Computer Simulation in Chemical Physics

NATO ASI Series

Advanced Science Institutes Series

A Series presenting the results of activities sponsored by the NATO Science Committee, which aims at the dissemination of advanced scientific and technological knowledge, with a view to strengthening links between scientific communities.

The Series is published by an international board of publishers in conjunction with the NATO Scientific Affairs Division

A Life Sciences **B Physics**	Plenum Publishing Corporation London and New York
C Mathematical and Physical Sciences **D Behavioural and Social Sciences** **E Applied Sciences**	Kluwer Academic Publishers Dordrecht, Boston and London
F Computer and Systems Sciences **G Ecological Sciences** **H Cell Biology** **I Global Environmental Change**	Springer-Verlag Berlin, Heidelberg, New York, London, Paris and Tokyo

NATO-PCO-DATA BASE

The electronic index to the NATO ASI Series provides full bibliographical references (with keywords and/or abstracts) to more than 30000 contributions from international scientists published in all sections of the NATO ASI Series.
Access to the NATO-PCO-DATA BASE is possible in two ways:

– via online FILE 128 (NATO-PCO-DATA BASE) hosted by ESRIN, Via Galileo Galilei, I-00044 Frascati, Italy.

– via CD-ROM "NATO-PCO-DATA BASE" with user-friendly retrieval software in English, French and German (© WTV GmbH and DATAWARE Technologies Inc. 1989).

The CD-ROM can be ordered through any member of the Board of Publishers or through NATO-PCO, Overijse, Belgium.

Series C: Mathematical and Physical Sciences - Vol. 397

Computer Simulation in Chemical Physics

edited by

M.P. Allen

H.H. Wills Physics Laboratory,
University of Bristol,
Bristol, U.K.

and

D.J. Tildesley

Department of Chemistry,
University of Southampton,
Southampton, U.K.

Springer Science+Business Media, B.V.

Proceedings of the NATO Advanced Study Institute on
New Perspectives in Computer Simulation in Chemical Physics
Alghero, Sardinia, Italy
September 14–24, 1992

A C.I.P. Catalogue record for this book is available from the Library of Congress.

ISBN 978-94-010-4734-0 ISBN 978-94-011-1679-4 (eBook)
DOI 10.1007/978-94-011-1679-4

Printed on acid-free paper

Contents

Preface

This book contains the Proceedings of a NATO Advanced Study Institute which was held at *CORISA* near Alghero in Sardinia, in September 1992. The school focused on recent progress in applying the methods of Computer Simulation to problems in Chemical Physics. It was organised and directed by Professor M. L. Klein and Dr. P. A. Madden, who reasoned that 1992 would be an appropriate time in the history of the field to mark recent developments with an Advanced Study Institute and an associated text.

Progress before 1987 had been summarized in the book "Computer Simulation of Statistical Mechanical Systems", edited by G. Ciccotti and W. G. Hoover, which contains the proceedings of an Enrico Fermi Summer School held in Varenna in 1985; in the reprint collection "Simulation of Liquids and Solids", edited by G. Ciccotti, D. Frenkel and I. R. McDonald, published in 1987; and in our own book, "Computer Simulation of Liquids", which is an introductory text, also published in 1987.

Over the subsequent five years there have been a number of remarkable advances in the techniques of computer simulation which have significantly expanded the scope of application of the methods. Good examples are the Car-Parrinello method, which allows the study of materials with itinerant electrons (such as metals and semiconductors); the Gibbs technique for the direct simulation of liquid-vapour phase equilibria; the transfer of scaling concepts from simulations of spin models to more complex systems; and the development of the configurational-biased Monte Carlo methods for studying dense polymers.

The Directors resolved to focus the school on giving a general, pedagogical account of these innovations, and to stress the continuous evolution of the simulation methodology. In the event, the scientific programme of the school exceeded expectations. Lecturer after lecturer delivered a beautifully constructed series of talks, focusing on a clear exposition of the subject, and not dwelling solely on their recent research results. The students responded by playing an active part in the discussions and presenting their own research in three excellent poster sessions.

In addition to these significant advances in basic techniques, the field has been stimulated by an enormous increase in available computing power and the provision of new applications software (particularly in the area of data visualisation). The school was used as a forum to discuss computer architectures and visualisation, and in particular the use of highly parallel machines. We were fortunate in that the manufacturers Hewlett-Packard, IBM and Silicon Graphics provided hardware and graphics software to allow the participants to gain hands-on experience of powerful workstations.

The Advanced Study Institute was held at the *CORISA* research centre, on the north side of the bay of Alghero, and participants were accommodated there and in the neighbouring Capo Caccia hotel. There was time to relax in the busy programme, to enjoy excursions to a nearby grotto, to see the local bird life (the 'Grifoni'), and to go out into the countryside to sample a traditional Sardinian shepherds' picnic. The beautiful setting provided opportunities to walk and swim, and the excellent facilities contributed greatly to the success of the school.

We are all indebted to Professor M. L. Klein and Dr. P. A. Madden for their skill in constructing the ASI and their tact and patience in supervising the progress of the event. The Directors were greatly helped in their planning by discussions with the Scientific Committee: G. Ciccotti, D. Frenkel, M. Gillan and J.-P. Ryckaert. A particular vote of thanks goes to the local organisers, Professor P. Suffritti and Dr. P. Demontis from the University of Sassari, for the tremendous efforts they made in meeting the needs of the participants. Support of various kinds was provided by the University of Sassari, the Italian Physical Society, Shell Research, Amsterdam, and, of course, NATO. The Editors would like to thank the Lecturers who cooperated closely in the process of producing this book, and who devoted much time and energy to preparing chapters of such high quality. Lastly, one of us (MPA) would like to thank all those who helped entertain the most exuberant member of the party, young Charles Allen.

We hope that the spirit of the meeting will pervade these Proceedings, that they will serve to remind all the participants of a stimulating time at Alghero, and that they will be enjoyed by a much wider audience than was able to attend the original lectures.

M. P. Allen and D. J. Tildesley, Editors

The Monte Carlo Method

D. J. Tildesley
Department of Chemistry
The University
Southampton SO9 5NH
United Kingdom

ABSTRACT. This Chapter describes the basic Metropolis Monte Carlo algorithm for simulating atoms in the canonical ensemble. The extension of the method to other common ensembles is outlined in a description of the Gibbs ensemble method for the direct simulation of fluid phase equilibria. The extension of the basic algorithm to include the case of a non-symmetrical underlying matrix is outlined with reference to some studies of variable polydispersity. In the simulation of molecular systems, the distinction is drawn between the classical rigid and flexible models, and the use of a metric potential for simulating a rigid model of butane is discussed. The kink-jump and reptation algorithms for the Monte Carlo simulation of polymer chains are described and a number of recent alternative algorithms are briefly reviewed.

1. The Metropolis algorithm

The Metropolis Monte Carlo method [1] is an importance-sampling technique for generating states of a system in a particular statistical ensemble. This section contains a description of the basic Metropolis algorithm for simulating N atoms in a volume V, at a temperature T. Readers who are unfamiliar with the method will find a gentler introduction in Chapter 4 of Ref. [2].

The method generates a Markov chain of states of the system whose limiting distribution is the probability density in the canonical ensemble,

$$\varrho_{NVT} = Q_{NVT}^{-1} \exp(-\beta\mathcal{H}(\mathbf{\Gamma})) \tag{1}$$

where Q_{NVT} is the canonical partition function, $\beta = 1/k_{\mathrm{B}}T$ and $\mathcal{H} = \mathcal{K} + \mathcal{V}$ is the hamiltonian of state $\mathbf{\Gamma}$ characterized by the $6N$ positions and momenta of the atoms.

Two states $\mathbf{\Gamma}_m$ and $\mathbf{\Gamma}_n$ of the system are linked by a transition probability π_{mn}: the probability that the system will move from state m to state n. The

M.P. Allen and D.J. Tildesley (eds.), Computer Simulation in Chemical Physics, 1–22.

Metropolis solution for the transition matrix $\boldsymbol{\pi}$ is

$$\begin{aligned} \pi_{mn} &= \alpha_{mn} \qquad\qquad \varrho_n \geq \varrho_m \quad m \neq n \\ \pi_{mn} &= \alpha_{mn}(\varrho_n/\varrho_m) \;\; \varrho_n < \varrho_m \quad m \neq n \end{aligned} \tag{2}$$

where ϱ_n is the probability density of the nth state of the system. The system also has a probability of remaining in the same state

$$\pi_{mm} = 1 - \sum_{n \neq m} \pi_{mn} \,. \tag{3}$$

In this solution, $\boldsymbol{\alpha}$ is the underlying matrix of the Markov chain. It is a symmetrical stochastic matrix and this property can be used with eqn (2) to show that

$$\sum_n \pi_{mn} = 1 \tag{4}$$

and

$$\varrho_m \pi_{mn} = \varrho_n \pi_{nm} \,. \tag{5}$$

Equation (5) is the principle of microscopic reversibility and it can be used with eqn (4) to show that the limiting distribution of the Markov chain generated by eqn (2) is the probability density, ϱ. The Metropolis solution only requires a knowledge of the ratio (ϱ_n/ϱ_m) and does not assume a knowledge of the partition function. Equation (2) is not unique, and there are other solutions for the transition matrix which rely on the symmetrical nature of $\boldsymbol{\alpha}$ [3]. In addition, there are possible solutions in which $\boldsymbol{\alpha}$ is non-symmetrical and these are discussed in section 3.

The Metropolis solution to the transition matrix is normally realized in the following way. A new trial state, n, is chosen by selecting an atom, i, in order or at random. The atom is given a uniform random displacement to a new position within a cube, R, which is centred at the old position of the atom. On the computer, there is a finite but large number of positions, N_R, in the cube which are sampled with equal probability. The underlying stochastic matrix is

$$\begin{aligned} \alpha_{mn} &= 1/N_R \qquad \mathbf{r}_i \in R \\ \alpha_{mn} &= 0 \qquad\qquad \mathbf{r}_i \notin R \,. \end{aligned} \tag{6}$$

The configurational energy of the old state, $\mathcal{V}(\boldsymbol{\Gamma}_m)$, and the trial state, $\mathcal{V}(\boldsymbol{\Gamma}_n)$, are calculated.

(i) If $\delta\mathcal{V}_{nm} = \mathcal{V}(\boldsymbol{\Gamma}_n) - \mathcal{V}(\boldsymbol{\Gamma}_m) < 0$ then $\varrho_n \geq \varrho_m$. In this case the trial move is immediately accepted.

(ii) If $\delta\mathcal{V}_{nm} > 0$, then $\varrho_n < \varrho_m$ and the move is accepted with a probability of $\varrho_n/\varrho_m = \exp(-\beta\delta\mathcal{V}_{nm})$. A trial state is accepted with a probability $\exp(-\beta\delta\mathcal{V}_{nm})$ using a standard rejection technique. A uniform random number, ξ, is generated between 0 and 1. If $\xi < \exp(-\beta\delta\mathcal{V}_{nm})$ the trial move is accepted.

(iii) If the move is rejected then the old state m is recounted as the new state in the chain.

In summary, a trial move is accepted with a probability, $\mathcal{P}$, given by

$$\mathcal{P} = \min(1, \exp(-\beta\delta\mathcal{V}_{nm})) . \tag{7}$$

It is clear that eqn (6) leads to a symmetrical $\boldsymbol{\alpha}$ since the elements of the underlying matrix are independent of n and m.

Once a technique, such as the Metropolis method, has been established for generating states with the canonical probability density, then the thermodynamic average of a configurational property $\mathcal{A}(\mathbf{r})$ can be estimated as an unweighted average over the states in the simulation.

$$\langle\mathcal{A}\rangle = \langle\mathcal{A}\rangle_{NVT} = (1/N_{\mathrm{states}}) \sum_{\mathrm{states}} \mathcal{A}(\mathbf{r}) \tag{8}$$

where $\mathbf{r}$ is the set of $3N$ positions $\{\mathbf{r}_i\}$ which describe a configuration of the system. There are two additional points to be made about the method. Firstly, the size of the trial displacement is a disposable parameter. If the maximum displacement of an atom is too small, then a large number of small moves will be accepted. If the maximum displacement is too large, then a small number of large moves will be accepted. It unlikely that either of these two extremes will lead to an efficient sampling of phase space and the maximum trial displacement is normally adjusted so that 50% of the trial moves are accepted. Secondly, the complete separation of the ideal and excess properties in classical statistical mechanics means that the Metropolis method need only sample states from the configurational manifold of phase space. The ideal gas contribution arising from integrals over the conjugate momenta can be calculated analytically and added to the configurational average, eqn (8), after the simulation.

2. Other ensembles

The methods described in section 1 can be readily extended to sample other common ensembles such as the isothermal-isobaric and the grand canonical ensembles. To illustrate these developments, we consider the Gibbs Monte Carlo method developed by Panagiotopoulos for the direct study of phase equilibria [4, 5]. A more detailed and formal development of the Gibbs ensemble is given in Chapter 6.

The Gibbs Monte Carlo method was originally developed to simulate two coexisting fluids by considering coupled moves in two simulation boxes. The first contains N^{I} atoms in a volume V^{I} and is buried deep in phase I. The second contains N^{II} atoms in a volume V^{II} buried deep in phase II. Each of these boxes is surrounded by periodic images of itself and there are no real interfaces between phases. The two boxes together form a representative system from a canonical ensemble at constant $N = N^{\mathrm{I}} + N^{\mathrm{II}}$, constant volume $V = V^{\mathrm{I}} + V^{\mathrm{II}}$ and constant T. There are three distinct trial moves, as illustrated in Fig. 1: an atom displacement; a volume rearrangement; an atom exchange.

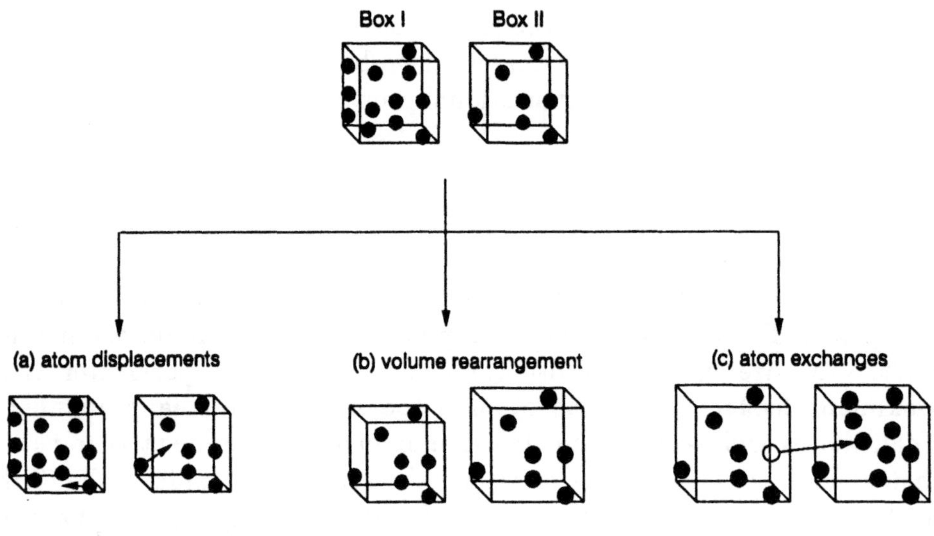

Fig. 1. The three types of moves in the Gibbs Monte Carlo simulation. (a) atom displacements, (b) volume rearrangement, (c) atom exchanges.

Atom displacements. In this step each box is independent and a trial state is generated by displacing an atom in its own periodic box and using the Metropolis algorithm to accept or reject the move.

Volume rearrangement. In this step the trial moves in each box are perfectly correlated. Consider box I to be part of a constant-$N^{\mathrm{I}}PT$ ensemble. A trial move consists of making a uniform random change in the volume of size ΔV and scaling the atom coordinates accordingly. In box I, $\mathcal{P}^{\mathrm{I}}_{\mathrm{vol}} = \varrho_n/\varrho_m$ is given by

$$\mathcal{P}^{\mathrm{I}}_{\mathrm{vol}} = \frac{\exp[-\beta P(V^{\mathrm{I}} + \Delta V) - \beta \mathcal{V}^{\mathrm{I}}(\mathbf{\Gamma}_n) + N^{\mathrm{I}} \ln(V^{\mathrm{I}} + \Delta V)]}{\exp[-\beta P V^{\mathrm{I}} - \beta \mathcal{V}^{\mathrm{I}}(\mathbf{\Gamma}_m) + N^{\mathrm{I}} \ln V^{\mathrm{I}}]} \tag{9}$$

which simplifies to

$$\mathcal{P}_{\text{vol}}^{\text{I}} = \exp[-\beta P\Delta V - \beta\delta\mathcal{V}_{nm}^{\text{I}} + N^{\text{I}}\ln(V^{\text{I}} + \Delta V) - N^{\text{I}}\ln(V^{\text{I}})] . \qquad (10)$$

The first term in the exponential is the work associated with the volume change against an external pressure P. The second term is the potential energy change on scaling the atom positions. The third and fourth terms are the jacobians associated with the volumes available to atoms in the trial state and the old state.

A simulation in the isothermal-isobaric ensemble would proceed by accepting the trial volume change with a probability given by $\min(1, \mathcal{P}_{\text{vol}}^{\text{I}})$. This type of move would need to be combined with trial moves involving atom displacements in box I.

In the Gibbs simulation the volume change in box I is coupled with a simultaneous change of $-\Delta V$ in the volume of box II. For box II

$$\mathcal{P}_{\text{vol}}^{\text{II}} = \exp[+\beta P\Delta V - \beta\delta\mathcal{V}_{nm}^{\text{II}} + N^{\text{II}}\ln(V^{\text{II}} - \Delta V) - N^{\text{II}}\ln V^{\text{II}}] . \qquad (11)$$

Since the Markov chains in the two boxes are independent, we can obtain the ratio of probabilities of the trial and old states in a combined move as

$$\mathcal{P}_{\text{vol}} = \mathcal{P}_{\text{vol}}^{\text{I}} \times \mathcal{P}_{\text{vol}}^{\text{II}} . \qquad (12)$$

At equilibrium, the pressure P in the two boxes is identical and

$$\begin{aligned}\mathcal{P}_{\text{vol}} = {} & \exp(-\beta[\delta\mathcal{V}_{nm}^{\text{I}} + \delta\mathcal{V}_{nm}^{\text{II}} - N^{\text{I}}k_{\text{B}}T\ln(1 + \Delta V/V^{\text{I}}) \\ & -N^{\text{II}}k_{\text{B}}T\ln(1 - \Delta V/V^{\text{II}})]) \qquad (13)\end{aligned}$$

and a joint volume rearrangement is accepted with a probability given by $\min(1, \mathcal{P}_{\text{vol}})$.

Atom exchanges. Again trial moves in both boxes are perfectly correlated. In this case box I is a representative system from a grand canonical ensemble at constant-$\mu V^{\text{I}}T$. A trial move consists of an atom creation or destruction in box I. If we concentrate on an atom being created in box I, then the ratio of the probabilities of the old and the new state is

$$\mathcal{P}_{\text{ex}}^{\text{I}} = \exp[-\beta[\delta\mathcal{V}_{nm}^{\text{I}} - k_{\text{B}}T\ln(zV^{\text{I}}/(N^{\text{I}} + 1))]] . \qquad (14)$$

Here the activity coefficient $z = \exp(\beta\mu)/\lambda^3$, where μ is the chemical potential and λ is the de Broglie wavelength of an atom.

In a normal grand canonical Monte Carlo simulation, this type of move would be accepted with a probability given by $\min(1, \mathcal{P}_{\text{ex}}^{\text{I}})$ and it would need to be combined with an equal number of trial atom-destructions and some trial atom-displacements.

In the Gibbs simulation method a corresponding trial destruction is attempted in box II. We can imagine the atom being transferred from box I to box II through an imaginary pipeline. At coexistence, μ is the same in both boxes and

$$\mathcal{P}_{\mathbf{ex}}^{\mathrm{II}} = \exp[-\beta[\delta\mathcal{V}_{nm}^{\mathrm{II}} - k_{\mathrm{B}}T\ln(N^{\mathrm{II}}/zV^{\mathrm{II}})]] \,. \tag{15}$$

The ratio of the probabilities of the trial and old states in the combined exchange step is

$$\begin{aligned} \mathcal{P}_{\mathbf{ex}} &= \mathcal{P}_{\mathbf{ex}}^{\mathrm{I}} \times \mathcal{P}_{\mathbf{ex}}^{\mathrm{II}} \\ &= \exp\left(-\beta\left[\delta\mathcal{V}_{nm}^{\mathrm{I}} + \delta\mathcal{V}_{nm}^{\mathrm{II}} + k_{\mathrm{B}}T\ln\left(\frac{V^{\mathrm{II}}(N^{\mathrm{I}}+1)}{V^{\mathrm{I}}N^{\mathrm{II}}}\right)\right]\right) \end{aligned} \tag{16}$$

and the joint creation-destruction attempt is accepted with a probability given by $\min(1, \mathcal{P}_{\mathbf{ex}})$. Trial creation and destruction attempts should be made in a particular box with equal probability to ensure microscopic reversibility.

There is some freedom in the choice of the relative frequencies of moves of these different types. Normally a cycle of trial displacements in each box is followed by one attempted volume rearrangement and several attempted atom exchanges.

The initial density and temperature of both simulation boxes are chosen to lie within a liquid-vapour two-phase region in the $\rho - T$ section of the one-component phase diagram. After a few hundred cycles, the volumes and the number of atoms in each box change so that the boxes contain coexisting liquid and gas at the same pressure and chemical potential. A recent example of this technique is a study of the orthobaric curve for chlorine [6]. The model of the intermolecular forces is a diatomic Lennard-Jones potential with $\varepsilon_{\mathrm{ClCl}}/k_{\mathrm{B}} = 178.3\,\mathrm{K}$, $\sigma_{\mathrm{ClCl}} = 3.332\,\mathrm{Å}$ and $L/\sigma = 0.63$ [7]. This simple model includes no representation of the permanent electrostatic interaction between molecules and no anisotropy in the repulsion-dispersion interaction. Both these aspects of the potential would need to be included to describe the experimentally measured liquid structure [8]. The Gibbs simulation method has been used to calculate the phase envelope for the liquid/vapour coexistence curve and this is shown in Fig. 2.

The agreement between the experimental and the simulation results for this simple model is remarkably good. The critical density and temperature have been estimated using the approximation of rectilinear diameters and assuming a critical exponent $\beta = 5/16$ from the 3-D Ising model. This analysis gives $\{T_c^*, \rho_c^*\}$ of $\{2.349, 0.1821\}$ compared to the experimental values of $\{2.338, 0.1812\}$ [9]. The use of more sophisticated techniques for studying phase equilibria and critical phenomena by simulation will be discussed in Chapter 3. The Gibbs technique has now been used to study phase equilibria for many different systems including pure fluids, mixtures, membrane equilibria [5], liquid-liquid equilibria [10] and polymers [11].

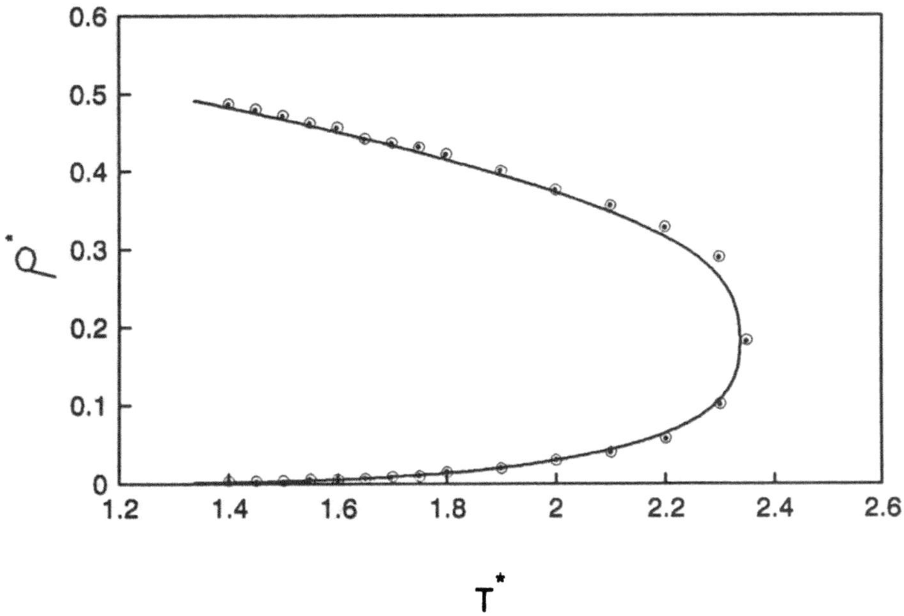

Fig. 2. The liquid/vapour coexistence curve for chlorine. The solid line is the experimental curve [9], the circles are the results of the Gibbs simulations [6].

Interestingly, the volume rearrangement and atom-exchange steps do not require an *a priori* knowledge of P and μ. The calculation of mechanical properties, such as P, in a Monte Carlo simulation is straightforward. However, the calculation of the statistical properties such as μ or the excess Helmholtz free energy F^{ex} is more difficult. In principle, F^{ex} can be represented as a simple canonical average.

$$F^{ex} = k_B T \ln\langle \exp(\beta \mathcal{V})\rangle_{NVT} \, . \tag{17}$$

However, the Metropolis Monte Carlo technique is designed to sample states which make an important contribution to $\langle \mathcal{V} \rangle$. It does not accurately sample the states which make important contributions to $\langle \exp(\beta\mathcal{V})\rangle$ and eqn (17) cannot be used to estimate F^{ex} accurately. Thermodynamic integration, overlapping distribution and umbrella sampling techniques which do allow for an accurate estimate of free energy differences are discussed in Chapter 4.

3. A non-symmetrical underlying matrix

There are many extensions to the basic Metropolis algorithm in which the underlying stochastic matrix is not symmetrical. Perhaps, the most common of these applications is the force-bias Monte Carlo method [12] in which the atom displacement is weighted in the direction of the force acting on it. In this section we will illustrate a simulation in this class with reference to some studies of variable polydispersity [13, 14].

In a polydisperse system some characteristic parameter, σ, of the atoms is described by a normalized distribution, $F(\sigma)$. σ might be the atom diameter and $F(\sigma)\mathrm{d}\sigma$ the fraction of atoms with diameters in the range σ to $\sigma + \mathrm{d}\sigma$. In the case of variable polydispersity, $F(\sigma)$ can change with density and temperature. An example of this type of system would be the non-ionic micelle formed from a poly-oxyethylene surfactant such as C_8E_4 in water. Theoretical and experimental studies indicate that these micelles are polydisperse in size, with the degree of polydispersity determined by the head-group repulsion, the amount of bound water and the isomerization of the hydrocarbon chains. An isolated micelle, that is one at low concentration, has this polydispersity described by an underlying distribution function, $F_{\mathrm{u}}(\sigma)$.

$$F_{\mathrm{u}}(\sigma) = C \exp(-\beta \mathcal{V}_{\mathrm{intra}}(\sigma)) \tag{18}$$

where C is a constant and $\mathcal{V}_{\mathrm{intra}}$ is the intraparticle free energy of the micelle. At a different temperature and density when the intermicellar interactions are important, the average distribution of micelle sizes will change to $F(\sigma)$. Other systems which exhibit this type of behaviour are oil/water emulsions, liquid foams and blood, all of which contain flexible particles which can distort under the influence of their neighbours because of the low interfacial tension at the boundary between the particle and the solvent.

Simulations of a fluid of this type can be performed in the canonical ensemble. A single trial move involves trying to change the particle's size and position. It is simpler to think of these steps separately and to note that particle displacement can be accomplished using the standard Metropolis algorithm. In changing the particle size the following underlying matrix is used

$$\begin{aligned} \alpha_{mn} &= F_{\mathrm{u}}(\sigma_n) \\ \alpha_{nm} &= F_{\mathrm{u}}(\sigma_m) . \end{aligned} \tag{19}$$

Since $F_{\mathrm{u}}(\sigma)$ is not generally uniform, $\boldsymbol{\alpha}$ is not symmetrical and the appropriate transition matrix is

$$\begin{aligned} \pi_{mn} &= \alpha_{mn} & \alpha_{nm}\varrho_n \geq \alpha_{mn}\varrho_m \\ \pi_{mn} &= \alpha_{mn}(\alpha_{nm}\varrho_n/\alpha_{mn}\varrho_m) & \alpha_{nm}\varrho_n < \alpha_{mn}\varrho_m . \end{aligned} \tag{20}$$

In this case the ratio of the probabilities of the old and the new states is given by

$$\varrho_n/\varrho_m = \exp(-\beta\delta\mathcal{V}_{nm}^{\mathrm{inter}}) \exp(-\beta\delta\mathcal{V}_{nm}^{\mathrm{intra}})$$

$$= \exp(-\beta\delta\mathcal{V}_{nm}^{\text{inter}})\,[F_{\text{u}}(\sigma_n)/F_{\text{u}}(\sigma_m)] \tag{21}$$

For a non-symmetrical $\boldsymbol{\alpha}$, the condition of microscopic reversibility is satisfied if the trial move is accepted with a probability given by

$$\min[1,\ \alpha_{nm}\varrho_n/\alpha_{mn}\varrho_m] \tag{22}$$

or substituting eqn (21) into eqn (22)

$$\min[1,\ \exp(-\beta\delta\mathcal{V}_{nm}^{\text{inter}})]\,. \tag{23}$$

To apply these ideas, a particle is picked at random and a new value of σ is chosen by randomly sampling from $F_{\text{u}}(\sigma)$ for $a \leq \sigma \leq b$ using a von Neumann rejection technique [15]. (a and b are the minimum and maximum size of the particle at low density). Only the change in the *interparticle* energy is used in eqn (23) to determine whether the trial move is accepted. The change in the *intraparticle* free energy has already been accounted for in the underlying matrix. Comparison with the standard Metropolis method shows that sampling directly from $F_{\text{u}}(\sigma)$ produces a markedly faster convergence of the Markov chain.

Simulations have been performed using a Schultz function as the model for the underlying distribution function

$$F_{\text{u}}(\sigma) = \begin{cases} (1/z!)[(z+1)/\sigma_0]^{z+1}\sigma^z \exp(-(z+1)\sigma/\sigma_0) & a \leq \sigma \leq b \\ 0 & \text{otherwise} \end{cases} \tag{24}$$

where z is an integer which determines the width of the underlying distribution and σ_0 is a characteristic particle diameter. Fig. 3 shows the average size of the particles as a function of the density. For the repulsive hard-sphere interaction the particle size decreases with increasing fluid density. For an attractive intermolecular potential the particles initially grow in size with increasing density before they shrink.

Defining an underlying distribution function fixes the chemical potential difference between the component particles in the mixture. This chemical potential difference function can be used in the semi-grand ensemble method developed by Kofke and Glandt [16]. (This method is described in more detail in Chapter 4, and for this problem is equivalent to the techniques described in this section).

Gibbs simulations of this model can also be performed to determine the relative sizes of particles in two coexisting phases [14]. Figure 4 shows the distribution of particle sizes for polydisperse Lennard-Jones atoms in a liquid/vapour (dense phase/dilute phase) equilibrium. The solid line is the underlying distribution function, which in this case is a Gaussian function. The dashed lines are the size distributions of particles in the coexisting dilute and dense phases. (Gaussian functions have been fitted through the

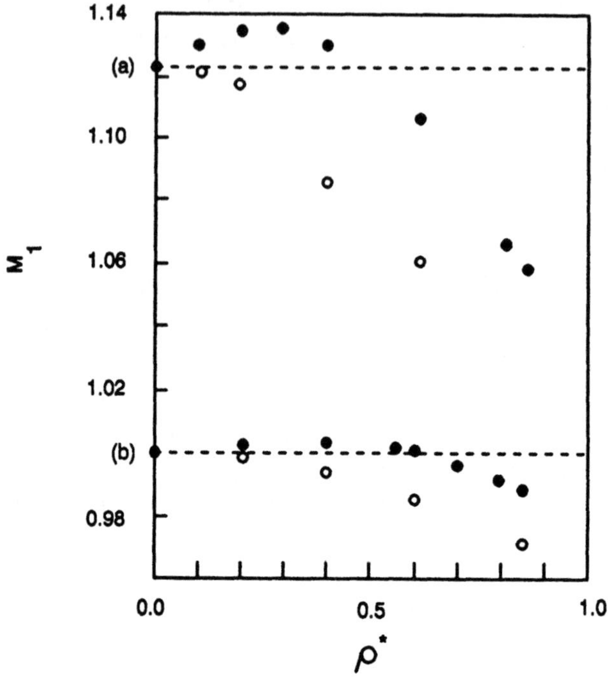

Fig. 3. The average size of the particles, M_1 as a function of density: (a) $z = 50$, $a = 1.0$, $b = 1.5$; (b) $z = 500$, $a = 0.9$, $b = 1.1$. A hard sphere intermolecular potential, ∘ ∘ ∘; a Lennard-Jones intermolecular potential, • • •.

histograms of the distributions obtained from the Gibbs simulation.) As the temperature approaches the critical temperature, the distributions, $F_u(\sigma)$, for the two phases converge.

4. Molecular systems

4.1. RIGID MOLECULES

The Monte Carlo simulation of a system of molecules is most easily achieved by transforming the hamiltonian of the system from a function of the atomic positions and momenta to a function of the generalized coordinates and their conjugate momenta. It is straightforward to make random changes in the generalized configurational variables which satisfy the principle of microscopic reversibility.

For example, the hamiltonian of a single N_2 molecule can be written in

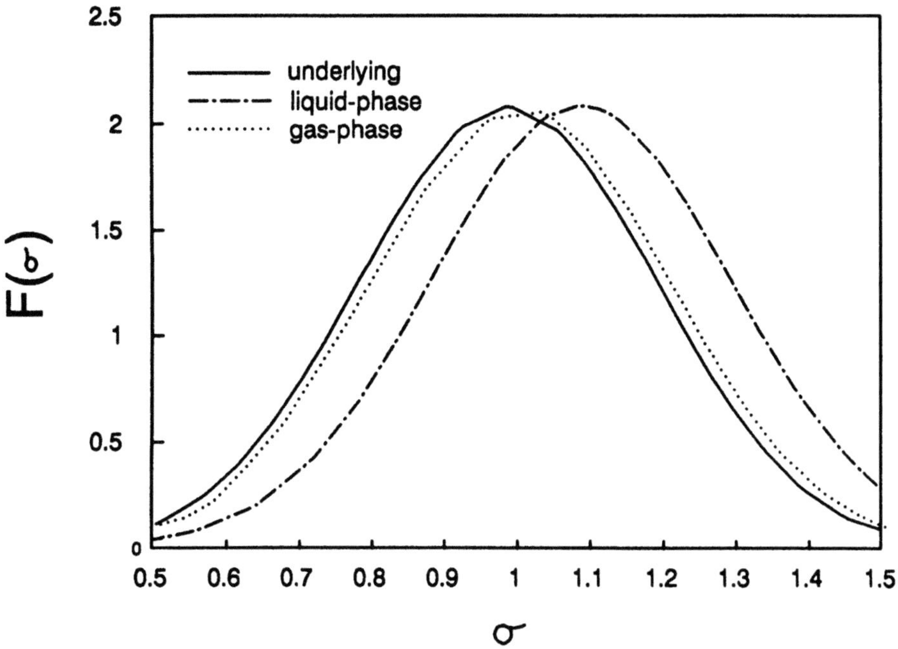

Fig. 4. The size distribution of particles in two coexisting phases of a polydisperse Lennard-Jones fluid at $T^* = 1.0$: ——, the underlying distribution function; ·····, the coexisting dilute phase; – · – · –, the coexisting dense phase.

terms of the atom coordinates, $\mathbf{r}_1$ and $\mathbf{r}_2$ and the corresponding momenta

$$\mathcal{H}(\mathbf{r},\mathbf{p}) = \tfrac{1}{2}\sum_{i=1}^{2} m_i \dot{\mathbf{r}}_i^2 + \mathcal{V}(\mathbf{r})\,. \tag{25}$$

The hamiltonian can be rewritten in terms of six generalized coordinates, $\mathbf{q}$, and their conjugate momenta $\mathbf{p}_q$. These coordinates are the three components of the molecular centre of mass, the bond length ℓ and the two Euler angles θ and ϕ which describe the orientation of the bond.

$$\mathcal{H}(\mathbf{q},\mathbf{p}_q) = \tfrac{1}{2}\mathbf{p}_q{}^{\mathrm{T}}\,\mathbf{G}^{-1}(\mathbf{q})\,\mathbf{p}_q + \mathcal{V}(\mathbf{q}) \tag{26}$$

where the superscript T means the transpose of a vector and $\mathbf{G}$ is the metric tensor. For a molecule of N atoms, $\mathbf{G}$ is a $3N \times 3N$ metric tensor

$$\mathbf{G}(\mathbf{q}) = \sum_{i=1}^{N} m_i \frac{\partial \mathbf{r}_i}{\partial \mathbf{q}} \cdot \frac{\partial \mathbf{r}_i}{\partial \mathbf{q}}\,. \tag{27}$$

The cartesian coordinates of atom i are functions of $\mathbf{q}$. The configurational average of a variable $\mathcal{A}$ in this flexible model is

$$\langle \mathcal{A} \rangle_{\mathrm{F}} = \frac{\int \cdots \int \mathrm{d}\mathbf{q}\, \mathcal{A}(\mathbf{q}) \exp(-\beta \mathcal{V}(\mathbf{q}))\, |\mathbf{G}|^{1/2}}{\int \cdots \int \mathrm{d}\mathbf{q}\, \exp(-\beta \mathcal{V}(\mathbf{q}))\, |\mathbf{G}|^{1/2}} \tag{28}$$

where the square-root of the determinant of the metric tensor comes from the integration over the conjugate momenta. For the flexible diatomic molecule $|\mathbf{G}|$ is the determinant of a 6×6 matrix. In this model, the potential $\mathcal{V}(\mathbf{q})$ might contain a term of the form $\frac{1}{2} k_\ell (\ell - \bar{\ell}\,)^2$ to control the bond length.

In contrast, in the rigid model of the diatomic, the bond length is constrained to its equilibrium value. In this case

$$\mathcal{H}(\mathbf{q}_\mathrm{s}, \mathbf{p}_\mathrm{s}) = \tfrac{1}{2} \mathbf{p}_\mathrm{s}{}^{\mathrm{T}} \mathbf{G}_\mathrm{s}^{-1}(\mathbf{q}_\mathrm{s}) \mathbf{p}_\mathrm{s} + \mathcal{V}'(\mathbf{q}_\mathrm{s}) \tag{29}$$

where the subscript s denotes the "soft variables" ($\mathbf{r}_{\mathrm{cm}}$, θ and ϕ for the diatomic) in contrast to the "hard variable" ℓ. $\mathbf{G}_\mathrm{s}$ is a sub-matrix of $\mathbf{G}$

$$\mathbf{G}_\mathrm{s}(\mathbf{q}_\mathrm{s}) = \sum_{i=1}^{N} m_i \frac{\partial \mathbf{r}_i}{\partial \mathbf{q}_\mathrm{s}} \cdot \frac{\partial \mathbf{r}_i}{\partial \mathbf{q}_\mathrm{s}} \tag{30}$$

associated with the transformation of the independent atomic coordinates to the generalized coordinates. The configurational average of the variable $\mathcal{A}$ in the rigid model is

$$\langle \mathcal{A} \rangle_{\mathrm{R}} = \frac{\int \cdots \int \mathrm{d}\mathbf{q}_\mathrm{s}\, \mathcal{A}(\mathbf{q}_\mathrm{s}) \exp(-\beta \mathcal{V}'(\mathbf{q}_\mathrm{s}))\, |\mathbf{G}_\mathrm{s}|^{1/2}}{\int \cdots \int \mathrm{d}\mathbf{q}_\mathrm{s}\, \exp(-\beta \mathcal{V}'(\mathbf{q}_\mathrm{s}))\, |\mathbf{G}_\mathrm{s}|^{1/2}} . \tag{31}$$

The potential $\mathcal{V}'$ is independent of ℓ and the generalized velocity associated with this variable is set to zero.

It is instructive to consider the relationship between the flexible and rigid models in the limit that the force constants associated with the internal soft variables of the flexible model tend to infinity, that is in the limit that $\mathcal{V}(\mathbf{q}_\mathrm{s}) = \mathcal{V}'(\mathbf{q}_\mathrm{s})$. Indeed throughout the rest of this Chapter, we will define the flexible model in this limit. For the linear diatomic, the canonical averages, eqns (28) and (31) are equivalent, because $|\mathbf{G}_\mathrm{s}|^{1/2}$ and $|\mathbf{G}|^{1/2}$ are both proportional to $\ell^2 \sin\theta$. Averages from the two models will always be identical if $|\mathbf{G}_\mathrm{s}|/|\mathbf{G}|$ is independent of the soft variables. The equivalence of these models is lost in the cases discussed in section 4.2.

For a system of N linear diatomic molecules, the average of $\mathcal{A}$ is

$$\langle \mathcal{A} \rangle = \frac{\int \mathrm{d}\mathbf{r} \int \mathrm{d}\boldsymbol{\Omega}\, \mathcal{A}(\mathbf{r}, \boldsymbol{\Omega}) \exp(-\beta \mathcal{V}(\mathbf{r}, \boldsymbol{\Omega}))}{\int \mathrm{d}\mathbf{r} \int \mathrm{d}\boldsymbol{\Omega}\, \exp(-\beta \mathcal{V}(\mathbf{r}, \boldsymbol{\Omega}))} \tag{32}$$

where $\mathrm{d}\boldsymbol{\Omega} = \prod_{i=1}^{N} \mathrm{d}\boldsymbol{\Omega}_i$ and $\mathrm{d}\boldsymbol{\Omega}_i = \sin\theta_i \mathrm{d}\theta_i \mathrm{d}\phi_i = \mathrm{d}(\cos\theta_i)\mathrm{d}\phi_i$. In a Monte Carlo simulation of this system, a trial move from state m to state n might consist of a random displacement of the centre of mass of the molecule and a random change in the orientation

$$\begin{aligned}\phi_i^n &= \phi_i^m + (2\xi_1 - 1)\,\delta\phi_{\max} \\ \theta_i^n &= \theta_i^m + (2\xi_2 - 1)\,\delta\theta_{\max}\end{aligned} \tag{33}$$

where ξ_1 and ξ_2 are uniform random numbers in the range 0 to 1, and $\delta\phi_{\max}$ and $\delta\theta_{\max}$ are the maximum allowed changes in the angles. A trial move would be accepted with a probability given by $\min(1,\, \varrho_n/\varrho_m)$ where

$$\frac{\varrho_n}{\varrho_m} = \exp(-\beta\delta\mathcal{V}_{nm})\,\frac{\sin\theta_i^n}{\sin\theta_i^m}\,. \tag{34}$$

This is not a useful description of the problem since ϱ_n/ϱ_m diverges as θ_i^m approaches 0. Normally a random change is made in $\cos\theta_i$

$$\cos\theta_i^n = \cos\theta_i^m + (2\xi_2 - 1)\,\delta(\cos\theta_{\max}) \tag{35}$$

and the move is accepted with a probability given by $\min(1, \exp(-\beta\delta\mathcal{V}_{nm}))$. The generalization to rigid, non-linear molecules involves making uniform random changes in the orientational variables ϕ, $\cos\theta$ and ψ. The trial move in orientational space can also be defined in terms of the four quaternion parameters that have been widely used in the molecular dynamics simulation of rigid molecules [17].

4.2. FLEXIBLE MOLECULES

The extension of these ideas to flexible molecules can be illustrated by considering the butane molecule shown in Fig. 5. This is a united-atom model containing four equivalent sites of equal mass to represent the methyl and methylene groups. For the flexible model, defined in section 4.1, the atomic coordinates $\{\mathbf{r}_1 \ldots \mathbf{r}_4\}$ can be represented as a function of 12 generalized coordinates. The generalized coordinates can be divided into two categories: the seven soft variables including the three coordinates of atom 1, the Euler angles (ϕ, θ, ψ) describing the orientation of the fragment 123, and the torsional angle (χ) describing the angle between the planes 123 and 234; and the five hard variables comprising the two valence angles α_1 and α_2 and the three bond lengths ℓ_1, ℓ_2 and ℓ_3. The determinant of the metric tensor $|\mathbf{G}|^{1/2}$ for this model is a product of a term $\sin\theta$ associated with the soft variables and a term ${\ell_1}^2{\ell_2}^2{\ell_3}^2 \sin\alpha_1 \sin\alpha_2$ which depends on the hard variables [18]. In the limit that the potentials governing the hard variables are very strong the hard-variable factor is a constant and the soft-variable factor is independent of the torsional angle χ.

The flexible model of butane can be simulated in the following way:

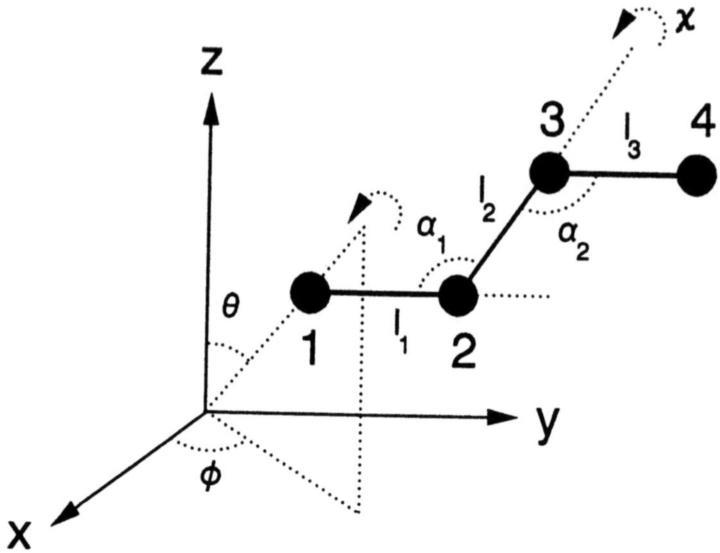

Fig. 5. The orientational variables used to describe the butane molecule.

(i) a trial configuration is generated by displacing atom 1, choosing a new orientation for the molecule by making uniform random changes in ϕ, $\cos\theta$ and ψ and finally changing the conformation by making a uniform random change in χ in the range 0 to 2π;

(ii) the trial configuration is accepted with a probability given by the $\min(1, \exp(-\beta\delta\mathcal{V}_{nm}))$, where $\delta\mathcal{V}_{nm}$ is the change in the intramolecular and intermolecular potential energies.

This scheme can be generalized to chains of arbitrary length without difficulty.

The simulation of the rigid model is more complicated. The metric tensor $|\mathbf{G}_s|$ is a 7×7 matrix. The determinant $|\mathbf{G}_s|^{1/2}$ depends on the torsional angle χ and is difficult to evaluate. Fixman [19] has presented an elegant solution to this problem by defining a matrix $|\mathbf{H}|$ such that

$$|\mathbf{G}_s| = |\mathbf{G}|\,|\mathbf{H}| \tag{36}$$

where

$$\mathbf{H} = \sum_{i=1}^{N} \frac{1}{m_i} \frac{\partial \mathbf{q}_h}{\partial \mathbf{r}_i} \cdot \frac{\partial \mathbf{q}_h}{\partial \mathbf{r}_i} \,. \tag{37}$$

$\mathbf{H}$ has the dimensions of the number of constrained degrees of freedom and is a 5×5 matrix for the rigid model of butane.

Details of the evaluation of $\mathbf{H}$ for butane are given elsewhere [20]. For the case where the bond lengths are fixed at 1.53 Å and $\alpha_1 = \alpha_2 = 109.47\,^\circ$

$$|\mathbf{H}| \propto 475 - 28\cos\chi - 222\cos^2\chi - 8\cos^3\chi + 8\cos^4\chi\,. \tag{38}$$

Substitution of eqn (36) into eqn (31) gives

$$\langle \mathcal{A} \rangle_{\mathrm{R}} = \frac{\int \cdots \int \mathrm{d}\mathbf{q}_{\mathrm{s}}\, \mathcal{A}(\mathbf{q}_{\mathrm{s}}) \exp(-\beta(\mathcal{V}'(\mathbf{q}_{\mathrm{s}}) + \mathcal{V}_{\mathrm{c}}))\, |\mathbf{G}|^{1/2}}{\int \cdots \int \mathrm{d}\mathbf{q}_{\mathrm{s}} \exp(-\beta(\mathcal{V}'(\mathbf{q}_{\mathrm{s}}) + \mathcal{V}_{\mathrm{c}}))\, |\mathbf{G}|^{1/2}}\,. \tag{39}$$

where

$$\mathcal{V}_{\mathrm{c}} = -\tfrac{1}{2} k_{\mathrm{B}} T \ln |\mathbf{H}|\,. \tag{40}$$

The rigid model of butane can be simulated in the following way:

(i) a trial configuration is generated by displacing atom 1, choosing a new orientation for the molecule by making uniform random changes in ϕ, $\cos\theta$ and ψ and finally changing the conformation by making a uniform random change in χ in the range 0 to 2π;

(ii) the trial configuration is accepted with a probability given by the $\min(1, \exp(-\beta\delta\mathcal{V}''_{nm}))$, where $\delta\mathcal{V}''_{nm}$ is the change in the intramolecular, the intermolecular and the metric potential (eqn (40)).

Intuitively, the flexible and rigid models of molecules like butane should be identical. However, the application of classical statistical mechanics to these models gives different results. In the case of an isolated butane molecule, the rigid model leads to an enhanced population of the *gauche* conformers over the percentage of these states found for the flexible model. There is no reason to prefer either model *a priori*. Both models can be simulated using the Monte Carlo method and the treatment of the flexible model is somewhat simpler. A detailed discussion of the differences between the rigid and flexible models has been presented by Ryckaert [21].

5. Polymers

The ideas developed in section 4 can be used to simulate simple models of polymers, but the technique of making an arbitrary change in a randomly chosen torsional angle is likely to lead to a high rejection ratio. Consider a freely jointed chain of hard-sphere beads in a solvent of hard spheres. A torsional angle is chosen in the middle of the chain and a small random change is made to the angle χ, while all the other bond lengths, valence and torsional angles are kept constant. If the chain is of a significant length this small rotation at the centre of the chain will produce a large translational displacement of one of the terminal beads which is likely to lead to an overlap and a rejection of the trial state. A number of simple methods for generating new trial states in a Monte Carlo simulation have been developed and two of these are discussed in this section.

The kink-jump algorithm. The kink-jump algorithm was first used by Baumgärtner and Binder [22] in Monte Carlo studies of a freely jointed chain with excluded volume interactions. Consider a freely-jointed linear polymer chain, consisting of $p+1$ monomers and p links:
(i) randomly choose a monomer unit i of the chain in state m;
(ii) rotate the monomer i in a circle around the vector joining $i-1$ to $i+1$ by choosing an angle $\delta\phi$ uniformly in the range $-\delta\phi_{\max}$ to $\delta\phi_{\max}$ (see Fig. 6). This move generates a trial state n;
(iii) accept the trial state with a probability given by $\min(1, \exp(-\beta\delta\mathcal{V}_{nm}))$.
If $i=1$ or $i=p+1$ then a trial state is generated by choosing a new random orientation for the first or the pth link respectively.

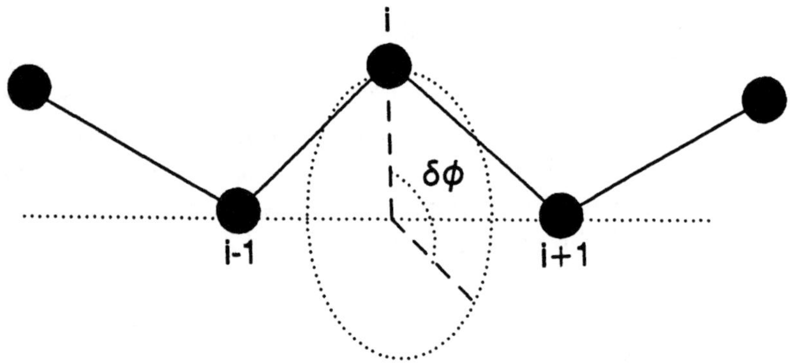

Fig. 6. The trial move in the kink jump algorithm used in the simulation of polymer chains.

The reptation algorithm. The reptation algorithm was first suggested by Wall and Mandel [23] in a simulation of a polymer on a square lattice. It has also been extensively used in off-lattice simulations. The algorithm consists of the following steps:
(i) choose one end of the chain at random to be the head, i.e. atom 1;
(ii) remove the tail link and the corresponding atom, $p+1$; attach the tail link to the head of the chain by choosing a new orientation for the link randomly and uniformly in space;
(iii) calculate the energy change $\delta\mathcal{V}_{nm} = \mathcal{V}_1^n - \mathcal{V}_{p+1}^m$ where $\mathcal{V}_1^n$ is the potential energy of the head atom in the trial state;

(iv) accept the trial state with a probability given by $\min(1, \exp(-\beta\delta\mathcal{V}_{nm}))$. The polymer moves by reptating along its own length like a slithering snake (see Fig. 7).

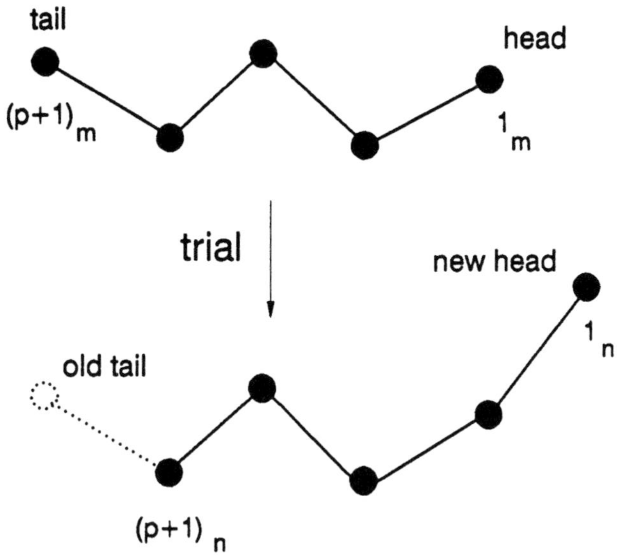

Fig. 7. A trial move in the reptation algorithm used in the simulation of polymer chains.

These two algorithms can be illustrated by simulating a chain without excluded volume effects. The chain consists of $p+1$ monomer beads linked by bonds of length ℓ. Beads and links in the chain do not interact with one another and can occupy the same region of space. For this type of chain, the mean squared end-to-end distance is

$$\langle R_p^2 \rangle = \ell^2 p \,. \tag{41}$$

All linear chain models correspond to the non-excluded-volume chain as $T \to \infty$. Figure 8 shows the results of two Metropolis Monte Carlo simulations of a chain with $p = 100$ and $\ell = 1.0$. The cumulative average of R_p^2 is plotted as a function of the number of cycles for the kink-jump and reptation algorithm. A cycle consists of p attempted moves. The reptation algorithm gives a result which agrees with the analytical average, $\langle R_p^2 \rangle = 100$, within 1×10^6 cycles. The kink-jump algorithm only reaches the analytical result after 2.5×10^6 cycles. The kink-jump algorithm generates successive configurations that are strongly correlated since only four torsional angles change

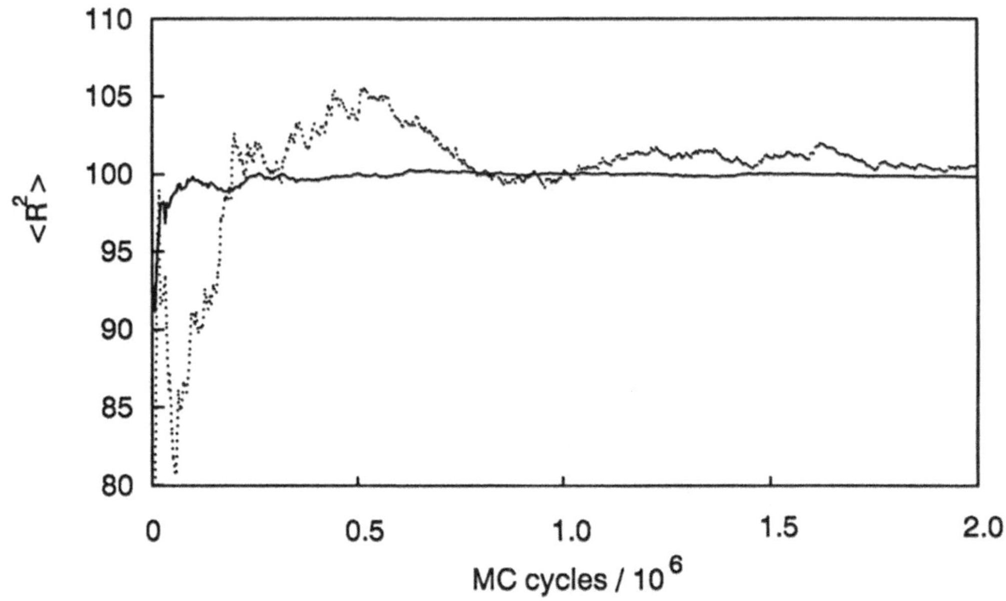

Fig. 8. The mean-squared end-to-end distance of a non-excluded volume polymer of 100 links: ——, the reptation algorithm; ······, the kink-jump algorithm.

in each trial move. In addition the moving bead is confined to a circle in space. In the reptation algorithm all of the $p-2$ torsional angles of the polymer change but only by label; one entirely new torsional angle is created. However, the orientation for the new link is chosen at random on the surface of a sphere. Using reptation a single chain can achieve a new configuration, on average, after p^2 moves, whereas it requires $p^{2.2}$ moves of the individual beads to achieve an equivalent change [24]. The computing time associated with the kink-jump algorithm is significantly higher than that required for the reptation algorithm and reasonable simulations of polymer fluids and moderate density can be successfully simulated using a cycle of trial reptation moves followed by one trial kink-jump.

Recently Siepmann and Frenkel [25] have introduced an algorithm based on the Rosenbluth weighting technique [26] which moves a polymer in a dense phase by effectively growing a chain from a random position or regrowing parts of a chain. This configurational-bias Monte Carlo method can be used to simulate polymers on lattices and in continuum systems. It can also be used to make estimates of the chemical potential of a polymer chain in dense solution [27, 28] and these developments will be discussed in Chapter 4.

An interesting recent advance has been the development of a concerted-rotation algorithm [29]. In this technique the trial move consists of a multiple rotation around seven adjacent skeletal bonds. A change in the first torsional angle is used to drive the move and the subsequent six torsions are determined by the six constraints that ensure that the moving portion of the chain rejoins the static residue at the end of the move. The exhaustive determination of all the possible solutions to these geometrical constraints is a difficult but tractable numerical problem and it is clear that for some initial choices of χ_1 there will not be a physically reasonable solution. For other choices of χ_1 there will be a number of possible solutions, with different values of $\{\chi_2 \dots \chi_6\}$. Once all the possibilities have been identified, then one of them can be selected at random to be the trial move. This method also requires an evaluation of the metric terms associated with the old and new states. It is apparently more rapidly convergent than the reptation algorithm for simulating certain realistic polymer models.

Finally in this section, it is worth questioning our initial premise that a high rejection ratio leads to a slow convergence in the calculation of the properties of the polymer. Simulations of self-avoiding random walks on a lattice have used the pivot algorithm originally suggested by Lal [30]. A trial move involves choosing a bead of the polymer to act as a pivotal point. Using this point as an origin, a random symmetry operation of the lattice (such as a rotation or a reflection) is performed on the part of the chain beyond the pivot point. This trial move would seem to be a disaster as it is almost certain to lead to an overlap between parts of the chain. However, the pivot moves are very radical and only a handful of successful trials are required to produce an "essentially new" configuration. In fact the detailed analysis of Madras and Sokal [30] shows the pivot algorithm is a factor of p times more efficient than many of the more standard algorithms used to generate self-avoiding random walks. The continuum analogue of this algorithm [31], which involves random changes in the torsional angles at a random point on the polymer backbone, certainly leads to a high rejection ratio. By analogy with the analysis for the self-avoiding random walk, it may still be an efficient algorithm for studying single polymer chains. However, the Monte Carlo simulation of polymer melts will require one of the techniques introduced in this section and discussed in more detail in Chapter 11.

6. Conclusions

In the first twenty years of the development of the Monte Carlo method, the important advantage that it offered over the molecular dynamics technique was the ease with which it could be extended to simulate a variety of different ensembles. This edge has effectively disappeared with the development of a raft of new molecular dynamics techniques in different ensembles which will be reviewed in Chapter 5. For example, a molecular dynamics implementation of the Gibbs method described in section 2 is feasible, and one would gain the important advantage of having the time dependent properties of the coexisting fluids.

The Monte Carlo simulation technique does have a number of advantages. Firstly for the realistic potentials which are increasingly used to describe the interactions between atoms within molecules [32], the analytical derivatives of the potential required for the forces are often complicated and time-consuming to compute. In these cases the Monte Carlo method, which only requires the potential energy, is significantly faster the molecular dynamics method. This is despite the fact that a Monte Carlo cycle requires twice the number of potential energy evaluations compared to the corresponding molecular dynamics time-step. (This problem can be mitigated by the use of table look-ups for the forces but these are difficult to implement for potentials which are functions of both the relative orientations and separation of the interaction sites). This advantage disappears if a force-biased Monte Carlo simulation is required to negotiate bottlenecks in the phase space of the system.

A clearer advantage of the Monte Carlo method is the flexibility in the choice of sampling functions. In studying an equilibrium system in the canonical ensemble, one most naturally samples states from the appropriate Boltzmann distribution. However, some properties have important contributions from states of low Boltzmann-probability, such as those involving significant atomic overlap. It is possible to sample from biased or weighted distributions which enhance the exploration of these states. It also possible to recover the normal Boltzmann average from the weighted average calculated in this simulation. These techniques are at the root of the configurational bias method for simulating polymers and the umbrella sampling method for calculating free energy differences.

One simple example of changing the sampling function is the partitioning of parts of the hamiltonian into the underlying stochastic matrix as illustrated in the account of variable polydispersity in section 3. This technique is also widely used in the study of flexible molecules where a change in a torsional angle, χ can be made by sampling from an underlying distribution function $\propto \exp(-\beta \mathcal{V}_{\mathrm{t}}(\chi))$ where $\mathcal{V}_{\mathrm{t}}$ is the torsional potential [20]. Clearly $\mathcal{V}_{\mathrm{t}}$ is not included in the calculation of $\delta \mathcal{V}_{nm}$ which is used in the acceptance criterion.

Regardless of the sampling function, the Monte Carlo method offers considerable flexibility in the choice of the underlying matrix or trial move, which must satisfy the principle of microscopic reversibility. An interesting example of this is a recent study of a dense monolayer comprising two chains of different length [33]. The system was a simple model for the adsorption of a mixture of $HS(CH_2)_9CH_3$ and $HS(CH_2)_{19}CH_3$ chain molecules on a gold surface. There is a strong tendency in this system for chains of different length to segregate into domains on the surface. This phenomenon would be extremely difficult to study by molecular dynamics because of the timescales for the diffusion of the thiol head-groups in the plane. However, the successful Monte Carlo method introduces a trial move which attempts to exchange the identities of the short and long chains. A long and a short chain are chosen at random and the terminal $-CH_3$ of the short chain is replaced by the $-(CH)_{10}CH_3$ fragment of the longer chain and *vice-versa*. This move, which is completely unphysical, allows for a rapid equilibration of the monolayer on the surface.

One important challenge for the Metropolis Monte Carlo method is the simulation of realistic models of natural and synthetic polymers. The development of the new configurational-bias sampling algorithms and the use of new trial moves, which should defy nature, bring this goal within reach.

Acknowledgement

I would like to thank my student Mr. Giulio Galassi for providing the material for Fig. 2 and Fig. 8 and for reading the manuscript.

References

[1] N. Metropolis, A. W. Rosenbluth, M. N. Rosenbluth, A. H. Teller, and E. Teller, *J. Chem. Phys.* **21**, 1087 (1953).

[2] M. P. Allen and D. J. Tildesley, *Computer Simulation of Liquids*, Clarendon Press, Oxford, 1987.

[3] A. A. Barker, *Aust. J. Phys.* **18**, 119 (1965).

[4] A. Z. Panagiotopoulos, *Molec. Phys.* **61**, 813 (1988).

[5] A. Z. Panagiotopoulos, N. Quirke, M. R. Stapleton, and D. J. Tildesley, *Molec. Phys.* **63**, 527 (1988).

[6] G. Galassi and D. J. Tildesley, *Molec. Phys.* (1993), submitted for publication.

[7] K. Singer, A. Taylor, and J. V. L. Singer, *Molec. Phys.* **33**, 1757 (1977).

[8] P. M. Rodger, A. J. Stone, and D. J. Tildesley, *Molec. Phys.* **63**, 173 (1988).

[9] S. Angus, B. Armstrong, and K. M. de Reuck, *International Thermodynamic Tables of the Fluid State, Vol. 8, Chlorine*, Pergamon Press, 1982.

[10] M. R. Stapleton and A. Z. Panagiotopoulos, *J. Chem. Phys.* **92**, 1285 (1990).

[11] G. C. A. M. Mooij and B. Smit, Direct simulation of phase equilibria for chain molecules, Poster presented at NATO ASI Alghero, 1992.

[12] C. Pangali, M. Rao, and B. J. Berne, *Chem. Phys. Lett.* **55**, 413 (1978).

[13] M. R. Stapleton, D. J. Tildesley, T. J. Sluckin, and N. Quirke, *J. Phys. Chem.* **82**, 4788 (1989).

[14] M. R. Stapleton, D. J. Tildesley, and N. Quirke, *J. Chem. Phys.* **92**, 4456 (1990).

[15] R. Y. RUBINSTEIN, *Simulation and the Monte Carlo method*, Wiley: New York, 1981.

[16] D. A. KOFKE and E. D. GLANDT, *J. Chem. Phys.* **87**, 4881 (1987).

[17] F. J. VESELY, *J. Comput. Phys.* **47**, 291 (1982).

[18] N. GO and H. A. SCHERAGA, *Macromolecules* **9**, 535 (1976).

[19] M. FIXMAN, *Proc. Nat. Acad. Sci.* **71**, 3050 (1974).

[20] N. G. ALMARZA, E. ENCISO, J. ALFONSO, F. J. BERMEJO, and M. ALVAREZ, *Molec. Phys.* **70**, 485 (1990).

[21] J. P. RYCKAERT, Molecular dynamics of polyatomic systems, in *Computer Simulation in Material Science*, edited by M. MEYER and V. PONTIKIS, Kluwer Academic Publishers, 1991.

[22] A. BAUMGARTNER and K. BINDER, *J. Chem. Phys.* **71**, 2541 (1979).

[23] F. T. WALL and F. MANDEL, *J. Chem. Phys.* **63**, 4592 (1975).

[24] M. BISHOP, D. CEPERLEY, H. L. FRISCH, and M. H. KALOS, *J. Chem. Phys.* **72**, 3228 (1980).

[25] J. I. SIEPMANN and D. FRENKEL, *Molec. Phys.* **75**, 59 (1992).

[26] M. N. ROSENBLUTH and A. W. ROSENBLUTH, *J. Chem. Phys.* **23**, 356 (1955).

[27] J. I. SIEPMANN, *Molec. Phys.* **70**, 1145 (1990).

[28] D. FRENKEL, G. C. A. M. MOOIJ, and B. SMIT, *J. Phys. Cond. Mat.* **4**, 3053 (1992).

[29] L. R. DODD, T. D. BOONE, and D. N. THEODOROU, *Molec. Phys.* (1993), accepted for publication.

[30] N. MEAL and A. D. SOKAL, *J. Stat. Phys.* **50**, 109 (1988).

[31] J. P. FREIRE and A. HORTA, *J. Chem. Phys.* **65**, 4049 (1976).

[32] S. L. PRICE, *Molec. Simul.* **1**, 135 (1988).

[33] J. I. SIEPMANN and I. R. MCDONALD, *Molec. Phys.* **75**, 255 (1992).

The Molecular Dynamics Method

D. J. Tildesley
Department of Chemistry
The University
Southampton SO9 5NH
United Kingdom

ABSTRACT. This Chapter describes the basic molecular dynamics method for simulating atoms in the microcanonical ensemble. The extension of the technique for studying molecules of arbitrary complexity using constraint dynamics is outlined and three algorithms for solving the equations of motion of the constraints are discussed. The fast multipole algorithm for calculating the forces in a system of charges or dipoles in periodic boundary conditions is described and the Leckner and Hautman and Klein methods for calculating long-range forces in a lamellar geometry are compared for some simple model systems.

1. Introduction

1.1. THE EQUATIONS OF MOTION

The molecular dynamics technique solves the classical equations of motion of N molecules interacting through a potential $\mathcal{V}$. The Euler-Lagrange equations of motion are of the form

$$\frac{\mathrm{d}}{\mathrm{d}t}(\partial \mathcal{L}/\partial \dot{\mathbf{q}}_i) - \partial \mathcal{L}/\partial \mathbf{q}_i = 0 \tag{1}$$

where the lagrangian $\mathcal{L}$ is defined in terms of the kinetic and potential energies

$$\mathcal{L} = \mathcal{K} - \mathcal{V} \, . \tag{2}$$

$\mathcal{L}$ is a function of the generalized coordinates $\mathbf{q}$ and their time derivatives $\dot{\mathbf{q}}$. If we consider a set of spherical atoms, then the generalized coordinates are the positions of the centres of mass, $\mathbf{r}$, and the normal definitions of $\mathcal{K}$ and $\mathcal{V}$ give the Newtonian equations of the system

$$m_i \ddot{\mathbf{r}}_i = \mathbf{f}_i \tag{3}$$

M.P. Allen and D.J. Tildesley (eds.), Computer Simulation in Chemical Physics, 23–47.

where m_i is the mass of atom i and $\mathbf{f}_i$ is the force on the atom. These $3N$ equations of motion are coupled since $\mathbf{f}_i$ depends on the positions of all of the atoms in the system through the gradient of the potential.

$$\mathbf{f}_i = -\partial\mathcal{V}/\partial\mathbf{r}_i \,. \tag{4}$$

It is also possible to define a generalized momentum conjugate to $\mathbf{q}_i$,

$$\mathbf{p}_i = \partial\mathcal{L}/\partial\dot{\mathbf{q}}_i \,. \tag{5}$$

The generalized coordinates and their conjugate momenta define a hamiltonian, which can be used to obtain two first-order equations of motion [1]. For a system of atoms these are

$$\begin{aligned} \dot{\mathbf{r}}_i &= \mathbf{p}_i/m_i \\ \dot{\mathbf{p}}_i &= -\partial\mathcal{V}/\partial\mathbf{r}_i \,. \end{aligned} \tag{6}$$

The calculation of the trajectories of the atoms is obtained by a numerical solution of either of the equivalent eqns (3) or (6).

The hamiltonian, $\mathcal{H} = \mathcal{K} + \mathcal{V}$, is a conserved variable of the motion. The conservation of $\mathcal{H}$ is a useful test of the accuracy of the numerical algorithm used to solve the equations of motions. For a system of atoms in a simulation which uses periodic boundary conditions, there is a translational invariance in the three coordinate directions and the linear momentum, $\mathbf{P}$, is a conserved variable. In this kind of geometry, the total angular momentum of the simulated system is not conserved [2].

1.2. ALGORITHMS FOR MOLECULAR DYNAMICS

There are two types of algorithms which are commonly used to solve eqns (3) and (6): the Gear predictor-corrector algorithms [3] and the Verlet-like algorithms. Both types of algorithm require only one evaluation of the forces at each time step.

The predictor-corrector algorithms usually store a significant number of time derivatives of the atomic positions at a particular step. Taylor series expansions are then used to advance each derivative forward from time t to $t + \delta t$, producing a series of predicted positions and their derivatives. The predicted positions can then be used to calculate the forces at $t + \delta t$ and, through eqn (3), the 'correct' accelerations. The difference between the corrected and predicted accelerations is used with a set of Gear constants to correct all of the predicted derivatives of the motion. The Gear coefficients are chosen to optimize the stability and accuracy of the trajectories. They depend on the order of the differential equation and the number of derivatives used in the Taylor series prediction of the positions (i.e. the order of the algorithm).

In contrast, the Verlet-like algorithms use information from the current and previous time steps to advance the atomic positions. The Verlet algorithm itself can be constructed by considering the Taylor expansions from t to $t + \delta t$ and $t - \delta t$.

$$\begin{aligned} \mathbf{r}_i(t + \delta t) &= \mathbf{r}_i(t) + \delta t \mathbf{v}_i(t) + \frac{1}{2}\delta t^2 \mathbf{a}_i(t) \\ \mathbf{r}_i(t - \delta t) &= \mathbf{r}_i(t) - \delta t \mathbf{v}_i(t) + \frac{1}{2}\delta t^2 \mathbf{a}_i(t) \end{aligned} \tag{7}$$

where $\mathbf{v}_i$ and $\mathbf{a}_i$ are the velocity and acceleration of atom i respectively. Adding these equations produces the equation for advancing the positions.

$$\mathbf{r}_i(t + \delta t) = 2\mathbf{r}_i(t) - \mathbf{r}_i(t - \delta t) + \delta t^2 \mathbf{f}_i(t)/m_i \,. \tag{8}$$

The velocity at a particular step can be obtained by subtracting the equations.

$$\mathbf{v}_i(t) = [\mathbf{r}_i(t + \delta t) - \mathbf{r}_i(t - \delta t)]/2\delta t \,. \tag{9}$$

In this algorithm the positions are correct to $\mathcal{O}(\delta t^4)$ whereas the velocities are correct to $\mathcal{O}(\delta t^2)$. A number of simple modifications to the basic Verlet scheme such as the 'leap-frog' and 'velocity-Verlet' algorithms have been suggested to try to improve the method of handling the velocities [2].

The Verlet-like algorithms are simpler to use then the Gear algorithms, and a fifth-order predictor-corrector method requires at least three times more memory than the Verlet algorithm. More importantly, the Gear algorithms become rapidly unstable and inaccurate with increasing time step whereas the root-mean-square fluctuation in $\mathcal{H}$ is almost a linear function of δt^2 for the Verlet algorithm. When the Verlet algorithm is used to simulate liquid argon, it produces an energy conservation of better than 1 part in 10^4 with a time step of 10^{-14} s. Most of the modified dynamical equations of motion which are used to simulate at constant temperature and pressure can now be solved using Verlet-like algorithms and there seems little advantage to considering other techniques.

The precise amount of time required for the integration algorithm is not particularly important because the overwhelming proportion of the computing effort is spent in the calculation of the force on each atom. This force calculation is simple if the intermolecular potential is pairwise additive and short range (i.e. it falls off more rapidly than r^{-d} where d is the dimensionality of the system). For large systems, geometrical decomposition algorithms have been developed [4] so that this calculation scales with the number of atoms N rather than N^2. If the force is long range, as in the interaction between two ions, then the use of the minimum-image convention is inappropriate. This problem is addressed in section 3.

1.3. THE MICROCANONICAL ENSEMBLE

If the equations of motion are accurately solved then the system stays on a constant-energy hypersurface in phase space, and the method generates states in a subset of the microcanonical ensemble at constant-$NVE\mathbf{P}$. The average of some property $\mathcal{A}$ in this ensemble is equal to the time average over the molecular dynamics trajectory in phase space.

$$\langle \mathcal{A} \rangle = \langle \mathcal{A} \rangle_{\mathbf{NVEP}} = \lim_{t' \to \infty} \frac{1}{t'} \int_0^{t'} \mathrm{d}t \; \mathcal{A}(\mathbf{\Gamma}(t)) \,. \tag{10}$$

The condition that the system stays on a constant-energy hypersurface, to say 1 part in 10^4, is a much weaker condition than insisting that each atom follows its exact classical trajectory from a given starting position and momentum. The weaker condition is all that is required for the calculation of the static averages such as $\mathcal{A}$. However, an important advantage of the molecular dynamics method is that it is possible to calculate time correlation functions and their corresponding spectra and transport coefficients. These calculations require 'exact' solutions of the atomic trajectories over the correlation times of interest. The balance in any molecular dynamics simulation is to increase the time step so that the phase space of the system can be widely explored without degrading the quality of the energy conservation.

1.4. MOLECULAR SYSTEMS

The Lagrangian development discussed in section 1.1 can be applied to a system of rigid molecules using the centre of mass of a molecule and the three Euler angles describing the molecular orientation as the generalized coordinates. The Euler-Lagrange equations for the body-fixed angular velocities and for the Euler angles are well-known (see eqns (3.29) and (3.32) of Ref. [2]). However, even for a small rigid molecule such as carbon tetrachloride, the rotational equations of motion expressed in terms of the Euler angles contain singularities. The most general and elegant solution to this problem involves the introduction of the four quaternion parameters as generalized orientational coordinates [5], and Fincham has developed a leap-frog algorithm for solving the quaternion equations of motion [6].

It is not sensible to solve these generalized equations of motion for a complicated molecule such as a cyclic hexapeptide. Finding such a set of coordinates for a large, flexible molecule is difficult and the resulting equations of motion can be awkward to solve numerically. (For a successful application of generalized coordinates to the simulation of solid butane see Ref. [7].) In contrast, the method of constraints can be applied with equal ease to small rigid molecules and to larger flexible molecules. This generality makes the technique particularly powerful and the method is outlined in the following section.

2. Constraint dynamics

2.1. CONSTRAINTS

The technique of constraint dynamics focuses on the component atoms of a molecule. These atoms are subject to the normal Newtonian forces in the system which arise from the intramolecular and intermolecular potentials, and to a number of holonomic constraints (i.e. a set of algebraic equations connecting the atomic coordinates which do not change with time).

For example, the rigid model of butane, discussed in Chapter 1, contains four united atoms representing the methylene or methyl groups of the chain. A torsional potential describes the angle between the planes 123 and 234, and a quadratic bond-angle potential is used to control the valence angles, 123 and 234. The bond lengths 12, 23 and 34 are normally fixed rigidly at their equilibrium value, ℓ_{CC}, by applying constraints of the form

$$\sigma_1(\mathbf{r}_1, \mathbf{r}_2) = r_{12}^2 - \ell_{\mathrm{CC}}^2 = 0\,. \tag{11}$$

In reality, the bond lengths are changing and this variation could be modelled crudely using an harmonic potential of the form

$$v_\ell(r_{12}) = \frac{1}{2}k_\ell(r_{12} - \ell_{\mathrm{CC}})^2 \tag{12}$$

where k_ℓ is the force constant of the vibration. However, the bond stretching potential, eqn (12), is usually very stiff and a small time step of the order of 10^{-16} s would be required with this model. A time step of this size is required to track the changes in the bond length and to ensure equipartition of energy in this degree of freedom. For very strong bonds such as the triple-bond in N_2, the harmonic oscillator is uncoupled from the orientational and translational degrees of freedom in a simulation of the liquid even when using a time step of 10^{-16} s. There is effectively no energy transfer between the bath and the oscillator. This phenomenon causes uncertainty in the definition of the number of degrees of freedom in the system and hence the temperature in a molecular dynamics simulation. If, on the other hand, these stiff bonds are rigidly constrained, then these degrees of freedom are completely removed from the system and a longer time step of approximately 2.5×10^{-15} s can be used.

More complicated geometrical constraints are frequently encountered in molecular problems. Figure 1 shows a methylene fragment from a longer chain in which the hydrogen atoms are explicitly represented. In a molecular dynamics simulation, the $\mathrm{C}_{i-1}\mathrm{C}_i\mathrm{C}_{i+1}$ bond angle would normally be controlled by a harmonic potential and it would be appropriate to freeze all of the other degrees of freedom in the methylene unit. This can be accomplished using six additional constraints [8].

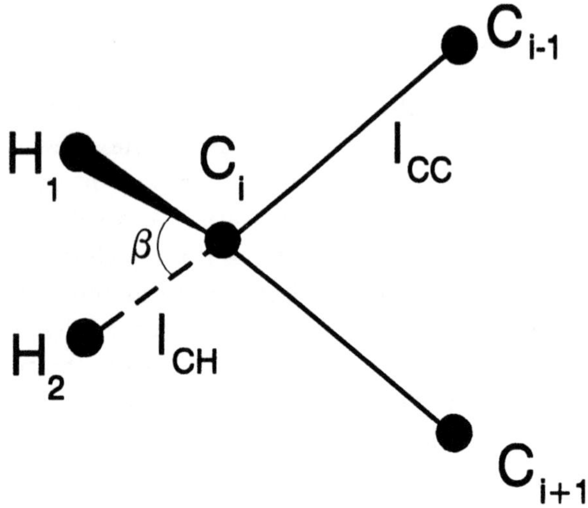

Fig. 1. A methylene fragment from an n-alkane chain, in which the hydrogen atoms are explicitly represented.

As in the united-atom model, the carbon-carbon bonds between adjacent methylene groups are fixed at their equilibrium bond lengths, ℓ_{CC}. In addition, the two carbon-hydrogen bonds in the methylene group are constrained to their equilibrium values, ℓ_{CH}. For example

$$\sigma_1(\mathbf{r}_1, \mathbf{r}_i) = |\mathbf{r}_1 - \mathbf{r}_i|^2 - \ell_{\mathrm{CH}}^2 = 0 \,. \tag{13}$$

The plane $H_1C_iH_2$ is constrained to be perpendicular to the plane $C_{i-1}C_iC_{i+1}$

$$\sigma_3(\mathbf{r}_1, \mathbf{r}_2, \mathbf{r}_{i+1}, \mathbf{r}_{i-1}) = (\mathbf{r}_1 - \mathbf{r}_2) \cdot (\mathbf{r}_{i-1} - \mathbf{r}_{i+1}) = 0 \,. \tag{14}$$

Finally, the vector from the mid-point of $C_{i-1}C_{i+1}$ to C_i is constrained to bisect the vector H_1H_2 at a distance $\ell_{\mathrm{CH}} \cos(\beta/2)$ from C_i, where β is the angle $H_1C_iH_2$

$$\sigma_4(\mathbf{r}_1, \mathbf{r}_2, \mathbf{r}_{i+1}, \mathbf{r}_i, \mathbf{r}_{i-1}) = \frac{1}{2}(\mathbf{r}_1 + \mathbf{r}_2) - \mathbf{r}_i + \ell_{\mathrm{CH}} \cos(\beta/2)\, \hat{\mathbf{u}} = 0 \tag{15}$$

where

$$\hat{\mathbf{u}} = \frac{\frac{1}{2}(\mathbf{r}_{i+1} + \mathbf{r}_{i-1}) - \mathbf{r}_i}{|\frac{1}{2}(\mathbf{r}_{i+1} + \mathbf{r}_{i-1}) - \mathbf{r}_i|} \,. \tag{16}$$

This representation of the methylene groups contains two additional atoms compared to the united-atom model, and is characterized by six additional constraints: three scalar constraints and a three-component vector constraint.

2.2. THE CONSTRAINT EQUATIONS OF MOTION

Consider one molecule composed of N atoms and subject to M constraints. This notation and the corresponding equations used in this section can be straightforwardly generalized to a system containing many molecules. The $3N$ Euler-Lagrange equations for the molecule are

$$\frac{\mathrm{d}}{\mathrm{d}t}(\partial\mathcal{L}/\partial\dot{\mathbf{r}}_i) - \partial\mathcal{L}/\partial\mathbf{r}_i = \mathbf{g}_i \tag{17}$$

where $\mathbf{g}_i$ is the constraint force on atom i

$$\mathbf{g}_i = \sum_{\alpha=1}^{M} \lambda_\alpha(\partial\sigma_\alpha/\partial\mathbf{r}_i) \tag{18}$$

and $\mathcal{L}$ is the lagrangian of the system of N atoms

$$\mathcal{L} = \tfrac{1}{2}\sum_{i=1}^{N} m_i\dot{\mathbf{r}}_i(t)^2 - \mathcal{V}(\mathbf{r}) \tag{19}$$

where $\mathbf{r}$ represents the positions of all the atoms. The λ_α are the undetermined multipliers associated with the α constraints. Physically they are proportional to the magnitudes of the forces of constraint. The $3N$ equations, (17), and the constraint equations

$$\sigma_\alpha(\mathbf{r}) = 0 \qquad \alpha = 1, \ldots, M \tag{20}$$

uniquely determine the $3N$ atomic positions and the undetermined multipliers at all times.

de Leeuw *et al.* [9] have shown that the constrained lagrangian can be used to construct the $6N$ Hamiltonian equations of motion

$$\begin{aligned} \dot{\mathbf{r}}_i &= m_i^{-1}\left(\mathbf{p}_i - \sum_{\alpha=1}^{M}\gamma_\alpha(\partial\sigma_\alpha/\partial\mathbf{r}_i)\right) \\ \dot{\mathbf{p}}_i &= -\frac{\partial\mathcal{V}}{\partial\mathbf{r}_i} + \sum_{\alpha=1}^{M}\sum_{j=1}^{N}\gamma_\alpha\frac{\partial^2\sigma_\alpha}{\partial\mathbf{r}_i\partial\mathbf{r}_j}\dot{\mathbf{r}}_j\,. \end{aligned} \tag{21}$$

The undetermined multipliers λ_α of eqn (18) are the first time derivatives of multipliers γ_α in eqn (21), i.e. $\dot{\gamma}_\alpha = \lambda_\alpha$. These first-order equations can be eliminated to produce $3N$ second-order differential equations

$$\ddot{\mathbf{r}}_i = m_i^{-1}\left(\mathbf{f}_i - \sum_\alpha \lambda_\alpha(\partial\sigma_\alpha/\partial\mathbf{r}_i)\right)\,, \tag{22}$$

where $\mathbf{f}_i$ is the force on atom i. An equation for the undetermined multipliers is obtained by differentiating the constraint equation, (20), twice

$$\begin{aligned} \frac{d\sigma_\alpha}{dt} &= \sum_i \dot{\mathbf{r}}_i \frac{\partial \sigma_\alpha}{\partial \mathbf{r}_i} = 0 \\ \frac{d^2\sigma_\alpha}{dt^2} &= \sum_i \ddot{\mathbf{r}}_i \frac{\partial \sigma_\alpha}{\partial \mathbf{r}_i} + \sum_i \sum_j \left(\dot{\mathbf{r}}_i \frac{\partial}{\partial \mathbf{r}_i}\right)\left(\dot{\mathbf{r}}_j \frac{\partial}{\partial \mathbf{r}_j}\right) \sigma_\alpha = 0 \,. \end{aligned} \tag{23}$$

Multiplying eqn (22) by $(\partial\sigma_\alpha/\partial\mathbf{r}_i)$, summing over i and substituting eqn (23) gives the equations for the multipliers. In a matrix notation [9] these are

$$\mathbf{\Lambda} = (\mathbf{M})^{-1}(\mathbf{F} + \mathbf{T}) \,. \tag{24}$$

$\mathbf{M}$ is an $M \times M$ matrix which is positive-definite and non-singular. Its inverse is also a positive-definite, symmetric matrix. The elements of $\mathbf{M}$ are

$$(\mathbf{M})_{\alpha\beta} = \sum_i m_i^{-1} \left(\frac{\partial \sigma_\alpha}{\partial \mathbf{r}_i}\right) \cdot \left(\frac{\partial \sigma_\beta}{\partial \mathbf{r}_i}\right) \,. \tag{25}$$

$\mathbf{\Lambda}$, $\mathbf{F}$ and $\mathbf{T}$ are vectors of length α;

$$\begin{aligned} (\mathbf{\Lambda})_\alpha &= \lambda_\alpha \\ (\mathbf{F})_\alpha &= \sum_i m_i^{-1} \mathbf{f}_i \cdot (\partial\sigma_\alpha/\partial\mathbf{r}_i) \\ (\mathbf{T})_\alpha &= \sum_i \sum_j \dot{\mathbf{r}}_i \frac{\partial^2 \sigma_\alpha}{\partial \mathbf{r}_i \partial \mathbf{r}_j} \dot{\mathbf{r}}_j \,. \end{aligned} \tag{26}$$

Once the Newtonian forces on the atoms and the required derivatives of the constraints have been evaluated, the matrix $\mathbf{M}$ can be inverted to provide the the undetermined multipliers. Equation (22) for the constrained atomic positions can then be solved using a standard differential equation solver.

The *exact* solution of eqn (22) will ensure that the constraints are satisfied at each time step. However, the error associated with the numerical solution of the differential equations means that the constraints are only satisfied to the order of the algorithm and the discrepancy with the exact solution grows with time. A number of numerical methods have been developed to avoid this problem and they are outlined in section 2.3.

2.3. ALGORITHMS FOR CONSTRAINT DYNAMICS

2.3.1. The SHAKE algorithm. The original algorithms for solving the equations of motion of constrained systems were designed to work with particular integration schemes such as the Verlet method described in section 1.2 [10]. The method begins with a time step in which the N atoms move subject to the Newtonian forces, $\mathbf{f}_i$, but without any consideration of the forces of constraint.

$$\mathbf{r}_i'(t+\delta t) = 2\mathbf{r}_i(t) - \mathbf{r}_i(t-\delta t) + \frac{\delta t^2}{m_i}\mathbf{f}_i \,. \tag{27}$$

The true positions at the end of the time step are

$$\mathbf{r}_i(t+\delta t) = 2\mathbf{r}_i(t) - \mathbf{r}_i(t-\delta t) + \frac{\delta t^2}{m_i}(\mathbf{f}_i - \mathbf{g}_i) \,. \tag{28}$$

Thus, for any atom i

$$\mathbf{r}_i(t+\delta t) = \mathbf{r}_i'(t+\delta t) - \frac{\delta t^2}{m_i}\sum_{\alpha=1}^{M} \lambda_\alpha(\partial\sigma_\alpha(\mathbf{r}(t))/\partial\mathbf{r}_i) \tag{29}$$

where $\mathbf{r}(t)$ represents the positions of the N atoms at time t. The atomic coordinates at $t+\delta t$ must also obey the constraint equation (20)

$$\sigma_\alpha(\mathbf{r}(t+\delta t)) = 0 \qquad \alpha = 1,\ldots,M \,. \tag{30}$$

In general, eqn (30) is a set of α non-linear equations for the multipliers, λ_α. These equations can be solved by a matrix inversion method [2] or by an iterative procedure known as the SHAKE algorithm [8].

In the SHAKE algorithm, the unconstrained atomic positions $\mathbf{r}_i'(t+\delta t)$ are used as a starting guess and the atomic coordinates are adjusted iteratively until the constraint equations are eventually satisfied to the required tolerance. All of the independent constraints are considered in a cyclic fashion at each iteration. Consider a particular constraint, κ, involving a subset of N_κ atoms. At the Ith iterative loop the atomic coordinates are $\mathbf{r}_i^{\text{old}}$. The forces of constraint act on these atoms at this iteration to produce new positions:

$$\mathbf{r}_i^{\text{new}} = \mathbf{r}_i^{\text{old}} - \frac{\delta t^2}{m_i}\lambda_\kappa^I\left[\frac{\partial\sigma_\kappa(\mathbf{r}(t))}{\partial\mathbf{r}_i}\right] \qquad i = 1,\ldots,N_\kappa \,. \tag{31}$$

The Lagrange multiplier is evaluated by solving $\sigma_\kappa(\mathbf{r}^{\text{new}}) = 0$. Expanding $\sigma_\kappa(\mathbf{r}^{\text{new}})$ about $\sigma_\kappa(\mathbf{r}^{\text{old}})$ using eqn (31) and retaining terms to $\mathcal{O}(\delta t^2)$ gives

$$\lambda_\kappa^I = \delta t^{-2}\sigma_\kappa(\mathbf{r}^{\text{old}}) \Big/ \sum_{i=1}^{N_\kappa} m_i^{-1}\left[\frac{\partial\sigma_\kappa(\mathbf{r}^{\text{old}})}{\partial\mathbf{r}_i}\right]\cdot\left[\frac{\partial\sigma_\kappa(\mathbf{r}(t))}{\partial\mathbf{r}_i}\right] \,. \tag{32}$$

Equations (31) and (32) are the central equations of the method. The multiplier for the κth constraint at the Ith iteration is evaluated and used in eqn (31) to obtain a new guess for the atomic positions. After some tens of iterations all of the constraints are satisfied to a relative accuracy of say 10^{-6} and the constrained atomic positions $\mathbf{r}(t+\delta t)$ are known.

The method is fast and robust and can be used with general constraints. Although it has been described for the Verlet algorithm, it can be used with predictor-corrector solvers [11], and an adaptation known as RATTLE has been developed for use with the velocity-Verlet algorithm [12]

2.3.2. Trajectory correction using a penalty function. Edberg *et al.* [13] have developed an algorithm which uses an equation of motion equivalent to eqns (22) and (24). At some point in the trajectory the positions and velocities are $\mathbf{r}_0$ and $\dot{\mathbf{r}}_0$. The subscript 0 indicates that the trajectory has moved some significant distance from the constraint surface. The trajectory can now be relaxed back on to the constraint surface by minimising two penalty functions

$$\begin{aligned} \Phi_\alpha &= [\sigma_\alpha(\mathbf{r})]^2 \\ \Psi_\alpha &= [\dot{\sigma}_\alpha(\mathbf{r}, \dot{\mathbf{r}})]^2 . \end{aligned} \tag{33}$$

For the linear constraint given by eqn (11)

$$\Psi_\alpha = (\mathbf{r}_{12} \cdot \dot{\mathbf{r}}_{12})^2 \tag{34}$$

is a positive function which is zero when the constraint distance is no longer changing with time.

The two penalty functions can be summed over all of the constraints in a molecule to produce the constraint potentials Φ and Ψ for each molecule. When the values of these potentials become sufficiently large, a non-linear minimization is performed to find the $6N$ values of $\mathbf{r}_i$ and $\dot{\mathbf{r}}_i$ at the minimum. The penalty function surfaces are positive-definite and nearly quadratic and the minimizations are rapid. Using a time step of 1.93 fs and simulating 64 molecules of butane, Edberg *et al.* [13] minimized Φ or Ψ if either exceeded 10^{-7}. The tolerances are reduced with respect to the Lennard-Jones parameters of a united-atom methylene group. The mean time between a minimization for any molecule in this simulation was 86 integration steps and the functions were minimized to a tolerance of 10^{-12} and 10^{-13} respectively. Similar results were obtained in a simulation of liquid decane and the method can be easily adapted for use with arbitrary constraints.

2.3.3. Trajectory correction by matrix inversion. A more elegant solution to this problem has recently been proposed by de Leeuw *et al.* [9]. The solution of eqn (22) leads to a set of coordinates $\mathbf{r}_0$, which have drifted some distance from the constraint surface at a particular time step. If the constraints were satisfied at the previous time step then

$$\sigma_\alpha(\mathbf{r}_0) = \mathcal{O}(\delta t^k) \tag{35}$$

where k is the order of the differential equation solver. The corrections required to produce the properly constrained coordinates, $\mathbf{r}$, from the unconstrained coordinates, $\mathbf{r}_0$, are

$$\mathbf{r}_i = \mathbf{r}_{i0} + \delta t^k \sum_{\alpha=1}^{M} \epsilon_\alpha \left(\partial\sigma_\alpha(\mathbf{r})/\partial\mathbf{r}_i\right). \tag{36}$$

These coordinates can be substituted into the constraint equation, (20), to obtain the correction parameters, ϵ_α. Using Taylor expansions of $\sigma(\mathbf{r})$ and $(\partial\sigma_\alpha/\partial\mathbf{r}_i)$ about $\mathbf{r}_0$ gives

$$\sigma_\alpha(\mathbf{r}) = \sigma_\alpha(\mathbf{r}_0) + \delta t^k \sum_{\beta=1}^{M}\sum_{i=1}^{N} \epsilon_\beta \left(\frac{\partial\sigma_\alpha(\mathbf{r}_0)}{\partial\mathbf{r}_i}\right)\cdot\left(\frac{\partial\sigma_\beta(\mathbf{r}_0)}{\partial\mathbf{r}_i}\right) \tag{37}$$

to $\mathcal{O}(\delta t^{2k})$. Since the higher-order terms neglected in eqn (37) are negligible, a non-singular set of M equations is obtained for the correction coefficients

$$\boldsymbol{\epsilon} = -\delta t^{-k}\,(\mathbf{M}_0)^{-1}\cdot\boldsymbol{\sigma} \tag{38}$$

where

$$(\mathbf{M}_0)_{\alpha\beta} = \sum_{i=1}^{N} \left(\frac{\partial\sigma_\alpha(\mathbf{r}_0)}{\partial\mathbf{r}_i}\right)\cdot\left(\frac{\partial\sigma_\beta(\mathbf{r}_0)}{\partial\mathbf{r}_i}\right). \tag{39}$$

and

$$\begin{aligned}(\boldsymbol{\sigma})_\alpha &= \sigma_\alpha(\mathbf{r}_0)\\ (\boldsymbol{\epsilon})_\alpha &= \epsilon_\alpha\,.\end{aligned} \tag{40}$$

The ϵ_α are used in eqn (36), with the approximation

$$(\partial\sigma_\alpha(\mathbf{r})/\partial\mathbf{r}_i) = (\partial\sigma_\alpha(\mathbf{r}_0)/\partial\mathbf{r}_i)$$

to correct the atomic positions to $\mathcal{O}(\delta t^{2k})$. de Leeuw *et al.* [9] performed simulations of a rigid model of CCl_4 using this technique with a fifth-order predictor corrector method. The correction algorithm was applied every time that $(r_{12}^2/\ell_{\mathrm{CCl}} - 1)$ exceeded 10^{-5}. Using this tolerance the correction algorithm was applied about every 20th integration step. The method has also been applied successfully to flexible molecules.

3. Long-range forces

The interaction between two charges

$$v_{\mathrm{ion}}(r_{ij}) = q_i q_j / 4\pi\epsilon_0 r_{ij} \tag{41}$$

decays to zero so slowly with distance, that the minimum-image convention with a spherical cut-off is inappropriate for calculating the forces on ions in the simulation of an ionic melt. Even in the simulation of an isolated cluster of ions, the calculation of this interaction requires $\mathcal{O}(N^2)$ operations. In a system with periodic boundary conditions, the Ewald sum is a technique for calculating the interaction between a charge i, all of its neighbouring charges j in the central simulation box and all of the images of i and j in the real space lattice of simulation boxes. Detailed theoretical studies have revealed the precise nature of the simulation when this type of lattice sum is employed [14] and this area has been reviewed elsewhere [2]. In this section we consider an alternative approach developed by Greengard and Rokhlin [15] based on a multipole expansion of regions of charge. In subsection 3.2 two technique for handling long-range forces in simulations with surfaces are described and the results of these techniques for some simple charge systems are compared.

3.1. THE GREENGARD-ROKHLIN ALGORITHM

The Greengard-Rokhlin algorithm is a technique for the evaluation of the potential and force fields. It can be used for any potential of the form r^{-n} but is particularly important for Coulombic and dipolar interactions because of its scaling behaviour. In this section the method is described for a fluid of charges. The algorithm is based upon two important results in classical electrostatics: the multipole and local expansions.

Consider N charges of strength q_i, $i = 1 \ldots N$. The charges are located at spherical polar coordinates $\{\rho_i, \alpha_i, \beta_i\}$ within a sphere $|\rho_i| < a$. The potential due to these charges at a point $P = (r, \theta, \phi)$ with $r > a$ is given by the multipole expansion

$$\varphi(P) = \sum_{l=0}^{\infty} \sum_{m=-l}^{m=l} \frac{M_{l,m}}{r^{l+1}} Y_{l,m}(\theta, \phi) \tag{42}$$

where $Y_{l,m}$ is a modified spherical harmonic (i.e. one that does not include the normalization factor $\sqrt{(2l+1)/4\pi}$) and $M_{l,m}$ is an expansion coefficient:

$$M_{l,m} = \sum_{i=1}^{N} q_i \, \rho_i^n \, Y_{l,-m}(\alpha_i, \beta_i) \,. \tag{43}$$

It is a simple operation to change the origin of this expansion and formulae for the new coefficients are obtained from the set of $M_{l,m}$ (see eqn (2.22) of Ref. [16]).

Consider N charges of strength q_i, $i = 1 \ldots N$, in a sphere of radius a centred at $Q = (\rho, \alpha, \beta)$. Q is a distance greater than $2a$ from the origin. The multipole expansion of these charges converges in a sphere of radius a

centred at the origin. The potential at some point P inside this sphere is given by the local expansion

$$\varphi(P) = \sum_{j=0}^{\infty} \sum_{k=-j}^{+j} R_{j,k} Y_{j,k}(\theta, \phi) r^j . \tag{44}$$

The coefficients in the local expansion are

$$R_{j,k} = \sum_{l=0}^{\infty} \sum_{m=-l}^{m=l} \frac{O_{l,m} J_{k,m} A_{l,m} A_{j,k} Y_{(l+j),(m-k)}(\alpha, \beta)}{A_{(l+j),(m-k)} \rho^{l+j+1}} \tag{45}$$

where

$$A_{l,m} = \frac{(-1)^l}{\sqrt{(l-m)!(l+m)!}} \tag{46}$$

and

$$J_{k,m} = \begin{cases} (-1)^l & \text{if } k.m \leq 0 \\ (-1)^l (-1)^{\min(|m|,|k|)} & \text{if } k.m > 0 \end{cases} . \tag{47}$$

$O_{l,m}$ are the coefficients in the multipole expansion of the charge around Q. Equation (2.29) of Ref. [16] can be used to calculate the change in the coefficients R given a change of the origin of the local expansion.

The fast-multipole method is most easily described for a free-standing cubic box containing charges. The problem of periodic boundary condition will be addressed presently. The algorithm begins by dividing the central simulation box into octants. The partition of space continues and the mesh at level $C+1$ is obtained by dividing the mesh at C into octants. A tree structure is imposed on this mesh, so that when a particular box at level C is divided, the resulting eight boxes at level $C+1$ are its children and the box itself is described as a parent. The number of distinct boxes at each mesh level is 8^C. This division is difficult to illustrate in three dimensions and Fig. 2 shows a two-dimensional example of a box divided to level $C = 3$. The top right-hand corner of the box is at level $C = 2$. The nearest-neighbour and next-nearest-neighbour boxes of b are hatched and shaded respectively. In later versions of the algorithm [17] different levels of mesh are used in different parts of the box depending on the local number density (the adaptive algorithm). However, the principles of the method can be described using the simpler uniform partition of space (the non-adaptive algorithm).

For a box b at level C, a nearest-neighbour box shares a boundary point with b. A second-nearest-neighbour box shares a boundary point with one of the nearest-neighbour boxes of b. The use of a local expansion is only valid if the external charges contributing to the expansion and those inside the

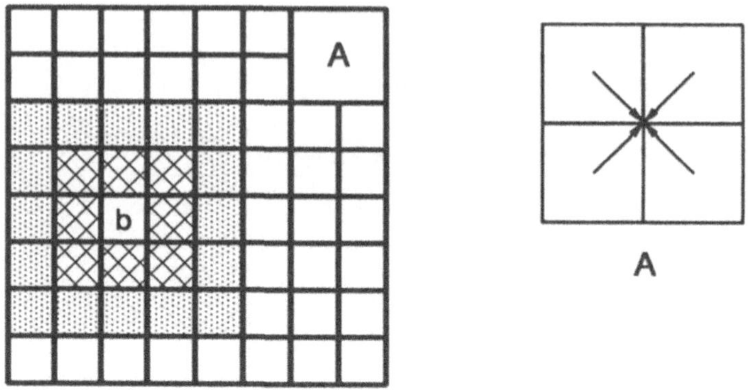

Fig. 2. The division of the computational box in the Greengard-Rokhlin algorithm to mesh level $C = 3$.

target box are well-separated. In other words the local expansion must arise from charges beyond the second-nearest-neighbour boxes if the field in b is to be accurately represented.

A number of basic definition are required to describe the algorithm:

- $\varphi_{C,b}$ is the Pth-order multipole expansion of the potential created by the charges contained inside box b at level C.
- $\psi_{C,b}$ is the Pth-order local expansion about the centre of box b at level C. This describes the potential due to all charges outside the box, its nearest-neighbour boxes and second-nearest-neighbour boxes.
- $\tilde{\psi}_{C,b}$ is the Pth-order local expansion about the centre of box b at level C from all the charges outside b's parents box and the parent's neighbours and the parent's second-nearest-neighbours.
- the *interaction list* for box b at level C is the set of boxes which are children of the nearest-neighbour and second-nearest-neighbour of b's parents and which are not nearest-neighbours and second-nearest-neighbours of b.

Suppose that at a particular level, $C-1$, of the algorithm the local expansion, $\psi_{C-1,b}$, has been obtained for all the boxes. The origin for the local expansion for a box at this level can be shifted so that the expansion acts at the centre of that box's children. For a particular child, j at level C, this shift will provide a representation of the potential due all the charges outside j's parent's neighbours and next-nearest-neighbours, that is $\tilde{\psi}_{C,j}$. The interaction list details the boxes whose contribution to the potential must be added to $\tilde{\psi}_{C,j}$ to produce $\psi_{C,j}$.

The fast-multipole algorithm involves an upward pass through the mesh followed by a downward pass.

The upward pass. The multipole expansion of the potential about the centre of each box is formed at the finest mesh level. The origin of the multipole expansion for each child is shifted to the centre of the parent box and the total potential is obtained at this point by addition. This process, shown in the inset to Fig. 2, is repeated at all mesh levels.

The downward pass. At mesh level $C = 1$, $\tilde{\psi}_{1,1} \ldots \tilde{\psi}_{1,8} = 0$. Consider a particular box, b, at $C = 1$ and identify the boxes on its interactions list. The multipole expansions for these boxes are transformed to local expansions at the centre of b and added to $\tilde{\psi}_{1,b}$ to form $\psi_{1,b}$. This operation is performed for each of the eight boxes.

At the next level of the downward pass, the local expansions $\psi_{1,1} \ldots \psi_{1,8}$ are shifted to the children to form $\tilde{\psi}_{2,1} \ldots \tilde{\psi}_{2,64}$. These local expansions are corrected using the interaction list at level 2 to form $\psi_{2,1} \ldots \psi_{2,64}$. This process is repeated at all mesh levels.

Once the local expansions at the finest mesh level are available, they can be used to generate the potential and force on all the charges from beyond the second-nearest-neighbour boxes at this level. The potential and forces from the nearest-neighbour and second-nearest-neighbour boxes at this mesh level are calculated directly.

The algorithm can be readily extended to include periodic boundary conditions. At the end of the upward pass the multipole expansion, $\varphi_{0,1}$ for the entire cell is available. In a system with periodic boundary conditions, this is also the expansion for each of the image boxes around their centres. Now image-boxes which are not nearest-neighbour or second-nearest-neighbour images are well-separated from the central box and it is straightforward to form a local expansion from these images, $\psi_{0,1}$ to act at the centre of the central box. In the notation of the algorithm

$$\boldsymbol{\psi}_{0,1} = \mathbf{T} \cdot \boldsymbol{\varphi}_{0,1} \tag{48}$$

where $\mathbf{T}$ is a $P \times P$ matrix which contains an analytic sum over the well-separated images [18]. Christiansen *et al.* [19] have shown that the order of the addition of the image contributions to this sum is important. The calculation of $\boldsymbol{\psi}_{0,1}$ can be thought of as the first step of the downward pass.

The important advantage of this algorithm is that the running time is of $\mathcal{O}(N)$. Greengard and Rokhlin [15] have shown that the time goes as

$$N(e\,(\log_2 \epsilon)^2 + f) \tag{49}$$

where e and f are constants which depend on the computer system and implementation and ϵ is the precision of the calculation of the potential. The precision fixes the order of the multipole and local expansions, i.e. $P \approx \log_2 \epsilon$. The storage requirements of the algorithm also scale with N. The estimates also apply to the adaptive algorithm which is faster and requires less storage. In a simulation of an isolated neutral system of 25,600 charges in two dimensions, the adaptive fast multipole algorithm is about 100 times faster than the direct calculation of all of the interactions. These calculations are performed with $P = 20$.

So far there have been few comparisons between the fast multipole algorithm and the Ewald summation. For large enough systems the fast multipole method will certainly be faster, but some preliminary results for dipolar lattices indicate that the method is slower than the Ewald-like technique for systems smaller than 30000 dipoles [19]. In contrast, some recent results of Ding *et al.* [20, 21], which use an elegant reduced cell multipole algorithm to handle the long-range forces, indicate the fast multipole method is 38 times faster than the the Ewald method for a box containing 4816 ions. The algorithm has been parallelized using a message-passing paradigm [22]. Recently, an alternative strategy for parallelization across networked workstations has been reported by Board *et al.* [23].

3.2. LONG-RANGE FORCES IN SURFACE SIMULATIONS

3.2.1. An Ewald method. The Ewald method has recently been adapted by Hautman and Klein to model a set of charges in a quasi-two-dimensional environment [24]. This method is appropriate for ions and dipoles adsorbed on a surface and for lamellar and multilayer systems such as those found in a lipid bilayer and Langmuir Blodgett films. It takes advantage of the small spread in the z-coordinates of the charges in the direction normal to the surface or to the bilayer. The system is infinitely periodic in the plane of the surface or the bilayer but not in the orthogonal direction. This kind of geometry is shown in Fig. 3. s_{ij} is the ion separation in the plane of the surface and the ions are at heights z_i and z_j. This technique is most appropriate if $z_{ij} \ll s_{ij}$. The in-plane contribution to $1/r_{ij}$ can be isolated by using the identity

$$\frac{1}{r_{ij}} = \left[\frac{1}{r_{ij}} - \sum_{n=0}^{P} \frac{a_n z_{ij}^{2n}}{s_{ij}^{2n+1}}\right] + \left[\sum_{n=0}^{P} \frac{a_n z_{ij}^{2n}}{s_{ij}^{2n+1}}\right] \tag{50}$$

where the sums are binomial expansions of $1/r_{ij}$ in powers of z_{ij}/s_{ij}, with $a_0 = 1$, $a_1 = -1/2$, $a_2 = 3/8$ etc. Equation (50) is an identity if both the sums are truncated at the same term P.

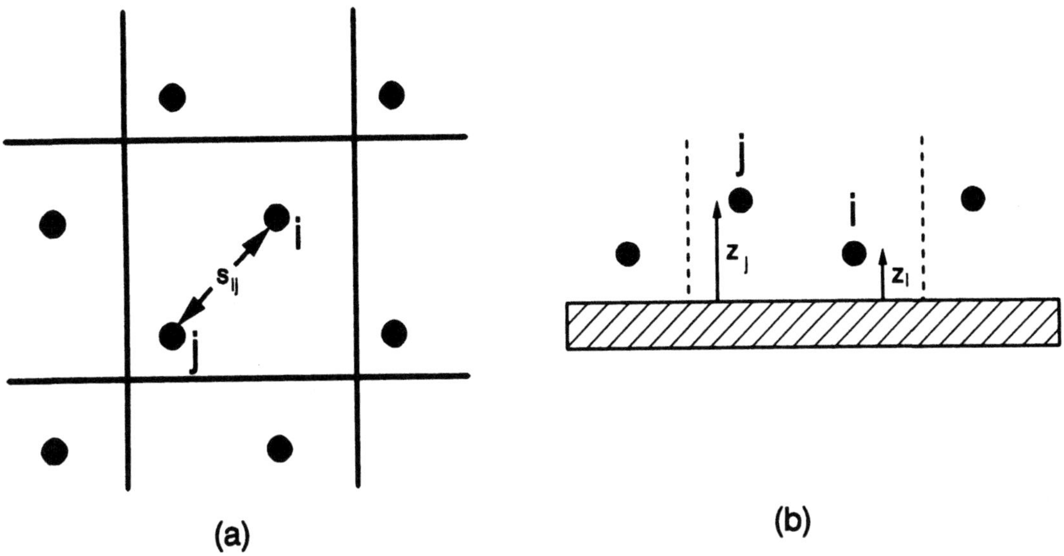

Fig. 3. Two charges close to a surface in a system with periodic boundary condition parallel to the surface.

In the Ewald method [2] a convergence function is used to rewrite the final summation in eqn (50).

$$\frac{1}{s_{ij}^{2n+1}} = \frac{1 - h_n(s_{ij},\, \alpha)}{s_{ij}^{2n+1}} + \frac{h_n(s_{ij},\, \alpha)}{s_{ij}^{2n+1}} \tag{51}$$

where the convergence function $h_n(s_{ij},\, \alpha)$ depends on a scalar convergence factor α and is different for each term of the P terms in the summation.

The total configurational energy of the infinitely periodic system is

$$\mathcal{V}_{\text{ion}} = \frac{1}{2} \sum_i \sum_j q_i q_j {\sum_{\boldsymbol{\nu}}}'' \frac{1}{r_{ij\nu}} \tag{52}$$

where

$$r_{ij\nu} = |\mathbf{r}_i - \mathbf{r}_j - \boldsymbol{\nu}| \tag{53}$$

and $\boldsymbol{\nu}$ is a two-dimensional lattice vector describing the array of simulation cells in the surface plane. The double prime on the sum over lattice vectors indicates that it is necessary to exclude the terms with $i = j$ when $|\boldsymbol{\nu}| = 0$. (The factor of $4\pi\epsilon_0$ has been included in the charges.) Substituting eqns (51) and (50) in eqn (52) gives

$$\mathcal{V}_{\text{ion}} = \mathcal{V}_{\text{s}} + \mathcal{V}_{\text{l}}$$

$$\mathcal{V}_s = \frac{1}{2}\sum_{i,j} q_i q_j \sum_{\boldsymbol{\nu}}{}'' \left[\frac{1}{r_{ij\nu}} - \sum_n \frac{a_n z_{ij}^{2n} h_n(s_{ij\nu},\, \alpha)}{s_{ij\nu}^{2n+1}}\right]$$

$$\mathcal{V}_l = \frac{1}{2}\sum_{i,j} q_i q_j \sum_{\boldsymbol{\nu}}{}'' \sum_n \frac{a_n z_{ij}^{2n} h_n(s_{ij\nu},\, \alpha)}{s_{ij\nu}^{2n+1}} \tag{54}$$

where the potential has been divided into a short-range part, $\mathcal{V}_s$, which is evaluated in real space, and a long-range part, $\mathcal{V}_l$, which is evaluated in reciprocal space.

It is necessary to choose P (the maximum term in the binomial expansion) to be sufficiently large that the real-space potentials converge rapidly. For examples in which the charges are sufficiently confined to a plane, such as the simulation of a Langmuir Blodgett film, the choice $P = 1$ allows the real-space sum to be truncated at $|\boldsymbol{\nu}| = 0$ and real-space interactions can be handled in the minimum-image convention. In this case

$$\mathcal{V}_s = \sum_{i<j} q_i q_j \left[\frac{1}{r_{ij}} - \frac{h_0(s_{ij},\, \alpha)}{s_{ij}} + \frac{z_{ij}^2 h_1(s_{ij},\, \alpha)}{2 s_{ij}^3}\right]. \tag{55}$$

Hautman and Klein have shown that the error functions and their derivatives are suitable convergence functions. The first two such functions are

$$h_0(s_{ij},\, \alpha) = \operatorname{erf}(t)$$

$$h_1(s_{ij},\, \alpha) = \operatorname{erf}(t) - \frac{2t}{\pi^{1/2}}(1 + 2t^2)\exp(-t^2) \tag{56}$$

where $t = s_{ij}/2\alpha$.

$\mathcal{V}_l$ is evaluated in reciprocal space to give

$$\mathcal{V}_l = \frac{\pi}{A}\sum_i \sum_j q_i q_j \sum_{\mathbf{G}\neq\mathbf{0}} \frac{\operatorname{erfc}(G\alpha)}{G}\left(1 + \frac{z_{ij}^2 G^2}{2}\right)\exp(-i\mathbf{G}\cdot\mathbf{s}_{ij}) - \left(\frac{1}{2\alpha\pi^{1/2}}\right)\sum_i q_i^2 . \tag{57}$$

$\mathbf{G}$ is a two-dimensional reciprocal lattice vector corresponding to the real-space lattice vector $\boldsymbol{\nu}$. The self-term, $i = j, \boldsymbol{\nu} = 0$, has to be included in the sum for $\mathcal{V}_l$ in eqn (54) to perform the Fourier transform and then subtracted from the final result; it is the last term in eqn (57). The double sum over i and j can be replaced by the products of single sums over i which can be rapidly evaluated. This factorization distinguishes the method from early techniques such as Ref. [25]. Finally, the long-range part is

$$\mathcal{V}_l = \frac{\pi}{A}\sum_{\mathbf{G}\neq\mathbf{0}} \frac{\operatorname{erfc}(G\alpha)}{G} A_{\mathbf{G}} A_{\mathbf{G}}^{*}$$

$$+\frac{\pi}{A}\sum_{\mathbf{G}\neq\mathbf{0}} G\,\mathrm{erfc}(G\alpha)\left[\frac{C_{\mathbf{G}}A_{\mathbf{G}}^* + C_{\mathbf{G}}^*A_{\mathbf{G}}}{2} - B_{\mathbf{G}}B_{\mathbf{G}}^*\right]$$

$$-\left(\frac{1}{2\alpha\pi^{1/2}}\right)\sum_i q_i^2 \tag{58}$$

where

$$\begin{aligned} A_{\mathbf{G}} &= \sum_i q_i \exp(-i\mathbf{G}\cdot\mathbf{s}_i) \\ B_{\mathbf{G}} &= \sum_i q_i z_i \exp(-i\mathbf{G}\cdot\mathbf{s}_i) \\ C_{\mathbf{G}} &= \sum_i q_i z_i^2 \exp(-i\mathbf{G}\cdot\mathbf{s}_i)\,. \end{aligned} \tag{59}$$

The calculation of the forces arising from this potential is straightforward. Hautman and Klein suggest that the method can also be used for nearly cubic cells between parallel walls such as water in a pore. In this case z_{ij} can be of $\mathcal{O}(L_z)$. The expansion $P = 1$ can still be used but the real-space simulation should include the direct interaction with images in cells $\nu_x = \pm 1$ and $\nu_y \pm 1$ as well as the central box.

3.2.2. Leckner's method. An interesting alternative to the use of the Ewald method is a summation technique developed by Leckner [26,27] which avoids the use of an arbitrary convergence parameter α. This technique is targeted at the simulation of charges in a pore geometry, although it can be used for completely periodic systems.

Consider a simulation involving Coulombic interactions in a cell of dimension L, L, D in the x, y and z directions. (Once again the cell is periodic the x and y directions but not in the z direction.) The method concentrates on calculating the force on an atom i from atom j and all of its images.

$$\mathbf{f}_{ij} = q_i q_j \sum_{\boldsymbol{\nu}} \frac{\mathbf{r}_i - \mathbf{r}_j - \boldsymbol{\nu}}{|\mathbf{r}_i - \mathbf{r}_j - \boldsymbol{\nu}|^3}\,. \tag{60}$$

Reduced distances, x, y and z, between atoms can be defined in terms of the box-length L

$$x_i - x_j = xL, \quad y_i - y_j = yL, \quad z_i - z_j = zL, \tag{61}$$

and the force on atom i can be written as a factor $q_i q_j / L$ times the following dimensionless functions

$$F_x(x, y, z) = \sum_{l=-\infty}^{\infty}\sum_{m=-\infty}^{\infty} \frac{x + l}{\left[(x+l)^2 + (y+m)^2 + z^2\right]^{3/2}}$$

$$F_y(x,y,z) = \sum_{l=-\infty}^{\infty}\sum_{m=-\infty}^{\infty} \frac{y+m}{\left[(x+l)^2+(y+m)^2+z^2\right]^{3/2}}$$

$$F_z(x,y,z) = \sum_{l=-\infty}^{\infty}\sum_{m=-\infty}^{\infty} \frac{z}{\left[(x+l)^2+(y+m)^2+z^2\right]^{3/2}} \,. \tag{62}$$

Clearly $F_y(x, y, z) = F_x(y, x, z)$ and it is only necessary to evaluate one of these two sums. Three exotic transformations are required to make these sums rapidly convergent. The first of these is the transformation of the sum to an integral using the Euler transformation

$$[(x+l)^2+(y+m)^2+z^2]^{-3/2} = \frac{2}{\sqrt{\pi}}\int_0^\infty dt\; t^{1/2}\exp(-t[(x+l)^2+(y+m)^2+z^2])\,. \tag{63}$$

The summations over the exponential in eqn (63) resulting from eqn (62) can be transformed using the Poisson-Jacobi identity

$$\sum_{l=-\infty}^{\infty}\exp[-(x+l)^2 t] = \left[\frac{\pi}{t}\right]^{1/2}\sum_{l=-\infty}^{\infty}\exp(-\pi^2 l^2/t)\cos(2\pi l x)\,. \tag{64}$$

The integral obtained after this substitution is a representation of the Bessel function of the second kind, K_ν

$$\int_0^\infty dt\; t^{\nu-1}\exp\left(-\frac{\pi^2 l^2}{t} - m^2 t\right) = 2\left(\pi\left|\frac{l}{m}\right|\right)^\nu K_\nu(2\pi|lm|)\,. \tag{65}$$

These three transformations can be used to produce the following convergent series

$$\begin{aligned} F_z(x,y,z) &= \frac{2\pi\sinh 2\pi z}{\cosh 2\pi z - \cos 2\pi y} \\ &\quad + 8\pi z\sum_{l=1}^{\infty} l\cos 2\pi l x \sum_{m=-\infty}^{\infty}[(y+m)^2+z^2]^{-1/2} \\ &\quad \times K_1(2\pi l\,[(y+m)^2+z^2]^{1/2}) \end{aligned} \tag{66}$$

and

$$\begin{aligned} F_x(x,y,z) &= \frac{2\pi\sin 2\pi x}{\cosh 2\pi z - \cos 2\pi x} \\ &\quad + 8\pi\sum_{l=-\infty}^{\infty}\frac{x+l}{[(x+l)^2+z^2]^{1/2}}\sum_{m=1}^{\infty} m\cos 2\pi m y \\ &\quad \times K_1(2\pi m[(x+l)^2+z^2]^{1/2})\,. \end{aligned} \tag{67}$$

The function $F_x(x, y, z)$ can be transformed to give

$$F_x(x, y, z) = 8\pi \sum_{l=1}^{\infty} l \sin 2\pi l x \sum_{m=-\infty}^{\infty} K_0(2\pi l[(y+m)^2 + z^2]^{1/2}) \,. \tag{68}$$

The potential energy between any two charges can be calculated indirectly by integrating the force. The result is

$$v_{\text{ion}}(r_{ij}) = \frac{q_i q_j}{L} V \tag{69}$$

where

$$\begin{aligned} V(x, y, z) \;=\; & 4 \sum_{l=1}^{\infty} \cos 2\pi l x \sum_{m=-\infty}^{\infty} K_0(2\pi l[(y+m)^2 + z^2]^{1/2}) \\ & - \log[\cosh 2\pi z - \cos 2\pi y] + C_2 \,. \end{aligned} \tag{70}$$

C_2 is an integration constant given by

$$C_2 = (1 + \sqrt{2}) M_s - \log 2 = 3.20711\ldots \tag{71}$$

where M_s is the Madelung constant for a square lattice of unit length with alternating unit charges,

$$M_s = 2\log 2 + 8 \sum_{l=1}^{\infty} \sum_{m=1}^{\infty} (-)^{l+1} K_0(\pi l(2m+1)) = 1.61554\ldots \,. \tag{72}$$

3.2.3. A comparison of the two techniques. These techniques can be tested by calculating the potential and forces associated with an infinite array of charges in the slab geometry. In the simplest case we consider two unit charges of opposite sign in a cubic box of unit length. A lattice of these boxes extends through space in the x and y direction and there are no images in the z direction. Charge 1 interacts with charge 2, all of the images of 1 and all of the images of 2. (The net force on 1 from all of the images of 1 is zero). Initially the separation $z = z_{ij}/L$ in the direction perpendicular to the surface is set to the small value of 0.1, while the charges are separated by $x = y = 0.1$ in the plane of the surface. The results for the potential energy are shown in Table I. For all of the calculations in this section which use Leckner's method, the first summation in eqn (70) runs from 1 to l_{max}: for $l = 3$, m runs from -3 to $+3$; for $l = 2$, m runs from -2 to $+2$; for $l = 1$, m run from -1 to $+1$; for all other l values only the $m = 0$ term is included. Leckner's method gives an accuracy of 1 part per million with $l_{\text{max}} = 10$. Hautman and Klein's method agrees with Leckner's method to within 200 parts per million.

As z increases to 0.25 (see Table II), Leckner's method converges with $l_{\text{max}} = 10$. Hautman and Klein's method produces an energy which is within

TABLE I
The energy between a charge +1.0 at (0.0, 0.0, 0.0) and −1.0 at (0.1, 0.1, 0.1) in an infinitely periodic array of cubic cells in a slab geometry using the methods of Leckner and Hautman and Klein.

Method		$-\mathcal{V}_{\text{ion}}$
Leckner	$l_{\text{max}} = 11$	5.77211
	$l_{\text{max}} = 12$	5.77212
	$l_{\text{max}} = 13$	5.77212
	$l_{\text{max}} = 14$	5.77211
HK	$\lvert\boldsymbol{\nu}_{\text{max}}\rvert = 0$	
	$\lvert\mathbf{G}_{\text{max}}\rvert = 15$	5.77191

TABLE II
The energy between a charge +1.0 at (0.0, 0.0, 0.0) and −1.0 at (0.0, 0.0, 0.25) in an infinitely periodic array of cubic cells in a slab geometry.

Method		$-\mathcal{V}_{\text{ion}}$
Leckner	$l_{\text{max}} = 4$	3.72467
	$l_{\text{max}} = 10$	3.72483
HK	$\lvert\boldsymbol{\nu}_{\text{max}}\rvert = 0$	
	$\lvert\mathbf{G}_{\text{max}}\rvert = 15$	3.71773
Heyes		3.7248

0.2% of the Leckner result. Leckner's result is in good agreement with an earlier result obtained by Heyes [25]. The Ewald-like method can be improved by including additional boxes in the real-space calculation. The $|\boldsymbol{\nu}_{\text{max}}| \leq 1$ calculation includes real-space interactions with the charges in the four nearest-neighbour cells. The $|\boldsymbol{\nu}_{\text{max}}| \leq 2$ calculation includes the additional real-space interactions with the eight next-nearest-neighbour cells. Table III shows that for two charges with $z = 0.2$, the Hautman-Klein method requires $|\boldsymbol{\nu}_{\text{max}}| \leq 2$ to produce results that are as accurate as Leckner's method. It is also clear that significant changes in α have little effect on the energy and the forces. Table III demonstrates that both methods can be used to calculate the forces as well as the potential energy. A test on pairs of charges with $z \approx 1$ indicates that the Hautman-Klein method at $P = 1$ requires $|\boldsymbol{\nu}_{\text{max}}| = 5$ to calculate forces with a reasonable accuracy and Leckner's method is the preferred choice for these geometries. This test is for two isolated charges on either side of the pore. The convergence of the Hautman-Klein method improves when the cell is filled with charges, so

TABLE III
The energy and forces between a charge +1.0 at (0.0, 0.0, 0.0) and −1.0 at (−0.25, −0.15, −0.2) in an infinitely periodic array of cubic cells in a slab geometry.

Method	α	$-\mathcal{V}_{\text{ion}}$	$-f_z$	$-f_x$	$-f_y$
Leckner ($l_{\text{max}} = 10$)		2.82156	6.44748	4.60495	2.91052
HK ($\lvert\nu_{\text{max}}\rvert = 0$)	0.16	2.82177	6.47830	4.56173	2.91496
HK ($\lvert\nu_{\text{max}}\rvert \leq 1$)	0.08	2.82132	6.45249	4.60519	2.91066
HK ($\lvert\nu_{\text{max}}\rvert \leq 1$)	0.14	2.82132	6.45250	4.60508	2.91059
HK ($\lvert\nu_{\text{max}}\rvert \leq 2$)	0.16	2.82151	6.44865	4.60493	2.91053

that ions on the opposite face are shielded from one another.

4. Conclusions

The molecular dynamics technique can be used to study the coupled Newtonian equations of motion of hundreds of thousands of atoms in an isolated cluster or in the bulk using periodic boundary conditions. Predictor-corrector methods or Verlet-like algorithms can be used to solve the equations of motion accurately so that the system stays on a constant-energy hypersurface in phase space and the method generates states in the microcanonical ensemble at constant-$NVE\mathbf{P}$. The average of some property $\mathcal{A}$ in this ensemble is equal to the time average over the molecular dynamics trajectory in phase space.

The technique now encompasses a variety of different ensembles. Extended lagrangians have been developed which allow for molecular dynamics at constant temperature, constant pressure and constant chemical potential and some of these are discussed in Chapter 5.

Constraint dynamics offers a general technique for solving the equations of motion of molecules of arbitrary complexity. The constrained Hamiltonian equations of motion have been identified, and for a system with M holonomic constraints an $M \times M$ matrix inversion is required to solve the equations of motion of the Lagrange multipliers. The error associated with the numerical solution of the differential equations means that the constraints are only satisfied to the order of the algorithm and the discrepancy with the exact solution grows with time. Three numerical methods which solve this problem have been discussed in section 2.3. Constraint dynamics plays an important rôle in the Car-Parrinello technique and this is discussed in Chapter 8.

For simulations of particles interacting through short-range forces such as the dispersion interaction, techniques such as the link-list algorithm mean that the computing time is of $\mathcal{O}(N)$, where N is the number of atoms in the simulation. For systems of charges and dipoles, the interactions are long range and the number of operations is of $\mathcal{O}(N^2)$. The Ewald method scales with $\mathcal{O}(N^{3/2})$ since additional terms have to be included in the k-space sum for larger systems. The fast-multipole algorithm described in section 3.1 provides a technique where the computing time is of $\mathcal{O}(N)$ and which will certainly be useful for simulations with large N. More detailed tests of the performance of this algorithm in comparison with the more established Ewald technique are required to determine the minimum system size at which it becomes cost-effective.

The Ewald and fast-multipole methods require that the simulation cell is extended by its images in the three coordinate directions. There is increasing interest in the simulations of charges in pores and thin films where the system is only periodic in two of the three coordinate directions. An extension of the Ewald method for pore geometries due to Hautman and Klein is compared with a direct summation technique due to Leckner. For a simple model system, the two techniques give very similar results for systems in which the z-separation of charges is $\approx 0.1L$. The Leckner technique provides an accurate and simple method of calculating the forces for simulations in which the z-separation of the charges is $\approx L$.

The molecular dynamics method is now a routine and well-established technique for studying systems of arbitrary molecular complexity. A small time step of $\mathcal{O}(10^{-15}\,\mathrm{s})$ is used to solve the equations of motion accurately and runs of 10^6 time steps are required to study a range of interesting phenomena on the nanosecond timescale. The necessary increases in speed are likely to come from parallelization of all parts of the code and this is an active area of research.

Acknowledgements

The work described in section 3.2.3 is carried out in collaboration with Dr. A. Clark of Unilever Research Port Sunlight, Dr D. Adolf and Dr. M.R.S. Pinches of Southampton University. I would like to thank Professor J. Perram for illuminating discussions on constraint dynamics and the fast-multipole method, and Dr. D. Adolf for reading this Chapter.

References

[1] H. GOLDSTEIN, *Classical Mechanics*, Addison-Wesley, Reading MA, second edition, 1980.

[2] M. P. ALLEN and D. J. TILDESLEY, *Computer Simulation of Liquids*, Clarendon Press, Oxford, 1987.

[3] C. W. GEAR, *The numerical integration of ordinary differential equations of various orders. Report ANL 7126*, Argonne National Laboratory, 1966.

[4] M. R. S. PINCHES, D. J. TILDESLEY, and W. SMITH, *Molec. Simul.* **6**, 51 (1991).

[5] D. J. EVANS, ***Molec. Phys.*** **34**, 317 (1977).

[6] D. FINCHAM, ***Molec. Simul.*** **8**, 165 (1992).

[7] K. REFSON and G. S. PAWLEY, ***Molec. Phys.*** **61**, 669 (1987).

[8] G. CICCOTTI and J. P. RYCKAERT, *Comput. Phys. Rep.* **4**, 345 (1986).

[9] S. W. DE LEEUW, J. W. PERRAM, and H. G. PETERSEN, *J. Stat. Phys.* **61**, 1203 (1990).

[10] J. P. RYCKAERT, G. CICCOTTI, and H. J. C. BERENDSEN, *J. Comput. Phys.* **23**, 327 (1977).

[11] W. F. VAN GUNSTEREN and H. J. C. BERENDSEN, ***Molec. Phys.*** **34**, 1311 (1977).

[12] H. C. ANDERSEN, *J. Comput. Phys.* **52**, 24 (1983).

[13] R. EDBERG, D. J. EVANS, and G. P. MORRISS, *J. Chem. Phys.* **84**, 6933 (1986).

[14] S. W. DE LEEUW, J. W. PERRAM, and E. R. SMITH, *Proc. Roy. Soc. Lond. A* **373**, 27 (1980).

[15] L. GREENGARD and V. ROKHLIN, *J. Comput. Phys.* **73**, 325 (1987).

[16] L. GREENGARD and V. ROKHLIN, *Lecture notes in Mathematics* **1360**, 121 (1988).

[17] J. CARRIER, L. GREENGARD, and V. ROKHLIN, *S.I.A.M. J. Sci. Stat. Comput.* **9**, 669 (1988).

[18] J. AMBROSIANO, L. GREENGARD, and V. ROKHLIN, *Comput. Phys. Commun.* **48**, 117 (1988).

[19] D. CHRISTIANSEN, J. W. PERRAM, and H. G. PETERSEN, *J. Comput. Phys.* (1993), to appear.

[20] H. Q. DING, N. KARASAWA, and W. A. GODDARD III, *J. Chem. Phys.* **97**, 4309 (1992).

[21] H. Q. DING, N. KARASAWA, and W. A. GODDARD III, *Chem. Phys. Lett.* **196**, 6 (1992).

[22] L. GREENGARD and W. D. GROPP, *Computers and Mathematics with Applications* **20**, 63 (1990).

[23] J. A. BOARD JR., J. W. CAUSEY, J. F. LEATHRUM JR., A. WINDEMUTH, and K. SCHULTEN, *Chem. Phys. Lett.* **198**, 89 (1992).

[24] J. HAUTMAN and M. L. KLEIN, ***Molec. Phys.*** **75**, 379 (1992).

[25] D. M. HEYES, ***Surface Science*** **110**, L619 (1987).

[26] J. LECKNER, ***Physica*** **A157**, 826 (1989).

[27] J. LECKNER, ***Physica*** **A176**, 485 (1991).

Back to Basics

M. P. Allen
H. H. Wills Physics Laboratory
Royal Fort
Tyndall Avenue
Bristol BS8 1TL
United Kingdom

ABSTRACT. An introduction is given to the statistical ensembles and distribution functions relevant to computer simulation methods. The effects of finite size, methods of extrapolating results to neighbouring state points, and behaviour near phase transitions, are all examined. Methods of computing transport coefficients in computer simulations are also discussed.

1. Introduction

The aim of this Advanced Study Institute is to spread some of the ideas that have developed in the field of condensed matter simulation over the last five years, and to look to the future. The danger of writing an introductory Chapter is that one either wastes space reviewing techniques that are familiar already, or one simply anticipates the material of later authors. Instead, I want to reprise some quite old ideas that should not be forgotten, and to give some pointers to the literature, old and new. The idea is to go into a little more depth, picking specific areas which I believe need close attention. Mostly, the motivation comes from current work: the rediscovery of old ideas in a new context, or the transfer of well-established methods from certain systems to others. I shall look at three basic aims of simulation: calculating thermodynamic properties, examining phase transitions, and investigating dynamics.

In chemical physics, we are generally most interested in atomic or molecular systems, rather than spin systems. Here, as elsewhere [1], we focus our attention on these models, in which the atoms move around, and we further restrict attention to single-component systems, for simplicity. However, many of the ideas come from the study of simpler spin systems, where the simulator has the luxury of extremely efficient simulation methods, per-

M.P. Allen and D.J. Tildesley (eds.), Computer Simulation in Chemical Physics, 49–92.

mitting long runs, to compensate for the sacrifice of realism. It may seem exacting to compare the quality of results from atomistic simulations with those from lattice simulations: however, the same principles and 'facts of life' apply to both. As computers become more powerful year by year, such comparisons become more meaningful.

It is impossible to give an introduction like this without covering some ground that is already well described elsewhere. Let me simply acknowledge my debt to the authors of the following reference works [2–6], and texts [7–14].

I start with some fairly simple questions, dealing with practical matters of carrying out simulations in different ensembles, run lengths, system sizes and so on. These will rely on general ideas of scaling, functional forms and asymptotic behaviour. Many of the underlying ideas were established years ago, when people had less powerful computers and, accordingly, more time and motivation to think about what they were doing. Then, I shall briefly discuss problems in 'real' physics: scaling behaviour near phase transitions and so on, but I stop short of attempting a detailed (and impossibly lengthy) exposition on the subject. There is, in any case, already an extensive literature in this area. Towards the end of this section, I give a case study: the critical point in the two-dimensional Lennard-Jones fluid. However, the development of specific applications is more the province of other contributors: see especially Chapters 4, 6, 11 and 10. Finally, I review some recent work concerning the calculation of transport coefficients by computer simulation, which again takes us back to basic principles established long ago.

2. Average values, fluctuations and errors

Here we revise something about average values and fluctuations *far away from* phase transitions.

2.1. MECHANICAL VARIABLES

Consider a set of atoms $i = 1 \ldots N$, with positions $\mathbf{r}_i$ and momenta $\mathbf{p}_i$. We use the shorthand $\mathbf{\Gamma}$ for these. Most of the following discussion applies equally to purely configurational simulations and properties, where $\mathbf{\Gamma}$ really means just the set of positions. Suppose we are looking at the average value of a quantity $\mathrm{a}(\mathbf{\Gamma})$, which is a well-defined function of $\mathbf{\Gamma}$. $\mathrm{a}(\mathbf{\Gamma})$ is *intensive*, i.e. does not scale with system size. It may be a sum, or pair sum, over all particles, divided by the number of particles:

$$\mathrm{a}(\mathbf{\Gamma}) = \frac{\mathcal{A}(\mathbf{\Gamma})}{N} = \frac{1}{N}\sum_i \mathrm{a}_i(\mathbf{\Gamma}),$$

an energy per molecule or magnetization per site, for example. Assuming that we have adopted sensible units based on molecular lengths, masses and energies, a and its average value $a = \langle \mathrm{a} \rangle$ in the ensemble being simulated will have values of $\mathcal{O}(1)$. The corresponding extensive variables $\mathcal{A}$ and $A = \langle \mathcal{A} \rangle$ will be of $\mathcal{O}(N)$. We define the fluctuation value $\delta \mathrm{a}(\mathbf{\Gamma}) = \mathrm{a}(\mathbf{\Gamma}) - \langle \mathrm{a} \rangle$, so $\langle \delta \mathrm{a} \rangle = 0$; $\left\langle \delta \mathrm{a}^2 \right\rangle$ is a measure of the fluctuations observed in a simulation. Define $\Delta(\mathrm{a}) = \sqrt{\left\langle \delta \mathrm{a}^2 \right\rangle}$, the root-mean-square fluctuation. We may expect $\left\langle \delta \mathrm{a}^2 \right\rangle \sim \mathcal{O}(N^{-1})$ and $\Delta(\mathrm{a}) \sim \mathcal{O}(N^{-1/2})$. We shall say much more about this later.

2.2. CORRELATION FUNCTIONS

We need to define two other quantities. a will vary with time: real time (in molecular dynamics) or fictitious time (in Monte Carlo). We can choose our own units for this: timesteps or Monte Carlo sweeps (that is, one attempted move *per particle*), as appropriate. Then, there will be a time correlation function for a, defined as $\langle \delta \mathrm{a}(0) \delta \mathrm{a}(t) \rangle$, which takes a value $\left\langle \delta \mathrm{a}^2 \right\rangle$ at $t = 0$ and decays to zero as $t \to \infty$ (because $\delta \mathrm{a}(t)$ and $\delta \mathrm{a}(0)$ become uncorrelated, and $\langle \delta \mathrm{a}(0) \delta \mathrm{a}(t) \rangle \to \langle \delta \mathrm{a}(0) \rangle \langle \delta \mathrm{a}(t) \rangle$). The timescale for this decay is called the *correlation time* τ_{a}. It may formally be defined as

$$\tau_{\mathrm{a}} = \int_0^\infty \mathrm{d}t \, \langle \delta \mathrm{a}(0) \delta \mathrm{a}(t) \rangle / \left\langle \delta \mathrm{a}^2 \right\rangle,$$

or it may be obtained from a long-time exponential decay $\exp\{-t/\tau_{\mathrm{a}}\}$ of the correlation function. The details are not crucial. In a similar way, a will vary from place to place. We assume that we can define a, $\langle \mathrm{a} \rangle$ and $\delta \mathrm{a}$ in a sub-region of our simulation box, just as we can for the whole box, ignoring (hopefully small) effects at the sub-region boundary. Then there will be a spatial correlation function $\langle \delta \mathrm{a}(0) \delta \mathrm{a}(r) \rangle$ relating values computed around points r apart. As above, this correlation will decay at large distances. The appropriate length scale is the *correlation length* ξ_{a}

$$\xi_{\mathrm{a}} = \int_0^\infty \mathrm{d}r \, \langle \delta \mathrm{a}(0) \delta \mathrm{a}(r) \rangle / \left\langle \delta \mathrm{a}^2 \right\rangle,$$

or again it may be associated with an asymptotically exponential decay $\exp\{-r/\xi_{\mathrm{a}}\}$ of the correlation function. (More precisely, $(1/r)\exp\{-r/\xi_{\mathrm{a}}\}$, see Chapter 10). Close to critical points both τ_{a} and ξ_{a} diverge in ways that are interesting (to some), but away from critical points they are typically only a few molecular units in magnitude. It is good practice, indeed almost essential, for simulation box sizes L to be large compared with ξ_{a} (for all properties of interest a), and for simulation run lengths t to be large compared with τ_{a}. Only then can we guarantee that reliably-sampled statistical properties are obtained. Close to critical points, we are forced to consider the converse situation.

To give an immediate application, we turn to the estimation of statistical errors in $\langle \mathrm{a} \rangle$. We highlight the fact that we are computing a finite time average by writing, temporarily, the simulation average as $\overline{\mathrm{a}}^t$ or a_t for short. We imagine this being a sample from an infinite ensemble (for example, a hypothetical infinite-time simulation). The standard result is that the rms error in the mean is

$$\Delta(\overline{\mathrm{a}}^t) = \Delta(\mathrm{a})/\sqrt{\mathcal{N}} \tag{1}$$

where $\mathcal{N}$ is the number of statistically independent samples in the run that go into $\overline{\mathrm{a}}^t$. We need to estimate $\mathcal{N}$. It is fairly clear that $\mathcal{N} \approx t/\tau_{\mathrm{a}}$. The result, eqn (1), follows from the 'central limit theorem' of probability. Giving a bit more detail, we write the time average and ensemble average separately. The time average of interest is

$$\delta a_t = \frac{1}{t}\int_0^t \mathrm{d}t' \; \delta \mathrm{a}(t')$$

so

$$\delta {a_t}^2 = \frac{1}{t^2}\int_0^t \mathrm{d}t' \int_0^t \mathrm{d}t'' \; \delta \mathrm{a}(t')\delta \mathrm{a}(t'').$$

Ensemble averaging, we obtain the standard result [15]

$$\begin{aligned} \left\langle \delta {a_t}^2 \right\rangle &= \frac{1}{t^2}\int_0^t \mathrm{d}t' \int_0^t \mathrm{d}t'' \; \langle \delta \mathrm{a}(t')\delta \mathrm{a}(t'') \rangle \\ &= \frac{2}{t}\int_0^t \mathrm{d}t' \; (1 - t'/t) \; \langle \delta \mathrm{a}(0)\delta \mathrm{a}(t') \rangle. \end{aligned}$$

If we average over only a short time compared with τ_{a}, this gives

$$\left\langle \delta {a_t}^2 \right\rangle = \left\langle \delta \mathrm{a}^2 \right\rangle$$

as expected: averaging makes no difference. On the other hand, if $t \gg \tau_{\mathrm{a}}$ we obtain, asymptotically,

$$\left\langle \delta {a_t}^2 \right\rangle = \left\langle \delta \mathrm{a}^2 \right\rangle \times 2\tau_{\mathrm{a}}/t \tag{2}$$

or, in other words,

$$\Delta(\overline{\mathrm{a}}^t) = \Delta(\mathrm{a})/\sqrt{t/2\tau_{\mathrm{a}}}. \tag{3}$$

So $\mathcal{N} = t/2\tau_{\mathrm{a}}$. If τ_{a} is not known *a priori* then the data can be analyzed to yield it, or $\mathcal{N}$, directly. One method is to select subruns of the full run, treating the latter as the 'underlying' ensemble, and estimating $\left\langle \delta {a_t}^2 \right\rangle$ for a range of shorter subrun times t. A plot of $t\left\langle\delta a_t{}^2\right\rangle\left\langle\delta \mathrm{a}^2\right\rangle^{-1}$ versus t^{-1} should yield an intercept $2\tau_{\mathrm{a}}$ in the limit $t^{-1} \to 0$. This is sometimes called the 'statistical inefficiency' $\mathcal{S}$, when computed in units of timesteps or Monte Carlo sweeps:

$$\mathcal{S} = 2\tau_{\mathrm{a}}/\delta t = t/\mathcal{N}\delta t = n_{\mathrm{step}}/\mathcal{N}.$$

It measures the ratio of 'effort', n_{step}, the number of steps, to 'achievement', $\mathcal{N}$, the number of statistically independent samples. A statistical inefficiency $\mathcal{S} = 20$ means that only every 20th step is truly independent. To make $\mathcal{S}$ as low as possible, evidently we should work with as long a timestep as possible, in MD, or investigate different algorithms empirically, in MC. We return to the question of errors shortly.

3. Ensembles and distributions

Molecular dynamics and Monte Carlo methods have been devised for each of the common statistical ensembles: the microcanonical, constant-NVE ensemble, the canonical, constant-NVT ensemble, the isothermal-isobaric, constant-NPT ensemble, and the grand canonical, constant-μVT ensemble [1, 16, 17]. There are eight such basic ensembles (basically a choice of two thermodynamically conjugate variables in each of three positions) and there has been some interest in developing the statistical mechanics of the four less common ones [18]. No doubt simulations of these will appear in due course.

3.1. ENSEMBLE EQUIVALENCE

The following paragraphs should be familiar from undergraduate and graduate statistical mechanics courses and texts [9, 10, 12]. The averages of a variable like a, computed in different ensembles, agree to within terms of $\mathcal{O}(1/N)$ or of $\mathcal{O}(L^{-d})$ where L is the box length and d the dimensionality. This is because the distributions of variables observed in many-particle simulations are very sharply peaked. Recall that the distribution function in the microcanonical ensemble is

$$\varrho_{NVE}(\mathbf{\Gamma}) = C\delta(E - \mathcal{E}(\mathbf{\Gamma}))$$

where we distinguish between E (the parameter of the ensemble) and $\mathcal{E}(\mathbf{\Gamma})$ (the energy of the system calculated as a function of positions and momenta $\mathbf{\Gamma}$). The phase-space volume is

$$\Omega_{NVE} = \int \mathrm{d}\mathbf{\Gamma}\ \varrho_{NVE}(\mathbf{\Gamma}) = C\int \mathrm{d}\mathbf{\Gamma}\ \delta(E - \mathcal{E}(\mathbf{\Gamma}))$$

where $\delta(x)$ is the Dirac delta function. If it helps, one can think of this as a 'surface area' rather than a volume, the surface being defined by the constraint of constant $\mathcal{E}(\mathbf{\Gamma})$; however, this 'surface' is actually of very high dimensionality. The entropy is $S = k_{\mathrm{B}} \ln \Omega_{NVE}$. However we can also define

$$\tilde{\Omega}_{NVE} = \int^{E} \mathrm{d}E'\, \Omega_{NVE'} = C \int \mathrm{d}\mathbf{\Gamma}\ \theta(E - \mathcal{E}(\mathbf{\Gamma}))$$

where $\theta(x)$ is the unit step function, and set $\tilde{S} = k_{\mathrm{B}} \ln \tilde{\Omega}_{NVE}$. The intensive quantities $s = S/N$ and $\tilde{s} = \tilde{S}/N$ differ only by terms of $\mathcal{O}(\ln N/N)$ at large N [10]. This is because the accessible volume of phase space rises so rapidly with E, that at the point of cutoff by the step function, only the region at $\mathcal{E}(\mathbf{\Gamma}) = E$ contributes. For an ideal gas, for example, $\tilde{\Omega}_{NVE} \propto E^{dN/2}$ and $\Omega_{NVE} \propto E^{dN/2-1}$. (To get a feel for these sorts of quantities, calculate the value of x^{500} for $x = 0.99, 1.00, 1.01$.) In the canonical ensemble we have

$$Q_{NVT} = \int \mathrm{d}\mathbf{\Gamma}\ \exp\{-\mathcal{E}(\mathbf{\Gamma})/k_{\mathrm{B}}T\}$$

so now the decaying exponential cuts off the rapid rise in phase space volume. The distribution of energies is still extremely sharply peaked around the maximum value $\hat{E}$, which will be close to the average $\langle \mathcal{E} \rangle_{NVT}$. Writing this as

$$Q_{NVT} = \int^{E} \mathrm{d}E'\, \Omega_{NVE'} \exp\{-E'/k_{\mathrm{B}}T\} \approx \Omega_{NVE} \exp\{-E/k_{\mathrm{B}}T\}$$

where we use the delta-function-like behaviour, and set $\langle \mathcal{E} \rangle_{NVT} = \hat{E} = E$, we see how this leads to the Helmholtz free energy

$$F = -k_{\mathrm{B}}T \ln Q_{NVT} = E - TS.$$

Also in this kind of work we learn never to try to Taylor-expand $\Omega_{NVE} \equiv \Omega(E)$ about a value of E, but instead to work with its logarithm. If $\Omega \sim E^N$ (just for illustration) then

$$\begin{aligned}
\Omega(E + \delta E) &\sim E^N[1 + \delta E/E]^N \\
&= E^N \left[1 + N(\delta E/E) + \tfrac{1}{2}N(N-1)(\delta E/E)^2 + \ldots\right] \\
&= E^N \left[1 + (\delta E/\epsilon) + \tfrac{1}{2}(\delta E/\epsilon)^2 + \ldots\right]
\end{aligned}$$

where $\epsilon = E/N$. The mounting powers of N prevent convergence; in practice, the expansion will be useless for δE even as small as the energy per particle, ϵ. (Again, to get a feel for this, imagine trying to expand x^{500} in a Taylor series around $x = 1$.) Instead, expanding the logarithm, $\ln \Omega \sim N \ln E$,

$$\begin{aligned}
\ln \Omega(E + \delta E) &\sim N \ln E + N \ln[1 + \delta E/E] \\
&= N \ln E + N \left[(\delta E/E) + \tfrac{1}{2}(\delta E/E)^2 + \ldots\right]
\end{aligned}$$

and the series converges rapidly. So it should, for $S = k_{\mathrm{B}} \ln \Omega$ and the derivatives of this (e.g. $\partial S/\partial E = 1/T$) are well-defined thermodynamic quantities.

3.2. SADDLE-POINT EXERCISE

The conversion between ensembles is almost a standard exercise in what are called 'saddle-point' methods [19]. We are interested in an integral of the form

$$\int_0^\infty dx\; x^n e^{-x} \equiv \Gamma(n+1)$$

where n is a large number ($n \sim N$ and $x \sim E$). This is actually the definition of the factorial or gamma function $\Gamma(n+1)$. The integrand has a (sharp) peak; it is easy to see that the maximum value is at $x = n$ (see Fig. 1).

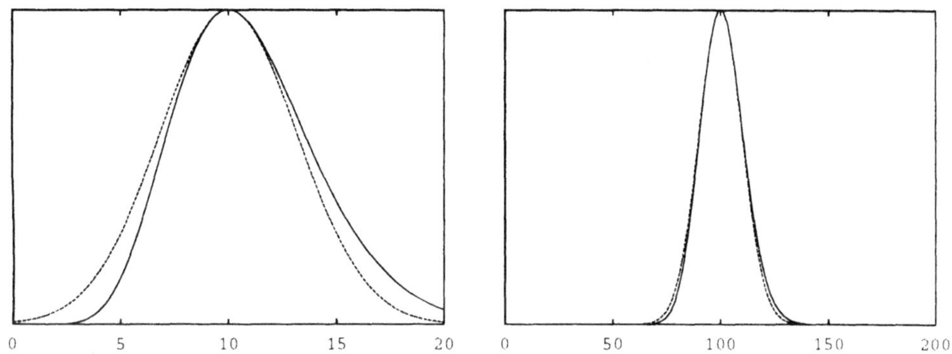

Fig. 1. Functions $x^{10}e^{-x}$ (left) and $x^{100}e^{-x}$ (right) illustrated with the Gaussian approximations (dashed lines) obtained by expanding the exponent. The vertical scales are arbitrary.

Writing the integrand as $\exp\{n \ln x - x\} \equiv \exp f(x)$ we expand about this value:

$$\left.\begin{array}{rcccc} f'(x) &=& (n/x) - 1 &=& 0 \\ f''(x) &=& -(n/x^2) &=& -1/n \end{array}\right\} \quad \text{for } x = n.$$

Then

$$\begin{aligned} \Gamma(n+1) &\approx \int_0^\infty dx\; \exp\left\{n \ln n - n - \frac{1}{2n}(x-n)^2\right\} \\ &\approx \exp\{n \ln n - n\} \int_{-\infty}^\infty dx\; \exp\left\{-\frac{1}{2n}(x-n)^2\right\}. \end{aligned}$$

Only a small error is made by extending the lower limit to $-\infty$ and the integral is just $\sqrt{2\pi n}$. The final result is

$$\Gamma(n+1) \approx \sqrt{2\pi n}\,(n^n e^{-n}).$$

The term in parentheses is the value of the integrand at its maximum; the prefactor is the correction term. This is perhaps more familiar as Stirling's approximation

$$\ln n! \approx n \ln n - n + \ln \sqrt{2\pi n} + \ldots .$$

3.3. ENSEMBLE CONVERSION

The example just treated was very simple, since there was only one independent variable, and so we sought a maximum of the function, not a saddle point. The same is true in our first example of ensemble conversion. We write

$$Q(\beta) = \exp\{-\beta F\} = \int \mathrm{d}E \; \Omega(E) \exp\{-\beta E\} \tag{4}$$

$$= \int \mathrm{d}E \; \exp\{S(E)/k_\mathrm{B} - \beta E\} \tag{5}$$

where we use the shorthand $Q(\beta) \equiv Q_{NVT}$, $\Omega(E) = \Omega_{NVE}$, and $\beta = 1/k_\mathrm{B}T$. To evaluate this, we need to identify the maximum. Let it be at $\hat{E}$ defined by

$$\left.\frac{\partial\{S(E)/k_\mathrm{B} - \beta E\}}{\partial E}\right|_{E=\hat{E}} = \left.\frac{\partial S(E)/k_\mathrm{B}}{\partial E}\right|_{E=\hat{E}} - \beta = 0$$

thus establishing the link $\partial S/\partial E = 1/T$ mentioned above. Then we evaluate the double derivative (at constant β) which will form the exponent of a Gaussian integral. Thermodynamics tells us that

$$\frac{\partial^2 S/k_\mathrm{B}}{\partial E^2} = \frac{\partial \beta}{\partial E} = -\beta^2 (k_\mathrm{B}/C_V) = -\frac{1}{k_\mathrm{B}T^2 C_V}.$$

This is negative, involves the constant-volume heat capacity $C_V = \partial E/\partial T$, and is of $\mathcal{O}(N^{-1})$, all of which is welcome news. The Gaussian integral gives $\sqrt{2\pi k_\mathrm{B}T^2 C_V}$ so

$$Q(\beta) = \Omega(\hat{E}) \exp\{-\beta \hat{E}\} \sqrt{2\pi k_\mathrm{B}T^2 C_V}$$

or

$$-\beta F = S/k_\mathrm{B} - \beta E + \ln \sqrt{2\pi k_\mathrm{B}T^2 C_V}. \tag{6}$$

We see that the corrections are small. The last term is only $\mathcal{O}(\ln N)$, so it can be neglected in comparison with the others, which are $\mathcal{O}(N)$; the relative error is $\mathcal{O}(\ln N/N)$.

This procedure may be run in reverse [13, 14] recognizing eqn (4) as a Laplace transform, and applying the saddle point method to its inverse

$$\Omega(E) = \frac{1}{2\pi i} \int_{\beta'-i\infty}^{\beta'+i\infty} \mathrm{d}\beta \; Q(\beta) \exp\{+\beta E\}$$

where $\beta' > 0$. This involves selecting a suitable path in the complex plane, $\beta' = \hat{\beta}$ defined by $(\partial \ln Q(\beta)/\partial\beta)_{\hat{\beta}} = -E$, but the procedure is straightforward and gives the same result

$$\begin{aligned}\ln \Omega(E) &\approx \ln Q(\hat{\beta}) + \hat{\beta}E + \ln\left[\left(2\pi\frac{\partial^2 \ln Q}{\partial\beta^2}\right)_{\hat{\beta}}^{-1/2}\right] \\ &= \ln Q(\hat{\beta}) + \hat{\beta}E - \ln\sqrt{2\pi k_{\mathrm{B}}T^2 C_V}.\end{aligned}$$

The saddle-point method is used by Smit, in discussing the Gibbs ensemble in Chapter 6.

3.4. DISTRIBUTIONS AND FLUCTUATIONS

We continue to discuss distributions in more detail. $\Omega_{NVE} \equiv \Omega(E)$, multiplied by the Boltzmann weight $\exp\{-\beta E\}$, is the distribution function for energies in the canonical ensemble. We can now write down our first approximation for this, around the average energy $\langle \mathcal{E} \rangle_{NVT}$ (which will be very close to the maximum value $\hat{E}$). Let $\delta E = E - \langle \mathcal{E} \rangle_{NVT} = E - \hat{E}$ and expand (just as we did a moment ago):

$$\mathcal{P}_{NVT}(\delta E) \propto \exp\{S(\hat{E} + \delta E)/k_{\mathrm{B}}\} \propto \exp\left\{\frac{\delta E^2}{2k_{\mathrm{B}}T^2 C_V}\right\}. \tag{7}$$

This is a familiar-looking Gaussian. We see the standard canonical ensemble result that

$$\left\langle \delta E^2 \right\rangle_{NVT} = k_{\mathrm{B}}T^2\left(\frac{\partial E}{\partial T}\right)_{NV} = k_{\mathrm{B}}T^2 C_V$$

or

$$\left\langle \delta\epsilon^2 \right\rangle_{NVT} = \frac{k_{\mathrm{B}}T^2}{N}\left(\frac{\partial \epsilon}{\partial T}\right)_{NV} = \frac{k_{\mathrm{B}}T^2}{N} c_V$$

where $c_V = C_V/N$ is the constant-volume heat capacity per particle. This is consistent with the correction term given above.

3.5. CONVERTING AVERAGES

It only requires a slight extension to obtain the lowest-order correction for any property, and this was essentially the work of Lebowitz *et al.* [20]. Simply Taylor-expanding about $\hat{E} \approx \langle \mathcal{E} \rangle_{NVT}$ gives the result

$$\langle \mathrm{a} \rangle_{NVT} = \langle \mathrm{a} \rangle_{NVE} + \tfrac{1}{2}\left(\frac{\partial^2 a}{\partial E^2}\right)\left\langle \delta E^2 \right\rangle_{NVT} = \langle \mathrm{a} \rangle_{NVE} + \tfrac{1}{2}\left(\frac{\partial^2 a}{\partial \epsilon^2}\right)\left\langle \delta\epsilon^2 \right\rangle_{NVT}$$

where the partial derivatives are evaluated at $\hat{E}$. Since $\left\langle \delta\epsilon^2 \right\rangle_{NVT} = k_B T^2 c_V / N$, the $\mathcal{O}(N^{-1})$ dependence of the difference between ensembles is obvious. Lebowitz *et al.* [20] rearrange this expression to conveniently express $\langle \mathrm{a} \rangle_{NVE}$ entirely in terms of constant-T averages and derivatives

$$\langle \mathrm{a} \rangle_{NVE} = \langle \mathrm{a} \rangle_{NVT} - \tfrac{1}{2} k_B T^2 \frac{\partial}{\partial T} \left((N c_V)^{-1} \frac{\partial \mathrm{a}}{\partial T} \right).$$

They also obtain the corresponding formulae for fluctuations, an example being

$$\left\langle \delta \mathrm{a}^2 \right\rangle_{NVE} = \left\langle \delta \mathrm{a}^2 \right\rangle_{NVT} - \frac{k_B T^2}{N c_V} \left(\frac{\partial a}{\partial T} \right)^2 .$$

All the terms here are of $\mathcal{O}(N^{-1})$. Quite generally, we see that fluctuations are smaller in the microcanonical ensemble than in the canonical ensemble: not surprising, since an extra constraint is applied in the former case. For $\mathrm{a} = \epsilon$ we obtain $\left\langle \delta\epsilon^2 \right\rangle_{NVT} = k_B T^2 c_v / N$ as before. For $\mathrm{a} = k$, the kinetic energy per particle, we obtain the well-known expression for calculating the specific heat in molecular dynamics simulations

$$\left\langle \delta k^2 \right\rangle_{NVE} = \frac{3 {k_B}^2 T^2}{2N} \left(1 - \frac{3 k_B}{2 c_v} \right).$$

3.6. DIRECT METHODS

Surprisingly (in retrospect) this seems to have been the main method for deriving statistical mechanical results in ensembles like the microcanonical one. Only in the last few years have more direct methods become popular. Pearson *et al.* [21] have derived several results at constant NVE. In most of their paper they adopt the $\tilde{S}$ definition of the entropy, but in one section they compare with the $S = k_B \ln \Omega_{NVE}$ definition, based on $\Omega_{NVE} = C \int \mathrm{d}\mathbf{\Gamma}\ \delta(E - \mathcal{E}(\mathbf{\Gamma}))$. This is the one most appropriate for molecular dynamics, where energy is strictly conserved. The momentum variables can be integrated out leaving (apart from an arbitrary additive constant)

$$S = k_B \ln \Omega_{NVE} = k_B \ln \int \mathrm{d}\mathbf{r}\ \frac{2}{3N} (E - \mathcal{V}(\mathbf{r}))^f$$

where $f = (3N/2) - 1$ is the number of degrees of freedom. (Actually, if the total momentum is conserved as well as the energy, $N \to N - 1$ here, and $f = [3(N-1)/2] - 1$.) The average of a configurational quantity $\mathrm{a}(\mathbf{r})$ is given by

$$\langle \mathrm{a} \rangle_{NVE} = \frac{\int \mathrm{d}\mathbf{r}\ \mathrm{a}(\mathbf{r})(E - \mathcal{V}(\mathbf{r}))^f}{\int \mathrm{d}\mathbf{r}\ (E - \mathcal{V}(\mathbf{r}))^f} = \frac{\int \mathrm{d}\mathbf{r}\ \mathrm{a}(\mathbf{r}) \mathcal{K}(\mathbf{r})^f}{\int \mathrm{d}\mathbf{r}\ \mathcal{K}(\mathbf{r})^f}.$$

The kinetic energy $\mathcal{K}$ is written as a function of *coordinates* through $\mathcal{K}(\mathbf{r}) = E - \mathcal{V}(\mathbf{r})$. Laplace transform techniques are used to tackle the integrals. A weight function of the kind appearing here has already been used (replacing the familiar Boltzmann factor) in Monte Carlo simulations of configurational properties [16], and the direct method has been used to show that the microcanonical ensemble velocity distribution is non-Maxwellian (for small systems) [22]. The temperature is defined by

$$(k_{\mathrm{B}}T)^{-1} = \left(\frac{\partial S}{\partial E}\right)_{NV} = f\left\langle \mathcal{K}^{-1}\right\rangle_{NVE} = \left((3/2) - N^{-1}\right)\left\langle k^{-1}\right\rangle_{NVE}$$

where $k = \mathcal{K}/N$ as usual. Further differentiation gives an expression for the heat capacity c_v which agrees (to within small-size correction terms) with that of Lebowitz *et al.*. Finally, we should recall that Frenkel [23] derived (directly) the expression for determining the excess chemical potential by test-particle insertion in this ensemble.

4. Distribution functions

4.1. MANIPULATING DISTRIBUTIONS

We have already started discussing *distributions* of the energy and other quantities. These are easily accumulated in computer simulations. Making a histogram of the frequency of occurrence of all values of a is equivalent (to the accuracy of the histogram bin width) to determining the distribution function:

$$\mathcal{P}(a) = C\langle \delta(\mathrm{a}(\mathbf{\Gamma}) - a)\rangle.$$

The constant is determined by normalization. We already noted that the energy distribution function in the canonical ensemble is simply the weighted density of states in the microcanonical ensemble

$$\begin{aligned} \mathcal{P}_{NVT}(E) &= C\langle \delta(\mathcal{E}(\mathbf{\Gamma}) - E)\rangle_{NVT} = C\Omega_{NVE} \exp\{-E/k_{\mathrm{B}}T\} \\ &= \frac{\Omega_{NVE} \exp\{-E/k_{\mathrm{B}}T\}}{\int \mathrm{d}E'\, \Omega_{NVE'} \exp\{-E'/k_{\mathrm{B}}T\}} \\ &= \frac{\Omega_{NVE} \exp\{-E/k_{\mathrm{B}}T\}}{Q_{NVT}}. \end{aligned}$$

Since $S = k_{\mathrm{B}} \ln \Omega_{NVE}$ and $T^{-1} = (\partial S/\partial E)$, this suggests that the entire equation of state $T(E)$ could be obtained from a single canonical ensemble distribution. Suppose that we have conducted a simulation at temperature T_0, and accumulated $\mathcal{P}_{NVT_0}(E)$. We define $\mathcal{F}_{NVT_0}(E)$ by

$$\beta_0\mathcal{F}_{NVT_0}(E) = -\ln \mathcal{P}_{NVT_0}(E) = \beta_0 E - S(E)/k_{\mathrm{B}} - \beta_0 F_0$$

where $\beta_0 = 1/k_B T_0$. The constant necessary to determine the absolute value of F_0 cannot be obtained from $\mathcal{P}_{NVT_0}(E)$, but if we differentiate with respect to E we obtain

$$\frac{\partial(-\beta_0 \mathcal{F}_{NVT_0}(E))}{\partial E} = \frac{\partial \ln \mathcal{P}_{NVT_0}(E)}{\partial E} = \frac{\partial \mathcal{P}_{NVT_0}(E)/\partial E}{\mathcal{P}_{NVT_0}(E)} = \beta(E) - \beta_0$$

where $\beta(E) = 1/k_B T(E)$. Thus, $\mathcal{P}_{NVT_0}(E)$ gives us the equation of state in the neighbourhood of T_0. We can see that this makes sense very close to T_0. $\mathcal{F}_{NVT_0}(E)$ clearly is the Helmholtz free energy, relative to F_0, associated with energy fluctuations, and it varies thermodynamically according to the above expression. Differentiating one more time gives

$$\frac{\partial^2(-\beta \mathcal{F}_{NVT_0}(E))}{\partial E^2} = \frac{\partial^2 \ln \mathcal{P}_{NVT_0}(E)}{\partial E^2} = \frac{\partial \beta(E)}{\partial E} = -\frac{1}{k_B T^2 C_V}$$

If we truncate at this point, effectively doing a (familiar-looking) Taylor expansion of $\ln \mathcal{P}_{NVT_0}(E)$ about its peak value $\hat{E}$, we obtain the usual Gaussian form for $\mathcal{P}_{NVT_0}(\delta E)$, eqn (7) with a mean-squared value $\left\langle \delta E^2 \right\rangle_{NVT_0} = k_B T_0^2 C_V$ as seen before. At the same time, we see that this is consistent with an equation of state with $\mathcal{F}_{NVT_0}(E)$ quadratic in E, and with β linear in E, the coefficient being determined by C_V. Further expansion of the probability distribution would yield non-Gaussian corrections, and higher-order derivatives of the equation of state.

In a similar way, the volume distribution in the isobaric ensemble is easily accumulated in a histogram,

$$\begin{aligned}
\mathcal{P}_{NPT}(V) &= C\langle \delta(V(\mathbf{\Gamma}) - V) \rangle_{NPT} = C Q_{NVT} \exp\{-\beta PV\} \\
&= \frac{Q_{NVT} \exp\{-\beta PV\}}{\int \mathrm{d}V' \, Q_{NV'T} \exp\{-\beta PV'\}} \\
&= \frac{Q_{NVT} \exp\{-\beta PV\}}{Q_{NPT}}.
\end{aligned}$$

Since $F = -k_B T \ln Q_{NVT}$ and $P = -(\partial F/\partial V)$, one might think that one could obtain $P(V)$ completely from a single determination of $\mathcal{P}_{NPT}(V)$. Define $\mathcal{G}_{NP_0T}(V)$ by

$$\beta \mathcal{G}_{NP_0T}(V) = -\ln \mathcal{P}_{NP_0T}(V) = \beta P_0 V + \beta F(V) - \beta G_0.$$

Again, our histogram cannot give us G_0, but differentiation with respect to V gives

$$\frac{\partial(-\beta \mathcal{G}_{NP_0T}(V))}{\partial V} = \frac{\partial \ln \mathcal{P}_{NP_0T}(V)}{\partial V} = \frac{\partial \mathcal{P}_{NP_0T}(V)/\partial V}{\mathcal{P}_{NP_0T}(V)} = \beta P(V) - \beta P_0.$$

So $\mathcal{P}_{NP_0T}(V)$ gives us the equation of state in the neighbourhood of P_0. $\mathcal{G}_{NP_0T}(V)$ is the Gibbs free energy for volume fluctuations, relative to G_0, and it varies thermodynamically according to the above expression. Differentiating one more time gives

$$\frac{\partial^2(-\beta\mathcal{G}_{NP_0T}(V))}{\partial V^2} = \frac{\partial^2 \ln \mathcal{P}_{NP_0T}(V)}{\partial V^2} = \frac{\partial \beta P(V)}{\partial V} = -\frac{1}{V k_B T \beta_T}$$

where β_T is the isothermal compressibility: $V\beta_T = -(\partial V/\partial P)$. Truncating gives the Gaussian form

$$\begin{aligned} \ln \mathcal{P}_{NP_0T}(\delta V) &= \text{constant} - \frac{\delta V^2}{2V k_B T \beta_T} \\ \mathcal{P}_{NP_0T}(\delta V) &\propto \exp\left\{-\frac{\delta V^2}{2V k_B T \beta_T}\right\} \end{aligned}$$

with a mean-squared value $\left\langle \delta V^2 \right\rangle_{NPT} = V k_B T \beta_T$. This is consistent with an equation of state with $\mathcal{G}_{NP_0T}(V)$ quadratic in V, and with V and P being linearly related through the isothermal compressibility.

At the risk of exhausting the story, the same approach applies to *any* variable a. We define the free energy function by

$$\begin{aligned} \mathcal{P}_{NPT}(a) &= C\langle\delta(\mathrm{a}(\mathbf{\Gamma}) - a)\rangle_{NPT} \\ &= \frac{\int \mathrm{d}\mathbf{\Gamma} \int \mathrm{d}V \ \exp\{-\beta(\mathcal{E}(\mathbf{\Gamma}) + PV)\}\delta(\mathrm{a}(\mathbf{\Gamma}) - a)}{Q_{NPT}} \\ &\equiv \exp\{-\beta\mathcal{G}_{NPT}(a)\}. \end{aligned}$$

There will be a thermodynamic variable conjugate to $\mathrm{a}(\mathbf{\Gamma})$, defined by

$$f a \equiv \left(\frac{\partial(-\beta\mathcal{G}_{NPT}(a))}{\partial a}\right) = \left(\frac{\partial \ln \mathcal{P}_{NPT}(a)}{\partial a}\right)$$

driving the system towards the minimum free energy (maximum probability). This is related to the free energy through the thermodynamic relation $fa \equiv (\partial(-\beta G)/\partial a)$. These functions govern fluctuations of a [11].

Even in their application to simulations, these ideas are very old [24] and are still being developed today. We return to them below. Suffice it to say now that, from what we know already, our determination of these distribution functions will be statistically very poor outside the very narrow range ($\Delta(\mathrm{a}) \sim \mathcal{O}(N^{-1})$) over which they adopt significant values. To proceed *beyond* the lowest (Gaussian) approximation will mean accurate measurements out in the wings, and this may not be so easy. We turn to this now.

4.2. NON-GAUSSIAN BEHAVIOUR AND FINITE SIZE

Now we look in more detail at the distribution functions for quantities, in a finite-size simulation. This material appears in many places [25, 26]. We concentrate on variables a which are effectively sums of contributions from all over the system. We could write them as a sum over particles,

$$\mathrm{a} = (1/N) \sum_i \mathrm{a}_i,$$

where a_i is a value per particle, or as an average over the box volume

$$\mathrm{a} = (1/V) \int d\mathbf{r}\ \mathrm{a}(\mathbf{r}) = L^{-d} \int d\mathbf{r}\ \mathrm{a}(\mathbf{r}).$$

If we define

$$\mathrm{a}(\mathbf{r}) = (V/N) \sum_i \mathrm{a}_i \delta(\mathbf{r} - \mathbf{r}_i)$$

we recover the first form, so it makes no difference. It is not necessary for a_i to depend only on the coordinates of particle i: just that it depend on particles in the locality of i. For example, we could write the potential energy this way, dividing pairwise terms (assumed short in range) equally between the participating atoms. Note that a has the form of an average, over the box volume, even before any ensemble averaging has taken place. We highlight this by writing it as $\overline{\mathrm{a}}^L$, or simply as a_L for short. What follows is just a generalization of the analysis of time averages $\overline{\mathrm{a}}^t$ seen earlier. Define

$$a_L = \overline{\mathrm{a}}^L = L^{-d} \int d\mathbf{r}\ \mathrm{a}(\mathbf{r}).$$

These finite-size averages are sampled from an ensemble: denote averages over the members of the ensemble by $\langle \ldots \rangle$. We may be considering a subsystem, size L^d, of an infinite system, or a complete system simulated in the grand canonical or some other ensemble, with prescribed boundary conditions. There will be differences between these cases in general, but we neglect these for simplicity here.

Generally, we expect the mean value $\langle a_L \rangle$ to exist, to be of $\mathcal{O}(1)$, and to be within $\mathcal{O}(L^{-d})$ of the thermodynamic limit. The static fluctuations will be smaller, scaling with L^{-d}, but when this scaling is divided out we are left with a thermodynamic quantity (like the specific heat). Call this c_{a}, i.e. define

$$\left\langle \delta a_L{}^2 \right\rangle = \langle a_L{}^2 \rangle - \langle a_L \rangle^2 \equiv L^{-d} c_{\mathrm{a}}. \tag{8}$$

c_{a} can be regarded as independent of L. This is consistent with a Gaussian probability distribution function,

$$\mathcal{P}(\delta a_L) = L^{d/2} (2\pi c_{\mathrm{a}})^{-1/2} \exp\{-L^d \delta \mathrm{a}^2 / 2c_{\mathrm{a}}\} \tag{9}$$

The Gaussian form will be correct if $L \gg \xi_{\mathrm{a}}$, the correlation length for the variable a. We believe this because we have been told about the central limit theorem of probability: any quantity which is the sum of statistically independent contributions will be distributed in a Gaussian way. We have the idea that a is the sum of almost independent contributions from different regions (of size ξ_{a}) of the fluid. Here we show this in (slightly) more detail, and discuss non-Gaussian corrections. The approach is similar to the one applied earlier to time correlations and statistical errors.

We shall need to discuss higher moments of the distribution, $\langle \delta a_L{}^n \rangle$, $n = 1, 2, 3, \ldots$, and *cumulants* $\langle a_L{}^n \rangle^{\mathrm{c}}$. The latter are more useful because they vanish for a Gaussian distribution. They are discussed in any text on probability theory, and also in Kubo's review article [27], which gives a range of applications. Cumulants are defined by the following expansion which, apart from a factor i in the exponent, is the *characteristic function*:

$$\langle \exp\{\kappa \mathrm{a}\} \rangle = \sum_{n=0}^{\infty} \frac{\kappa^n \langle \mathrm{a}^n \rangle}{n!} \equiv \exp\left\{ \sum_{n=1}^{\infty} \frac{\kappa^n}{n!} \langle \mathrm{a}^n \rangle^{\mathrm{c}} \right\}.$$

Note that we are expanding an *exponent*, which, as we have seen, is a good idea. The first few cumulants are

$$\begin{aligned}
\langle a_L \rangle^{\mathrm{c}} &= \langle a_L \rangle \\
\langle a_L{}^2 \rangle^{\mathrm{c}} &= \left\langle \delta a_L{}^2 \right\rangle \\
\langle a_L{}^3 \rangle^{\mathrm{c}} &= \left\langle \delta a_L{}^3 \right\rangle \\
\langle a_L{}^4 \rangle^{\mathrm{c}} &= \left\langle \delta a_L{}^4 \right\rangle - 3\left\langle \delta a_L{}^2 \right\rangle^2 \\
\langle a_L{}^5 \rangle^{\mathrm{c}} &= \left\langle \delta a_L{}^5 \right\rangle - 10\left\langle \delta a_L{}^3 \right\rangle\left\langle \delta a_L{}^2 \right\rangle \\
\langle a_L{}^6 \rangle^{\mathrm{c}} &= \left\langle \delta a_L{}^6 \right\rangle - 15\left\langle \delta a_L{}^4 \right\rangle\left\langle \delta a_L{}^2 \right\rangle - 10\left\langle \delta a_L{}^3 \right\rangle^2 + 30\left\langle \delta a_L{}^2 \right\rangle^3
\end{aligned}$$

The first term, $\langle a_L \rangle^{\mathrm{c}}$, is of $\mathcal{O}(1)$, and the second is $\langle a_L{}^2 \rangle^{\mathrm{c}} \sim \mathcal{O}(N^{-1}) \sim \mathcal{O}(L^{-d})$, as we have seen earlier. It turns out that $\langle a_L{}^3 \rangle^{\mathrm{c}} \sim \mathcal{O}(L^{-2d})$ and $\langle a_L{}^4 \rangle^{\mathrm{c}} \sim \mathcal{O}(L^{-3d})$. The important point is that an extra power of L^{-d} comes in with each higher-order cumulant: in general

$$\langle a_L{}^n \rangle^{\mathrm{c}} \sim \mathcal{O}(N^{n-1}) \sim \mathcal{O}(L^{d(n-1)}).$$

This can be understood as follows. Write δa_L as a volume integral,

$$\delta a_L = V^{-1} \int \mathrm{d}\mathbf{r}\ \delta \mathrm{a}(\mathbf{r})$$

where $\delta \mathrm{a}(\mathbf{r}) = \mathrm{a}(\mathbf{r}) - \langle \mathrm{a} \rangle$. Then write

$$\left\langle \delta a_L{}^2 \right\rangle = \frac{1}{V^2} \int \mathrm{d}\mathbf{r}_1 \int \mathrm{d}\mathbf{r}_2\ \langle \delta \mathrm{a}(\mathbf{r}_1) \delta \mathrm{a}(\mathbf{r}_2) \rangle = \frac{1}{V} \int \mathrm{d}\mathbf{r}_{12}\ \langle \delta \mathrm{a} \delta \mathrm{a}(\mathbf{r}_{12}) \rangle$$

We have eliminated one of the integrals, by assuming translational invariance: we simply choose one point as the origin. Then we use the definition of the pair correlation function to write $\langle \delta a \delta a(\mathbf{r}) \rangle = \left\langle \delta a^2 \right\rangle c^{(2)}(\mathbf{r})$, where $\left\langle \delta a^2 \right\rangle$ is the average over a small distance (or the single-particle average): it will have a value of $\mathcal{O}(1)$. We do not need the precise form of $c^{(2)}(\mathbf{r})$. All we need to know is that it will decay to zero on a length scale ξ_a, so the d-dimensional integration will yield a volume of order $\xi_a{}^d$. The factor V^{-1}, of course, is just L^{-d}. Thus

$$\left\langle \delta a_L{}^2 \right\rangle \sim \left\langle \delta a^2 \right\rangle \times (\xi_a/L)^d \equiv L^{-d} c_a$$

where we compare with eqn (8).

For the next cumulant, a similar derivation goes through. We will need to integrate over three coordinates; one of the integrations will be redundant. Thus

$$\begin{aligned} \langle \delta a(\mathbf{r}_1) \delta a(\mathbf{r}_2) \delta a(\mathbf{r}_3) \rangle &= \left\langle \delta a^3 \right\rangle V^{-3} \int d\mathbf{r}_1 \int d\mathbf{r}_2 \int d\mathbf{r}_3 \; c^{(3)}(\mathbf{r}_1, \mathbf{r}_2, \mathbf{r}_3) \\ &= \left\langle \delta a^3 \right\rangle V^{-2} \int d\mathbf{r}_{12} \int d\mathbf{r}_{13} \; c^{(3)}(\mathbf{r}_{12}, \mathbf{r}_{13}) \end{aligned}$$

where we introduce a three-particle correlation function. This will vanish whenever any particle strays outside a range of ξ_a of the other two, so the integral can only give rise to a value of $\mathcal{O}(\xi_a{}^{2d})$. The result is

$$\left\langle \delta a_L{}^3 \right\rangle \sim \left\langle \delta a^3 \right\rangle \times (\xi_a/L)^{2d} \equiv L^{-2d} c_a'$$

by analogy with eqn (8). c_a' will turn out to be another thermodynamic derivative like c_a.

For the fourth cumulant the derivation has one more twist. In the crucial integral

$$\langle \delta a(\mathbf{r}_1) \delta a(\mathbf{r}_2) \delta a(\mathbf{r}_3) \delta a(\mathbf{r}_4) \rangle = \left\langle \delta a^4 \right\rangle V^{-4} \int d\mathbf{r}_1 \ldots \int d\mathbf{r}_4 \; c^{(4)}(\mathbf{r}_1, \mathbf{r}_2, \mathbf{r}_3, \mathbf{r}_4)$$

a nonzero integrand does not necessarily depend on *all* the coordinates lying within the same region of space, of dimension ξ_a. There will be a contribution whenever pairs are close, let's say $\{\mathbf{r}_1, \mathbf{r}_2\}$ and $\{\mathbf{r}_3, \mathbf{r}_4\}$. We can see that we get terms involving $\left\langle \delta a_L{}^2 \right\rangle$ again:

$$\begin{aligned} \left\langle \overbrace{\delta a(\mathbf{r}_1) \delta a(\mathbf{r}_2)} \overbrace{\delta a(\mathbf{r}_3) \delta a(\mathbf{r}_4)} \right\rangle &= \left[\left\langle \delta a^2 \right\rangle V^{-2} \int d\mathbf{r}_1 \int d\mathbf{r}_2 \; c^{(2)}(\mathbf{r}_1, \mathbf{r}_2) \right] \\ &\quad \times \left[\left\langle \delta a^2 \right\rangle V^{-2} \int d\mathbf{r}_3 \int d\mathbf{r}_4 \; c^{(2)}(\mathbf{r}_3, \mathbf{r}_4) \right] \\ &\sim \left\langle \delta a_L{}^2 \right\rangle^2 \sim (\xi_a/L)^{2d} \end{aligned}$$

where we indicate the proximity constraint by braces. This spoils the steady reduction in the order of magnitude of the moments, which is why the cumulants are defined with these terms subtracted off. There are three possible ways of pairing the coordinates, so the modified average

$$\begin{aligned}\langle\delta \mathrm{a}(\mathbf{r}_1)\delta \mathrm{a}(\mathbf{r}_2)\delta \mathrm{a}(\mathbf{r}_3)\delta \mathrm{a}(\mathbf{r}_4)\rangle \; &- \; \langle\delta \mathrm{a}(\mathbf{r}_1)\delta \mathrm{a}(\mathbf{r}_2)\rangle\langle\delta \mathrm{a}(\mathbf{r}_3)\delta \mathrm{a}(\mathbf{r}_4)\rangle \\ &- \; \langle\delta \mathrm{a}(\mathbf{r}_1)\delta \mathrm{a}(\mathbf{r}_3)\rangle\langle\delta \mathrm{a}(\mathbf{r}_2)\delta \mathrm{a}(\mathbf{r}_4)\rangle \\ &- \; \langle\delta \mathrm{a}(\mathbf{r}_1)\delta \mathrm{a}(\mathbf{r}_4)\rangle\langle\delta \mathrm{a}(\mathbf{r}_2)\delta \mathrm{a}(\mathbf{r}_3)\rangle\end{aligned}$$

genuinely does involve four points *all* confined to a region of space of linear dimension ξ_a, and

$$\left\langle \delta a_L{}^4 \right\rangle - 3\left\langle \delta a_L{}^2 \right\rangle^2 \sim \left\langle \delta \mathrm{a}^4 \right\rangle \times (\xi_\mathrm{a}/L)^{3d} \equiv L^{-3d} c_\mathrm{a}{}''$$

(defining yet another quantity $c_\mathrm{a}{}''$) as required. The derivation goes on in similar manner. It is essentially the same as the derivation of the virial expansion in terms of cluster integrals.

A similar derivation results if we write down the partition function for a set of subsystems of the original system [13]. As an exercise we can see it by roughly modelling interparticle correlations by considering blocks of length $2\xi_\mathrm{a}$: within each block, the variable a is *perfectly* correlated, i.e. the same everywhere, but between different blocks there is no correlation. A slightly more sophisticated assumption is that all correlations decay as $\exp\{-r/\xi_\mathrm{a}\}$, and this also gives results of the above kind.

If we wish to discuss the time-averaging aspect, $a_{Lt} = \overline{\mathrm{a}}^{Lt}$, we simply include a factor τ_a/t along with the factors of $(\xi_\mathrm{a}/L)^d$. The same observations about integrals and cumulants apply: we require proximity in space *and* time for the integrand not to vanish. Thus,

$$\left\langle \delta a_{Lt}{}^2 \right\rangle \sim \left\langle \delta \mathrm{a}^2 \right\rangle \times (\xi_\mathrm{a}/L)^d(\tau_\mathrm{a}/t) \sim \left\langle \delta \mathrm{a}^2 \right\rangle \times (\xi_\mathrm{a}/L)^d/\mathcal{N}$$

and the square root of this is the rms error in $\langle a_{Lt}\rangle$. We have already seen this (without the L-dependence) in eqn (2).

Now we can answer the question: given double the computer time, is it better to run for twice as long or to increase the system size? Most simulation algorithms are, optimally, of $\mathcal{O}(N)$ or $\mathcal{O}(L^d)$, that is (provided interaction ranges are reasonably short), it only requires twice the cpu time to simulate twice the number of molecules for the same number of steps or sweeps. (Of course, depending on circumstances, things may be worse!) In other words, cpu $\propto L^d t$. However, we see the combination $L^d t$ appearing above, so it makes no difference whether we do longer runs or increase the system size. Of course, if the simulation algorithm is worse than $\mathcal{O}(N)$, it is best to keep the system size small and to run for longer.

How about the fluctuation quantities? Consider the fluctuations in $\left\langle \delta a_L{}^2 \right\rangle$. We can write

$$\left\langle \left(\overline{\delta a_L{}^2}^t \right)^2 \right\rangle = \left(\left\langle \delta a_L{}^4 \right\rangle - \left\langle \delta a_L{}^2 \right\rangle^2 \right) / \mathcal{N}$$

and neglect non-Gaussian corrections, so

$$\Delta(\delta a_L{}^2) = \sqrt{2 \left\langle \delta a_L{}^2 \right\rangle^2 / \mathcal{N}}$$

or $\Delta(c_{\mathrm{a}})/c_{\mathrm{a}} = \sqrt{2/\mathcal{N}}$. Now there is no point in increasing the system size: it will have no effect on the relative error in c_{a}. It is best to use the smallest system for which L is safely larger than all the ξ_{a}, and then run for as long as possible. It might be better, in certain cases, to calculate quantities like c_{a} by numerical differentiation of averages rather than from fluctuation expressions [6,25]. Static linear response theory relates $\left\langle \delta \mathrm{a}^2 \right\rangle$ to the response of a to an extra term in the hamiltonian coupling to a. This amounts to changing the state point in some way; the change in $\langle \mathrm{a} \rangle$ will be measurable, and provides an alternative route to c_{a}. The effect must be small enough to lie within the linear regime. If it is too small, statistical errors on $\langle \mathrm{a} \rangle$ will wash out the response and the fluctuation method will be superior.

Relative errors in the higher cumulants become progressively worse. For example, we can write

$$\left\langle \left(\overline{\delta a_L{}^3}^t \right)^2 \right\rangle = \left(\left\langle \delta a_L{}^6 \right\rangle - \left\langle \delta a_L{}^3 \right\rangle^2 \right) / \mathcal{N}$$

and again make the Gaussian approximation giving

$$\Delta(\delta a_L{}^3) = \sqrt{15 \left\langle \delta a_L{}^2 \right\rangle^3 / \mathcal{N}}$$

or $\Delta(c_{\mathrm{a}}')/c_{\mathrm{a}}' \sim \sqrt{L^d/\mathcal{N}}$. Now increasing the system size is positively to be avoided!

4.3. HISTOGRAMS AND CUMULANTS

It has been well-known for many years [28] that it is possible to use simulation results obtained at one state point to calculate properties at another. Recently this approach has been rediscovered [29, 30] and applied to the precise determination of phase transitions, and refined [31–33] by expanding in cumulants. Here we set out the definitions and make some general comments. We return to phase transitions themselves shortly.

We start from the relation linking averages at different temperatures

$$\langle \mathrm{a}\rangle_{NVT} = \frac{\langle \mathrm{a}\exp\{-\delta\beta\mathcal{E}\}\rangle_{NVT_0}}{\langle \exp\{-\delta\beta\mathcal{E}\}\rangle_{NVT_0}}$$

where $\delta\beta = \beta - \beta_0$, $\beta = 1/k_BT$ and $\beta_0 = 1/k_BT_0$ where the simulations have been conducted at temperature T_0, and the temperature of interest is T. Here we shall concentrate on the simplest cases: the average energy as a function of T and, later, the density (or volume) as a function of P. (This will also give us the free energy differences between different state points.) However, if histograms of any property a are accumulated at the same time, their averages can be reconstructed and manipulated in the same way. Recall that when we accumulate a histogram of energies $\mathcal{E}$ in a *canonical* simulation, we are essentially recording the *microcanonical* ensemble density of states $\Omega(\mathcal{E})$ (with a known Boltzmann weighting, but this is an incidental factor) at energies in the neighbourhood of $\langle\mathcal{E}\rangle_{NVT}$.

It is convenient to choose $\mathrm{a} = \delta(\mathcal{E}(\mathbf{\Gamma}) - E)$ to obtain a relation between the probability distributions at two different temperatures:

$$\mathcal{P}_{NVT}(E) \equiv \langle\delta(\mathcal{E}(\mathbf{\Gamma}) - E)\rangle_{NVT} = \mathcal{P}_{NVT_0}(E)\frac{\exp\{-\delta\beta E\}}{\langle\exp\{-\delta\beta\mathcal{E}\}\rangle_{NVT_0}}. \tag{10}$$

More general quantities of interest $\mathrm{a}(E)$ (functions of energy) can be obtained by completing the integration over E

$$\langle \mathrm{a}(E)\rangle_{NVT} = \int \mathrm{d}E\ \mathcal{P}_{NVT}(E)\mathrm{a}(E)$$

and using eqn (10) to relate them to the distribution at T_0. We can sample $\mathcal{P}_{NVT_0}(E)$ by constructing a histogram in the simulation. As we have discussed at length, it will be sharply peaked. It will be approximately Gaussian, with successive cumulants decaying by factors inversely proportional to the system size. Accordingly, it may not be possible to extend the results too far either side, since the appropriate regions of energy space will be poorly sampled.

Phillpot and Rickman [31–33] propose calculating the cumulants of the distribution directly, rather than constructing a histogram. Indeed, this is a useful exercise, and it highlights the limitations of the method both in principle and in practice. The cumulant expansion is seen to be particularly appropriate when one writes

$$\begin{aligned}\exp\{-\delta(\beta F)\} &= \frac{Q_{NVT}}{Q_{NVT_0}} = \langle\exp\{-\delta\beta\mathcal{E}\}\rangle_{NVT_0}\\ &\equiv \exp\left\{\sum_{n=1}^{\infty}\frac{(-\delta\beta)^n}{n!}\langle\mathcal{E}^n\rangle^{\mathrm{c}}_{NVT_0}\right\}.\end{aligned}$$

Here $\delta(\beta F) = \beta F - \beta_0 F_0$. We recognize a Taylor-Maclaurin expansion here

$$-\delta(\beta F) = \sum_{n=1}^{\infty} \frac{(-\delta\beta)^n}{n!} \langle \mathcal{E}^n \rangle^c_{NVT_0}$$

so that

$$\langle \mathcal{E}^n \rangle^c_{NVT_0} = \left(\frac{\partial^n(-\beta F)}{\partial(-\beta)^n} \right)$$

or

$$\langle \varepsilon^n \rangle^c_{NVT_0} = (1/N)^n \left(\frac{\partial^n(-\beta F)}{\partial(-\beta)^n} \right).$$

The decreasing system-size dependence is obvious. The cumulants are directly given by successive derivatives of the free energy with respect to β, or indeed, recognizing the first one as $\langle \varepsilon \rangle_{NVT_0} = \epsilon$, by successive derivatives of the energy

$$\langle \varepsilon^n \rangle^c_{NVT_0} = (1/N)^{n-1} \left(\frac{\partial^{n-1}\epsilon}{\partial(-\beta)^{n-1}} \right).$$

Let us follow this through for volume fluctuations in the isothermal-isobaric ensemble. Consider making a small pressure change $\delta P = P - P_0$. We write

$$\begin{aligned} \exp\{-\beta\delta G\} &= \frac{Q_{NPT}}{Q_{NP_0T}} = \langle \exp\{-\beta\delta P V\} \rangle_{NP_0T} \\ &\equiv \exp\left\{ \sum_{n=1}^{\infty} \frac{(-\beta\delta P)^n}{n!} \langle V^n \rangle^c_{NP_0T} \right\}. \end{aligned}$$

Here $\delta G = G - G_0$. The Maclaurin expansion is

$$-\beta\delta G = \sum_{n=1}^{\infty} \frac{(-\beta\delta P)^n}{n!} \langle V^n \rangle^c_{NP_0T}$$

so that

$$\langle V^n \rangle^c_{NP_0T} = \left(\frac{\partial^n(-\beta G)}{\partial(-\beta P)^n} \right)$$

or

$$\langle v^n \rangle^c_{NP_0T} = (1/N)^n \left(\frac{\partial^n(-\beta G)}{\partial(-\beta P)^n} \right).$$

The decreasing system size dependence is again explicit. The cumulants are directly given by successive derivatives of the Gibbs free energy with respect to P, or, since we have $V = \partial G/\partial P$,

$$\langle v^n \rangle^c_{NP_0T} = (1/N)^{n-1} \left(\frac{\partial^{n-1} v}{\partial(-\beta P)^{n-1}} \right) .$$

Is this technique going to work? That is, can we calculate the cumulants directly in a simulation (instead of building up a histogram) and then use them to work out properties at other state points? Essentially we are fitting our distribution function using its first few cumulants. It may well be true that it can be represented quite accurately this way, and it appears that only a small error results from truncating the expansion at some point. The cumulants decrease in magnitude very rapidly as the order n goes up. However this is *bad* news, not good news! The decrease is due to the powers of N above: the crucial thermodynamic derivatives are multiplied by these factors. There are two basic sources of error in the method to consider: systematic and statistical errors.

The systematic error comes simply from truncating the cumulant expansion. Phillpot and Rickman truncated at the third term; statistics on the fourth cumulant were so poor that they could not even determine its sign, and this is likely to be a common problem. Even if the first three cumulants are computed exactly, this amounts to approximating the equation of state $\epsilon(\beta)$ or $v(P)$, locally, by a quadratic equation, setting to zero all higher derivatives. This is analogous to truncating the gas phase virial expansion at B_3, a fairly limited approximation. We can estimate the likely errors immediately for some test cases.

Firstly we use the Nicolas equation of state [34] for the Lennard-Jones liquid, fixing the density at about the triple-point value, $\rho = 0.85$ (in the usual reduced units) and picking a temperature $k_B T = 1.0$ approximately midway between the triple and critical-point values ($k_B T_t \approx 0.75$ and $k_B T_c \approx 1.35$). The cumulants can be obtained exactly from the assumed equation of state, at this state point. In Fig. 2, we compare the model equation of state $\epsilon(\beta)$ with the cumulant expansion truncated at various levels, over the entire liquid range. The results are quite encouraging: the systematic error is of the order of 0.3% for the best approximation at the extremes of this range.

Secondly, we adopt the Carnahan-Starling equation of state for hard spheres, picking a pressure $\beta P = 1.6$ ($\beta = 1$, formally) corresponding to a density roughly half that at which freezing begins. The appropriate $v(P)$ curves are shown in Fig. 3, together with the results of a cumulant expansion. Over a modest range of P ($\sim \pm 30\%$) very accurate results can be achieved; however if we try to extend the range too much, dramatic divergence sets in.

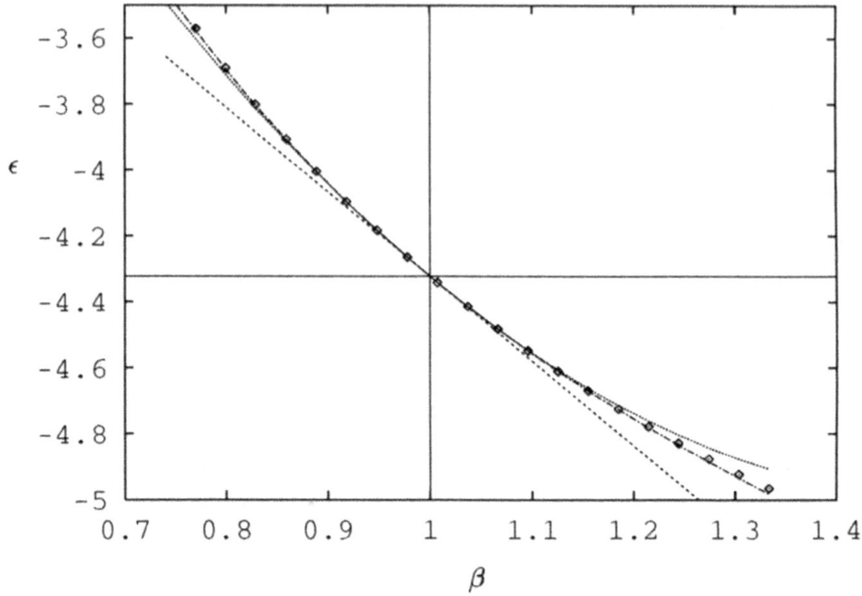

Fig. 2. The Nicolas equation of state for energy ϵ as a function of inverse temperature β (points), with the curves obtained from energy cumulant expansions about the indicated state point (solid lines). We show expansions to order 2 (dashed line), 3 (dotted line) and 4 (dot-dashed line).

The second problem is that statistical errors will be very high on the higher cumulants. The effect of this can be calculated using the techniques described earlier. Clearly, problems of poor sampling in the wings of the underlying distribution will affect whatever extrapolation is attempted, whether the cumulant approach is used or not.

In conclusion, it is doubtful that the histogram and cumulant methods are well suited to mapping out large regions of the equation of state. The cumulant method is certainly no more accurate than the accumulation of histograms: it is simply a convenient approximate representation of the distribution function. For a featureless equation of state, in the neighbourhood of the simulated state point, the systematic errors in using cumulants may not be too severe, and the general histogram technique is very useful. Any "interesting" features of the equation of state will, however, be difficult to represent using cumulants. This is especially true near a phase transition. However, it is in exactly in this region that the histogram reweighting method is useful, as we shall see.

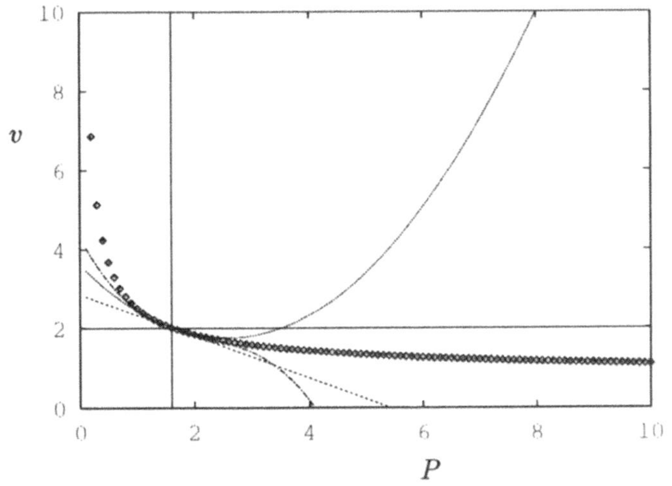

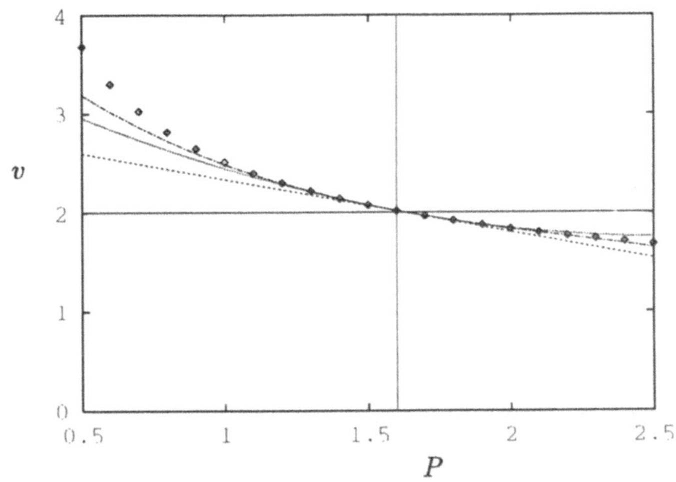

Fig. 3. The Carnahan-Starling equation of state for volume v as a function of pressure P, with the curves obtained from volume cumulant expansions about the indicated state point (solid lines), over a narrow and a wide range of P. We show expansions to order 2 (dashed line), 3 (dotted line) and 4 (dot-dashed line).

5. Finite-size scaling and phase transitions

5.1. INTRODUCTION

Very roughly speaking, we can classify phase transitions into two types: first-order and continuous. At a continuous phase transition, some correlation length ξ_a in the system diverges, and an order parameter (expressible as the spatial integral over the corresponding correlation function $\langle \delta a(0) \delta a(r) \rangle$) becomes macroscopically large. The approach to the phase transition is heralded by this divergent behaviour, which may well be describable in terms of universal exponent relations. The effects of finite size on such systems close to continuous phase transitions are well studied. By contrast, a system undergoes a first-order phase transition abruptly, when one phase becomes thermodynamically more stable than another. There are no transition precursors, but again it is possible to describe the effects of finite size in this case.

In both cases, the effects of finite size are to *shift* the transition and to *round* it in some way. The shift is certainly expected for first order transitions (because properties like the free energy differ from their infinite system limits by terms of $\mathcal{O}(1/N)$) and for continuous transitions (because the correlation length ξ_a diverges at the transition in the infinite system limit, but need only reach a value $\sim L$ for the finite system). The rounding is expected in both cases as well, because the partition function only has singularities (and hence produces discontinuous or divergent properties) in the limit $L \to \infty$; otherwise, it is analytic.

In either case, it is useful to consider the probability distribution function for variables averaged over a spatial block of side L; this may be the complete simulation box (in which case we must specify the ensemble and boundary conditions) or it may be a subsystem. For purposes of illustration we shall not distinguish these possibilities.

5.2. FIRST-ORDER TRANSITIONS

Physicists traditionally consider the case of a spin system [6, 25] in which one can define an order parameter a (typically a net magnetization variable) equally likely to take positive or negative values in the ordered phase, and zero in the disordered phase. Generally one specifies the state point by setting the temperature and magnetic field. More generally, say for the melting or vapourization transition, or an order-disorder transition of some kind, there will be no obvious microscopic symmetry. The relevant thermodynamic parameters may be the temperature and chemical potential. To begin with, we concentrate on varying the temperature and examine energy distributions.

Consider simulating a system in the canonical , and in the microcanonical ensemble, close to a phase transition. It is possible to describe what happens in a self-consistent way [24, 35]. In one phase, $\mathcal{P}_{NVT}(E)$ is essentially a Gaussian centred around a value E^-, while in the other phase the peak is around E^+. Far from the transition, one or other of these will apply. Close to the phase transition we may see contributions from both Gaussians, and a doubly-peaked distribution results. The weight of each Gaussian changes as the temperature is varied. Thus a smooth crossover occurs from one branch of the equation of state $E(T)$ to the other. In the transition region we may expect to see anomalies such as an increased specific heat: the double-peak distribution is wider than its constituent single-peaked ones. All of these phenomena will be influenced by the system size, because the peaks become narrower (and overlap less) as the system becomes larger.

In the microcanonical ensemble, the signature of a first-order phase transition is often said to be 'hysteresis' in the equation of state, now written as $T(E)$ or $\beta(E)$, or the appearance of a 'van der Waals loop'. We already have the equations necessary to predict the form of the crossover from one branch of the equation of state $\beta(E)$ to the other. We saw earlier that differentiation of the $\mathcal{P}(E)$ curve and use of $T^{-1} = k_B\beta = (\partial S/\partial E)$ gives

$$\frac{\partial(-\beta\mathcal{F}_{NVT_0}(E))}{\partial E} = \frac{\partial \ln \mathcal{P}_{NVT_0}(E)}{\partial E} = \frac{(\partial\mathcal{P}_{NVT_0}(E)/\partial E)}{\mathcal{P}_{NVT_0}(E)} = \beta(E) - \beta_0$$

At the transition temperature T_0 let $E^-(T_0) = E_0^-$ and $E^+(T_0) = E_0^+$. If $\mathcal{P}_{NVT_0}(E)$ consists of two peaks, centred on these values, then $\beta(E)$ will have the shape shown in Fig. 4.

At the maxima, and at the minimum, of $\mathcal{P}_{NVT_0}(E)$, we must have $\beta(E) = \beta_0$, from the above equation. It is easy to see that the traditional 'equal area' construction applied to the $\beta(E) - \beta_0$ curve corresponds to equal peak heights for $\mathcal{P}_{NVT_0}(E)$. Draw a line at constant $\beta = \hat{\beta}$ intersecting the $\beta(E)$ curve at E_0^-, E_0^{mid} and E_0^+, such that

$$\int_{E_0^-}^{E_0^+} \mathrm{d}E'(\beta(E') - \hat{\beta}) = 0.$$

Then, from the definitions above, $\mathcal{P}_{NVT_0}(E_0^-) = \mathcal{P}_{NVT_0}(E_0^+)$ where $k_B T = 1/\beta$, and we associate this with equal Helmholtz free energies, so $\hat{\beta} = \beta_0$. The equal-area construction is equivalent to a common-tangent construction for the entropy curve $S(E)$ (see Fig. 4). The gradient of the tangent determines a common value of β at both state points, and the tangent line itself is given by $\Delta S = (S^+ - S^-) = k_B\beta\Delta E = k_B\beta(E^+ - E^-)$ which implies equal Helmholtz free energies $F^+ = F^-$. Consequently, it seems that exactly at the transition we can model the distribution function thus:

$$\mathcal{P}_{NVT_0}(E) \propto \exp\left\{\frac{(E - E_0^+)^2}{2k_B T_0^2 C^+}\right\} + \exp\left\{\frac{(E - E_0^-)^2}{2k_B T_0^2 C^-}\right\} \tag{11}$$

with equal peak heights. However, there is a small correction to make for ensemble conversion. According to eqn (6), the Helmholtz free energy measured in a constant-NVT Monte Carlo simulation will be related to the microcanonical ensemble entropy and energy by

$$-\beta F = S/k_B - \beta E + \ln\sqrt{2\pi k_B T^2 C}.$$

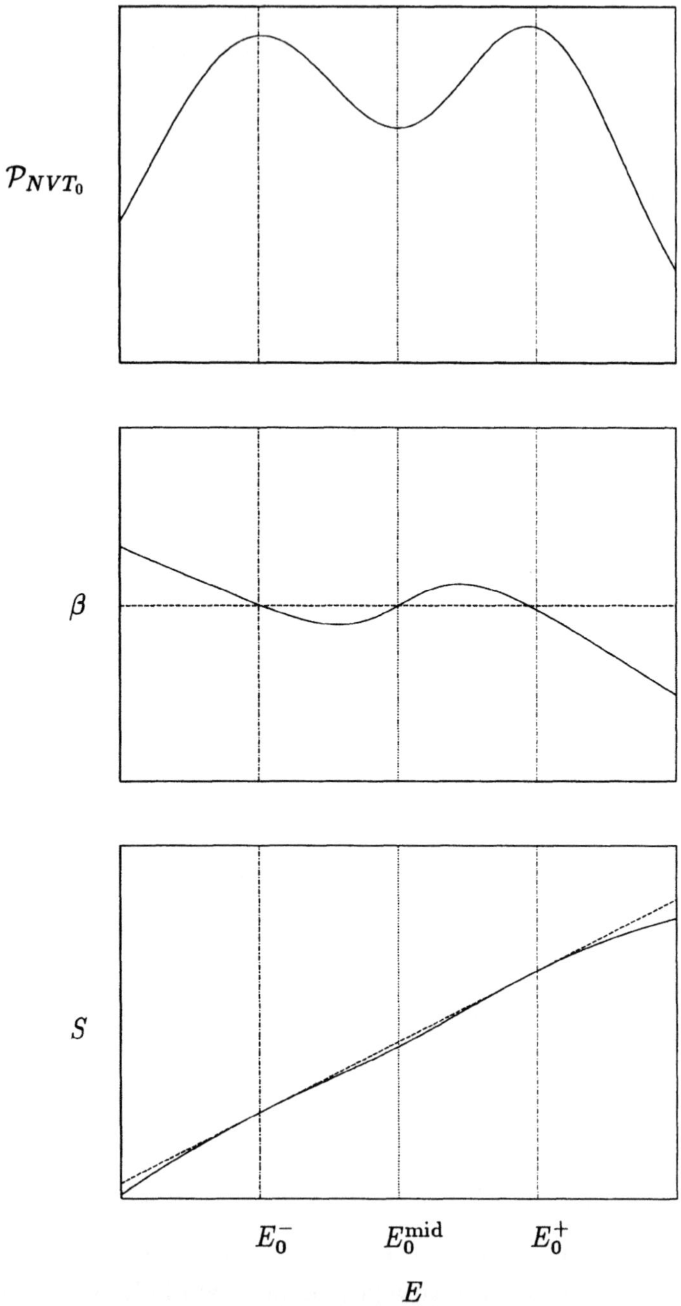

Fig. 4. The probability distribution function $\mathcal{P}_{NVT_0}(E)$, equation of state $\beta(E)$ and entropy $S(E)$ close to a first-order phase transition. We show the equal-area construction applied to $\beta(E)$ and the equivalent double-tangent construction applied to $S(E)$. Vertical lines mark E_0^-, E_0^{mid} and E_0^+. Note that the specific heats of the two phases are chosen to be different: $C^- > C^+$.

The correction term arises because we are *averaging* over the energy fluctuations corresponding to the two peaks, we are not simply constrained to the values E_0^+ and E_0^-. Now if we set $F^+ = F^-$ and apply this relation we obtain

$$\sqrt{C^-}\mathcal{P}_{NVT_0}(E_0^-) = \sqrt{C^+}\mathcal{P}_{NVT_0}(E_0^+).$$

Recalling that the peak width is proportional to $\sqrt{C}$, we see that this implies equal peak *areas* at coexistence. The correct (NVT ensemble) two-Gaussian form at coexistence is therefore

$$\mathcal{P}_{NVT_0}(E) \propto \frac{1}{\sqrt{C^+}} \exp\left\{\frac{(E-E_0^+)^2}{2k_\mathrm{B}T_0^2C^+}\right\} + \frac{1}{\sqrt{C^-}} \exp\left\{\frac{(E-E_0^-)^2}{2k_\mathrm{B}T_0^2C^-}\right\} \quad (12)$$

The above argument echoes Privman's discussion (p84 of Ref. [36]) of the methods adopted in two frequently-cited papers [37, 38].

The effects of increasing system size and moving away from the phase transition temperature are shown in Fig. 5 and Fig. 6. For a larger system, the peaks in eqn (11) scale in the usual way with system size, becoming sharper, and the variation of $\beta(E)$ (and other thermodynamic functions) is more abrupt. In the thermodynamic limit we obtain a vertical straight line joining the two $\beta(E)$ branches. Moving away from the transition by even a small amount quickly changes the peak heights in $\mathcal{P}_{NVT}(E)$, as well as shifting the peak positions. The effect is directly obtained from eqn (10). When we set $\mathcal{P}_{NVT}(E) \propto \exp\{-\delta\beta E\}\mathcal{P}_{NVT_0}(E)$, where $\delta\beta = \beta - \beta_0$, to a first approximation, the peaks shift as they must according to the equation of state, i.e. $E^\pm = E_0^\pm + C^\pm \delta T$ (where $\delta T = T - T_0$). Moreover, peaks in the neighbourhood of E_0^+ are scaled by $\exp\{-\delta\beta E_0^+\}$, while those around E_0^- are multiplied by $\exp\{-\delta\beta E_0^-\}$. This is easily associated with the shift in free energy difference δF between the two phases which accompanies the shift in temperature. If we adopt the double-Gaussian form we see

$$\mathcal{P}_{NVT}(E) \propto \frac{A^+}{\sqrt{C^+}} \exp\left\{\frac{(E-E^+)^2}{2k_\mathrm{B}T_0^2C^+}\right\} + \frac{A^-}{\sqrt{C^-}} \exp\left\{\frac{(E-E^-)^2}{2k_\mathrm{B}T_0^2C^-}\right\} \quad (13)$$

where E^+ and E^- are given above, and $A^+/A^- = \exp\{-\delta\beta(E_0^+ - E_0^-)\}$. Notice that as the system size grows, the distinction between equal peak heights and equal peak areas as signatures of coexistence becomes less important. For a large system, the appearance of two peaks with amplitudes in any way comparable with each other indicates close proximity to the phase transition.

These general ideas are well established, and apply to other variables. If we are simulating a fluid-solid transition, where the density changes, at constant pressure, we expect $\mathcal{P}(V)$ to show double peaks [24]. Again, the link between equations of state in different ensembles is fairly clear. At constant

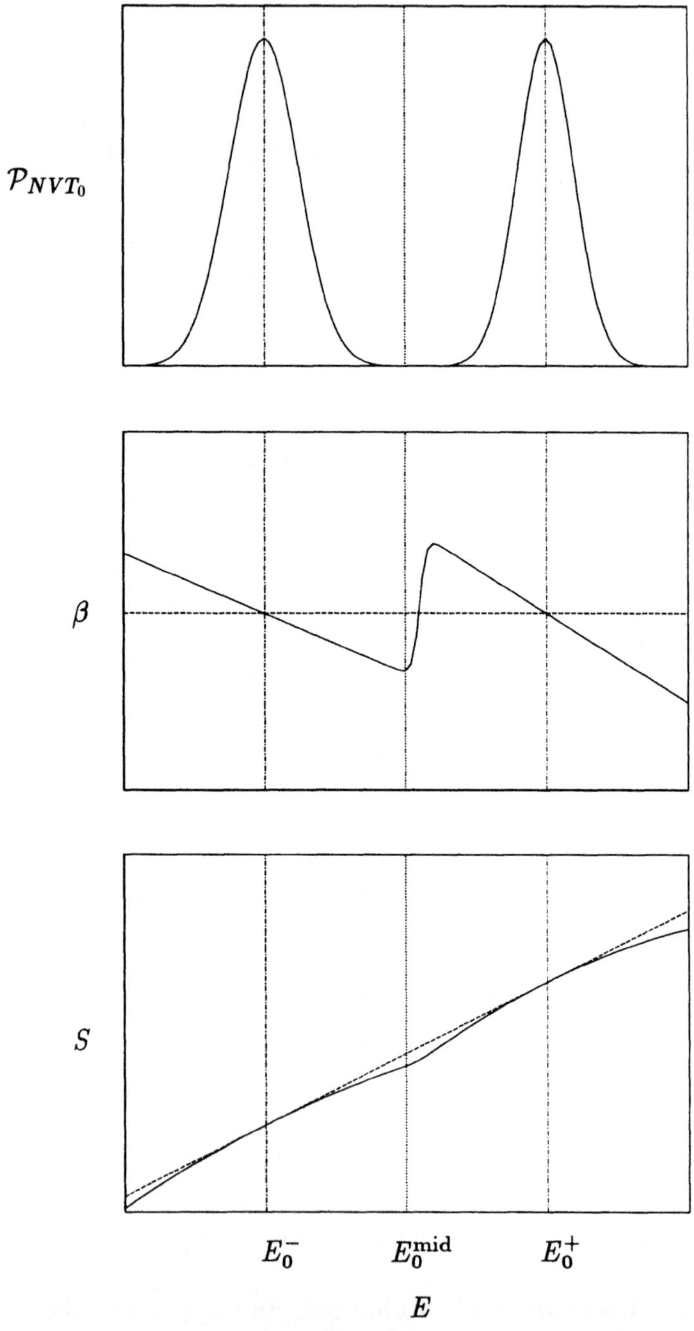

Fig. 5. The probability distribution function $\mathcal{P}_{NVT_0}(E)$, equation of state $\beta(E)$ and entropy $S(E)$ close to a first-order phase transition. Vertical lines mark E_0^-, $E_0^{\rm mid}$ and E_0^+. Larger system size than that of Fig. 4. Again, note that the specific heats of the two phases are chosen to be different: $C^- > C^+$.

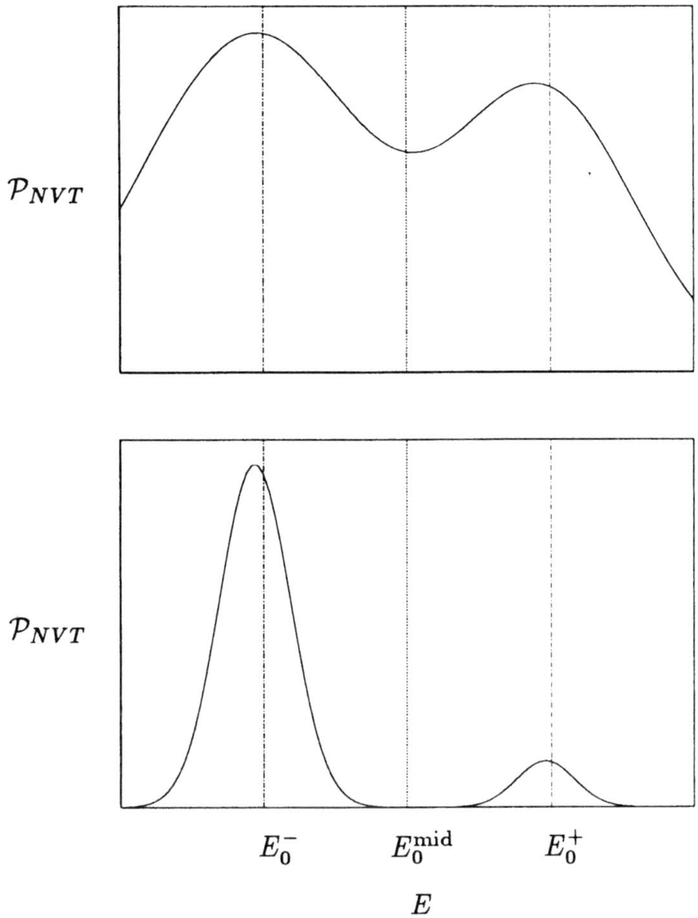

Fig. 6. The probability distribution function $\mathcal{P}_{NVT}(E)$, with β slightly higher than the transition value β_0 (T slightly lower than T_0). The upper curve corresponds to the system of Fig. 4; the lower curve to the system of Fig. 5.

volume, a 'loop' is seen in the equation of state $P(V)$. Choose a pressure P in the loop region, intersecting the constant-NVT curve at three places, $V^- < V^{\text{mid}} < V^+$. Then it can be proved that $\mathcal{P}_{NPT}(V)$ has two peaks, with maxima at V^- and V^+ and a minimum between them at V^{mid}. The equal area construction for $V(P)$, or common tangent construction for F, correspond to equal peak heights in $\mathcal{P}_{NPT}(V)$. Once more, converting back to the isobaric ensemble, this translates to equal peak areas.

We should note in passing that the doubly-peaked distribution is not well represented by the double-Gaussian form in the middle (where the probability is lowest, and the free energy barrier is found). Actually, $\mathcal{P}_{NVT}(E^{\text{mid}})$

(or $\mathcal{P}_{NPT}(V^{\mathrm{mid}})$) is larger than expected, and the free energy barrier correspondingly lower, because the mid-region corresponds to two phases in coexistence. The free energy associated with this is dominated by the interfacial contribution [25, 39]. One phase appears as a spherical droplet of diameter roughly L within the other phase, or as a slab spanning the periodic system. In either case, the free energy difference

$$\Delta\mathcal{F} = \mathcal{F}(E^{\mathrm{mid}}) - \mathcal{F}(E^{\pm})$$

is just an interfacial area multiplied by a surface tension. The free energy 'barrier height' is of $\mathcal{O}(L^{d-1})$, not of $\mathcal{O}(L^{d})$ as the simple two-Gaussian form might suggest. This situation is discussed in a little more detail by Smit in Chapter 6.

If there is any doubt as to the first-order character of a transition, the serious student should really examine distributions of this kind as a function of system size. Doubts may arise because the two Gaussians overlap. The solution is simply to simulate a larger system and observe the narrowing and separating phenomenon. (Of course, this may be expensive). Lee and Kosterlitz [30] propose the growth in free energy barrier $\Delta\mathcal{F} \sim \mathcal{O}(L^{d-1})$, as a signature of first-order character. A recent example of this technique applied to a weak transition is the investigation of the isotropic liquid $\leftrightarrow$ nematic liquid crystal transition for a simple spin model [40]. This paper contains free energy curves calculated as functions of order parameter, showing, in an exemplary way, the expected variation with system size and with temperature.

What has to be watched out for, especially for atomistic simulations, is the danger of confusion between metastability effects (insufficient equilibration) and the genuine consequences of the phase transition on the equation of state. From what we have just seen, there is a free energy barrier $\Delta\mathcal{F}$ to get over, between one phase and the other. If this is significant, we may not be able to sample $\mathcal{P}(E)$ sufficiently well to determine $\Delta\mathcal{F}$. Thus, Catch-22 operates: we need to observe a high barrier to decide if the transition is first order, but a high barrier will inhibit equilibration. A possible way round this is to use narrow windows, to simulate a restricted range of energies at one time. This has recently been carried out at the melting transition [41]. Simulations in windows near the top of the energy barrier will correspond to two phases in equilibrium. The same is true of microcanonical simulations designed to reveal the shape of the $\beta(E)$ equation of state. In all cases, timescales appropriate for two-phase equilibration are necessary. Of course, if the transition is known to be first order, and there is no interest in the two-phase region, then simple free energy determination on the two branches of the equation of state may be sufficient to locate it.

5.3. CONTINUOUS TRANSITIONS

Now we consider continuous transitions. As made clear before, we cannot go into too much detail. The interested reader is advised to consult articles in the reprint collection of Cardy [42], the volume edited by Privman [36], and references therein.

Suppose that the transition is characterized by an order parameter a and a divergent correlation length ξ_{a}. We shall be interested in the block average value $a_L = \overline{\mathrm{a}}^L$. In a magnetic system a is the magnetization; in a fluid it is the density. The basic idea of finite size scaling analysis is that the values of properties of the system are dictated by the ratio ξ_{a}/L, and that no other length scales enter the problem. The *position* of the phase transition, and hence the finite-size shift, are determined by $\xi_{\mathrm{a}} \to L$. Any property can be written

$$\langle a_L \rangle = a \times \Phi(\xi_{\mathrm{a}}/L)$$

where a is the average value in the infinite system limit. There will be exponent laws dictating the behaviour of a in the vicinity of the phase transition, and more scaling laws stating how ξ_{a} behaves inside the function Φ. In fact, the correlation length exponent ν is defined such that

$$\xi_{\mathrm{a}} \sim |T - T_{\mathrm{c}}|^{-\nu}.$$

Remember that a may be singular at the transition, but $\langle a_L \rangle$ will not be: therefore it is part of the job of Φ to remove this singularity. For example, the magnetic susceptibility χ diverges like

$$\chi \sim \xi_{\mathrm{a}}{}^{\gamma/\nu}$$

so we can write, in the finite system,

$$\langle \chi_L \rangle \sim L^{\gamma/\nu} \tilde{\Phi}(L^{1/\nu}(T - T_{\mathrm{c}}))$$

In these equations, if we take the $L \to \infty$ limit, the usual power-law scaling is recovered. In all of the above we have not included the dependence on magnetic field (for a spin system) or chemical potential (for a fluid). This is easily done, but not necessary here.

Scaling analysis of this kind was applied to the block distribution function $\mathcal{P}(a_L)$, discussed earlier, by Binder [26] and Bruce [43]. Again, we mention that we should carefully distinguish between a variable averaged over a block which is a *subsystem* of the simulated system, and one constituting the entire system, in a specified ensemble with specified boundary conditions. However, we can neglect these points for the purposes of illustration. The scaling form is

$$\mathcal{P}(a_L) \sim L^{\beta/\nu} \tilde{\mathcal{P}}(L^{\beta/\nu} a_L, \xi_{\mathrm{a}}/L).$$

or

$$\mathcal{P}(a_L) \sim L^{\beta/\nu}\tilde{\mathcal{P}}(L^{\beta/\nu}a_L, L^{1/\nu}(T - T_c)).$$

The $L^{\beta/\nu}$ scaling guarantees that $\left\langle \delta a_L{}^2 \right\rangle \sim L^{-2\beta/\nu}$ as required by the finite-size-limited behaviour of the correlation function. The $L^{1/\nu}(T - T_c)$ scaling factor reflects the dependence on ξ_a/L referred to above. If we define a scaled variable $\tilde{a_L} = L^{\beta/\nu}a_L$ then, close to the critical point, we should have

$$\mathcal{P}(a_L) \sim L^{\beta/\nu}[\tilde{\mathcal{P}}_0(\tilde{a_L}) + L^{1/\nu}(T - T_c)\tilde{\mathcal{P}}_1(\tilde{a_L}) + \ldots]$$

where both $\tilde{\mathcal{P}}_0(\tilde{a_L})$ and $\tilde{\mathcal{P}}_1(\tilde{a_L})$ are universal functions, characterizing the class of the transition. When the correlation length is small compared with L, above the critical temperature, a single Gaussian results; well below it we get a double Gaussian. Actually at the critical point, the distribution can be calculated by simulation, or renormalization group theory [26, 43, 44].

The results are quite different for different universality classes. For the Ising/fluid $d = 3$ systems, $\tilde{\mathcal{P}}_0(\tilde{a_L})$ has a single well-formed central peak, characterized by a cumulant ratio

$$G \equiv -\langle \tilde{a_L}^4 \rangle^c / 2\langle \tilde{a_L}^2 \rangle^{c2} \approx 0.33 \qquad d = 3.$$

(For a perfect single-Gaussian distribution we would have $G = 0$, while for a function consisting of two equal, sharp peaks, $G = 1$.) In the analogous $d = 2$ system, $\tilde{\mathcal{P}}_0(\tilde{a_L})$ has a very distinct double-peaked form shown in Fig. 7. For this curve the cumulant ratio is

$$G \approx 0.87 \qquad d = 2.$$

Thus, for a finite system of this kind, the critical point does *not* correspond to the coalescence of the two peaks which represent the densities of the co-existing phases. Examination of these functions is a powerful way of locating and characterizing critical points. For example, the prewetting critical point has been shown to lie in the $d = 2$ Ising universality class in this way [45].

These methods have been used to locate and characterize the critical point of the $d = 2$ Lennard-Jones fluid [46–49]. This problem has a long history. The transition is believed to be $d = 2$ Ising-like, which means that the coexistence curve in the ρ, T phase diagram is exceptionally flat at the top, having a critical exponent $\beta = 0.125$. Analysis of equation-of-state data [50] could give only a very rough idea of the critical parameters, while Gibbs ensemble simulations, aimed at locating the peak coalescence point [51–55] must still be extrapolated to infinite system size in order to avoid systematically overestimating the critical temperature. Rovere *et al.* [46, 47] attempted a block distribution function analysis for subsystems within a canonical ensemble simulation; Wilding and Bruce [48, 49] did something

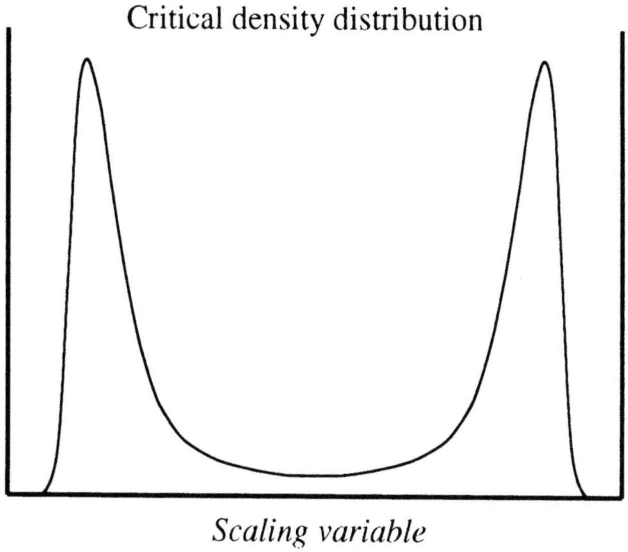

Fig. 7. The universal density scaling function at criticality, $\tilde{\mathcal{P}}_0(\tilde{a_L})$, for a $d = 2$ Ising-like system.

similar, using grand canonical simulations (which allow the whole system density to vary). This last work represents the current state-of-the-art study for this system. It demonstrated that extremely slow density fluctuations occur, requiring runs of order $10^6 - 10^8$ sweeps! *Exactly* at the critical point the density distribution scales in a simple way

$$\mathcal{P}(\delta\rho_L) \sim L^{\beta/\nu}\tilde{\mathcal{P}}_0(L^{\beta/\nu}\delta\rho_L)$$

where $\delta\rho_L = \rho_L - \rho_L(\text{crit})$. (Away from criticality there is mixing of the two relevant fields (temperature and chemical potential) and loss of (Ising-like) symmetry, but this can still be analyzed satisfactorily.) Wilding and Bruce studied two system sizes corresponding to $\langle N \rangle \approx 100$ and $\langle N \rangle \approx 400$ at criticality. They tracked the system along the coexistence line in the μ, β plane: this is defined by the existence of two *equal* peaks in the density distribution. They plotted the scaled distribution at each state point. This is illustrated in Fig. 8 for two system sizes. Practically, the density scaling factor is defined by making the distribution have unit variance (as well as being normalized), so it is not necessary to feed in the value of β/ν. (A value $\beta/\nu = 0.125$ arises from the fit: this is exactly correct for the Ising universality class). The key is to compare results for the two system sizes. Exactly at criticality, the scaled density distribution does not change with system size: as discussed above, it is *universal*. For T slightly below T_c, the distribution is similar in appearance, *but* increasing the system size causes a sharpening of the side peaks and a drop of the central region. As the system size increases, or as

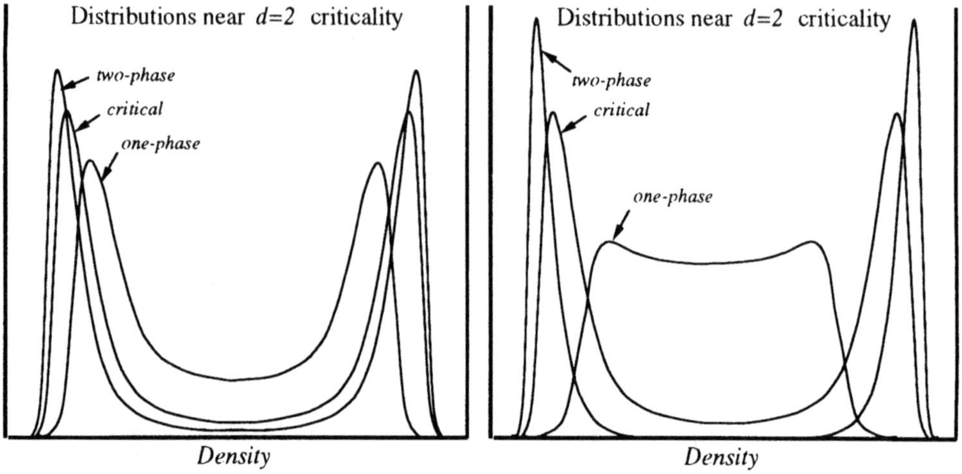

Fig. 8. The density scaling function $\tilde{\mathcal{P}}_0(\tilde{a_L})$ (schematic) illustrating the change on increasing the system size (left - small system; right - large system) at three temperatures: in the two-phase region ($T < T_c$), exactly at criticality ($T = T_c$) and in the one-phase region ($T > T_c$). The shape of the critical curve does not change.

we move further back down the coexistence curve, the distribution smoothly goes over into two separate Gaussians, as we would expect. For T slightly above T_c, again the appearance of the distribution is similar to the critical distribution, but increasing the system size results in a flattening, and a broadening of the side peaks. Further progress along the extrapolation of the true coexistence curve, or simply a further increase in system size at fixed state point, results in a completely flat 'mesa' appearance, followed by the growth of a single, central Gaussian.

The approach is similar to that of Lee and Kosterlitz [30], except that they look only at the side-peak heights relative to the central minimum of $\tilde{\mathcal{P}}_0$ (and hence the 'free energy' barrier) while Wilding and Bruce examine the complete distribution. Note that, throughout the above discussion, T_c means the true, infinite-system, critical temperature. It is not necessary to introduce a finite-size pseudocritical temperature $T_c(L)$, namely the temperature at which the peaks appear to coalesce. One could do this, and extrapolate $T_c(L) \to T_c$ as $L \to \infty$, but much of the information in the distribution function would be thrown away by this approach. It is interesting to note that Wilding and Bruce's results show that for the Lennard-Jones system, and for $N \sim$ few hundred particles, the two peaks persist until the temperature has risen about 3% above the critical value. This is the magnitude of finite-size effects on $T_c(L)$.

The questions any serious 'critical-point seeker' must answer are as follows.

(i) For your chosen ensemble and boundary conditions, is there a finite-size scaling form for the appropriate distribution function, giving a universal prediction at criticality? If not, you cannot hope to locate the critical point accurately, (to within a few percent) because this is the way it is characterized. Let's assume that this has been done, and that it is similar to the scaling form discussed above.

(ii) Do you have plenty of cpu time available? Even for systems of a few hundred atoms, very slow fluctuations, $\sim 10^6$ sweeps, are seen 1% away from criticality, and exactly at the critical point, runs of length $\sim 5 \times 10^7$ seem to be necessary.

(iii) Do your measured distributions look like the correct scaling form? If not, you cannot be close to the critical point. You cannot extrapolate to the critical point until you have entered the asymptotic regime!

(iv) How do the scaled distributions change with increasing system size? Only at the critical point will they be invariant. For $2d$ Ising-like systems they have the two-peaked structure discussed above: in the coexistence region, the side peaks will become 'peakier', while in the one-fluid region they will broaden and the central region will grow. This system-size-dependence is the essential ingredient. For $3d$ Ising-like systems the shape is quite different at T_c, but the general pattern of behaviour is similar.

Of course, if one simply wants a *rough guide* to the position of the critical point, it may be sufficient to make a fit to the coexistence curve far away from T_c (possibly using Gibbs ensemble simulations to determine the curve) and assume the law of rectilinear diameters, as outlined in Chapter 2. A somewhat more accurate guide may be provided by examining the form of distribution functions in the Gibbs ensemble, as described in Chapter 6. However, for accurate determination of T_c, a full asymptotic analysis of the distribution functions is required.

The histogram reweighting techniques seem to have their most useful applications here, when a simulation close by a phase transition can be reweighted as a function (say) of T to locate the transition point precisely. In doing this one must beware of unphysical peaks appearing in the extrapolated distribution [56], a possible consequence of poor sampling in the wings of the distribution. The cumulant approximation method cannot be recommended here, as we need to model the shape of the distribution quite accurately.

6. Transport coefficients

Now we have a complete change of gear. I want to discuss the way in which we compute transport coefficients in computer simulations using periodic boundary conditions. This is partly stimulated by a recent paper [57] which considers the possibility of measuring the shear viscosity from the dynamics of *single* particles. However, we come to that in due course. First I want to review how we normally calculate transport coefficients, and I focus on the diffusion coefficient D as an example of a single-particle one, and the shear viscosity η as an example of a collective one.

6.1. DIFFUSION IN BOUNDED AND UNBOUNDED SYSTEMS

First, consider diffusion. There are two well-known expressions for D, one involving the Einstein relation

$$D = \tfrac{1}{2}\lim_{t\to\infty} \tfrac{\mathrm{d}}{\mathrm{d}t}\langle \Delta z_i(t)^2\rangle \qquad \text{(Einstein)}$$

and the other the infinite time integral of the velocity autocorrelation function, a so-called Green-Kubo relation

$$D = \textstyle\int_0^\infty \mathrm{d}t\ \langle v_{zi}(0) v_{zi}(t)\rangle \qquad \text{(Green-Kubo).}$$

Here $\Delta z_i(t) = z_i(t) - z_i(0)$, and $z_i(t)$ and $v_{zi}(t)$ are, respectively, the z-coordinate and velocity of a chosen particle i. These expressions are equivalent to one another if one sets $\Delta z_i(t) = \int_0^t \mathrm{d}t'\ v_{zi}(t')$ and manipulates the integrals in a straightforward way. However, they have slightly unusual properties.

In an infinite system, the coordinates are unbounded: a particle can travel indefinitely far from its starting position. In a simulation we mimic this by following atoms through the walls of the periodic simulation box: we do *not* replace an outgoing atom by its incoming periodic image when a box boundary is crossed. As a consequence, the averages appearing in the Einstein relation are *not* stationary. Expanding $\Delta z_i(t)$ we get

$$\langle \Delta z_i(t)^2\rangle = \langle z_i(t)^2\rangle - \langle z_i(0)^2\rangle - 2\langle z_i(0)\Delta z_i(t)\rangle. \tag{14}$$

In the above, each average is over particle positions at time $t = 0$ lying within the basic simulation box: $-L/2 \le z_i(0) < L/2$; $z_i(t)$ however is unrestricted. The last term in eqn (14) is evidently zero, on grounds of translational invariance: the displacement cannot be correlated with the initial position (assuming that the fluid is uniform). At long times, then,

$$\langle z_i(t)^2\rangle - \langle z_i(0)^2\rangle \to 2Dt.$$

This may look odd to those used to writing $\langle a(t)^2\rangle = \langle a(0)^2\rangle$ on the grounds of stationarity, but of course it is quite correct, and follows directly from the classical diffusion equation. In the simulation, $\langle z_i(0)^2\rangle$ is a simple average over the initial distribution, uniform within the basic simulation box, while $\langle z_i(t)^2\rangle$ involves the same ensemble of initial positions, but with the $z_i(t)$ values corresponding to particles that have diffused out into the wide blue yonder.

The situation is rather different if we consider a box of length L with real walls (and later take the limit $L \to \infty$). Now the system is bounded but no longer translationally invariant. Starting from a uniform initial distribution within the box (which is the equilibrium distribution, apart from some wall-induced structure which we neglect) it is easy to see from the diffusion equation that

$$\langle z_i(t)^2 \rangle = \langle z_i(0)^2 \rangle$$

and

$$\langle z_i(0) \Delta z_i(t) \rangle = -Dt.$$

This last equation is only true on a timescale

$$\tau_v \ll t \ll \tau_{\mathrm{box}}$$

where $\tau_{\mathrm{box}} = L^2/2D$, the time for a particle to diffuse the length of the box. This average is dominated by the particles that start very close to the walls, and which must, on average, move away from them. The walls cannot be ignored, *no matter how large the system is.* On this timescale we expect to see, from eqn (14), as a result of this term, $\langle \Delta z_i(t)^2 \rangle \sim 2Dt$; at long times ($t \to \tau_{\mathrm{box}}$) the mean-squared displacement curve tails off. An example of this is shown in Fig. 9. We show results for hard spheres using two different system sizes $N = 108, 216$ in single- and double-length boxes. Density is half the close-packed value, $T = 1$. Here, τ_{box} is quite long (~ 150 for the $N = 108$ system) and diffusive behaviour is established well before the box-limited plateau sets in. The lesson is that, in a finite, bounded system, we can obtain an intermediate-time ($\tau_v \ll t \ll \tau_{\mathrm{box}}$) value of D, but we cannot allow $t \to \infty$.

It is well known that one has to take care with the long-time and large-system limits. Remember that transport coefficients arise from conservation relations of the kind $\dot{\alpha}(k,t) = -ikj_\alpha(k,t)$, where we take k to lie in the z direction for convenience, and where $j_\alpha(k,t)$ is well-behaved in the low-k limit. This guarantees that $\alpha(k,t)$ is a slowly-evolving variable at low k. Variables a and $\dot{\mathrm{a}}$ appearing in the Einstein and Green-Kubo relations are defined by

$$\mathrm{a}(t) = \lim_{k \to 0} \alpha(k,t)/(-ik)$$

and

$$\dot{\mathrm{a}}(t) = \lim_{k \to 0} \dot{\alpha}(k,t)/(-ik) = \lim_{k \to 0} j_\alpha(k,t).$$

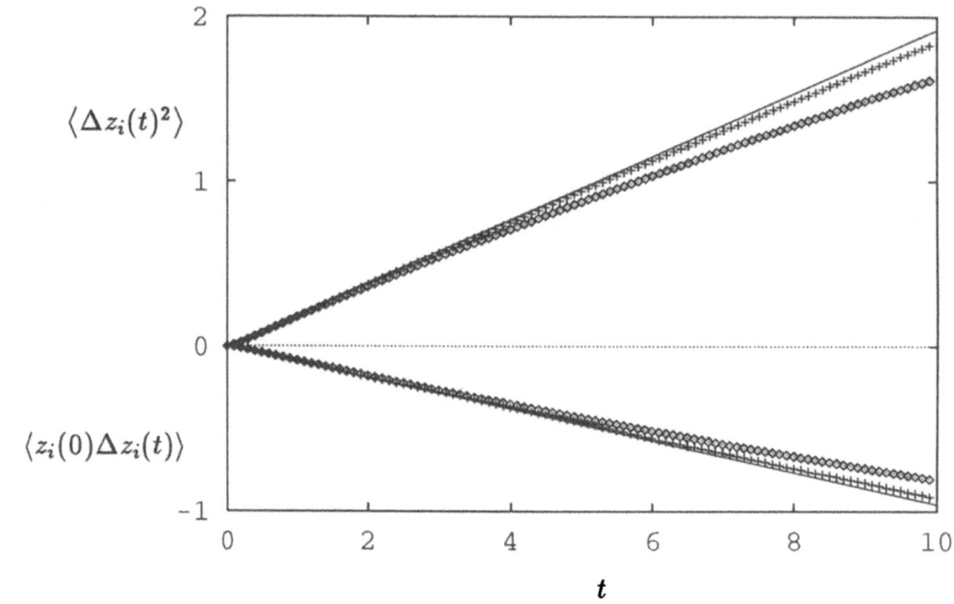

Fig. 9. Einstein relations for diffusion for systems with reflecting walls in the z direction. We plot a line with gradient $2D$, $\langle \Delta z_i(t)^2 \rangle$ for a box with reflecting walls and $N = 108$ atoms (diamonds) and the same quantity for a double-length box with reflecting walls and $N = 216$ atoms (pluses). Below the t axis are corresponding values of $\langle z_i(0)\Delta z_i(t)\rangle$ and a straight line with gradient $-D$.

This is the procedure by which the large-system limit is correctly taken before the long time limit; this is essential to avoid reaching the conclusion that the transport coefficient vanishes. For diffusion we have

$$\alpha(k,t) = \exp\{-ikz_i(t)\},$$

the Fourier transform of the single-particle density, and

$$j_\alpha(k,t) = -ikv_{zi}(t)\exp\{-ikz_i(t)\},$$

the single-particle current. Hence $\mathrm{a} = z_i(t)$, $\dot{\mathrm{a}} = v_{zi}(t)$.

6.2. SHEAR MODES IN BOUNDED AND UNBOUNDED SYSTEMS

Things become more tricky when we look at the analogous expressions for shear motion. Now we have

$$\alpha(k,t) = p_x(k,t) \equiv \sum_{i=1}^{N} p_{xi}(t)\exp\{-ikz_i(t)\},$$

the *collective* transverse momentum density, and

$$j_\alpha(k,t) = \Pi_{zx}(k,t),$$

the zx-component of the k-dependent pressure tensor. The proof that $\dot{p}_x(k,t) = -ik\Pi_{zx}(k,t)$, and the precise form of $\Pi_{zx}(k,t)$, are given in standard references [7, 58]. Then, zero-k variables can be defined by

$$\mathcal{A}(t) = \sum_{i=1}^{N} p_{xi}(t) z_i(t) = \sum_{i=1}^{N} a_i(t) \tag{15}$$

and

$$\dot{\mathcal{A}}(t) = \left(\sum_{i=1}^{N} \frac{p_{xi}(t) p_{zi}(t)}{m} + \sum_{i=1}^{N} f_{xi} z_i \right) = \sum_{i=1}^{N} \dot{a}_i(t) \tag{16}$$

$$= \left(\sum_{i=1}^{N} \frac{p_{xi}(t) p_{zi}(t)}{m} + \sum_{i=1}^{N} \sum_{j>i}^{N} f_{xij} z_{ij} \right) \tag{17}$$

where f_{xij} is the x-component of the force between i and j (assumed pairwise additive) and $z_{ij} = z_i - z_j$ is the interatomic separation. Helfand [59] has carried through the analysis of collective Einstein relations in this case. The appropriate equations are

$$\eta = \tfrac{1}{2} \lim_{t\to\infty} \tfrac{\mathrm{d}}{\mathrm{d}t} \frac{1}{V k_{\mathrm{B}} T} \langle \Delta \mathcal{A}(t)^2 \rangle$$

$$= \tfrac{1}{2} \lim_{t\to\infty} \tfrac{\mathrm{d}}{\mathrm{d}t} \frac{1}{V k_{\mathrm{B}} T} \left\langle \left[\textstyle\sum_i \Delta a_i(t) \right]^2 \right\rangle \qquad \text{(Einstein, Helfand)}$$

and

$$\eta = \frac{1}{V k_{\mathrm{B}} T} \int_0^\infty \mathrm{d}t \left\langle \dot{\mathcal{A}}(t) \dot{\mathcal{A}}(0) \right\rangle \qquad \text{(Green-Kubo)}.$$

Note that we can formally write $\mathcal{A}(t)$ and $\dot{\mathcal{A}}(t)$ as sums of single-particle quantities, as in eqns (15) and (16). There is no problem with this in the bounded case, for a system with reflecting walls in the z direction. However, in a fully periodic system the above equations are misleading. $\dot{\mathcal{A}}$ is re-written in pairwise form in eqn (17) making it independent of the origin of coordinates. This relies on the double summation, and momentum conservation in the form of Newton's third law $f_{xij} = -f_{xji}$:

$$\sum_{i=1}^{N} f_{xi} z_i = \sum_{i=1}^{N} \sum_{j\neq i}^{N} f_{xij} z_i = \tfrac{1}{2} \sum_{i=1}^{N} \sum_{j\neq i}^{N} f_{xij} z_{ij}.$$

The subtlety is that these pairwise quantities are always evaluated with the *minimum image convention* applied. The pairs involved are always minimum image pairs. This is a sleight of hand: $\dot{\mathcal{A}}$ is *not* simply related to the time derivative of $\mathcal{A}$ as defined above in single-particle form. What is the form of the Einstein relation, then? Well, if we define $\Delta\mathcal{A}(t) = \int_0^t \mathrm{d}t' \, \dot{\mathcal{A}}(t')$, there is no problem. The link between Einstein and Green-Kubo goes through. This is actually how things are done in simulations [60, 61]. This is *not* equivalent to simply writing down $\Delta\mathcal{A}(t) = \mathcal{A}(t) - \mathcal{A}(0)$ with $\mathcal{A}(t)$ given by the equation above.

Can $\mathcal{A}(t)$ be evaluated directly, in some consistent way, and used to calculate $\Delta\mathcal{A}(t)$ for the Einstein-Helfand relation? It seems not, in periodic systems at any rate. Clearly, it would be wrong to simply evaluate $\mathcal{A}(t)$ using periodically-imaged coordinates. The next alternative is to follow the particle coordinates as they diffuse out of the box, as we did for self diffusion. This also is easily shown to be wrong. Soon, the individual particles will diffuse apart and will become uncorrelated; the cross-terms in $\left\langle [\sum_i \Delta \mathrm{a}_i(t)]^2 \right\rangle$ will disappear. Then, the separate mean-squared values of $\Delta \mathrm{a}_i(t)$ will be dictated by single-particle diffusion, not shear viscosity. The *only* correct way, it seems, to handle $\Delta\mathcal{A}(t)$ is to write it as $\int_0^t \mathrm{d}t' \, \dot{\mathcal{A}}(t')$, and express $\dot{\mathcal{A}}$ in pairwise, minimum-image form.

In the closed-system picture, however, we can use Helfand's expression as written, just as we did for diffusion. Suppose we have real reflecting walls bounding the system in the z direction, but maintain periodicity in the x and y directions (so as not to jeopardize momentum conservation). Then the analysis we applied to diffusion applies exactly to shear modes; the kinematic viscosity $\eta/\rho m$ replaces the diffusion coefficient. This is illustrated in Fig. 10. Again, we get a plateau value, spoiling the growth in $\langle \Delta\mathcal{A}(t)^2 \rangle$. Unfortunately, though, $\tau_{\mathrm{box}} = \rho m L^2/2\eta$ is now quite short (~ 7.5) and more rapid departure from the infinite-system straight line occurs. Nonetheless, it is clear that, given a sufficiently large box, Helfand's formula would hold over a range of times $t \ll \tau_{\mathrm{box}}$. This observed behaviour agrees well with the solution of the 'diffusion equation' (using $\eta/\rho m$ in place of D) in the finite box.

6.3. THE MCQUARRIE EXPRESSION

These questions have resurfaced because of current interest in an expression appearing in McQuarrie's book on Statistical Mechanics [8]. Here, McQuarrie wrote down an expression involving *single-particle* variables only. Chialvo and Debenedetti [57] have suggested that this expression might be useful. In the event, it turns out that it is not. McQuarrie writes

$$\eta = \tfrac{1}{2}\lim_{t\to\infty}\tfrac{\mathrm{d}}{\mathrm{d}t} \frac{1}{Vk_{\mathrm{B}}T} \left\langle \sum_i [\Delta \mathrm{a}_i(t)]^2 \right\rangle \qquad \text{(Einstein, McQuarrie)}$$

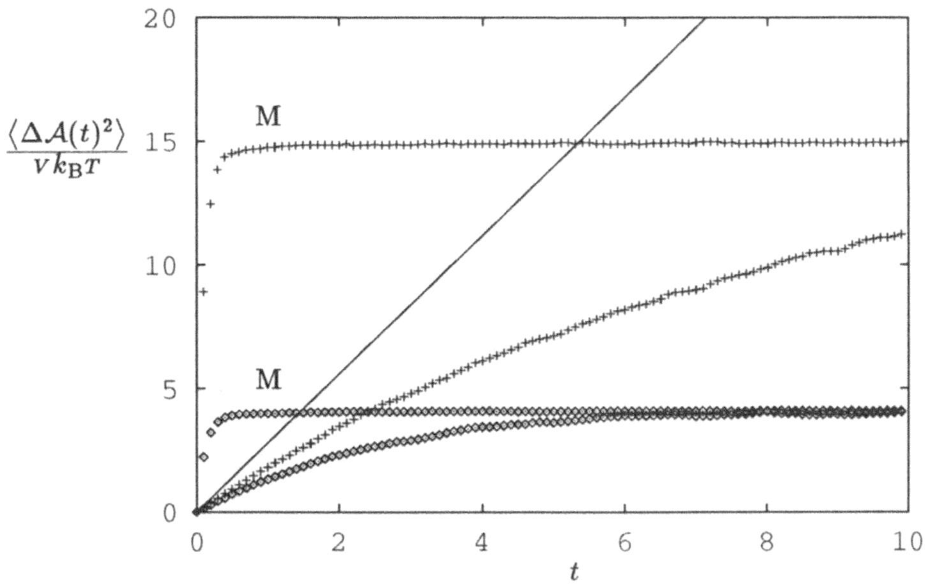

Fig. 10. Einstein relations for viscosity for systems with reflecting walls in the z direction. We plot a line with gradient 2η, Helfand's expression $(Vk_BT)^{-1}\langle \Delta \mathcal{A}(t)^2 \rangle$ for simulations in a box with reflecting walls and $N = 108$ atoms (diamonds) and the same quantity for a double-length box with $N = 216$ atoms (pluses). The corresponding McQuarrie expression $(Vk_BT)^{-1}\langle \sum_i [\Delta a_i(t)^2] \rangle$ is plotted using the same symbols in each case, and labelled with an 'M'.

which differs from the correct expression by taking the summation out of the square brackets. He leaves the proof to the reader. If this equation were true, it would be marvellous: greatly improved statistics could be expected in the computation of shear viscosities.

Unfortunately, this proves not to be the case. In a periodic box, it is impossible to write $\Delta a_i(t)$ in a way that is translationally invariant. Even if we write $\Delta a_i(t) = \int_0^t dt' \, \dot{a}_i(t')$, the time derivative $\dot{a}_i(t')$ is itself not invariant to a shift in the coordinates, and cannot be made consistent with periodic boundary conditions. If we sum over i, giving $\dot{\mathcal{A}} = \sum_i \dot{a}_i$ there is no problem, because the momentum conservation law causes all the translationally invariant terms to cancel. The difficulty arises directly because there is no conservation relation for the k-dependent *single-particle* variable $p_{xi}(k,t) \equiv p_{xi}(t) \exp\{-ikz_i(t)\}$: $\dot{p}_{xi}(k,t)$ is not simply proportional to k. Without such a conservation law, in fact, there is no transport coefficient! Chialvo and Debenedetti seemed to have calculated η in periodic boundaries using the McQuarrie prescription, but of course they had to use an algorithm which was consistent with periodic boundary conditions, and this entailed adopting a specific procedure to handle crossings of the periodic boundaries. We have shown [62] that their results are entirely due to this procedure, and are not related to the viscosity. In fact, Brownian Dynamics simulations of independent particles, carried out using the same procedure, give very similar-looking results for the 'shear viscosity', which can be explained completely in terms of *static* quantities, not dynamic ones.

In a finite box with reflecting walls, it is possible to write down the McQuarrie form directly, without any worries about translational invariance in the z direction, nor any need to adopt special prescriptions to handle periodic boundary crossings. The question now is whether or not a medium-time gradient in the Einstein curve can be seen ahead of the long-time limit at $t \sim \tau_{\text{box}}$, as was the case for the Helfand expression. We see the results [63] in Fig. 10. The McQuarrie form rises very quickly (within a velocity decay time) to its limiting constant value, which is equal to the $t \to \infty$ Helfand value. There is no sign of an intermediate gradient related to the viscosity. So, while it is true that the two expression are equivalent in this system at long times, this is only trivially so: the McQuarrie expression, unlike the Helfand expression, clearly has no Einstein-like behaviour.

In summary, there are many subtleties associated with the calculation of transport coefficients in computer simulations. These subtleties cannot be completely divorced from a consideration of the boundary conditions.

7. Conclusions

I hope that I have given some examples which illustrate the continuing development of computer simulation techniques, and the need, even today, not to forget some of the basic principles of statistical mechanics which have been applied since the earliest days of simulation. Simulation programs are not black boxes; to treat them in this way is to run the risk of (at best) throwing away useful information or (at worst) getting completely incorrect results. These basic principles are especially useful when new, more efficient, or more appropriate simulation algorithms need to be devised. In some of the later Chapters of this book, we see examples of this in action.

Acknowledgements

Thanks are due to Carl Mason and Doug Cleaver, for providing material related to the cumulant expansion technique, and to Andy Masters and Dave Brown, for collaborating on the transport coefficient material. Alastair Bruce gave useful advice on the Lennard-Jones critical point simulations.

References

[1] M. P. ALLEN and D. J. TILDESLEY, *Computer Simulation of Liquids*, Clarendon Press, Oxford, 1987.

[2] G. CICCOTTI and W. G. HOOVER, editors, *Molecular Dynamics Simulation of Statistical Mechanical Systems*, International School of Physics 'Enrico Fermi', Varenna, 1985, Course XCVII, North Holland, 1986.

[3] G. CICCOTTI, D. FRENKEL, and I. R. MCDONALD, editors, *Simulation of Liquids and Solids*, North Holland, 1987.

[4] K. BINDER and D. P. LANDAU, *Phys. Rev. B* **30**, 1477 (1984).

[5] K. BINDER, editor, *Monte Carlo Methods in Statistical Physics*, volume 7 of *Topics in Current Physics*, Springer, second edition, 1986.

[6] K. BINDER and D. W. HEERMANN, *Monte Carlo Simulation in Statistical Physics*, volume 80 of *Solid State Sciences*, Springer, 1988.

[7] J. P. HANSEN and I. R. MCDONALD, *Theory of Simple Liquids*, Academic Press, second edition, 1986.

[8] D. A. MCQUARRIE, *Statistical Mechanics*, Harper and Row, New York, 1976.

[9] T. L. HILL, *An Introduction to Statistical Thermodynamics*, Addison-Wesley, 1960.

[10] K. HUANG, *Statistical Mechanics*, Wiley, 1963.

[11] E. M. LIFSHITZ and L. P. PITAEVSKII, *Statistical Physics*, volume 5 of *Landau and Lifshitz Course of Theoretical Physics*, Pergamon, third edition, 1980.

[12] R. BECKER, *Theory of Heat*, Springer, 1967.

[13] R. KUBO, *Statistical Mechanics*, North Holland, 1965.

[14] S. K. MA, *Statistical Mechanics*, World Scientific, 1985.

[15] A. PAPOULIS, *Probability, Random Variables and Stochastic Processes*, McGraw-Hill, 1965.

[16] J. R. RAY, *Phys. Rev. A* **44**, 4061 (1991).

[17] T. CAGIN and B. M. PETTITT, *Molec. Phys.* **72**, 169 (1991).

[18] H. W. GRABEN and J. R. RAY, *Phys. Rev. A* **43**, 4100 (1991).

[19] J. MATHEWS and R. L. WALKER, *Mathematical Methods of Physics*, W. A. Benjamin, second edition, 1973.

[20] J. L. LEBOWITZ, J. K. PERCUS, and L. VERLET, *Phys. Rev.* **153**, 250 (1967).

[21] E. M. PEARSON, T. HALICIOGLU, and W. A. TILLER, *Phys. Rev. A* **32**, 3030 (1985).

[22] J. R. RAY and H. W. GRABEN, *Phys. Rev. A* **44**, 6905 (1991).

[23] D. FRENKEL, in *Molecular Dynamics Simulation of Statistical Mechanical Systems*, edited by G. CICCOTTI and W. G. HOOVER, p. 151, International School of Physics 'Enrico Fermi', Varenna, 1985, Course XCVII, North Holland, 1986.

[24] W. W. WOOD, in *Physics of Simple Liquids*, edited by H. N. V. TEMPERLEY, J. S. ROWLINSON, and G. S. RUSHBROOKE, chapter 5, North Holland, 1968.

[25] A. MILCHEV, K. BINDER, and D. W. HEERMANN, *Z. Phys. B. Cond. Mat.* **63**, 521 (1986).

[26] K. BINDER, *Z. Phys. B. Cond. Mat.* **43**, 119 (1981).

[27] R. KUBO, *J. Phys. Soc. Japan* **17**, 1100 (1962).

[28] I. R. MCDONALD and K. SINGER, *Discuss. Faraday Soc.* **43**, 40 (1967).

[29] A. M. FERRENBERG and R. H. SWENDSEN, *Phys. Rev. Lett.* **61**, 2635 (1988).

[30] J. LEE and J. M. KOSTERLITZ, *Phys. Rev. Lett.* **65**, 137 (1990).

[31] S. R. PHILLPOT and J. M. RICKMAN, *J. Chem. Phys.* **94**, 1454 (1991).

[32] J. M. RICKMAN and S. R. PHILLPOT, *Phys. Rev. Lett.* **66**, 349 (1991).

[33] S. R. PHILLPOT and J. M. RICKMAN, *Molec. Phys.* **75**, 189 (1992).

[34] J. J. NICOLAS, K. E. GUBBINS, W. B. STREETT, and D. J. TILDESLEY, *Molec. Phys.* **37**, 1429 (1979).

[35] F. R. BROWN and A. YEGULALP, *Phys. Lett. A* **155**, 252 (1991).

[36] V. PRIVMAN, editor, *Finite Size Scaling and Numerical Simulation of Statistical Systems*, World Scientific, 1990.

[37] K. BINDER, editor, *Applications of the Monte Carlo Method in Statistical Physics*, volume 36 of *Topics in Current Physics*, Springer, 1984.

[38] M. S. S. CHALLA, D. P. LANDAU, and K. BINDER, *Phys. Rev. B* **34**, 1841 (1986).

[39] J. E. MAYER and W. W. WOOD, *J. Chem. Phys.* **42**, 4268 (1965).

[40] Z. ZHANG, O. G. MOURITSEN, and M. J. ZUCKERMANN, *Phys. Rev. Lett.* **69**, 2803 (1992).

[41] D. FRENKEL and J. S. VAN DUIJNEVELDT, *J. Chem. Phys.* **96**, 4655 (1992).

[42] J. L. CARDY, editor, *Finite-Size Scaling*, volume 2 of *Current Physics - Sources and Comments*, North Holland, 1988.

[43] A. D. BRUCE, *J. Phys. C* **14**, 3667 (1981).

[44] D. NICOLAIDES and A. D. BRUCE, *J. Phys. A Math. Gen.* **21**, 233 (1988).

[45] D. NICOLAIDES and R. EVANS, *Phys. Rev. Lett.* **63**, 778 (1989).

[46] M. ROVERE, D. W. HEERMANN, and K. BINDER, *Europhys. Lett.* **6**, 585 (1988).

[47] M. ROVERE, D. W. HEERMANN, and K. BINDER, *J. Phys. Cond. Mat.* **2**, 7009 (1990).

[48] A. D. BRUCE and N. B. WILDING, *Phys. Rev. Lett.* **68**, 193 (1992).

[49] N. B. WILDING and A. D. BRUCE, *J. Phys. Cond. Mat.* **4**, 3087 (1992).

[50] M. R. REDDY and S. F. O'SHEA, *Can. J. Phys.* **64**, 677 (1986).

[51] B. SMIT, P. DE SMEDT, and D. FRENKEL, *Molec. Phys.* **68**, 931 (1989).

[52] B. SMIT and D. FRENKEL, *Molec. Phys.* **68**, 951 (1989).

[53] R. R. SINGH, K. S. PITZER, J. J. DE PABLO, and K. M. PRAUSNITZ, *J. Chem. Phys.* **92**, 5463 (1990).

[54] B. SMIT and D. FRENKEL, *J. Chem. Phys.* **94**, 5663 (1991).

[55] K. K. MON and K. BINDER, *J. Chem. Phys.* **96**, 6989 (1992).

[56] P. PECZAK, A. M. FERRENBERG, and D. P. LANDAU, *Phys. Rev. B* **43**, 6087 (1991).

[57] A. A. CHIALVO and P. G. DEBENEDETTI, *Phys. Rev. A* **43**, 4289 (1991).

[58] D. J. EVANS and G. P. MORRISS, *Statistical Mechanics of Nonequilibrium Liquids*, Academic Press, 1990.

[59] E. HELFAND, *Phys. Rev.* **119**, 1 (1960).

[60] B. J. ALDER, D. M. GASS, and T. E. WAINWRIGHT, *J. Chem. Phys.* **53**, 3813 (1970).

[61] J. J. ERPENBECK, *Phys. Rev. A* **38**, 6255 (1988).

[62] M. P. ALLEN, D. BROWN, and A. J. MASTERS, Submitted for publication, 1992.

[63] M. P. ALLEN and A. J. MASTERS, *Molec. Phys.* (1992), accepted.

Advanced Monte Carlo Techniques

D. Frenkel
FOM Institute for Atomic and Molecular Physics
Kruislaan 407
1098 SJ Amsterdam
The Netherlands

ABSTRACT. This Chapter starts with a description of Monte Carlo techniques to study many-body systems in the canonical (constant-NVT), isobaric-isothermal (constant-NPT) and grand-canonical (constant-μVT) ensembles. Subsequently, I give a unified description of two families of 'smart' Monte Carlo schemes, namely Swendsen-Wang-style cluster moves, and configurational-bias MC techniques for collective moves of pre-selected particles. Next, I briefly discuss free-energy calculations in the context of first-order phase transitions. I conclude with a discussion of recent advances in grand-canonical simulations of (hybrid) lattice models.

1. Introduction

In this Chapter, I will assume that the basic idea behind the Monte Carlo technique of Metropolis *et al.* [1], as described in Chapter 1, is known. Nevertheless, I briefly recall those aspects of the Metropolis method that are essential for what follows. The Metropolis method provides us with a recipe to perform a random walk through configuration space, in such a way that the probability of visiting a given configuration i (i.e. a given set of coordinates $\{\mathbf{r}_1, \mathbf{r}_2, \ldots, \mathbf{r}_N\}$), is proportional to the Boltzmann weight ϱ_i of that configuration. In order to construct such a random walk, one should specify the matrix π_{ij} that determines the probability to make a transition from state i to another state j. The matrix π_{ij} should satisfy the important condition that, if we prepare an ensemble of systems in equilibrium, then after one Monte Carlo step, this ensemble is still in equilibrium. One way to achieve this is to impose that π_{ij} satisfies the detailed-balance condition

$$\varrho_i \pi_{ij} = \varrho_j \pi_{ji} \; . \tag{1}$$

There are many possible forms of the transition matrix π_{ij} that satisfy eqn (1). Let us consider how π_{ij} is constructed in practice. We recall that

M.P. Allen and D.J. Tildesley (eds.), Computer Simulation in Chemical Physics, 93–152.

a Monte Carlo move consists of two stages: first we perform a trial move from state i to state j. We denote the transition matrix that determines the probability to perform a trial move from i to j by α_{ij}. $\boldsymbol{\alpha}$ is usually referred to as the underlying matrix of the Markov chain. The next stage is the decision to either accept or reject this trial move. Let us denote the probability of accepting a trial move from i to j by P_{ij}. Clearly, $\pi_{ij} = \alpha_{ij} P_{ij}$. In the original Metropolis scheme, $\boldsymbol{\alpha}$ is chosen to be a symmetric matrix. However, in later sections we shall see several examples where $\boldsymbol{\alpha}$ is *not* symmetric. If $\boldsymbol{\alpha}$ is symmetric, we can rewrite eqn (1) in terms of the P_{ij} only :

$$\varrho_i P_{ij} = \varrho_j P_{ji} \, . \tag{2}$$

From eqn (2) follows:

$$P_{ij}/P_{ji} = \varrho_j/\varrho_i = \exp(-\beta(\mathcal{V}_j - \mathcal{V}_i)) \, . \tag{3}$$

Again, there are many possible choices for P_{ij} that satisfy this condition (and the obvious condition that the probability P_{ij} cannot exceed 1). The choice of Metropolis *et al.* was

$$P_{ij} = \min(\varrho_j/\varrho_i, 1) \, . \tag{4}$$

Note that we have not specified the matrix $\boldsymbol{\alpha}$, but for the fact that it must be symmetric. This reflects the fact that there is a considerable freedom in the choice of our trial moves. We will come back to this point in subsequent sections. Thus far, we have not mentioned another important condition that π_{ij} should satisfy, namely that it is *ergodic*, i.e. that every accessible point in configuration space can be reached in a finite number of Monte Carlo steps from any other point. Although there are simple MC schemes that are guaranteed to be ergodic, these are often not the most efficient schemes. Conversely, there are many efficient Monte Carlo schemes that have either not been proven to be ergodic or, worse, been proven to be non-ergodic. If you are worried that purists may dislike your favorite non-ergodic scheme, the solution is to mix that scheme with a less efficient ergodic scheme. Then everybody is happy.

Before proceeding with a discussion of Monte-Carlo sampling methods, it is important to stress one limitation of the Metropolis-style MC method. The MC method *can* be used to estimate configuration-space averages of structural properties of a many-body system, i.e.

$$\langle \mathcal{A} \rangle_{NVT} = \frac{\int \mathrm{d}\mathbf{q}^N \, \mathcal{A}(\mathbf{q}^N) \exp(-\beta \mathcal{V}(\mathbf{q}^N))}{Z_{NVT}} \, , \tag{5}$$

with $Z_{NVT} = \int \mathrm{d}\mathbf{q}^N \exp(-\beta \mathcal{V}(\mathbf{q}^N))$. However, the MC method cannot be used to compute the configurational integral Z_{NVT}, as it depends on the accessible *volume* in configuration space. Knowledge of the configurational integral is required to compute "thermal" properties, such as the Helmholtz free energy F, the Gibbs free energy G, and the chemical potential μ. Techniques to compute the latter quantities (that play a crucial role in the calculation of phase equilibria), will be discussed in considerable detail in subsequent sections.

2. Other ensembles

In the previous section, I have briefly reviewed the Metropolis Monte Carlo simulation method. This technique was designed to compute the average value of an arbitrary function $\mathcal{A}(\mathbf{q}^N)$ of the coordinates of a system of N particles at constant temperature T and volume V (eqn (5)):

$$\langle \mathcal{A} \rangle_{NVT} = \frac{\int d\mathbf{q}^N \, \mathcal{A}(\mathbf{q}^N) \exp(-\beta \mathcal{V}(\mathbf{q}^N))}{Z_{NVT}} \, .$$

This average is usually referred to as the 'canonical' ensemble average. To be more precise, the true canonical ensemble average includes the particle momenta. However, in all cases of practical interest, the integration over momenta can be carried out analytically, and therefore does not require numerical sampling.

A large fraction of all Monte Carlo calculations to date have been performed in the constant-NVT ensemble. However, it is sometimes advisable to perform a simulation where thermodynamic variables other than N,V or T are kept fixed. In the present section we discuss such simulations. I will limit this discussion to those ensembles that have found most widespread use in the simulation of classical many-body systems, such as the constant-NPT and μVT ensembles. I refer readers interested in *microcanonical* MC simulations to the original paper by Creutz [2] and the book by Binder [3].

2.1. ISOBARIC-ISOTHERMAL ENSEMBLE

Apart from the canonical ensemble, the isobaric-isothermal (constant-NPT) ensemble is probably the one most widely used in Monte Carlo simulations. This is not surprising because most real experiments are also carried out under conditions of controlled pressure and temperature. Moreover, constant-NPT simulations can be used to measure the equation of state of a model system even if the virial expression for the pressure cannot be readily evaluated. This may, for instance, be the case for certain models of non-spherical hard-core molecules. But also for the increasingly important class of models where the (non-pairwise additive) potential energy function is computed *numerically* for each new configuration. Finally, it is often convenient to use constant-NPT MC to simulate systems in the vicinity of a first-order phase transition, because at constant pressure the system is free (given enough time, of course) to transform completely into the state of lowest (Gibbs) free energy, whereas in a constant-NVT simulation the system may be kept at a density where it would like to phase-separate into two bulk phases of different density, but is prevented from doing so by finite size effects.

Constant-pressure Monte Carlo simulations were first described by Wood [4] in the context of a simulation study of two-dimensional hard disks. Although the method introduced by Wood is very elegant, it is not readily applied to systems with arbitrary continuous potentials. McDonald [5] was the first to apply constant-NPT simulations to a system with continuous intermolecular forces (a Lennard-Jones mixture), and it is the constant-pressure method of McDonald that is now being used almost universally, and that is discussed below.

I will derive the basic equations of constant-pressure Monte Carlo in a way that may appear unnecessarily complicated. However, this derivation has the advantage that the same framework can be used to introduce some of the other non-NVT Monte Carlo methods to be discussed below. For the sake of convenience we shall initially assume that we are dealing with a system of N identical atoms. The partition function for this system is given by:

$$Q(N,V,T) = \frac{1}{\Lambda^{3N} N!} \int_0^L \cdots \int_0^L \mathrm{d}\mathbf{q}^N \exp(-\beta \mathcal{V}(\mathbf{q}^N)) \,, \tag{6}$$

where $\Lambda = \sqrt{h^2/2\pi m k_B T}$ is the thermal de Broglie wavelength. It is convenient to rewrite eqn (6) in a slightly different way. We have assumed that the system is contained in a cubic box with diameter $L = V^{1/3}$. We now define scaled coordinates $\mathbf{s}^N$, by:

$$\mathbf{q}_i = L\mathbf{s}_i$$

for $i = 1, 2, \ldots, N$. If we now insert these scaled coordinates in eqn (6) we obtain:

$$Q(N,V,T) = \frac{V^N}{\Lambda^{3N} N!} \int_0^1 \cdots \int_0^1 \mathrm{d}\mathbf{s}^N \exp(-\beta \mathcal{V}(\mathbf{s}^N; L)) \,. \tag{7}$$

In eqn (7), we have written $\mathcal{V}(\mathbf{s}^N; L)$ to indicate that $\mathcal{V}$ depends on the *real* rather than the *scaled* distances between the particles. The expression for the Helmholtz free energy of the system is:

$$\begin{aligned} F(N,V,T) &= -k_B T \ln Q \\ &= -k_B T \ln \left(\frac{V^N}{\Lambda^{3N} N!} \right) - k_B T \ln \left(\int \mathrm{d}\mathbf{s}^N \exp(-\beta \mathcal{V}(\mathbf{s}^N; L)) \right) \\ &= F_{\mathrm{id}}(N,V,T) + F_{\mathrm{ex}}(N,V,T) \,. \end{aligned} \tag{8}$$

In the last line of the above equation we have identified the two contributions to the Helmholtz free energy on the previous line as the ideal-gas expression

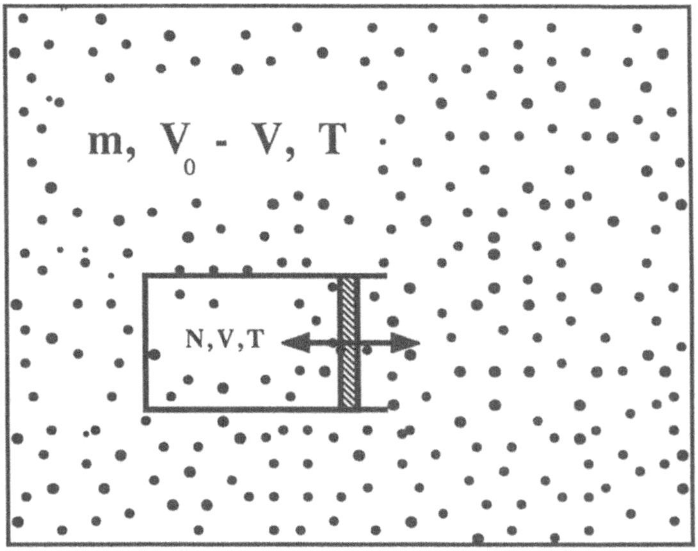

Fig. 1. Ideal gas (m particles, volume $V_0 - V$) can exchange volume with an N-particle system (volume V)

plus an excess part. Let us now assume that the system is separated by a 'piston' [6] from an ideal gas reservoir (see Fig. 1). The total volume of the system plus reservoir is fixed at a value V_0. The total number of particles is M. Hence the volume accessible to the $m \equiv M - N$ ideal gas molecules is $V_0 - V$. The partition function of the total system is simply the product of the partition functions of the constituent sub-systems:

$$Q(N, m, V, V_0, T) = \frac{V^N (V_0 - V)^m}{\Lambda^{3M} N! m!} \int d\mathbf{s}^m \int d\mathbf{s}^N \exp(-\beta \mathcal{V}(\mathbf{s}^N; L)) \,. \quad (9)$$

Note that the integral over the $\mathbf{s}^m$ scaled coordinates of the ideal gas yields simply 1. For the sake of compactness we have assumed that the thermal wavelength of the ideal gas molecules is also equal to Λ. The total free energy of this combined system is $F_{\text{tot}} = -k_B T \ln Q(N, m, V, V_0, T)$. Now let us assume that the 'piston' between the two subsystems is free to move, so that the volume V of the N-particle subsystem can fluctuate. Of course, the most probable value of V will be the one that minimizes the free energy of the combined system. The probability density $\mathcal{P}(V)$ that the N-particle subsystem has a volume V is given by:

$$\mathcal{P}(V) = \frac{V^N (V_0 - V)^m \int \mathrm{d}\mathbf{s}^N \exp(-\beta \mathcal{V}(\mathbf{s}^N; L))}{\int_0^{V_0} \mathrm{d}V' \, V'^N (V_0 - V')^m \int \mathrm{d}\mathbf{s}^N \exp(-\beta \mathcal{V}(\mathbf{s}^N; L'))} \tag{10}$$

We now consider the limit that the size of the reservoir tends to infinity ($V_0 \to \infty, M \to \infty, m/V_0 \to \rho$). In that limit, a small volume change of the small system does not change the pressure P of the large system. In other words, the large system works as a manostat for the small system. In that case we can simplify eqns (9) and (10). Note that in the limit $V/V_0 \to 0$ we can write:

$$(V_0 - V)^m = V_0^m (1 - (V/V_0))^m \to V_0^m \exp(-mV/V_0).$$

In the limit $m \to \infty$, $\exp(-mV/V_0) \to \exp(-\rho V)$. But, as the reservoir contains an ideal gas, ρ can be written as βP. With these substitutions, eqn (10) becomes:

$$\mathcal{P}(V) = \frac{V^N \exp(-\beta P V) \int \mathrm{d}\mathbf{s}^N \exp(-\beta \mathcal{V}(\mathbf{s}^N; L))}{\int_0^{V_0} \mathrm{d}V' \, V'^N \exp(-\beta P V') \int \mathrm{d}\mathbf{s}^N \exp(-\beta \mathcal{V}(\mathbf{s}^N; L'))} . \tag{11}$$

In the same limit, the difference in free energy between the combined system and the ideal-gas system in the absence of the N-particle subsystem is the well-known Gibbs free energy G :

$$G(N, P, T) = -k_B T \ln \int \mathrm{d}V \frac{V^N \exp(-\beta P V)}{\Lambda^{3N} N!} \int \mathrm{d}\mathbf{s}^N \exp(-\beta \mathcal{V}(\mathbf{s}^N; L)) \tag{12}$$

Equation (11) is the starting point for constant-NPT Monte Carlo simulations. The idea is that the probability density to find the small system in a particular configuration of the N atoms (as specified by $\mathbf{s}^N$) at a given volume V is given by:

$$\mathcal{P}(V; \mathbf{s}^N) = \frac{V^N \exp(-\beta P V) \exp(-\beta \mathcal{V}(\mathbf{s}^N; L))}{\int_0^{V_0} \mathrm{d}V' \, V'^N \exp(-\beta P V') \int \mathrm{d}\mathbf{s}^N \exp(-\beta \mathcal{V}(\mathbf{s}^N; L'))} .$$

We can now carry out Metropolis sampling on the reduced coordinates $\mathbf{s}^N$ *and* the volume V, with a weight-function $\varrho(\mathbf{s}^N, V)$ proportional to $\exp(-\beta\{\mathcal{V}(\mathbf{s}^N, V) + PV - N\beta^{-1} \ln V\})$. In the constant-$NPT$ MC method, V is simply treated as an additional coordinate, and trial moves in V must satisfy the same rules as trial moves in $\mathbf{q}$ (in particular, we should maintain the reversibility of the underlying Markov chain). Let us assume that our trial moves consist of an attempted change of the volume from V to $V' = V + \Delta V$, where ΔV is a random number, uniformly distributed over the interval $[-\Delta V_{\max}, +\Delta V_{\max}]$. In the Metropolis scheme such a random, volume-changing move will be accepted if:

$$\exp(-\beta[\mathcal{V}(\mathbf{s}^N, V') - \mathcal{V}(\mathbf{s}^N, V) + P(V' - V) - N\beta^{-1} \ln(V'/V)]) > \mathcal{R} \tag{13}$$

where $\mathcal{R}$ is a random number, uniformly distributed over the interval $[0, 1]$. Instead of attempting random changes in the volume itself, one might construct trial moves in the box-length L [5], or in the logarithm of the volume [7]. Such trial moves are equally legitimate, as long as the reversibility of the underlying Markov chain is maintained. However, such alternative schemes result in a slightly different form for eqn (13). The isobaric-isothermal MC method has been extended to simulate systems at constant stress. This problem was first in the context of molecular dynamics simulations by Parrinello and Rahman [8, 9] who developed an extension of the constant-pressure molecular dynamics technique introduced by Andersen [10]. The extension of the Parrinello-Rahman method to Monte Carlo simulations is straightforward [11].

2.2. GRAND CANONICAL ENSEMBLE

The Metropolis sampling scheme was introduced as a method to compute thermal averages of functions $\mathcal{A}(\mathbf{q}^N)$, that depend explicitly on the coordinates of the molecules in the N-body system under study. Examples of such 'mechanical' properties are the potential energy or the virial contribution to the pressure. However, the Metropolis method cannot be used to determine the configurational integral

$$Z_{NVT} \equiv \int \mathrm{d}\mathbf{q}^N \exp(-\beta\mathcal{V}(\mathbf{q}^N))$$

itself. The latter quantity measures the effective volume in configuration space that is accessible to the system. Hence the original Metropolis scheme cannot be used to determine those thermodynamic properties of a system that depend explicitly on the configurational integral. Examples of such 'thermal' properties are: the Helmholtz free energy

$$F = -k_\mathrm{B}T \ln Q_{NVT} ,$$

the entropy S and the Gibbs free energy G. However, although the Metropolis method cannot be used to measure, for instance, free energies directly, it can be used to measure the *difference* in free energy between two possible states of an N-body system. This fact is exploited in the grand-canonical Monte Carlo method (GCMC) first implemented for classical fluids by Norman and Filinov [12], and later extended and improved by a number of other groups [13–21]. The basic idea of the GCMC method is explained below.

In order to understand the statistical mechanical basis of the GCMC technique, let us return to eqn (9) of section 2.1. This equation gives the partition function of a combined system of N interacting particles in volume V and $m = M - N$ ideal gas molecules in volume $V_0 - V$:

$$Q(N, m, V, V_0, T) = \frac{V^N (V_0 - V)^m}{\Lambda^{3M} N!m!} \int \mathrm{d}\mathrm{s}^m \int \mathrm{d}\mathrm{s}^N \exp(-\beta\mathcal{V}(\mathrm{s}^N)) \,. \quad (9)$$

Now, instead of allowing the two systems exchange volume, let us see what happens if the systems can exchange particles (see Fig. 2). To be more

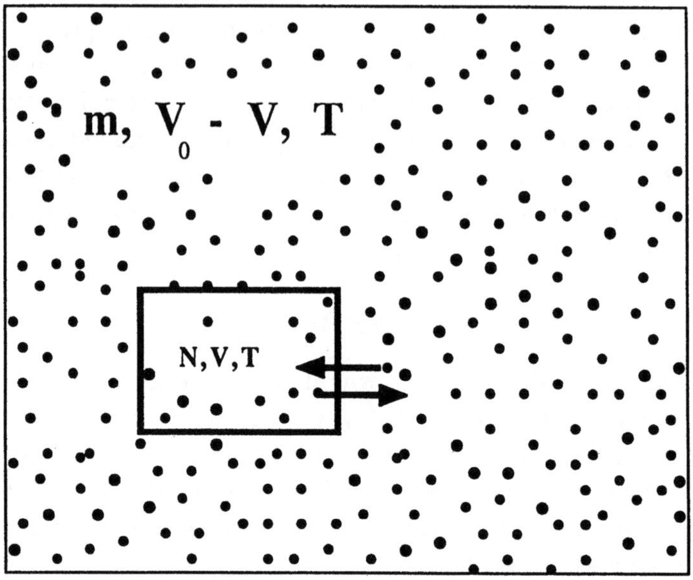

Fig. 2. Ideal gas (m particles, volume $V_0 - V$) can exchange particles with an N-particle system (volume V)

precise, we assume that the molecules in the two sub-volumes are actually identical particles. The only difference is that when they find themselves in volume V they interact and when they are in volume $V_0 - V$, they do not. If we transfer a molecule i from a reduced coordinate $\mathbf{s}_i$ in the volume $V_0 - V$ to the same reduced coordinate in volume V, then the potential energy function $\mathcal{V}$ changes from $\mathcal{V}(\mathbf{s}^N)$ to $\mathcal{V}(\mathbf{s}^{N+1})$. The expression for the total partition function of the system, including all possible distributions of the M particles over the two sub-volumes is:

$$Q(M,V,V_0,T) = \sum_{N=0}^{M} \frac{V^N (V_0 - V)^{M-N} \int d\mathbf{s}^{M-N} \int d\mathbf{s}^N \exp(-\beta \mathcal{V}(\mathbf{s}^N))}{\Lambda^{3M} N!(M-N)!} . \tag{14}$$

Following the approach of section 2.1, we now write the probability density to find a system with $M - N$ particles at reduced coordinates $\mathbf{s}^{M-N}$ in volume $V' \equiv V_0 - V$ and N particles at reduced coordinates $\mathbf{s}^N$ in volume V:

$$\mathcal{P}(\mathbf{s}^M; N) = \frac{V^N V'^{M-N}}{Q(M,V,V',T)\Lambda^{3M} N!(M-N)!} \exp(-\beta \mathcal{V}(\mathbf{s}^N)) . \tag{15}$$

Let us now consider a trial move in which a particle is transferred from V' to the same scaled coordinate in V. First we should make sure that we construct an underlying Markov chain that is symmetric. Symmetry, in this case, implies that the *a priori* probability to move a particle from V' to V should be equal to the *a priori* probability of the reverse move. The probability of acceptance of a trial move in which we move a particle *to* or *from* volume V is determined by the ratio of the corresponding probability densities (eqn (15)):

$$P_{N\rightarrow N+1} = \frac{V(M-N)}{V'(N+1)} \exp(-\beta(\mathcal{V}(\mathbf{s}^{N+1}) - \mathcal{V}(\mathbf{s}^{N}))) \tag{16}$$

$$P_{N+1\rightarrow N} = \frac{V'(N+1)}{V(M-N)} \exp(-\beta(\mathcal{V}(\mathbf{s}^{N}) - \mathcal{V}(\mathbf{s}^{N+1}))) \,. \tag{17}$$

Now let us consider the limit that the ideal gas system is very much larger than the interacting system: $M \rightarrow \infty, V' \rightarrow \infty, (M/V') \rightarrow \rho$. Note that for an ideal gas the chemical potential μ is related to the particle-density ρ by: $\mu = k_{\mathrm{B}}T \ln \Lambda^3 \rho$. Therefore, in the limit $(M/N) \rightarrow \infty$, eqns (16),(17) become:

$$P_{N\rightarrow N+1} = \frac{V}{N+1} \Lambda^{-3} \exp(\beta\mu) \exp(-\beta(\mathcal{V}(\mathbf{s}^{N+1}) - \mathcal{V}(\mathbf{s}^{N}))) \tag{18}$$

$$P_{N+1\rightarrow N} = \frac{(N+1)}{V} \Lambda^{3} \exp(-\beta\mu) \exp(-\beta(\mathcal{V}(\mathbf{s}^{N}) - \mathcal{V}(\mathbf{s}^{N+1}))) \,. \tag{19}$$

In the Metropolis scheme, an attempted particle addition to (removal from) volume V is accepted if $P_{N\rightarrow N+1} > \mathcal{R}$ ($P_{N+1\rightarrow N} > \mathcal{R}$), where $\mathcal{R}$ is a random number uniformly distributed in the interval $[0, 1]$.

Equations (18), (19) are the basic equations for grand canonical Monte Carlo simulations. Acceptable trial moves are: (i) the random insertion of an additional particle at any point in volume V; (ii) the random removal of any of the N particles in volume V. In addition, the particles in volume V can sample the accessible configuration space by conventional Monte Carlo sampling. Note that in eqns (18), (19) all explicit reference to the ideal-gas system has disappeared. The rules only refer to the system of volume V.

The most salient feature of the GCMC technique is that in such simulations the chemical potential μ is *imposed*, while the number of particles N is a fluctuating quantity. During the simulation we may measure other thermodynamic quantities, such as the pressure P, the average density $\langle\rho\rangle$ or the internal energy $\langle\mathcal{V}\rangle$. As we know μ, we can derive all other 'thermal' properties, such as the Helmholtz free energy or the entropy. This may seem surprising. After all, in section 1 we stated that Metropolis sampling *cannot* be used to sample absolute free energies and related quantities. Yet, with grand canonical MC we appear to be doing precisely that. The answer is that, in fact, we do not. What we measure is not an absolute but a relative free energy. In GCMC, we are equating the chemical potential of a molecule in an ideal gas at density ρ (for the ideal-gas case we know how to compute μ) and the chemical potential of the same species in an interacting system at density ρ'.

Grand canonical MC works best if the acceptance of trial moves by which particles are added or removed is not too low. For atomic fluids this condition effectively limits the maximum density at which the method can be used to about twice the critical density. Special tricks are needed to extend GCMC to somewhat higher densities [19]. GCMC is easily implemented for mixtures and inhomogeneous systems, such as fluids near interfaces. In fact, some of the most useful applications of the GCMC method are precisely in these areas of research. Although GCMC can be applied to simple models of non-spherical molecules, it converges very poorly for all but the smallest polyatomic molecules. However, in section 4.1, we shall discuss techniques that can be used to generalize GCMC simulations to chain molecules. For more details on the more conventional grand canonical Monte Carlo method, the reader is referred to the book by Allen and Tildesley [22] and a review paper by Frenkel [23].

2.3. GIBBS ENSEMBLE

One of the prime reasons why we are interested in simulations at constant P or constant μ, is that the condition for coexistence of two or more phases I, II,... is that the pressure of all coexisting phases must be equal ($P_{\mathrm{I}} = P_{\mathrm{II}} = \ldots = P$), as must be the temperature ($T_{\mathrm{I}} = T_{\mathrm{II}} = \ldots = T$) and the chemical potentials of all species ($\mu_{\mathrm{I}}^{\alpha} = \mu_{\mathrm{II}}^{\alpha} = \ldots = \mu^{\alpha}$). Hence one might be inclined to think that the best ensemble to study would be the 'constant-μPT ensemble'. The quotation marks around the name of this 'ensemble' are intentional, because, strictly speaking, no such ensemble exists. The reason is simple: if we specify only intensive parameters, such as P, T and μ, the extensive variables (such as V) are unbounded. Another way to say the same thing is that the set P, T, μ is linearly dependent. In order to get a decent ensemble, we must fix at least one extensive variable. In the case of constant-pressure MC this variable is the number of particles N, while in grand canonical Monte Carlo, the volume V of the system is fixed.

A few years ago, however, Panagiotopoulos [24] developed a new Monte Carlo scheme, called the 'Gibbs method', that comes very close to achieving the impossible: namely simulating phase equilibria under conditions where the pressure, temperature and chemical potential(s) of the coexisting phases are equal. The reason why this method can work is that, although the *difference* between chemical potentials in different phases is fixed (namely, at $\Delta\mu = 0$), the absolute values are still free to change. The Gibbs-ensemble method has developed into one of the most important techniques to study phase-equilibria (in particular, fluid-fluid equilibria) by Monte Carlo simulation. In view of the great importance of this method, Chapter 6 in this volume has been devoted to the ins and outs of the Gibbs ensemble method.

2.4. SEMI-GRAND ENSEMBLE

At present, the Gibbs ensemble method has become the technique *par excellence* to study fluid-fluid phase equilibria. However, like the grand-canonical ensemble scheme, the method does rely on a reasonable number of successful particle insertions to achieve compositional equilibrium. As a consequence, the Gibbs ensemble method cannot be used to study phase equilibria involving very dense phase or crystalline solids. Although there exist, to my knowledge, no numerical schemes that allow *direct* simulation of, say, solid-liquid phase coexistence in a multicomponent system, there exists a technique that greatly facilitates the numerical study of such phase equilibria. This is the so-called 'semi-grand ensemble method' of Kofke and Glandt [25]. In this section, I shall present a simple description of this method.

In order to understand why the semi-grand ensemble can be successful where other schemes fail, it is useful to borrow a result that will be derived in section 3.1. In that section, we show that the excess chemical potential of a molecule in a fluid is related to the average Boltzmann factor associated with the random addition of such a molecule to a fluid with N particles present (eqn (40)):

$$\mu_{\mathrm{ex}} = -k_{\mathrm{B}}T \ln \left\langle \int \mathrm{d}s_{N+1} \exp(-\beta \Delta \mathcal{V}) \right\rangle_N .$$

Let us assume that we wish to study phase equilibria in a binary mixture. In that case, we must compute the Gibbs free energy per mole of the mixture, as a function of the composition:

$$G(X_A) = X_A \mu_A + X_B \mu_B , \tag{20}$$

where X_A $(= 1 - X_B)$ denotes the mole fraction of species A, and μ_A (μ_B) denote the chemical potentials of the two components in the mixture. Let us assume that we have somehow succeeded in computing the Gibbs free energy of one of the pure phases (for instance, by one of the thermodynamic integration methods described in section 5). At first sight it might seem that, in order to compute G as a function of X_A, we have to repeat such a thermodynamic integration for a large number of X_A values. Fortunately, this is usually not the case. Rather than recomputing $G(X)$ for a number of compositions, we can study the *change* in $G(X)$ with X. In other words, we need to have a 'microscopic' expression for

$$\left(\frac{\partial G(X)}{\partial X} \right)_{P,T,N} = \mu_A - \mu_B \tag{21}$$

$$= (\mu_A - \mu_B)_{\rm id} + (\mu_A - \mu_B)_{\rm ex} \, .$$

In the first line of eqn (21), we have used the Gibbs-Duhem relation. I assume that the ideal-gas contributions to the chemical potential of both A and B are known. The quantity that we must compute is $\Delta\mu_{\rm ex} \equiv (\mu_A - \mu_B)_{\rm ex}$. Naively, one might try to measure this quantity by using the particle insertion method to obtain $\mu_{\rm ex}$ of species A and B separately and then subtracting the result. Although such an approach would be correct *in principle*, it is time consuming and not very accurate. Fortunately, $\Delta\mu_{\rm ex}$ can be obtained much more directly by measuring the Boltzmann factor associated with a virtual trial move where a randomly selected particle of type B is transformed into a particle of type A [26–28]. I leave it as an exercise to the reader to derive that the resulting expression for $\Delta\mu_{\rm ex}$ at constant pressure is:

$$\Delta\mu_{\rm ex} = -k_{\rm B}T \ln \left\langle \frac{N_B}{N_A+1} \exp(-\beta\Delta\mathcal{V}^{+-}) \right\rangle \tag{22}$$

where $\Delta\mathcal{V}^{+-}$ denotes the change in potential energy of the system if one particle of type B is changed into type A. $-k_{\rm B}T\ln(N_B/[N_A+1])$ is simply the ideal mixing contribution to the chemical potential. The point to note about eqn (22) is that for a perfect mixture (i.e. A and B have the same intermolecular interactions), $\ln\langle\exp(-\beta\Delta\mathcal{V}^{+-})\rangle$ is identically equal to zero. In other words, we may obtain very good statistics on $\Delta\mu_{\rm ex}$ even if the direct measurement of the excess chemical potential of the individual species would yield poor statistics.

The aim of this introduction to the semi-grand ensemble is twofold: first of all, eqn (22) shows that the Boltzmann factor associated with the change of identity of a particle in a mixture is related to the difference in excess chemical potential of the two species that take part in the interchange. And secondly, I have made it plausible that we can get good statistics on $\Delta\mu_{\rm ex}$ even when the particle insertion method to measure the excess chemical potential of the individual species fails, for instance in a crystalline solid [29]. I recall that the grand-canonical Monte Carlo method has about the same range of applicability as the particle insertion method. It is therefore logical to infer that it should be possible to construct a simulation scheme based on particle *interchanges* that should work under conditions where the GCMC scheme fails. The semi-grand canonical Monte Carlo (SGCMC) method is such a scheme.

How does the SGCMC method work? Let us first consider the expression for Ξ, the grand-canonical partition function for an n-component mixture:

$$\Xi = \sum_{N_1,N_2,\ldots,N_n}^{\infty} \prod_{i=1}^{n} \frac{q_i^{N_i}\exp(\beta\mu_i N_i)}{N_i!} V^N \int {\rm d}{\rm s}^N \exp(-\beta\mathcal{V}({\rm s}^N)) \, , \tag{23}$$

where $N \equiv \sum_i N_i$, $\mathcal{V}(\mathbf{s}^N)$ denotes the potential energy function of the n-component mixture and q_i is the 'kinetic' contribution to the partition function due to species i. Next, we consider a related partition function Ξ', identical to Ξ, but for the fact that we have imposed the constraint that $N = \sum_i N_i$ is fixed. If N is fixed, we can eliminate one of the N_i, for instance N_1, from the sum in eqn (23) and we obtain:

$$\Xi' = \sum_{N_2,\ldots,N_n}' q_1^N \exp(\beta\mu_1 N) \prod_{i=1}^{n} \left(\frac{q_i}{q_1}\right)^{N_i} \frac{\exp(\beta(\mu_i - \mu_1)N_i)}{N_i!} \times \left(V^N \int \mathrm{d}\mathbf{s}^N \exp(-\beta\mathcal{V}(\mathbf{s}^N))\right) . \quad (24)$$

We now multiply this equation on both sides by $\exp(-\beta\mu_1 N)$ and we define a new partition function $\mathcal{Y} \equiv \Xi' \exp(-\beta\mu_1 N)$:

$$\mathcal{Y} = \sum_{N_2,\ldots,N_n}' q_1^N \prod_{i=1}^{n} \left(\frac{q_i}{q_1}\right)^{N_i} \frac{\exp(\beta(\mu_i - \mu_1)N_i)}{N_i!} \times \left(V^N \int \mathrm{d}\mathbf{s}^N \exp(-\beta\mathcal{V}(\mathbf{s}^N))\right) . \quad (25)$$

The next step is subtle. We shall reinterpret the sum over all N_i in eqns (23)–(25). In these equations, we had assumed that to every composition $N_1, N_2, \ldots, N_n$, there corresponds *one* term in the sum. Let us now take a somewhat different point of view: we assume that all these different species are all manifestations of one and the same 'particle'. This sounds strange, so I shall use an analogy to explain what I mean. Let us consider that we have a group of 100 people, made up of 3 groups: eaters, drinkers and sleepers. In fact, we want to consider all possible combinations of these groups, with the constraint that the total number is fixed. One such combination would be: 30 eaters, 30 drinkers and 40 sleepers. Now we make a discovery: one and the same person can be an eater, a drinker or a sleeper, but not simultaneously. Now our sum over all combinations becomes different: we have 100 'persons' who can all take on any one out of the three possible identities. In that case, we have many more ways in which we can make a group of 30 eaters, 30 drinkers and 40 sleepers, namely 100!/(30!30!40!). If we wish to have the same total number of terms in our sum as before, we have to divide by this factor.

Let us now translate this example back to the sum over particles in eqn (25). We replace the sum over numbers of particles of species i, by a sum over all possible identities of all particles. But then we must correct for double counting by dividing by $N!/(N_1! \times \ldots \times N_n!)$. If we do that, eqn (25) becomes:

$$\mathcal{Y} = \sum_{\text{identities}} \frac{q_1^N}{N!} \prod_{i=1}^{n} \left(\frac{q_i}{q_1}\right)^{N_i} \exp(\beta(\mu_i - \mu_1)N_i) \times$$
$$\left(V^N \int \mathrm{d}\mathbf{s}^N \exp(-\beta\mathcal{V}(\mathbf{s}^N))\right) . \tag{26}$$

Finally, it turns out that it is more convenient to consider the corresponding ensemble at constant pressure. In that case, the partition function changes to

$$\mathcal{Y}' = \beta P \int \mathrm{d}V \exp(-\beta PV) \frac{(Vq_1)^N}{N!} \times \tag{27}$$
$$\left(\sum_{\text{identities}} \prod_{i=1}^{n} \left(\frac{q_i}{q_1}\right)^{N_i} \exp(\beta(\mu_i - \mu_1)N_i) \int \mathrm{d}\mathbf{s}^N \exp(-\beta\mathcal{V}(\mathbf{s}^N))\right) .$$

For cosmetic reasons, we rewite eqn (27) in terms of the *fugacities* f_i, rather than the chemical potentials μ_i. Recall that the fugacity of a species i is defined by the expression:

$$\mu_i(P, T, \{X_i\}) \equiv \mu_i^0(T) + k_B T \ln f_i , \tag{28}$$

where $\mu_i^0(T)$ is the chemical potential of the ideal gas reference state ($P = 1$) of species i. Using the expression for the chemical potential of an ideal gas at pressure P, and recalling that for an ideal gas $f = P$, it is easy to show that:

$$\mu_i^0(T) = -k_B T \ln(k_B T q_i) . \tag{29}$$

Inserting eqn (29) in eqn (27), we obtain:

$$\mathcal{Y}' = \beta P \int \mathrm{d}V \exp(-\beta PV) \frac{(Vq_1)^N}{N!} \sum_{\text{identities}} \prod_{i=1}^{n} \left(\frac{f_i}{f_1}\right)^{N_i} \times$$
$$\left(\int \mathrm{d}\mathbf{s}^N \exp(-\beta\mathcal{V}(\mathbf{s}^N))\right) . \tag{30}$$

By now the reader is probably thoroughly confused by all the algebra. In order to see what we have done, it is instructive to look at the characteristic thermodynamic function associated with the successive partition functions. We started with a grand-canonical ensemble. The link with thermodynamics is given by:

$$\beta PV = \ln \Xi(V, T, \{\mu_i\}) . \tag{31}$$

The transformation to $\mathcal{Y}$ (eqn (25)), corresponds to a change to the thermodynamic variable

$$\beta(PV - \mu_1 N) = \ln \mathcal{Y}(V, T, N, \{\mu_i | i \neq 1\}) . \tag{32}$$

Finally, we change to constant pressure, which means that the characteristic thermodynamic function becomes:

$$\beta\mu_1 N = -\ln \mathcal{Y}'(P, T, \{\ln(f_i/f_1)|i \neq 1\}) . \tag{33}$$

Rather than use $\ln(f_i/f_1)$ as the independent variable, it is more convenient to follow Ref. [25] and use the *fugacity fraction* ξ_i, defined as

$$\xi_i \equiv \frac{f_i}{\sum_{j=1}^{n} f_j} . \tag{34}$$

The advantage is that, while $\ln(f_i/f_1)$ varies between $-\infty$ and $+\infty$ as we go from pure 1 to pure i, ξ_i varies between 0 and 1. Clearly,

$$\begin{aligned}\beta\mu_1 N \; = \; -\ln \Bigg(\beta P \int \mathrm{d}V \; \exp(-\beta PV) \frac{(Vq_1)^N}{N!} \; \times \\ \sum_{\text{identities}} \prod_{i=1}^{n} \left(\frac{\xi_i}{\xi_1}\right)^{N_i} \int \mathrm{d}\mathbf{s}^N \exp(-\beta\mathcal{V}(\mathbf{s}^N)) \Bigg) .\end{aligned} \tag{35}$$

How does μ_1, the chemical potential of the reference species, change with the fugacity fractions of the other species? To see this we consider the derivative of eqn (35) with respect to ξ_i:

$$\left(\frac{\partial \beta\mu_1 N}{\partial \xi_i}\right)_{N,P,T,\{\xi_j|j\neq i\}} = -\left\langle \frac{N_i}{\xi_i} + \frac{N - N_1}{\xi_1} \right\rangle , \tag{36}$$

where we have used the fact that $\mathrm{d}\xi_1 = -\sum_{j\neq i} \mathrm{d}\xi_j$. Equation (36) tells us how we can measure μ_1 changes as we change the chemical potential difference between species 1 and the other species. Let us consider the application to phase coexistence in a binary mixture. In that case we only vary ξ_2. First we measure the excess chemical potential in phase I consisting of pure 1 and in phase II consisting of pure species 2 (for instance by thermodynamic integration). Next we compute the change in μ_1 in phase I as we increase ξ_2 from 0 and the corresponding change in μ_1 in phase II as we lower ξ_2 from 1. The point where $\mu_1^{(\mathrm{I})}(\xi_2) = \mu_1^{(\mathrm{II})}(\xi_2)$ is the coexistence point, because at that point $f_1^{(\mathrm{I})} = f_1^{(\mathrm{II})}$ and $f_2^{(\mathrm{I})} = f_2^{(\mathrm{II})}$. Note that we have not specified the nature of phase I and II. They could be liquid, solid or liquid crystalline.

The only practical problem that remains is the Monte Carlo sampling of $-\langle N_i/\xi_i + (N - N_1)/\xi_1 \rangle$ (eqn (36)). Note that the N_i are the dependent variables. The ξ_i are imposed during a given simulation. In addition to the usual particle moves and volume changes we must now also consider a move where a particle changes identity. To this end, we select one of the N particles at random and with equal probability assign it one of the n possible identities. The probability of accepting such a trial move is

$$P_{\text{acc}} = \min\left(1, \frac{\xi_{i'}}{\xi_i} \exp(-\beta\Delta\mathcal{V}(\mathbf{s}^N))\right) , \tag{37}$$

where $\Delta\mathcal{V}(\mathbf{s}^N)$ denotes the change in potential energy of the system if we change the identity of a randomly selected particle from i into i'.

I conclude this discussion of the semi-grand ensemble with three comments. First of all, SGCMC is very well suited to study phase equilibria in multicomponent systems that are also in chemical equilibrium. The reason is that every chemical equilibrium simply imposes a relation between the fugacities of the reacting species. Hence, the only effect of a chemical equilibrium is that the number of *independent* fugacities is reduced by one. In practice, the method cannot be used to study *any* chemical equilibrium: the total number of molecules should be conserved in the reaction.

Secondly, the SGCMC scheme can be used to simulate phase equilibria in truly polydisperse systems, including polydisperse solids. And finally, it can be quite advantageous to combine the SGCMC method with the Gibbs ensemble method for mixtures. In that case the fugacity ratios in both simulation boxes are kept the same. In other words, we allow particles in either box to change identity while remaining in the same box. But in addition we allow trial moves where we attempt to move a particle of the reference species 1 from one box to the other. Now the selection of the particle to be swapped goes as follows: first select box I or box II with equal probability. Next, select any molecule of type 1 in the selected box and try to insert it in the other box (see Chapter 6). (The implementation that I suggest here is slightly different from the one advocated in Ref. [25] and closer to the approach of Stapleton *et al.* [30]). The natural choice for the reference species 1 is clearly that species that can be swapped most efficiently, e.g. the smallest molecule in the system.

3. Virtual moves

In the previous sections we have been discussing several Monte Carlo schemes that all aimed to sample equilibrium averages in one ensemble or another. The procedure by which this was achieved was by some kind of random walk through configuration space, where the probability of acceptance of a trial move from one configuration to another was determined by the ratio of the probabilities of these configurations. In the present section we discuss sampling schemes in which certain trial moves are attempted *but never accepted!*

3.1. PARTICLE INSERTION METHOD

As an example of a Monte Carlo technique that employ virtual moves, we consider the so-called 'particle-insertion' method (often referred to as the Widom method [31]) to measure the chemical potential μ of a species in a pure fluid or in a mixture. The statistical mechanics that is the basis for this method is quite simple. Consider the definition of the chemical potential μ_α of a species α. From thermodynamics we know that μ is defined as:

$$\mu = \left(\frac{\partial G}{\partial N}\right)_{PT} = \left(\frac{\partial F}{\partial N}\right)_{VT} = -T\left(\frac{\partial S}{\partial N}\right)_{VE}, \tag{38}$$

where G, F and S are the Gibbs free energy, the Helmholtz free energy and the entropy, respectively. Here, and in the next few paragraphs, we focus on a one-component system, and hence we drop the subscript α. If we express the Helmholtz free energy of an N-particle system in terms of the partition function Q_N (eqn (8)), then it is obvious from eqn (38) that, for sufficiently large N the chemical potential is given by: $\mu = -k_BT\ln(Q_{N+1}/Q_N)$. If we use the explicit form (eqn (8)) for Q_N, we find:

$$\begin{aligned}\mu &= -k_BT\ln(Q_{N+1}/Q_N)\\ &= -k_BT\ln\left(\frac{V}{\Lambda^3(N+1)}\right) - k_BT\ln\left(\frac{\int d\mathbf{s}^{N+1}\exp(-\beta\mathcal{V}(\mathbf{s}^{N+1}))}{\int d\mathbf{s}^N\exp(-\beta\mathcal{V}(\mathbf{s}^N))}\right)\\ &= \mu_{\mathrm{id}}(V) + \mu_{\mathrm{ex}}\,. \end{aligned}\tag{39}$$

In the last line of eqn (39), we have indicated the separation into the ideal-gas contribution to the chemical potential, and the excess part. As $\mu_{\mathrm{id}}(V)$ can be evaluated analytically, we focus on μ_{ex}. We now separate the potential energy of the $(N+1)$-particle system into the potential energy function of the N-particle system, $\mathcal{V}(\mathbf{s}^N)$, and the interaction energy of the $(N+1)^{\mathrm{th}}$ particle with the rest: $\Delta\mathcal{V} \equiv \mathcal{V}(\mathbf{s}^{N+1}) - \mathcal{V}(\mathbf{s}^N)$. Using this separation, we can write μ_{ex} as:

$$\mu_{\mathrm{ex}} = -k_BT\ln\left\langle\int d\mathbf{s}_{N+1}\exp(-\beta\Delta\mathcal{V})\right\rangle_N, \tag{40}$$

where $\langle\ldots\rangle_N$ denotes canonical ensemble averaging over the configuration space of the N-particle system. The important point to note is that eqn (40) expresses μ_{ex} as an ensemble average that can be sampled by the conventional Metropolis scheme. There is only one aspect of this equation that makes it different from the averages that we considered before, namely the fact that we compute the average of an *integral* over the position of particle $N+1$. This last integral can be sampled by brute-force (unweighted) Monte Carlo sampling. In practice the procedure is as follows: we carry out a perfectly normal constant NVT Monte Carlo simulation on the system of N particles. At frequent intervals during this simulation (for instance, after every MC trial move) we randomly generate a coordinate $\mathbf{s}_{N+1}$, uniformly over the unit cube. With this value of $\mathbf{s}_{N+1}$, we then compute $\exp(-\beta\Delta\mathcal{V})$. By averaging the latter quantity over all generated trial positions, we obtain the average that appears in eqn (40). So, in effect, we are computing the average of the Boltzmann factor associated with the random insertion of an additional particle in an N-particle system, *but we never accept any such trial insertions*, because then we would no longer be sampling the average needed in eqn (40). The Widom method is a very powerful method to compute the chemical potential of (not too dense) atomic and simple molecular liquids. Its main advantage is its great simplicity, and the fact that it can be added on to an existing constant-NVT MC program, without any modifications to the original sampling scheme: we are simply computing one more thermal average. There is something else about the Widom method that makes it appealing, but that has nothing to do with computational efficiency: it really provides an insight into the *meaning* of the chemical potential, an insight that is often hard to extract from most text-books on statistical thermodynamics.

Extensive numerical studies of the excess chemical potential of finite periodic systems, have shown that this quantity is rather strongly dependent on the size of the system [32, 33]. Computer simulations are typically carried out for periodic systems in which the fundamental cell contains of order 10^2 to 10^3 particles, and the correction needed to give the infinite-system result can therefore be large. Of course, it is possible to estimate the finite-size correction empirically by carrying out simulations for different values of N (the number of particles), but this is very time consuming. It would clearly be much more convenient if the finite-size correction could be estimated directly, since it would then be possible to estimate the chemical potential in the thermodynamic limit on the basis of results obtained from simulations of relatively small systems. In fact, Siepmann, McDonald and Frenkel [34] have recently derived an expression for the leading ($\mathcal{O}(1/N)$) system-size dependence of the excess chemical potential:

$$\Delta\mu_{\mathrm{ex}}(N) = \frac{1}{2N}\left(\frac{\partial P}{\partial \rho}\right)\left[1 - k_{\mathrm{B}}T\left(\frac{\partial \rho}{\partial P}\right) - \rho k_{\mathrm{B}}T\,\frac{(\partial^2 P/\partial \rho^2)}{(\partial P/\partial \rho)^2}\right] . \tag{41}$$

As shown in Ref. [34], this expression agrees with the exact result for hard rods in one dimension, and is in excellent agreement with numerical results for hard-disks in two dimensions.

The particle insertion method can be modified to measure the *difference* in chemical potential between two species α and β in a mixture. In this case a trial move consists of an attempt to transform a particle of species α into species β (without, of course, ever accepting such trial moves). For more details, the reader is referred to Refs. [35, 28]. Let us finally mention that the particle insertion and swapping techniques are not limited to the measurement of chemical potentials. In fact, a wide class of partial molar quantities (such as, for example the partial molar enthalpy h_α or the partial molar volume v_α) can be measured in this way. For details, see Refs. [28, 36].

The particle insertion method fails under the same conditions where grand canonical Monte Carlo simulations become inefficient, and for the same reason: namely that the probability of 'accepting' a trial insertion becomes very small. One consequence is that the 'brute-force' particle insertion method is less suited for molecular than for atomic systems. Recently, however, Siepmann [37] has developed a scheme to measure the chemical potential of flexible molecules with a finite number of conformations. A similar approach, in the context of a lattice model, had been proposed earlier by Harris and Rice [38]. The next section discusses the basic idea behind these schemes.

3.2. CHEMICAL POTENTIAL OF MACRO-MOLECULES WITH DISCRETE CONFORMATIONS

In order to understand the methods that have been devised to calculate the chemical potential of chain molecules, it is instructive to first consider how we would compute μ_{ex} of a chain molecule with the Widom technique. To this end, I introduce the following notation: the position of the first segment of the chain molecule is denoted by $\mathbf{q}$ and the conformation of the molecule is described by $\mathbf{\Gamma}$. The configurational part of the partition function of a system of chain molecules can be written as

$$Q_{\text{chain}}(N,V,T) = \frac{1}{N!}\int \mathrm{d}\mathbf{q}^N \sum_{\mathbf{\Gamma}_1,\ldots,\mathbf{\Gamma}_n} \exp(-\beta \mathcal{V}(\mathbf{q}^N,\mathbf{\Gamma}^N)) \,. \tag{42}$$

The analogy with the previous sections suggests that the excess chemical potential of a chain molecule is obtained by considering the ratio

$$Q(N+1,V,T)/[Q(N,V,T)Q_{\text{non-interacting}}(1,V,T)],$$

where $Q(N+1,V,T)$ is the (configurational part of) the partition function of a system of $N+1$ interacting chain molecules and $Q(N,V,T) \times Q_{\text{non-interacting}}(1,V,T)$ is the partition function for a system consisting of N interacting chains and one chain that does not interact with the others. The latter chain plays the role of the ideal gas molecule in the previous sections. Note, however, that although this molecule does not interact with any of the other molecules it *does* interact with itself, both through bonded and through non-bonded interactions. Unfortunately, this is not a particularly useful reference state, as we do not, in general, know the partition function of an isolated self-avoiding chain.

We therefore use another reference state, namely that of the isolated non-selfavoiding chain. To be specific, let us consider the case of a molecule that consists of ℓ segments. Starting from segment 1, we can add segment 2 in b_2 equivalent directions, and so on. Clearly, the total number of non-selfavoiding conformations is $\Omega_{\text{id}} = \prod_{i=1}^{\ell} b_i$. For convenience, I have assumed that for a given i, all b_i directions are equally likely (i.e. I ignore *gauche-trans* potential energy differences and I even allow the ideal chain to fold back on itself). These limitations are not essential but they simplify the notation. Finally, I assume that all b_i are the same. Hence, for the simple model that we consider, $\Omega_{\text{id}} = b^{\ell}$. If we use such an ideal chain as our reference system, the expression for the excess chemical potential becomes

$$\begin{aligned}\beta\mu_{\text{ex}} &= -k_B T \ln\left(\frac{Q_{\text{chain}}(N+1,V,T)}{Q(N,V,T)Q_{\text{ideal}}(1,V,T)}\right) \\ &= -k_B T \ln\left\langle \exp[-\beta\Delta\mathcal{V}(\mathbf{q}^N,\mathbf{\Gamma}^N;\mathbf{q}_{N+1},\mathbf{\Gamma}_{N+1})]\right\rangle ,\end{aligned} \tag{43}$$

where $\Delta\mathcal{V}$ denotes the interaction of the test chain with the N chains that are already present in the system *and with itself*, while $\langle\ldots\rangle$ indicates averaging over all starting positions and all ideal-chain conformations of a randomly inserted chain.

The problem with the Widom approach to eqn (43) is that almost all randomly inserted ideal chain conformations will overlap either with particles already present in the system, or internally. The most important contributions to μ_{ex} will come from the extremely rare cases where the trial chain happens to be in just the right conformation to fit into the available space in the fluid. Clearly, it would be desirable if we could restrict our sampling to those conformations that satisfy this condition. If we do that, we introduce a bias in our computation of the insertion probability and we must somehow correct for that bias. The scheme developed by Harris and Rice and, independently, by Siepmann shows how that can be done.

In order to explain this scheme, I will first describe how the method works, and then show that it does indeed lead to the desired answer. The approach used in Refs. [37, 38] consists of two steps: in the first step a chain conformation is generated in such a way that 'acceptable' conformations are created with a high probability. The next step corrects for this bias by multiplying with a weight factor. A scheme that generates 'acceptable' chain conformations with a high probability was developed by Rosenbluth and Rosenbluth in the early fifties [39]. In the Rosenbluth scheme, a conformation of a chain molecule is constructed segment-by-segment. For every segment, we have a choice of b possible directions. In the Rosenbluth scheme, this choice is not random but favors the direction with the largest Boltzmann factor. To be specific, the probability (P) to generate a polymer with a conformation $\mathbf{\Gamma}$ using the Rosenbluth algorithm is given by

$$P_{\mathbf{\Gamma}} = \prod_{i=1}^{\ell} \frac{\exp\left[-\beta v^{(i)}(\mathbf{\Gamma}_i)\right]}{Z_i}, \tag{44}$$

where $v^{(i)}(\mathbf{\Gamma}_i)$ denotes the energy of segment i of the chain with conformation $\mathbf{\Gamma}$ (note that this energy excludes the contributions of segments $i+1$ to l, so the total energy of the chain is given by: $\mathcal{V}_{\mathbf{\Gamma}} = \sum_{i=1}^{\ell} v^{(i)}(\mathbf{\Gamma}_i))$. Z_i in eqn (44) is shorthand for

$$Z_i \equiv \sum_{j=1}^{b} \exp\left[-\beta v^{(i)}(\mathbf{\Gamma}_j)\right] .$$

where j enumerates all possible orientations from which the ith segment of the chain can be chosen and $v^{(i)}(\mathbf{\Gamma}_j)$ denotes the potential energy of the ith segment in orientation j. An important property of the probability given by eqn (44) is that it is normalized, i.e.

$$\sum_{\mathbf{\Gamma}} P_{\mathbf{\Gamma}} = 1 .$$

The Rosenbluth weight factor that corrects for the bias in the selection of conformation $\mathbf{\Gamma}$ is given by

$$W_{\mathbf{\Gamma}} \equiv \prod_{i=1}^{\ell} \frac{Z_i}{b}. \tag{45}$$

Now let us assume that we use the Rosenbluth scheme to generate a large number of chain conformations while keeping the coordinates of all other particles in the system fixed. For this set of conformations, we compute the average of the Rosenbluth weight factor W, $\overline{W}$. If we also perform an ensemble average over all coordinates and conformations of the N particles in the system, we obtain

$$\langle W \rangle = \left\langle \sum_{\mathbf{\Gamma}} P_{\mathbf{\Gamma}}(\mathbf{q}^N, \mathbf{\Gamma}^N) W_{\mathbf{\Gamma}}(\mathbf{q}^N, \mathbf{\Gamma}^N) \right\rangle \tag{46}$$

where the angular brackets denote the ensemble average over all configurations $\{\mathbf{q}^N\}$ of the 'solvent'. Note that the test polymer does not form part of the N-particle system. Therefore the probability to find the remaining particles in a configuration $\{\mathbf{q}^N, \mathbf{\Gamma}^N\}$ does not depend on the conformation $\mathbf{\Gamma}$ of the polymer.

In order to simplify the expression for the average in eqn (46), we first consider the average of the Rosenbluth factor for a given configuration $\{\mathbf{q}^N, \mathbf{\Gamma}^N\}$ of the solvent

$$\overline{W}(\{\mathbf{q}^N, \mathbf{\Gamma}^N\}) = \sum_{\mathbf{\Gamma}} P_{\mathbf{\Gamma}}(\mathbf{q}^N) W_{\mathbf{\Gamma}}(\{\mathbf{q}^N, \mathbf{\Gamma}^N\}). \tag{47}$$

Substitution of eqns (44) and (45) yields

$$\overline{W} = \sum_{\mathbf{\Gamma}} \left[\prod_{i=1}^{\ell} \frac{\exp\left[-\beta v^{(i)}(\mathbf{\Gamma}_i)\right]}{Z_i} \right] \left[\prod_{i=1}^{\ell} \frac{Z_i}{b} \right]$$

$$= \sum_{\mathbf{\Gamma}} \prod_{i=1}^{\ell} \frac{1}{b} \exp\left[-\beta v^{(i)}(\mathbf{\Gamma}_i)\right]$$

$$= \sum_{\mathbf{\Gamma}} \frac{1}{b^{\ell}} \exp\left[-\beta \mathcal{V}_{\mathbf{\Gamma}}\right] , \qquad (48)$$

where we have dropped all explicit reference to the solvent coordinates $\{\mathbf{q}^N, \mathbf{\Gamma}^N\}$. Note that eqn (48) can be interpreted as an average over all *ideal* chain conformations of the Boltzmann factor $\exp[-\beta\mathcal{V}_{\mathbf{\Gamma}}]$. If we now substitute eqn (48) in eqn (46) we obtain

$$\langle W \rangle = \frac{\sum_{\mathbf{\Gamma}} \left\langle \exp[-\beta\Delta\mathcal{V}(\mathbf{q}^N, \mathbf{\Gamma}^N; \mathbf{q}_{N+1}, \mathbf{\Gamma}_{N+1})]\right\rangle}{\sum_{\mathbf{\Gamma}}} . \qquad (49)$$

If we compare eqn (49) with eqn (43), we see that the ensemble average of the Rosenbluth factor is directly related to the excess chemical potential of the chain molecule.

$$\beta\mu_{\mathrm{ex}} = -k_{\mathrm{B}}T \ln \langle W \rangle . \qquad (50)$$

This completes our proof that a measurement of the average Rosenbluth factor of a trial chain can indeed be used to estimate the excess chemical potential of a polymer in a dense fluid.

Siepmann has compared this method with a brute-force application of the Widom method for a simple two-dimensional model of chain molecules on a lattice [37], and found the 'Rosenbluth' scheme to be very much superior, in particular at high densities. I should stress that the above method to measure the chemical potential is in no way limited to chain molecules on a lattice. What *is* essential is that the number of possible directions for each segment (b) relative to the previous one is finite. The special case of chain molecules on a lattice was studied by Mooij and Frenkel [40, 41] who found that, in that case, the calculation of the chemical potential can even be made 2–3 orders of magnitude more efficient than the original scheme of Refs. [37, 38].

3.3. EXTENSION TO CONTINUOUSLY DEFORMABLE MOLECULES

The numerical computation of the (excess) chemical potential of a flexible chain (with or without elastic forces that counteract bending), is rather different from the corresponding calculation for a chain molecule that has a large but fixed number of undeformable conformations.

Below, I shall consider the case of a flexible molecule *with* internal energy. The fully flexible case then follows trivially. Consider a 'worm-like' chain of ℓ linear segments. The potential energy of a given conformation has two contributions:

(i) The internal potential energy $\mathcal{V}_{\text{int}}$ is equal to the sum of the contributions of the individual joints. A joint between segments i and $i+1$ (say) has a potential energy $v(\theta_i)$ that depends on the angle θ_i between the successive segments. For instance, $v(\theta_i)$ could be of the form $v(\theta)=\alpha\theta^2$. For realistic models for poly-atomic molecules, $\mathcal{V}_{\text{int}}$ would account for all local internal potential energy changes due to bending and torsion.

(ii) The 'external' potential energy $\mathcal{V}_{\text{ext}}$. This energy accounts for all interactions with other molecules and for the non-bonded intra-molecular interactions. In addition, interactions with any external field that may be present are also included in $\mathcal{V}_{\text{ext}}$.

In what follows I shall refer to the chain in the absence of the 'external' interactions as the *ideal* chain. Clearly, the conformational partition function of the ideal chain is equal to

$$Z_{\text{id}} = c \int \cdots \int \mathrm{d}\mathbf{\Gamma}_1 \ldots \mathrm{d}\mathbf{\Gamma}_\ell \prod_{i=1}^{\ell} \exp(-\beta v_{\text{id}}(\theta_i)) \tag{51}$$

where c is a numerical constant. Our aim is to compute the effect of the external interactions on the conformational partition function. Hence, we wish to evaluate Z/Z_{id}, where Z denotes the partition function of the interacting chain. The excess chemical potential of the interacting chain is given by

$$\mu_{\text{ex}} = -k_{\text{B}}T \ln(Z/Z_{\text{id}}).$$

Below, I shall sketch an efficient procedure to evaluate this ratio by Monte Carlo sampling. However, before considering the 'smart' approach, let us briefly review two not-so-smart methods.

The simplest (and most stupid) way to compute the excess chemical potential of the interacting chain is to generate a very large number of completely random conformations of the freely jointed chain. For every conformation we compute both $\exp(-\beta\mathcal{V}_{\text{int}})$ and $\exp(-\beta[\mathcal{V}_{\text{int}}+\mathcal{V}_{\text{ext}}])$. The average of the former quantity is proportional to Z_{id}, while the average of the latter Boltzmann factor is proportional to Z. The ratio of these two averages should therefore yield Z/Z_{id}.

The problem with this approach is that the overwhelming majority of randomly generated conformations correspond to worm-like chains with a very high internal energy (and therefore a very small Boltzmann weight). Hence the statistical accuracy of this sampling scheme will be very poor.

The second scheme is designed to alleviate this problem. Rather than generating conformations of a freely jointed chain, we now sample the internal angles in the chain in such a way that the probability of finding a given angle θ_i is given by the Boltzmann weight

$$P(\theta_i) = \frac{\exp(-\beta v(\theta_i))}{\int \mathrm{d}\mathbf{\Gamma}_i \exp(-\beta v(\theta_i))}\,.$$

Here, and in what follows, I use the symbol $\mathbf{\Gamma}_i$ to denote the unit vector that specifies the orientation of the ith segment of the chain molecule. For every conformation thus generated, we compute the Boltzmann factor $\exp(-\beta\mathcal{V}_{\text{ext}})$. The average of this Boltzmann weight is then equal to

$$\begin{aligned}\langle\exp(-\beta\mathcal{V}_{\text{ext}})\rangle &= \frac{\int\prod \mathrm{d}\mathbf{\Gamma}\exp(-\beta[\mathcal{V}_{\text{id}}+\mathcal{V}_{\text{ext}}])}{\int\prod \mathrm{d}\mathbf{\Gamma}\exp(-\beta\mathcal{V}_{\text{id}})} \\ &= Z/Z_{\text{id}}. \end{aligned} \tag{52}$$

This approach is obviously superior to the first scheme. However, in many practical situations it will still yield poor statistics, because most ideal chain conformations will not correspond to energetically favourable situations for the interacting chain. Hence the Boltzmann weights will, again, be small for most conformations and the statistical accuracy will not be very good.

The problem with both schemes described above, is that neither allows us to focus on those conformations that should contribute most to Z, namely the ones for which both the internal and external potential energy are not much larger than a few $k_{\text{B}}T$ per degree of freedom. It would clearly be desirable to *bias* the sampling towards such favourable conformations. Let us now consider one possible solution to the sampling problem sketched above. The procedure is similar to the scheme to compute the excess chemical potential of a chain molecule with many fixed conformations (see section 3.2). Yet, there is an important difference precisely because the number of conformations is now, in principle, infinite. We can never hope to sample over *all* possible orientations of a new segment as we grow a chain. Hence, we generate a random sample of possible segment directions and use these in a modified Rosenbluth scheme. Below I shall first sketch how the sampling scheme works. Next I shall show that it does indeed yield the desired average.

To compute μ_{ex}, we apply the following 'recipe' to construct a conformation of a chain of ℓ segments. The construction of chain conformations proceeds segment by segment. Let us consider the addition of one such segment. To be specific, let us assume that we have already grown i segments, and that we are trying to add segment $i+1$. This is done as follows:

(i) Generate a fixed number (say b) trial segments. The orientations of the trial segments are distributed according to the Boltzmann weight associated with the internal energy $v(\theta)$. We denote the different trial segment by indices $1, 2, \ldots b$.

(ii) For all b trial segments, we compute the 'external' Boltzmann factor $\exp(-\beta v_{\text{ext}}(j))$.

(iii) Select one of the trial segments, say j, with a probability

$$P_j = \exp(-\beta v_{\text{ext}}(j))/Z_i\ ,$$

where we have defined

$$Z_i \equiv \sum_{j'=1}^{b} \exp(-\beta v_{\rm ext}(j')) \,.$$

(iv) Add this segment as segment $i+1$ to the chain and store the corresponding partial Rosenbluth weight $w_i = Z_i/b$.

The desired ratio $Z/Z_{\rm id}$ is than equal to the average value (over many trial chains) of the product of the partial Rosenbluth weights:

$$Z/Z_{\rm id} = \left\langle \prod_{i=1}^{\ell} w_i \right\rangle . \tag{53}$$

The advantage of this scheme is that step (iii) biases the sampling towards energetically favourable conformations. However, it still remains to be shown that eqn (53) is, in fact, correct. This, I shall now proceed to do. To this end let us consider the probability with which we generate a given chain conformation. This probability is the product of a number of factors. Let us first consider these factors for one segment, and then later extend the result to the complete chain. The probability to generate a given set of b trial segments with orientations $\mathbf{\Gamma}_1$ through $\mathbf{\Gamma}_b$ is

$$P_{\rm id}(\mathbf{\Gamma}_1).P_{\rm id}(\mathbf{\Gamma}_2)\ldots P_{\rm id}(\mathbf{\Gamma}_b)\mathrm{d}\mathbf{\Gamma}_1\ldots\mathrm{d}\mathbf{\Gamma}_b.$$

The probability of selecting any one of these trial segments, say segment j, is

$$\exp(-\beta v_{\rm ext}(j))/Z_i.$$

We wish to compute the average of a quantity, say w, over all possible sets of trial segments and all possible choices of the segment. To this end, we must sum over all j and integrate over all orientations $\prod_{j'=1}^{b}\mathrm{d}\mathbf{\Gamma}_{j'}$ (i.e. we average over the normalized probability distribution for the orientation of segment $i+1$):

$$\langle w\rangle = \int [\prod_{j'=1}^{b} \mathrm{d}\mathbf{\Gamma}_{j'} P_{\rm id}(\mathbf{\Gamma}_{j'})] \sum_{j''=1}^{b} \frac{\exp(-\beta v_{\rm ext}(j''))}{Z_i} w(1,2,\ldots,b) \,. \tag{54}$$

Now we make use of the fact that $w_i(1,2,\ldots,b)$ is equal to Z_i/b (see step (iv) of the 'recipe' above). Inserting this expression in eqn (54), we obtain:

$$\langle w\rangle = \int [\prod_{j'=1}^{b} \mathrm{d}\mathbf{\Gamma}_{j'} P_{\rm id}(\mathbf{\Gamma}_{j'})] \sum_{j''=1}^{b} \frac{\exp(-\beta v_{\rm ext}(j''))}{b} \,. \tag{55}$$

Now, as the labeling of the trial segments is arbitrary, all b terms in the sum in eqn (55) yield the same contribution, and eqn (55) simplifies to

$$\langle w \rangle = \int d\mathbf{\Gamma} P_{\rm id}(\mathbf{\Gamma}) \exp(-\beta v_{\rm ext}(\mathbf{\Gamma})) \tag{56}$$

$$= \frac{\int d\mathbf{\Gamma} \exp(-\beta[v_{\rm id}(\mathbf{\Gamma}) + v_{\rm ext}(\mathbf{\Gamma})])}{\int d\mathbf{\Gamma} \exp(-\beta v_{\rm id}(\mathbf{\Gamma}))} \tag{57}$$

$$= \frac{Z^{(1)}}{Z^{(1)}_{\rm id}}, \tag{58}$$

which is indeed the desired result, but for the fact that the expression in eqn (58) refers to 1 segment (as indicated by the superscript in $Z^{(1)}$. The extension to a chain of ℓ segments is straightforward, albeit that the intermediate expressions become a little unwieldy.

The above scheme to compute the chemical potential of continuously deformable molecules has recently been tested by Frenkel and Smit [42] and by de Pablo *et al.* [43]. The results of Refs. [42,43] indicate that the method can be applied directly to chain molecules of 20–30 segments. However, there is some evidence that, for longer chain lengths, the scheme runs in to sampling problems. In that case, there is an alternative scheme to compute the chemical potential of chain molecules, due to Kumar *et al.* [44]. This scheme is more robust, but also more expensive than the method described above, because it computes the *incremental* change in the chemical potential of a chain molecule upon sequential addition of segments. In principle, the computation of the chemical potential of a chain molecule consisting of ℓ segments, would require $\ell + 1$ simulations in the Kumar scheme. However, Kumar *et al.* argue that, in practice, reasonable estimates of the chemical potential may be obtained using far fewer runs. The idea behind the latter approach is that 'incremental' excess chemical potential can be related to the excess chemical potential of the polymer as a whole, when the latter increases linearly with the number of segments.

This assumption has been tested for an isolated chain by Smit *et al.* [45], using the present, rigorous scheme. The work of Ref. [45] suggests that, at least in this case, the assumption of a linear length dependence of the excess chemical potential of a chain molecule is, in fact, an oversimplification. It seems likely that a hybrid of the techniques of Refs. [42] and [44] will be preferable to either technique.

3.4. OVERLAPPING DISTRIBUTION METHOD

The reader may wonder why, in the previous sections, we have only been discussing a trial move that attempts to *add* a particle to the system, and not the reverse move. After all, the excess chemical potential can also be written as:

$$\begin{aligned}\mu &= k_B T \ln(Q_N/Q_{N+1}) \\ &= \mu_{id} + k_B T \ln \langle \exp(+\beta \Delta \mathcal{V}) \rangle_{N+1} \ , \end{aligned} \tag{59}$$

where $\Delta\mathcal{V}$ denotes the interaction energy of particle $N+1$ with the remaining N particles. It would seem that eqn (59) can be sampled by straightforward Metropolis Monte Carlo. In general, however, this is not true. The reason is that the function $\exp(\beta\Delta\mathcal{V})$ is, in principle not bounded. It can become arbitrarily large, as $\Delta\mathcal{V}$ grows. (Incidentally, this is not true for $\exp(-\beta\Delta\mathcal{V})$, because one of the conditions that a system must satisfy in order to be describable by classical statistical mechanics is that its potential energy function has a lower bound). The problem with eqn (59) is that very large values of the integrand coincide with very small values ($\mathcal{O}(\exp(-\beta\Delta\mathcal{V}))$ of the Boltzmann factor (that determines how often a configuration is sampled during a Monte Carlo run). As a consequence, an appreciable contribution to the average in eqn (59) comes from a part of configuration space that is hardly ever, or indeed never, sampled during a run. Hard spheres offer a good illustration. As the potential energy function of non-overlapping hard spheres is always zero, a simple Monte Carlo sampling of eqn (59) for a dense fluid of hard spheres would always yield the nonsensical estimate $\mu_{ex} = 0$ (whereas, in fact, at freezing, $\mu_{ex}/k_B T \sim 15$). The correct way to obtain chemical potentials from simulations involving both particle insertions and particle removals has been indicated by Shing and Gubbins [46, 47]. However, I find it convenient to discuss this problem in the context of a more general technique to measure free energy differences, first introduced by Bennett [48], called the overlapping distribution method.

Consider two N-particle systems, labeled 0 and 1 with partition functions Q_0 and Q_1. For convenience we assume that both systems have the same volume V, but this is not essential. From eqn (8) it follows that the free energy difference $\Delta F = F_1 - F_0$ can be written as:

$$\begin{aligned}\Delta F &= -k_B T \ln(Q_1/Q_0) \\ &= -k_B T \ln \left(\frac{\int d\mathbf{s}^N \exp(-\beta \mathcal{V}_1(\mathbf{s}^N))}{\int d\mathbf{s}^N \exp(-\beta \mathcal{V}_0(\mathbf{s}^N))} \right)\end{aligned} \tag{60}$$

Suppose that we are carrying out a (Metropolis) sampling of the configuration space of system 1. For every configuration visited during this sampling of system 1 we can compute the potential energy of system 0 ($\mathcal{V}_0(\mathbf{s}^N)$) for the same configuration, and hence the potential energy difference $\Delta\mathcal{V} = \mathcal{V}_1(\mathbf{s}^N) - \mathcal{V}_0(\mathbf{s}^N)$ (see Fig. 3). We use this information to construct a histogram that measures the probability density for the potential energy difference $\Delta\mathcal{V}$. Let us denote this probability density by $p_1(\Delta\mathcal{V})$. In the canonical

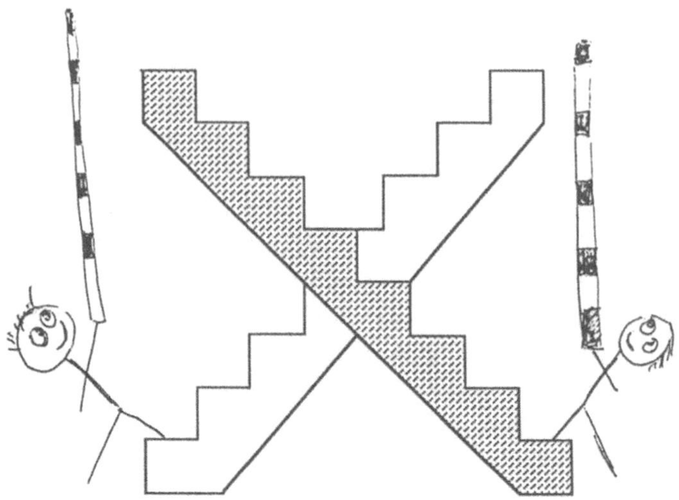

Fig. 3. Example of measurement of histogram of potential energy differences (here represented by height differences) between two systems 0 and 1. Note that the two histograms will, in general, be different.

ensemble, $p_1(\Delta\mathcal{V})$ is given by:

$$p_1(\Delta\mathcal{V}) = \frac{\int ds^N \exp(-\beta\mathcal{V}_1)\,\delta(\mathcal{V}_1 - \mathcal{V}_0 - \Delta\mathcal{V})}{q_1}, \tag{61}$$

where we have denoted the scaled, configurational part of the partition function by a lower-case q (e.g. $q_1 = \int ds^N \exp(-\beta\mathcal{V}_1(s^N))$). The δ-function in eqn (61) means that we can substitute $\mathcal{V}_0 + \Delta\mathcal{V}$ for $\mathcal{V}_1$ in the Boltzmann factor, hence:

$$\begin{aligned} p_1(\Delta\mathcal{V}) &= \frac{\int ds^N \exp(-\beta(\mathcal{V}_0 + \Delta\mathcal{V}))\,\delta(\mathcal{V}_1 - \mathcal{V}_0 - \Delta\mathcal{V})}{q_1} \\ &= \frac{q_0}{q_1}\exp(-\beta\Delta\mathcal{V})\frac{\int ds^N \exp(-\beta\mathcal{V}_0)\,\delta(\mathcal{V}_1 - \mathcal{V}_0 - \Delta\mathcal{V})}{q_0} \\ &= \frac{q_0}{q_1}\exp(-\beta\Delta\mathcal{V})p_0(\Delta\mathcal{V}), \end{aligned} \tag{62}$$

where $p_0(\Delta\mathcal{V})$ is the probability density to find a potential energy difference $\Delta\mathcal{V}$ between systems 1 and 0, while Boltzmann-sampling the available configurations of system 0. As the free energy difference between systems 1 and

0 is simply $\Delta F = -k_B T \ln(q_1/q_0)$, we find from eqn (62):

$$\ln p_1(\Delta\mathcal{V}) = \beta(\Delta F - \Delta\mathcal{V}) + \ln p_0(\Delta\mathcal{V}) \,. \tag{63}$$

In order to obtain ΔF from eqn (63) in practical cases, it is convenient to define two functions f_0 and f_1 by:

$$f_0(\Delta\mathcal{V}) = \ln p_0(\Delta\mathcal{V}) - \frac{\beta\Delta\mathcal{V}}{2}$$

and

$$f_1(\Delta\mathcal{V}) = \ln p_1(\Delta\mathcal{V}) + \frac{\beta\Delta\mathcal{V}}{2}$$

such that

$$f_1(\Delta\mathcal{V}) = f_0(\Delta\mathcal{V}) + \beta\Delta F \,.$$

Suppose that we have measured f_0 and f_1 in two separate simulations (one sampling system 0, the other system 1). We can then obtain ΔF by fitting the functions f_0 and f_1 to two polynomials in $\Delta\mathcal{V}$ that are identical but for the constant term. The constant offset between the two polynomials yields our estimate for ΔF. Note that, in order to perform such a fit, it is not even necessary that there exists a range of $\Delta\mathcal{V}$ where both f_0 and f_1 can be measured. However, in the absence of such a range of overlap, the statistical accuracy of the method is usually poor.

Now consider the particle insertion/removal problem. Let us assume that system 1 is a system with N interacting particles, while system 0 contains $N-1$ interacting particles and 1 ideal-gas particle. The difference in free energy between these two systems is obviously equal to μ_{ex}. Applying eqn (63) to this particular case, we find:

$$\Delta\mu_{ex} = f_1(\Delta\mathcal{V}) - f_0(\Delta\mathcal{V}) \,. \tag{64}$$

Equation (64) is equivalent to the result obtained by Shing and Gubbins. Using the overlapping distribution method it is possible to combine the results of simulations with trial insertions and trial removals to arrive at a more accurate estimate for the chemical potential. A generalization of the Shing-and-Gubbins method to chain molecules, has been proposed by de Pablo *et al.* [43].

3.5. UMBRELLA SAMPLING

In the previous section, we introduced the distibution functions $p_0(\Delta\mathcal{V})$ and $p_1(\Delta\mathcal{V})$ that measure the probability of finding system 0 (1) in an equilibrium configuration $\mathbf{s}^N$ where its potential energy difference with system 1 (0) equals $\Delta\mathcal{V}$. At first sight it would seem that knowledge of either p_0 or

p_1 is sufficient to estimate the free energy difference between systems 0 and 1. After all, eqn (63) states that

$$p_1(\Delta\mathcal{V}) = p_0(\Delta\mathcal{V}) \exp(\beta(\Delta F - \Delta\mathcal{V})) .$$

If we integrate over $\Delta\mathcal{V}$ on both sides of this equation, we obtain:

$$\begin{aligned} \int_{-\infty}^{\infty} p_1(\Delta\mathcal{V})\, \mathrm{d}\Delta\mathcal{V} &= \exp(\beta\Delta F) \int_{-\infty}^{\infty} p_0(\Delta\mathcal{V}) \exp(-\beta\Delta\mathcal{V})\, \mathrm{d}\Delta\mathcal{V} \\ 1 &= \exp(\beta\Delta F) \left\langle \exp(-\beta\Delta\mathcal{V}) \right\rangle_0 , \end{aligned} \tag{65}$$

or

$$\exp(-\beta\Delta F) = \left\langle \exp(-\beta\Delta\mathcal{V}) \right\rangle_0$$

Although eqn (65) is very useful to estimate free energy differences between two systems that are not too dissimilar, its applicability is limited. The problem is that, in many cases of practical interest, the largest contributions to the average $\langle \exp(-\beta\Delta\mathcal{V}) \rangle_0$ come from regions of configuration space where $p_0(\Delta\mathcal{V})$ is very small, while $\exp(-\beta\Delta\mathcal{V})$ is very large. As a result, the statistical error in ΔF is large.

One method to achieve a more accurate estimate of ΔF is the 'umbrella sampling' scheme suggested by Torrie and Valleau [49]. The basic idea behind umbrella sampling is that, in order to obtain an accurate estimate of the free energy difference between two systems (0 and 1), one should sample both the part of configuration space that is accessible to system 1 and the part that is accessible to 0. In order to achieve this in a single simulation, one should modify the Markov chain that governs the sampling of configuration space. This is achieved by multiplying the Boltzmann factor of system 0 by a (positive) weighting function w. As a result, the probability of visiting a point $\mathbf{q}^N$ in configuration space is now proportional to $\exp(-\beta\mathcal{V}_0(\mathbf{q}^N))w(\mathbf{q}^N)$. The expression for $\langle \exp(-\beta\Delta\mathcal{V}) \rangle_0$ now becomes:

$$\left\langle \exp(-\beta\Delta\mathcal{V}) \right\rangle_0 = \frac{\int \mathrm{d}\mathbf{q}^N \exp(-\beta\mathcal{V}_0(\mathbf{q}^N)) w(\mathbf{q}^N) [\exp(-\beta\Delta\mathcal{V}(\mathbf{q}^N))/w(\mathbf{q}^N)]}{\int \mathrm{d}\mathbf{q}^N \exp(-\beta\mathcal{V}_0(\mathbf{q}^N)) w(\mathbf{q}^N) w^{-1}(\mathbf{q}^N)} , \tag{66}$$

or, introducing the notation $\langle \ldots \rangle_w$ to denote an average over a probability distribution proportional to $\exp(-\beta\mathcal{V}_0(\mathbf{q}^N))w(\mathbf{q}^N)$:

$$\begin{aligned} \left\langle \exp(-\beta\Delta\mathcal{V}) \right\rangle_0 &= \frac{\left\langle \exp(-\beta\Delta\mathcal{V})/w \right\rangle_w}{\left\langle 1/w \right\rangle_w} \\ &= \left\langle \exp(-\beta\Delta\mathcal{V})/w \right\rangle_w \left\langle w \right\rangle_0 . \end{aligned} \tag{67}$$

The second line of eqn (67) shows that, in order for both terms on the right hand of this equation to be non-zero, $p_0(\Delta\mathcal{V})w$ should have an appreciable overlap with $p_1(\Delta\mathcal{V})$, while w itself should overlap with $p_0(\Delta\mathcal{V})$. This 'bridging' property of w is responsible for the name 'umbrella' sampling.

Although umbrella sampling is, in principle, a powerful technique, one drawback is that the function w is not known *a priori*. Rather, it must be constructed using the available information about the functions p_0 and p_1. This requires some skill. A second (though related) problem is that an unfortunate choice of w may result in estimates for ΔF that appear reasonable but are, in fact, wrong. Only by systematic testing is it possible to get a feeling for the statistical errors in an umbrella sampling calculation. At first sight, it might seem advantageous to refine the computation of w in such a way that *all* relevant configurations can be sampled in one run. Surprisingly, this is not the case. Rather, it is advantageous to perform several umbrella sampling runs in (partially overlapping) 'windows'. To see this, let us define an 'order parameter' Φ that is a measure for the 'location' of a given configuration between systems 0 and 1. For instance, Φ might be the potential energy, or the (in constant-pressure simulations) the density. Let us denote the average value of Φ in system 1 by $\Phi_{\max}$ and the value in 0, by $\Phi_{\min}$. Let us assume that we sample an interval $\Phi_{\max} - \Phi_{\min} \equiv \Delta\Phi$ in n umbrella-sampling simulations. The optimum choice of n is clearly the one that samples the complete Φ-interval in the minimum computing time. In order to estimate this time, let us assume that the system performs a random walk in Φ-space within the window $\Delta\Phi/n$. Associated with the random walk in Φ-space is a 'diffusion constant' D_Φ. The characteristic time needed to sample one interval $\Delta\Phi/n$ is

$$\tau_n = \frac{(\Delta\Phi/n)^2}{D_\Phi}.$$

Clearly, the total time to sample all n windows is

$$\tau_{\text{tot}} = n\tau_n = \frac{(\Delta\Phi)^2}{nD_\Phi}.$$

The important point to note is that the computing time *decreases* with increasing n. It would, however, be incorrect to assume that n should be chosen as large as possible. The actual equilibration time of a run in one of the Φ-windows also depends on the rate at which all coordinates 'orthogonal' to Φ are sampled. Let us denote this time by $\tau_\perp$. Clearly, once $\tau_\perp$ becomes appreciably larger than τ_n, the total computation will scale as $n \times \tau_\perp$. This suggests that the optimum choice of n is the one for which $\tau_n \approx \tau_\perp$. For a more detailed discussion, see Ref. [50]. A recent example of multi-stage umbrella sampling applied to the problem of homogeneous nucleation can be found in Ref. [51].

It has been suggested [52] that there is, in fact, a unique optimum choice for w that minimizes the statistical error in the measurement of the free-energy difference between two systems in a single simulation. In order to achieve this, the sampling of configuration space should be carried out with an overall probability density W, given by

$$W = \text{const.} \times \sqrt{\frac{\exp(-2\beta\mathcal{V}_0)}{Q_0^2} + \frac{\exp(-2\beta\mathcal{V}_1)}{Q_1^2}}\,. \tag{68}$$

Although the above equation is, in principle, correct, its practical use may be limited by two factors. One is not serious, the other one usually is. The not-serious problem is that we do not know Q_0/Q_1 *a priori* (if we did, there would be little need for umbrella sampling). In practice, the simulation will be done using a reasonable guess for this ratio. Any error in this guess will worsen the statistics of our simulation somewhat, but it will not systematically affect our estimate for the free-energy difference. The more serious restriction of eqn (68) is that it will only work if systems 0 and 1 are sufficiently close in configuration space that W is not vanishingly small in between. We can interpret $-k_{\mathrm{B}}T\ln W$ as an effective free-energy surface probed in the simulation. If the basins in configuration space belonging to systems 0 and 1 are separated by a region were W is very small, then there is a high 'free energy barrier' between the two systems. Our sampling scheme will then converge poorly. In fact, the equilibration time will diverge as $\exp(\beta F_{\mathrm{max}})$, where $F_{\mathrm{max}} = -k_{\mathrm{B}}T\ln W_{\mathrm{min}}$ is the height of the free-energy barrier separating the two regions in configuration space. As before, this problem may be alleviated by replacing a single run by a series of runs restricted to successive 'windows' that 'tile' the path in configuration space between 0 and 1.

4. Beyond single-particle moves

Thus far, I have described Monte Carlo procedures that were based on the original Metropolis method to sample the configurations of the system under study. In particular, I have assumed that a trial move consisted of an attempted displacement of a single, randomly selected, particle. In fact, there have been many attempts to construct Monte Carlo schemes that utilize 'collective' many-particle moves. Special techniques are required to construct such collective Monte Carlo moves, because a scheme where we would simply attempt a random displacement of all N particles simultaneously, would be *less* rather than *more* efficient than a sequence of single-particle moves, because a very small step-size is required to get a reasonable acceptance. In order to construct larger collective trial moves, several authors have constructed hybrid techniques that are, in a sense, intermediate between Monte Carlo and molecular dynamics (see e.g. Ref. [53] and the book of Allen and Tildesley [22]). The idea behind the earlier 'smart' MC schemes was to construct trial moves that were not completely random, but utilized information about the forces acting on the particles in the system. Recently, these techniques have been considerably extended [54] in such a way that an MD algorithm can be converted into a 'collective' MC algorithm. The difference is that, whereas for MD, we should use time increments that are sufficiently small to ensure energy conservation (see Chapter 2), for MC we should use large time steps, and accept or reject the position thus generated using (say) the Metropolis criterion. It should be stressed that this scheme can only be applied to MD algorithms that are time-reversible or, more precisely, that conserve volume in phase space (otherwise there would be a bias in the sampling for which we should correct). For more details, the reader is referred to Ref. [54]. In the remainder of this section, we shall discuss other 'collective' MC schemes of a completely different nature. The reason to include this discussion is that, during the past few years, much progress has been made in the development of Monte-Carlo schemes where a group of particles is displaced in a trial move. In fact, we should distinguish between two rather different schemes:

(i) The first scheme, or actually family of schemes, is based on the cluster algorithm of Swendsen and Wang [55]. In this method, stochastic rules are used to construct clusters, and these clusters are subsequently moved (or, in the case of spins, flipped) as a single unit.

(ii) In the second scheme, that I denote by the name 'Configurational Bias Monte Carlo', we know in advance which particles we shall attempt to move e.g. some or all particles belonging to a given chain molecule). The scheme provides a set of rules that make it possible to perform large-scale collective moves with a reasonable acceptance probability.

In the following sections, I describe first the Configurational-Bias MC (CBMC) method (section 4.1) and then the "Swendsen-Wang" method (section 4.2). In the latter case, my presentation will be a bit different from the one given in the recent review by Swendsen, Wang and Ferrenberg in the book by Binder [3]. The latter review emphasizes application to lattice systems. In contrast, I try to present the scheme in a way that facilitates extension to off-lattice problems. Both the SW and the CBMC methods are examples of Monte-Carlo schemes in which the *a priori* probability to make a transition from state i to state j (i.e. the matrix α_{ij} in section 1) is *not* symmetric. In this respect, both schemes are related to the earlier 'smart' or 'force-bias' Monte-Carlo schemes (see Refs. [22, 53]).

4.1. CONFIGURATIONAL-BIAS MONTE CARLO SCHEME

The configurational-bias MC method was originally developed in the context of simulations of chain molecules. The central idea is that the Rosenbluth trial insertion scheme described in sections 3.2 and 3.3, can be used as a starting point for a Monte Carlo scheme to sample equilibrium configurations of systems consisting of chain molecules [38, 56, 57]. At first sight, this may not appear to be a new result but a very old one. After all, the original Rosenbluth scheme itself was designed as a method to sample polymer conformations. However, the Rosenbluth scheme suffers from the drawback that it generates an unrepresentative sample of all polymer conformations: i.e. the probability to generate a particular conformation $\mathbf{\Gamma}$ using the Rosenbluth scheme, is *not* proportional to the Boltzmann weight of that conformation. The Rosenbluth weight W, discussed in the section 3.2, was introduced to correct for this bias in the computation of thermal averages. However, such a correction procedure only works for relatively short chains. This drawback of the Rosenbluth sampling scheme is, in fact, well known (see, e.g. Ref. [58]). The solution of this problem is to bias the Rosenbluth sampling in such a way that the correct (Boltzmann) distribution of chain conformations is generated in a Monte Carlo sequence. In the configurational bias scheme to be discussed below, the Rosenbluth weight is used to bias the *acceptance* of trial conformations that are generated with the Rosenbluth procedure. As a consequence, all conformations are generated with their *correct* Boltzmann weight. This removes the main drawback of the original Rosenbluth scheme. Moreover, I shall show that the configurational bias scheme can also be extended to continuously deformable molecules, such as worm-like chains.

4.1.1. Sampling discrete conformations. Consider a trial move from a chain-conformation $\mathbf{\Gamma}_1$ with Rosenbluth weight w_1 to some other conformation $\mathbf{\Gamma}_2$ with Rosenbluth weight w_2. Let us denote the probability of finding the chain in conformation $\mathbf{\Gamma}_1$ by P_1 and the corresponding probability for $\mathbf{\Gamma}_2$ by P_2. The number of chains in conformation 1(2), N_1 (N_2) is obviously proportional P_1 (P_2). We wish to perform a Monte-Carlo move such that detailed balance is satisfied, i.e. the rate K_{12} with which conformations of type 1 are transformed into type 2 equals the reverse rate:

$$K_{12} = K_{21}. \tag{69}$$

Let us try to compute K_{12} and K_{21}. K_{12} is equal to N_1 times $P_{\text{gen}}(\mathbf{\Gamma}_2)$, the probability of generating chain 2 by Rosenbluth sampling, times the acceptance probability P_{acc}. We assume that P_{acc} is an, as yet unspecified function of the ratio w_2/w_1, $g(w_2/w_1)$. $P_{\text{gen}}(\mathbf{\Gamma}_2)$ is equal to

$$P_{\text{gen}}(\mathbf{\Gamma}_2) = \prod_{i=1}^{\ell} \frac{\exp(-\beta v^{(i)}(\mathbf{\Gamma}_2(i)))}{Z_i}, \tag{70}$$

where, as before, we use the notation $Z_i \equiv \sum_{j=1}^{b} \exp(-\beta v^{(i)}(\mathbf{\Gamma}_2(j)))$. The sum in Z_i runs over all b possible directions of the ith segment of the polymer. With this notation, eqn (69) becomes

$$N_1 P_{\text{gen}}(\mathbf{\Gamma}_2) g(w_2/w_1) = N_2 P_{\text{gen}}(\mathbf{\Gamma}_1) g(w_1/w2). \tag{71}$$

Equation (71) can be simplified by using the result of the section 3.2 that the Rosenbluth weight w_2 times the generating probability $P_{\text{gen}}(\mathbf{\Gamma}_2)$ is proportional to N_2 (and similarly for 1). Hence multiplying eqn (71) on the left by w_2/w_2 and on the right by w_1/w_1 yields

$$N_1 N_2 \frac{g(w_2/w_1)}{w_2} = N_2 N_1 \frac{g(w_1/w_2)}{w_1} \tag{72}$$

or

$$\frac{g(w_2/w_1)}{g(w_1/w_2)} = \frac{w_2}{w_1} \,. \tag{73}$$

There are many choices of g that satisfy this condition. One obvious choice is the Metropolis form [1]

$$g(x) = \min(x, 1). \tag{74}$$

In words, the Configurational Bias Monte Carlo (CBMC) scheme works as follows:

(i) Generate a trial conformation by using the Rosenbluth scheme (i.e. eqn (70) to regrow the entire molecule, or part thereof.

(ii) Compute the Rosenbluth weights w_{trial} and w_{old} of the trial conformation *and of the old conformation.*

(iii) Accept the trial move with a probability $\min(w_{\text{trial}}/w_{\text{old}}, 1)$.

4.1.2. Continuously deformable chain. Next consider Configurational Bias Monte Carlo sampling of worm-like chains or, for that matter, any chain with flexible joints with or without internal bending energy. Again, we first consider the probability to generate a trial configuration $\mathbf{\Gamma}_j$ using the (extended) Rosenbluth scheme. As before, we only consider the expression for one of the ℓ segments, to keep the equations simple:

$$P_{\text{gen}}(j) = \mathrm{d}\mathbf{\Gamma}_j P_{\text{id}}(\mathbf{\Gamma}_j) \left[\prod_{j' \neq j}^{b} \mathrm{d}\mathbf{\Gamma}_{j'} P_{\text{id}}(\mathbf{\Gamma}_{j'}) \right] \frac{\exp(-\beta v_{\text{ext}}(j))}{\sum_{j'=1}^{b} \exp(-\beta v_{\text{ext}}(j'))} \,. \tag{75}$$

We now generate around the original conformation i the same set of additional directions (i.e. those conformations in eqn (75) with $j' \neq j$). Next we compute the ratio of the Rosenbluth weights $w_j^{(t)}/w_i^{(o)}$, where

$$w_j^{(t)} = \frac{\exp(-\beta v_{\text{ext}}^{(t)}(j)) + \sum_{j' \neq j}^{b} \exp(-\beta v_{\text{ext}}^{(t)}(j'))}{b}$$

and

$$w_i^{(o)} = \frac{\exp(-\beta v_{\text{ext}}^{(o)}(i)) + \sum_{j' \neq j}^{b} \exp(-\beta v_{\text{ext}}^{(o)}(j'))}{b}$$

The superscript (t) and (o) distinguish the trial conformation from the old conformation. Again, we shall devise a Metropolis criterion to determine the acceptance probability of a trial move from i to j. As before, this acceptance probability is determined by the ratio $x \equiv w_j^{(t)}/w_i^{(o)}$ (actually, for a molecule of ℓ segments, we should compute a product of such factors). Let us assume that $w_j^{(t)} < w_i^{(o)}$. In that case, $P_{\rm acc}(i \to j) = x$ while $P_{\rm acc}(j \to i) = 1$. Next, let us check whether detailed balance is satisfied. To do so, we write down the explicit expressions for K_{ij} and K_{ji}.

$$K_{ij} = N_i P_{\rm gen}(j) w_j^{(t)}/w_i^{(o)}$$

and

$$K_{ji} = N_j P_{\rm gen}(i) 1.$$

Here it should be stressed that there are many different routes to go from i to j and vice versa, depending on the selection of the trial orientations j'. What we show is that, as for *any* specific choice of this set, K_{ij} and K_{ji} are equal ('super-detailed balance'), this equality must also hold for the sum over all possible sets j'. In addition, we use the fact that

$$P_{\rm gen}(j) w_j \sim N_j \prod_{j' \neq j}^{b} {\rm d}\mathbf{\Gamma}_{j'} P_{\rm id}(\mathbf{\Gamma}_{j'})$$

and

$$P_{\rm gen}(i) w_i \sim N_i \prod_{j' \neq j}^{b} {\rm d}\mathbf{\Gamma}_{j'} P_{\rm id}(\mathbf{\Gamma}_{j'}).$$

For a given set j', we can therefore write:

$$K_{ij} = N_i P_{\rm gen}(j) w_j / w_i = \text{constant} \times N_i N_j \prod_{j' \neq j}^{b} {\rm d}\mathbf{\Gamma}_{j'} P_{\rm id}(\mathbf{\Gamma}_{j'}) \frac{1}{w_i}$$

while

$$K_{ji} = N_j P_{\rm gen}(i) 1 = \text{constant} \times N_j N_i \prod_{j' \neq j}^{b} {\rm d}\mathbf{\Gamma}_{j'} P_{\rm id}(\mathbf{\Gamma}_{j'}) \frac{1}{w_i}.$$

Hence, K_{ij} is indeed equal to K_{ji}. This completes the proof that the above Configurational Bias MC scheme satisfies detailed balance.

The CBMC scheme for completely flexible molecules has been applied by Frenkel *et al.* [59] to study conformational changes of chain molecules in solution. The method was found to be particularly efficient in generating large conformational changes of such a molecule. More interestingly, the CBMC scheme makes it possible to perform direct simulations of phase equilibria in polymer mixtures, using the Gibbs-ensemble method [60, 61].

4.1.3. From polymers back to atoms. Up to this point, I have introduced the configurational-bias MC scheme as a method to sample conformations of linear chain molecules. However, the method is applicable to branched polymers and ring polymers. In the latter case, the condition that the Nth segment is connected to the first, is accounted for by biasing the generation of trial directions. The *a priori* probability to generate a given trial direction $\mathbf{\Gamma}_i$ as the mth segment, is proportional to the number of ideal (i.e. non-(self)-avoiding) conformations that connect this segment to the first segment in $N - m$ steps.

Above, the configurational-bias scheme has been presented exclusively as a method to generate *polymer* conformations. In fact, the method is more general than that. It can be used as a scheme to perform collective rearrangements of *any* set of labeled coordinates. Such a scheme could, for instance, be used to carry out Monte Carlo moves to swap n small particles within a volume ΔV with one large particle that occupies the same (excluded) volume. Similarly, one could envisage a Monte Carlo move in which part of a polymer *plus the surrounding solvent* is regrown. Such an approach may be useful in the study of molecules of biological interest.

4.2. THE SWENDSEN-WANG SAMPLING SCHEME

The central idea behind the Swendsen-Wang (SW) scheme [55] and subsequent extensions and modifications (see Ref. [3]), is to generate trial configurations with a probability that is proportional to the Boltzmann weight of that configuration. As a result, the subsequent trial moves can be accepted with 100% probability. The easiest way to describe the SW scheme is to consider the expression for detailed balance in a Monte Carlo scheme to simulate a discrete spin model. Consider an 'old' configuration (labeled by a superscript o) and a 'new' configuration, denoted by a superscript n. Detailed balance is satisfied if the following equality holds:

$$\begin{aligned} &P_{\text{B}}^{o} P_{\text{Gen}}^{o}(\{\text{cluster}\}) P_{\text{sel}}^{\{\text{cl}\}}(\text{flip}) P_{\text{acc}}(o \to n) \; = \\ &P_{\text{B}}^{n} P_{\text{Gen}}^{n}(\{\text{cluster}\}) P_{\text{sel}}^{\{\text{cl}\}}(\text{flip}) P_{\text{acc}}(n \to o) \end{aligned} \tag{76}$$

where $P_{\text{B}}^{o(n)}$ is the Boltzmann weight of old (new) configuration, $P_{\text{Gen}}(\{\text{cluster}\})$ denotes the probability of breaking the system up into a specified set of clusters, each of which can be subsequently flipped at random, $P_{\text{sel}}^{\{\text{cl}\}}(\text{flip})$ is the probability to select, from a set of clusters $\{1, 2, \ldots, p\}$ a given subset (e.g. $\{1, 5, \ldots, p-1\}$) that is to be flipped, and finally $P_{\text{acc}}(o \to n)$ is the acceptance probability of a given trial move. We can simplify eqn (76) in two ways: first of all, we note that the probability $P_{\text{sel}}^{\{\text{cl}\}}(\text{flip})$ is the same for the forward and reverse moves (i.e. we consider the *same* subset of the *same* set of clusters). Moreover, we wish to impose $P_{\text{acc}} = 1$ for both forward and reverse moves. This may not always be feasible. However, for most of the cases that I discuss below, this is indeed possible. The detailed balance equation then becomes:

$$P_{\mathrm{B}}^{o} P_{\mathrm{Gen}}^{o}(\{\text{cluster}\}) = P_{\mathrm{B}}^{n} P_{\mathrm{Gen}}^{n}(\{\text{cluster}\}) \tag{77}$$

or

$$\frac{P_{\mathrm{Gen}}^{o}(\{\text{cluster}\})}{P_{\mathrm{Gen}}^{n}(\{\text{cluster}\})} = \frac{P_{\mathrm{B}}^{n}}{P_{\mathrm{B}}^{o}} = \exp(-\beta\Delta E)\,, \tag{78}$$

where ΔE is the difference in energy between the new and the old configurations. The trick is then to find a recipe for cluster generation that will satisfy eqn (78). In order to illustrate how this works, I consider two simple examples, namely the Ising model and the xy-model. The extension to many other models is straightforward.

4.2.1. Ising model. For the construction of the Ising SW algorithm, the dimensionality of the model is irrelevant. Consider a given configuration of the spin system, with N_{p} spin-pairs parallel and N_{a} spin-pairs anti-parallel. The total energy of that configuration is:

$$E = (N_{\mathrm{a}} - N_{\mathrm{p}})J$$

where J denotes the strength of the nearest-neighbor interaction. The Boltzmann weight of that configuration is

$$P_{\mathrm{B}}^{o} = \exp(-\beta J(N_{\mathrm{a}} - N_{\mathrm{p}}))/Z$$

where Z is the partition function of the system. In general, Z is unknown, but that is unimportant. We only use the fact that Z is a constant. Next, we construct clusters by creating bonds between spin pairs according to the following recipe:

- If nearest neighbors are *anti-parallel*, they are not connected.
- If nearest neighbors are parallel, they are connected with probability p and disconnected with probability $(1-p)$.

Here, I have assumed that J is positive. If J is negative (anti-ferromagnetic interaction), parallel spins are not connected, while anti-parallel spins are connected with a probability p.

In the case that we consider, there are N_{p} parallel spin pairs. The probability that n_{c} of these are connected and $n_{\mathrm{b}} = N_{\mathrm{p}} - n_{\mathrm{c}}$ are 'broken' is:

$$P_{\mathrm{Gen}}^{o}(\{\text{cluster}\}) = p^{n_{\mathrm{c}}}(1-p)^{n_{\mathrm{b}}}\ .$$

Note that this is the probability of connecting (*c.q.* breaking) a *specified* subset of all links between parallel spins. Once the connected bonds have

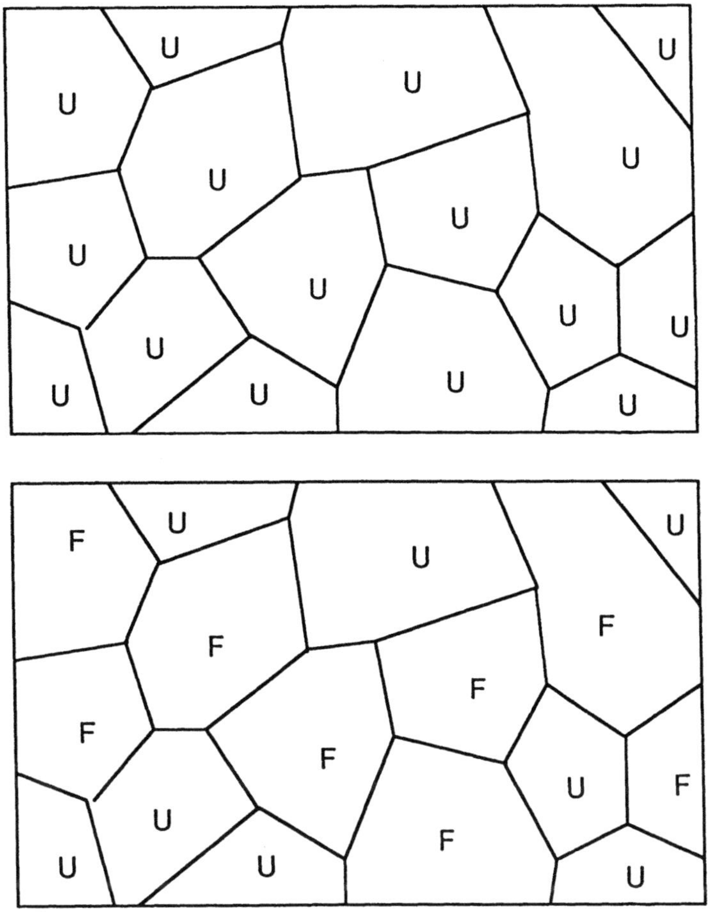

Fig. 4. Swendsen-Wang algorithm. Top figure: the system is divided into clusters. **U** denotes the "unflipped" state of the cluster. In the bottom figure, some clusters have been flipped (**F**). Note, however, that the cluster *structure* is unchanged.

been selected, we can define the clusters in the system. A cluster is a set of spins that are at least singly connected by bonds. We now choose our subset of clusters to flip (see Fig. 4). After the cluster flipping, the number of parallel and anti-parallel spin pairs will have changed, e.g.

$$N_{\mathbf{p}}^{\mathbf{new}} = N_{\mathbf{p}}^{\mathbf{old}} + \Delta$$

and (hence)

$$N_{\mathbf{a}}^{\mathbf{new}} = N_{\mathbf{a}}^{\mathbf{old}} - \Delta \, .$$

Therefore, the total energy of the system will have changed by an amount $-2J\Delta$:

$$E_{\text{new}} = E_{\text{old}} - 2J\Delta \ .$$

Let us now consider the probability to make the *reverse* move. In order to do this, we should generate the *same* cluster structure, but now starting from a situation where there are $N_{\text{p}} + \Delta$ parallel spin pairs and $N_{\text{a}} - \Delta$ anti-parallel pairs. As before, the bonds between anti-parallel pairs are assumed to be broken (this is compatible with the same cluster structure). We also know that the new number of connected bonds, n'_{c} must be equal to n_{c}, *because the same number of connected bonds is required to generate the same cluster structure.* The difference appears when we consider how many of the bonds between parallel spins in the new configuration should be broken (n'_{b}). Using

$$N_{\text{p}}^{\text{new}} = n'_{\text{c}} + n'_{\text{b}} = n_{\text{c}} + n'_{\text{b}} = N_{\text{p}}^{\text{old}} + \Delta = n_{\text{c}} + n_{\text{b}} + \Delta$$

we see that

$$n'_{\text{b}} = n_{\text{b}} + \Delta$$

If we insert this in eqn (78), we obtain

$$\frac{P_{\text{Gen}}^{o}(\{\text{cluster}\})}{P_{\text{Gen}}^{n}(\{\text{cluster}\})} = \frac{p^{n_{\text{c}}}(1-p)^{n_{\text{b}}}}{p^{n_{\text{c}}}(1-p)^{n_{\text{b}}+\Delta}} = (1-p)^{-\Delta} = \frac{P_{\text{B}}^{n}}{P_{\text{B}}^{o}} = \exp(2\beta J\Delta) \ . \quad (79)$$

In order to satisfy this equation, we must have:

$$1 - p = \exp(-2\beta J)$$

or

$$p = 1 - \exp(-2\beta J) \ ,$$

which is the Swendsen-Wang rule.

4.2.2. xy-model. In this section, I describe one application to a 'continuous' model, namely the planar Heisenberg model (xy-model). In this model, spins on neighboring lattice sites have an interaction

$$v_{ij} = -J\mathbf{u}_i.\mathbf{u}_j$$

where $\mathbf{u}_i$ denotes the unit vector that specifies the direction of a spin on site i, while J measures the strength of the interaction. The first step is to define the cluster moves. The essential requirement is that the relative orientation of particles within one cluster is not changed. One way to achieve this is to reflect the orientation of all particles within the selected cluster with respect to a *randomly* chosen plane. For convenience, I use coordinates such that this plane is perpendicular to the x-axis. Let us consider what happens to the interaction energy of two neighboring spins if one is reflected, but not the other. Initially, the interaction energy was

$$v_{ij}^{\text{old}} = -J(u_i^x u_j^x + u_i^y u_j^y)$$

If only one spin is reflected (either one), the product $u_i^x u_j^x$ changes sign. Therefore,

$$v_{ij}^{\text{new}} = v_{ij}^{\text{old}} + 2Ju_i^x u_j^x \equiv v_{ij}^{\text{old}} + \Delta v_{ij} \; .$$

Note that Δv_{ij} may be either positive or negative. Now we construct clusters by linking neighboring spins according to the following recipe:

- If $\Delta v_{ij} \leq 0$, i and j are always *disconnected.* We call these spin pairs 'anti-parallel', by analogy to the Ising case.
- If $\Delta v_{ij} > 0$, i and j are *connected* with a probability p_{ij} and disconnected with a probability $(1 - p_{ij})$. We call these spin pairs 'parallel', by analogy to the Ising case.

Note that the probability to break a bond between 'parallel' spins depends on the magnitude of Δv_{ij}.

Let us next consider how we can impose detailed balance. The equivalent of eqn (77) is

$$P_B^o \prod_{ij}^{n_c} p_{ij} \prod_{kl}^{n_b} (1 - p_{kl}) = P_B^n \prod_{ij}^{n'_c} p_{ij} \prod_{kl}^{n'_b} (1 - p_{kl}) \tag{80}$$

where the set of pairs $\{ij\}$ and $\{kl\}$ denote all pairs of neighboring spins that are 'parallel'. Note that, as before, the set $\{ij\}$ is the *same* for the old and new configuration, because we are considering the same cluster structure. As a consequence, $n_c = n'_c$. However, the number of disconnected parallel spin pairs will, in general differ in the old and new situations. We can write the change in energy between the old and new configurations due to cluster flips as

$$\Delta E = \sum_{ij} \Delta v_{ij}^+ + \sum_{kl} \Delta v_{kl}^-$$

where Δv^+ denotes a positive energy change (i.e. where a pair of spins goes from 'parallel' to 'anti-parallel') and Δv^- denotes a negative energy change. Note that, for a given pair ij, Δ is either positive or negative (or zero). Hence, the sets $\{ij\}$ and $\{kl\}$ that give non-zero contributions are *non-overlapping*. We can now re-express our condition for detailed balance (eqn (80)) as:

$$P_{\mathrm{B}}^{o} \prod_{ij}^{n_c} p_{ij} \prod_{kl}^{n_b}(1 - p_{kl}) =$$

$$P_{\mathrm{B}}^{o} \exp\left[-\beta\left(\sum_{\alpha\beta} \Delta v_{\alpha\beta}^{+} + \sum_{\gamma\delta} \Delta v_{\gamma\delta}^{-}\right)\right] \prod_{ij}^{n'_c} p_{ij} \prod_{kl}^{n'_b}(1 - p_{kl}) . \tag{81}$$

Clearly, as the same set of bonds are connected in the old and new configuration, we can simplify the above equation

$$\prod_{kl}^{n_b}(1 - p_{kl}) = \exp\left[-\beta\left(\sum_{\alpha\beta} \Delta v_{\alpha\beta}^{+} + \sum_{\gamma\delta} \Delta v_{\gamma\delta}^{-}\right)\right] \prod_{kl}^{n'_b}(1 - p_{kl}) . \tag{82}$$

Let us denote the number of bonds between parallel spins that are broken both in the old *and* in the new configuration by N_b. We now write $n_b = N_b + \Delta n_b^o$ and $n'_b = N_b + \Delta n_b^n$. Using this definition, we can rewrite our detailed balance condition as

$$\prod_{ij}^{\Delta n_b^o} (1 - p_{ij}) \exp\left[\beta\left(\sum_{ij} \Delta v_{ij}^{+}\right)\right] =$$

$$\exp\left[-\beta\left(\sum_{kl} \Delta v_{kl}^{-}\right)\right] \prod_{kl}^{\Delta n_b^n} (1 - p_{kl}) . \tag{83}$$

It is important to note that the pairs with a positive (negative) Δv are precisely the ones that are parallel in the old (new) configuration but not in the new (old) one. The above equality can be satisfied trivially if we impose that

$$1 - p_{ij} = \exp(-\beta \Delta v_{ij}^{+})$$

or

$$p_{ij} = 1 - \exp(-\beta \Delta v_{ij}^{+}) .$$

This, then, is the 'Swendsen-Wang' recipe for the xy-model. In fact, it will work for a wide class of d-dimensional spin models. It contains the Ising model as a special case.

4.2.3. Translationally disordered systems. Here the problem is more complex. Consider two 'molecules' i and j that either belong to the same cluster, or move independently. We should now distinguish three cases:

(i) The relative coordinates of i and j are unchanged (they belong to the same cluster). This yields $\Delta v_{ij}=0$.

(ii) i moves, but not j. This yields $\Delta v_{ij} \equiv v_1$.
(iii) j moves, but not i. This yields $\Delta v_{ij} \equiv v_2$.
In general, $v_1 \neq v_2$. There are probably many ways to define clusters in this situation. I suggest the following, scheme. Grow *a single cluster* starting from a randomly chosen site (say i). An atom k in the cluster is linked to another atom (l) with an, as yet unspecified, probability p_{kl}. We terminate the growth of the cluster as soon as we have exhausted all possibilities to form bonds to atoms outside the cluster. We then try to move this cluster over a distance Δ. The trial move is accepted with an (as yet unspecified) probability $P_{\mathrm{acc}}(0 \to 1)$, where 0 denotes the old position, and 1 the new position. The problem arises when we consider the reverse move: i.e. we attempt to move atom i over a distance $-\Delta$. For any atom j, moving i over a distance $-\Delta$ and keeping j fixed, is equivalent to moving j over $+\Delta$ and keeping i fixed. As a consequence, Δv_{ij} is now equal to v_2, rather than v_1. Clearly, whatever we do, the probability of forming a bond in a cluster should not depend on the sign of Δ. One solution would be to have p_{ij} depend on the current distance of i and j only (such an approach has been followed successfully by Wu, Chandler and Smit [62]). Alternatively, one may construct p_{ij} in such a way that it depends on Δ, but in such a way that it is symmetric in v_1 and v_2. A possible expression would be:

$$p_{ij} = 1 - \frac{\min(1, \exp(-\beta v_1)) + \min(1, \exp(-\beta v_2))}{2}$$

This expression reduces to the previous one for the Ising and xy models. The difference with these earlier examples is that, in those cases, once a cluster (or set of clusters) had been generated, the cluster moves were accepted with unit probability. In contrast, we must now explicitly compute the acceptance probability of a trial move. If we denote the potential energy of the old (new) configuration by $\mathcal{V}_0$ ($\mathcal{V}_1$), the detailed balance condition requires that

$$\exp(-\beta\mathcal{V}_0) \prod_{kl} (1 - p^{\mathrm{f}}_{kl}) P_{\mathrm{acc}}(0 \to 1) = \exp(-\beta\mathcal{V}_1) \prod_{kl} (1 - p^{\mathrm{r}}_{kl}) P_{\mathrm{acc}}(1 \to 0) \quad (84)$$

where k denotes a particle *on* the cluster, and l a particle outside it. The superscripts f and r denote forward and reverse moves. In writing eqn (84), we have used the fact that the probability to form intra-cluster bonds is the same for forward and reverse moves. Note, however, that for particles on different clusters, the probability to form a bond on the forward and reverse moves is not the same. From eqn (84), we derive an expression for the ratio of the acceptance probabilities:

$$R \equiv \frac{P_{\mathrm{acc}}(0 \to 1)}{P_{\mathrm{acc}}(1 \to 0)} = \exp(-\beta(\mathcal{V}_1 - \mathcal{V}_0)) \prod_{kl} \frac{(1 - p^{\mathrm{r}}_{kl})}{(1 - p^{\mathrm{f}}_{kl})} . \quad (85)$$

Note that all quantities on the right hand of eqn (85) can be evaluated once the cluster around i has been generated. As before, it is probably best to use the Metropolis rule

$$P_{\mathrm{acc}}(0 \rightarrow 1) = \min(R, 1) .$$

Clearly, many choices for p_{kl} are possible and there is, as yet, little experience with such schemes.

4.2.4. An alternative for 'Metropolis' sampling?. Finally, I wish to point out that the concept underlying the Swendsen-Wang scheme suggests an alternative to the Metropolis sampling for single-particle and collective moves in conventional Monte Carlo simulations. Before discussing *how* this can be achieved, I should first explain *why* such an alternative scheme is, in fact, desirable. To see this, consider the often quoted rule-of-thumb which states that it is advisable to choose the magnitude of a trial move such that the acceptance probability of such moves is approximately 50%. Of course, this is not a rigorous rule, but for systems that interact through a continuous potential, it is reasonable. In contrast, if we consider systems with hard-core interactions, it is better to work with much lower acceptance probabilities. The reason is the following: a trial move of particle i is accepted with a probability $P_{\mathrm{acc}}(i) = \exp(-\beta \Delta \mathcal{V}(i))$, where $\Delta \mathcal{V}(i)$ represents the change in the potential energy of the system due to the displacement of particle i. In what follows, I shall assume that all interactions are pairwise additive, i.e.

$$\Delta \mathcal{V} = \sum_{j \neq i} \Delta v_{ij} .$$

In systems with continuous intermolecular interactions, the usual procedure to determine whether or not a trial move is accepted, is to compute Δv_{ij} for all j and then, at the end, compare $\exp(-\beta \sum \Delta v_{ij})$ with a random number between 0 and 1. However, for hard-core systems, we can compute $\exp(-\beta \Delta v_{ij})$ for every j and reject the move *as soon as we find an overlap.* As a consequence, rejected moves are cheaper than accepted moves in a hard-core system, but not in a system with continuous interactions. We can now balance the amount of computational work spent on accepted and rejected moves in the hard core system, by increasing the magnitude of the trial move (and thereby reducing the acceptance probability). I shall now show that we can use the concept of 'bonds' to construct a MC scheme for continuous potentials that will allow us to reject *at an early stage* moves that are likely to be 'unfavorable'. For hard-core particles, this scheme reduces to the conventional Metropolis scheme (see, however, below).

As before, I consider the trial move of a single particle i. Particle i interacts with 'neighboring' particles $\{j\}$. Due to the trial displacement the interaction energy will change by an amount Δv_{ij}. I now construct *bonds* between particle i and its neighbors $\{j\}$ according to the following rule:

(i) If $\Delta v_{ij} \leq 0$, i and j are not connected.

(ii) If $\Delta v_{ij} > 0$, i and j are connected with a probability $p_{ij} = 1 - \exp(-\beta \Delta v_{ij})$ and disconnected otherwise.

The trial move is carried out if particle i is not connected with any other particle j. More importantly, the trial move is rejected *as soon as a bond is formed between i and any j*. Note that this scheme is different from the Metropolis scheme, because it will result in the rejection of some moves that *lower* the potential energy. The scheme does, however, satisfy detailed balance. To see this, consider the probability of accepting a trial move from the 'old' position of particle i to the new position:

$$P_{\mathrm{acc}}(i \to i') = \prod_{j \neq i} (1 - p_{ij}) = \exp\left[-\beta \sum_{j \neq i} \Delta v_{ij}^{+}\right] , \tag{86}$$

where the sum over j enumerates all interaction energies that *increase* as a result of the trial move. Next, consider the reverse move. I will assume that the underlying Markov chain of the Monte Carlo scheme is symmetric, i.e. the *a priori* probabilities of going from i to i' and from i' to i are equal. The acceptance probability for the trial move from i' to i is

$$P_{\mathrm{acc}}(i' \to i) = \prod_{j \neq i} (1 - p_{ij}) = \exp\left[-\beta \sum_{j \neq i} \Delta v_{ij}^{-}\right] . \tag{87}$$

But now the sum over j includes only those pairs $\{ij\}$ for which the energy increases when going from the trial position to the original one. The ratio of these acceptance probabilities is

$$\frac{P_{\mathrm{acc}}(i \to i')}{P_{\mathrm{acc}}(i' \to i)} = \exp\left[-\beta \sum_{j \neq i} \Delta v_{ij}\right] . \tag{88}$$

In eqn (88), the sum over j is unrestricted! Hence our scheme satisfies detailed balance.

A few remarks are in place. First of all, one should bear in mind that there is a trade-off between the possibility of rejecting a move at an early stage and the computation of a large number of quasi-random numbers. In practice, it will probably be best to use the bond-generating MC scheme only for those interactions that are likely to change appreciably in a trial move, i.e. the interactions with near neighbors. If this part of the test is passed, the change in long-range interactions can be summed in the usual way and the normal Metropolis rule can be used to test if the overall move will be accepted.

Secondly, the scheme described here, is in no way limited to single-particle moves. In fact, it is probably most useful to apply it to complex many-particle moves that have a very low acceptance. Examples are volume changes in constant pressure simulations of systems with strong, short-ranged attractions (e.g. the attractive, hard-core Yukawa system). A second interesting application is the *staging* approach to path-integral Monte Carlo in systems with *attractive* interactions. A third, useful application is the sampling of *configurational-bias* moves in the simulation of systems of chain molecules (see section 4.1). Here too, the average acceptance probability of trial moves may be very small. This suggests that we may often find that, after 'regrowing' only a small fraction of the chain molecule, the ratio of the Rosenbluth weights that we are considering becomes very small. This implies that, most likely, the move will be rejected. In the conventional CBMC scheme, we must nevertheless follow the regrowth process to the very end. In contrast, the present scheme allows us to terminate a 'doomed' trial move at an earlier stage. It should be noted that, in this application, the bond-generating MC scheme should be superior to the conventional 'Metropolis' scheme, *even for hard-core molecules.*

In this brief outline, I have only sketched a few obvious applications. However, there are clearly many others. In general, one may expect the present scheme to be most efficient in situations where conventional Monte Carlo moves (i) are expensive and (ii) have a very low acceptance probability. The present scheme allows one to quickly eliminate a large fraction of the 'doomed' trial moves. It should be stressed that we are free to decompose the contributions to the acceptance probability of trial moves in any way that we consider convenient (as long as this is done consistently). In particular, this means that there is no need to generate a quasi-random number for *every* interaction that we compute. However, as the best way of implementing the algorithm clearly depends on the system under consideration, I shall not discuss this point here.

5. Thermodynamic integration

In section 3.1 it was mentioned that knowledge of 'thermal' quantities, such as μ, is usually necessary to locate the coexistence line for a first-order phase-transition. At first sight knowledge of μ may appear superfluous. After all, a computer simulation mimics the behaviour of a real solid or liquid. If the simulation is ergodic it should spontaneously transform to whatever phase is thermodynamically most stable, and then we would know all there is to know. Unfortunately, this approach does not work. At least, not for phase transitions involving three-dimensional solids. At a solid-solid or solid-liquid phase transition very strong hysteresis effects are usually observed in a simulation (see Fig. 5). In fact, it is very difficult to nucleate a crystal

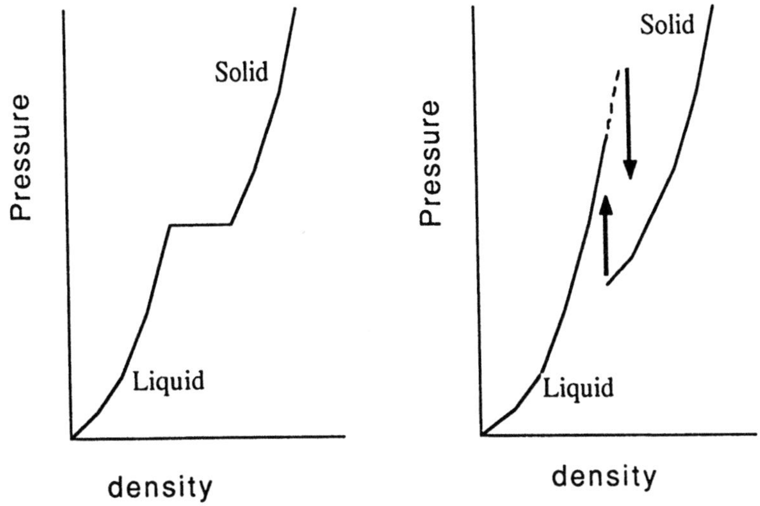

Fig. 5. First-order phase transitions in small model systems with periodic boundary conditions (right) tend to exhibit much stronger hysteresis effects than are usually observed in the real world (left)

from a liquid during a simulation. Hence, to locate the point where two phases coexist, we must compute the chemical potential of the homogeneous phases at the same temperature and pressure and find the point where the two μ's are equal. For not-too-dense fluids, one might use the Gibbs method (see Chapter 6) to allow the computer to find this point. However, the latter method does not work for phase transitions involving solids. In such cases we really must compute μ. General methods to compute μ and related thermal quantities by computer simulation are discussed in Ref. [23]. In the present section I wish to describe a simple, and usually reliable, method to compute the free energy F (and hence μ).

Let us first recall how free energies are measured in real experiments. In the real world free energies cannot be obtained from a single measurement either. What can be measured, however, is the derivative of the free energy with respect to volume V and temperature T:

$$\left(\frac{\partial F}{\partial V}\right)_{NT} = -P \tag{89}$$

and

$$\left(\frac{\partial F/T}{\partial 1/T}\right)_{VT} = E\,. \tag{90}$$

Here P is the pressure and E the energy of the system under consideration.

The trick is now to find a reversible path that links the state under consideration to a state of known free energy. The change in F along that path can then simply be evaluated by integration of eqns (89) and (90). There are only very few thermodynamic states where the free energy of a substance is known. One state is the ideal gas phase, the other is the perfectly ordered ground state at $T = 0K$.

In computer simulations, the situation is quite similar. In order to compute the free energy of a dense liquid, one may construct a reversible path to the very dilute gas phase. It is not really necessary to go all the way to the ideal gas. But at least one should reach a state that is sufficiently dilute that the free energy can be computed accurately, either from knowledge of the first few terms in the virial expansion of the compressibility factor PV/Nk_BT, or that the chemical potential can be computed by other means (see section 3.1 and 3.4). For the solid, the ideal gas reference state is less useful (although techniques have been developed to construct a reversible path from a dense solid to a dilute (lattice-) gas [63]). The obvious reference state for solids is the harmonic lattice. Computing the absolute free energy of a harmonic solid is relatively straightforward, at least for atomic and simple molecular solids. However, not all solid phases can be reached by a reversible route from a harmonic reference state. For instance, in molecular systems it is quite common to find a strongly anharmonic plastic phase just below the melting line. This plastic phase is not (meta-) stable at low temperatures.

Fortunately, in computer simulations we do not have to rely on the presence of a 'natural' reversible path between the phase under study and a reference state of known free energy. If such a path does not exist, we can construct an artificial path. This is in fact a standard trick in statistical mechanics (see e.g. Ref. [64]). It works as follows: Consider a case where we need to know the free energy $F(V,T)$ of a system with a potential energy function $\mathcal{V}_1$, where $\mathcal{V}_1$ is such that no 'natural' reversible path exists to a state of known free energy. Suppose now that we can find another model system with a potential energy function $\mathcal{V}_0$ for which the free energy *can* be computed exactly. Now let us define a generalized potential energy function $\mathcal{V}(\lambda)$, such that $\mathcal{V}(\lambda = 0) = \mathcal{V}_0$ and $\mathcal{V}(\lambda = 1) = \mathcal{V}_1$. The free energy of a system with this generalized potential is denoted by $F(\lambda)$. Although $F(\lambda)$ itself cannot be measured directly in a simulation, we can measure its derivative with respect to λ:

$$\left(\frac{\partial F}{\partial \lambda}\right)_{NVT\lambda} = \left\langle\frac{\partial \mathcal{V}(\lambda)}{\partial \lambda}\right\rangle_{NVT\lambda} . \tag{91}$$

If the path from $\lambda = 0$ to $\lambda = 1$ is reversible, we can use eqn (91) to compute the desired $F(V,T)$. We simply measure $\langle \partial\mathcal{V}/\partial\lambda \rangle$ for a number of values of λ between 0 and 1. Typically, 10 quadrature points will be sufficient to get the absolute free energy per particle accurate to within $0.01k_BT$.

It is however important to select a reasonable reference system. One of the safest approaches appears to be to choose as a reference system an Einstein crystal with the same structure as the phase under study [65]. This choice of reference system makes it extremely improbable that the path connecting $\lambda = 0$ and $\lambda = 1$ will cross an (irreversible) first-order phase transition from the initial structure to another, only to go back to its original structure for still larger values of λ. Nevertheless, it is important that the parametrization of $\mathcal{V}(\lambda)$ be chosen carefully. Usually, a linear parametrization (i.e. $\mathcal{V}(\lambda) = \lambda\mathcal{V}_1 + (1-\lambda)\mathcal{V}_0$) is quite satisfactory. But occasionally such a parametrization may lead to weak (and relatively harmless) singularities in eqn (91) for $\lambda \to 0$. More details about such free energy computations can be found in Ref. [23].

Similar techniques can be used to locate first-order phase transitions involving phases with partial order (e.g. liquid crystals). For details, the reader is referred to Refs. [66–68].

6. The grand canonical miracle

In this final section I discuss entropy-driven phase transitions. The main aim of the examples discussed here is to show that, in some cases, the grand-canonical ensemble greatly simplifies the description of solvent effects on the structure and phase behavior of complex fluids.

6.1. BINARY MIXTURES

Phase separation in binary mixtures is the example that is used in many textbooks to illustrate the competition between energy and entropy in a phase transformation. For a mixture at constant total volume V, the Helmholtz free energy F, should be minimal. As a first approximation, the entropy of mixing of a mixture of two species A and B, is replaced by the entropy of mixing of an ideal mixture

$$S_{\text{id}}(X) = -Nk_{\text{B}}[X \ln X + (1 - X)\ln(1 - X)] \tag{92}$$

where X denotes the mole fraction of one component (say A): $X = N_A/(N_A + N_B)$. The entropy of mixing given by eqn (92) is a convex function of X. As a consequence, $S_{\text{id}}(X)$ will always decrease if phase separation takes place. This implies that phase separation can only take place if the resulting decrease in energy E outweighs the increase in $-TS_{\text{id}}$. In a hard-core mixture, there is no energy change upon mixing. Hence, if eqn (92) were exact, we should never observe phase separation in a hard-core mixture. And, even though eqn (92) is known to be only an approximation, it was commonly thought that a fluid mixture of dissimilar hard spheres would not phase separate. This belief was, at least partly, based on the work of Lebowitz and

Rowlinson [69], who studied the phase-behavior of such binary mixtures of dissimilar hard spheres, using a more accurate approximation for the entropy of mixing, based on the Percus-Yevick (PY) integral equation. Lebowitz and Rowlinson found that, at least within that approximation, hard spheres of arbitrary size ratio will mix in all proportions in the fluid phase. The implication of this result was that entropic effects are not enough to cause a miscibility gap in a simple fluid mixture. Recently, however, Biben and Hansen [70] have applied a more sophisticated analytical theory for dense fluid mixtures to the same problem. Unlike the PY approximation, this theory is found to predict that an asymmetric binary hard-sphere mixture should phase separate if the sizes of the two spheres are sufficiently dissimilar (typically, if the size ratio is less than 0.2). However, as in the case of Ref. [69], it is based on an approximate theory for the hard-sphere mixture. Hence, one may wonder to what extent the result found in Ref. [70] depends on the approximations that are used to compute the equation-of-state of the mixture. Clearly, it would be highly desirable to have a model system of a binary mixture for which the existence of a purely entropic demixing transition can be proven exactly.

It turns out that, in a special case [71], this can indeed be done by constructing a simple lattice model of binary hard-core mixture. By transforming to a grand-canonical ensemble, this binary hard-core mixture can be mapped onto a one-component lattice gas with attractive nearest neighbor interactions. This mapping is of interest because the one-component lattice gases can, in turn, be mapped onto Ising-like models for which, in some cases, the phase behavior is known exactly. To illustrate our approach, first consider a trivial model-system, namely a square lattice with at most one particle allowed per square. Apart from the fact that no two particles can occupy the same square face, there is no interaction between the particles. For a lattice of N sites, the grand-canonical partition function is:

$$\Xi = \sum_{\{n_i\}} \exp\left[\beta\mu \sum_i n_i\right] . \tag{93}$$

The sum is over all allowed sets of occupation numbers $\{n_i\}$ and μ is the chemical potential. Next, we include "small" hard particles that are allowed to sit on the links of the lattice. (see Fig. 6). These small particles are excluded from the edges of a square that is occupied by a large particle.

For a given configuration $\{n_i\}$ of the large particles, one can then exactly calculate the grand canonical partition function of the small particles. Let $M = M(\{n_i\})$ be the number of free spaces accessible to the small particles. Then clearly:

$$\Xi_{\text{small}}(\{n_i\}) = \sum_{l=0}^{M} \frac{M!z_s^l}{l!(M-l)!} = (1+z_s)^{M(\{n_i\})}, \tag{94}$$

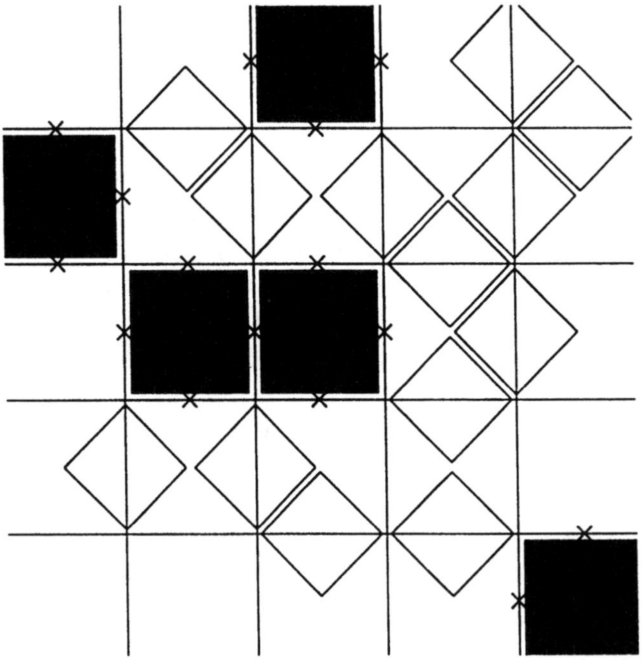

Fig. 6. Exactly solvable lattice-gas model for a mixture of large (black squares) and small (white squares) hard particles. The crosses indicate which small-particle sites are excluded by the presence of large particles.

where z_s is the fugacity of the small particles. M can be written as

$$M(\{n_i\}) = 2N - 4\sum_i n_i + \sum_{<ij>} n_i n_j \ , \tag{95}$$

where $2N$ is the number of links on the lattice and the second sum is over nearest-neighbor pairs, and comes from the fact that when two large particles touch, one link is doubly excluded [72]. The total partition function for the "mixture" is:

$$\Xi_{\text{mixture}} = \sum_{\{n_i\}} \exp\left\{[\beta\mu - 4\log(1+z_s)]\sum_i n_i + [\log(1+z_s)]\sum_{<ij>} n_i n_j\right\} \ , \tag{96}$$

where we have omitted a constant factor $(1+z_s)^{2N}$. Equation (96) is simply the expression for the grand-canonical partition function of a one-component lattice-gas Ising model with a shifted chemical potential and an effective nearest neighbor attraction with an interaction strength $\log(1+z_s)/\beta$. As is well known, this lattice model can again be transformed to a 2-D Ising spin model that can be solved in the zero field case [73, 74]. In the language of our mixture model, no external magnetic field means:

$$(1+z_s)^2 = z_l, \tag{97}$$

where $z_l = \exp \beta\mu$, the large particle fugacity. The order-disorder transition in the Ising model then corresponds to phase separation in the language of our model. This demixing is purely entropic, just like the transition predicted [70] for the hard sphere mixture. In fact, the mapping described above can also be carried through when energetic interactions between the large particles are included. However, for the sake of clarity, we will restrict ourselves to athermal hard-core mixtures.

Of course, there is a wide variety of lattice models for hard-core mixtures that can be mapped onto one-component systems with effective attraction. The model discussed above is only special in the sense that it can be mapped onto a model that is exactly solvable. In particular, from the known results for the 2D-Ising model (see e.g. Ref. [75]), it is straightforward to evaluate the composition of the coexisting phases along the entire coexistence line. This coexistence curve is shown in Fig. 7. One important question that is raised by the work of Biben and Hansen [70] is whether the demixing transition is of the fluid-fluid or the fluid-solid type. The phase transition in the square-lattice model that we discussed above provides no answer to this question, as there is no distinction between 'liquid' and 'solid' in a lattice gas on a square, or simple-cubic lattice. Yet, from the above mapping, it is immediately obvious that a fluid-fluid transition does, in fact, occur in another hard-core mixture, namely that of a mixture of hard-core monomers and polymers on a lattice. To this end, we consider, once again our model of a mixture of large and small hard squares (cubes, in 3-D) on a square (cubic) lattice, i.e. the model that could be mapped onto the 1-component lattice gas with nearest neighbor interactions. We now construct 'polymers' by connecting N large squares (cubes), while the solvent is represented by the small particles. The simplest polymers are made by connecting N large particles by nearest neighbor bonds only. A grand-canonical summation over all configurations of the small particles, yields a very simple expression for the (canonical) partition function of the polymers, namely

$$Z_{\text{polymers}} = \sum_{\{n_i\}} \exp\left[\frac{J}{k_B T} \sum_{<ij>} n_i n_j\right] \tag{98}$$

with $J \equiv k_B T \log(1+z_s)$ and where the sum is over all acceptable (i.e. non-overlapping) configurations of the hard-core polymers. Equation (98) is precisely the expression for the partition function of the Flory-Huggins lattice

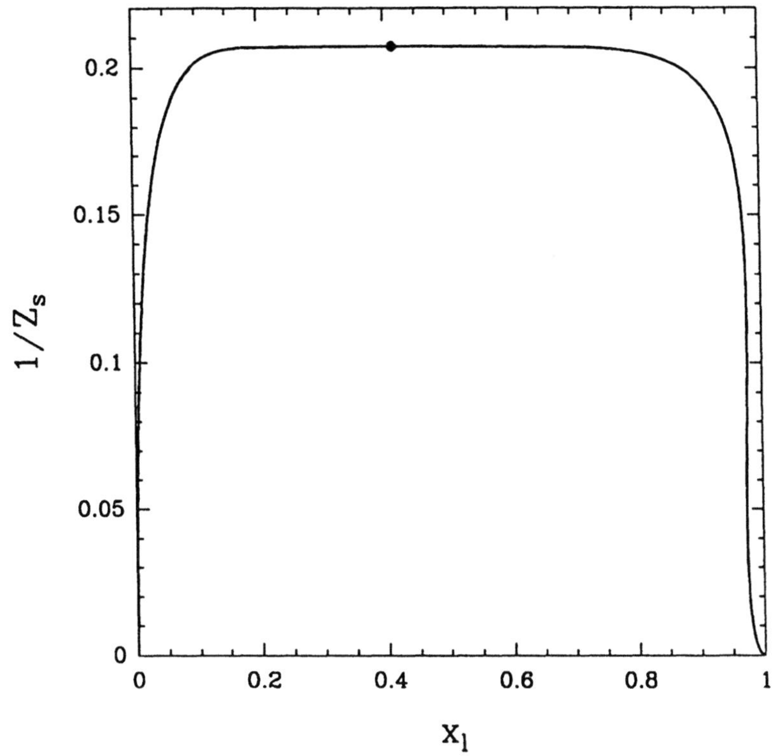

Fig. 7. Phase diagram of the model system shown in Fig. 6. X_l denotes the molar fraction of large squares, z_s is the fugacity of small squares. The drawn curve indicates the composition of coexisting phases. Note that this curve is asymmetric. In particular, the critical point (indicated by a black dot) is located at $X_l = 2^{1/2} - 1$. Along the coexistence curve the fugacities of the large and small particles are related through eqn (97).

model [76]. This model has been studied extensively both using approximate analytical theories, in particular the well-known Flory-Huggins theory and modifications thereof (for a critical review see [77]) and, more recently, by direct numerical simulations (see e.g. Refs. [58, 78]). Traditionally, the coupling constant J has been interpreted as a purely *energetic* interaction. In contrast, in the context of the model that we consider, J is of completely entropic origin. We can now translate the results that have been obtained for the Flory-Huggins lattice-gas model directly to our hard-core mixture. This means that many effects that in the past were interpreted as originating from energetic attractions, can also be obtained from purely entropic effects. In particular, the existence of a first-order fluid-fluid phase separation in this model provides direct proof that a purely entropic demixing transition exists in a hard-core polymer-solvent mixture. Our analysis also yields a very simple 'entropic' interpretation of other, closely related phenomena, such as polymer collapse in a 'poor' solvent. It is, perhaps, worth pointing out that in our model, the solvent molecules are smaller than the monomeric units of the polymer. In contrast, in the original Flory-Huggins lattice model, there is no such size-asymmetry. We can, however, construct a hard-core lattice model where solvent and monomeric units occupy the same volume. It can be shown that the solvent-induced polymer collapse will also occur in such a 'symmetric' model system.

Finally, we note that the Flory-Huggins theory for polymer solutions yields the following (approximate) expression for the free-energy of a polymer solution:

$$\frac{F_{\text{mix}}}{k_{\text{B}}T} = \frac{\phi}{N}\log\phi + (1-\phi)\log(1-\phi) + \chi\phi(1-\phi) \tag{99}$$

where ϕ is the fraction of the volume occupied by polymer, while the parameter χ is related to the coupling constant J of the original lattice model, by

$$\chi = \frac{1}{2}\frac{J}{k_{\text{B}}T}n_{\text{b}} = \frac{1}{2}n_{\text{b}}\ln(1+z_s), \tag{100}$$

where n_{b} is the coordination number of the lattice. If, as was assumed in the original Flory-Huggins theory, J is due to energetic interactions, then χ should vary as $1/T$. However, in the present (extreme) interpretation of the same lattice-gas model, the parameter χ would be *completely independent of temperature*. There is, in fact, a large body of experimental data that shows that, for many polymer solutions, χ has a large, if not dominant, entropic (i.e. temperature-independent) part.

6.2. DEPLETION FLOCCULATION

Let us next consider a slightly more complex example of an entropy-driven phase separation in a binary mixture, namely polymer-induced flocculation of colloids. Experimentally, it is well known that the addition of a small amount of free, non-adsorbing polymer to a colloidal suspension induces an effective attraction between the colloidal particles and may even lead to coagulation. This effect has been studied extensively [79] and is well understood, at least qualitatively. As in the example discussed above, the polymer-induced attraction between colloids is an *entropic* effect: when the colloidal particles are close together, the total number of accessible polymer conformations is larger than when the colloidal particles are far apart. However, although the physical mechanism responsible for polymer-induced coagulation is understood *qualitatively*, a quantitative description of this phenomenon is difficult. This is so because the polymer-induced attraction between the colloidal particles is non-pairwise additive. Moreover, it depends both on the osmotic pressure of the polymer and on the concentration of the colloid. Yet, in the theoretical description of polymer-induced clustering [80], the effect of the polymer is usually replaced by an effective, density-independent, pair-wise additive attraction between the colloidal particles. However, in this case, no analytical evaluation of the (grand-canonical) partition function is possible, even when one considers only the very simplest model, *viz.* that of a mixture of hard-core colloidal particles with ideal chain molecules with conformations that are restricted to a lattice. In this case, it would clearly be desirable to carry out 'exact' numerical simulations to investigate the phase behavior. Yet, the computational problems are still formidable. What is required is a numerical scheme that samples the positions of the colloidal particles while averaging over all possible conformations of a large (and fluctuating) number of chain molecules. The 'conventional' Monte Carlo schemes to simulate lattice models of polymer systems [58] would be woefully inadequate for such a calculation.

Fortunately, it is possible to construct a rigorous and efficient Monte Carlo scheme to study this model. Our approach relies on the fact that we can recursively compute the partition function of an ideal (non-self avoiding) chain on a lattice in an arbitrary external potential [81, 82]. This is most easily seen by considering a chain of length $\ell-1$ on a lattice. For convenience, we assume that the external potential is either zero or infinite. The total number of accessible ideal chain conformations that terminate on lattice site i is denoted by $\omega_{\ell-1}(i)$. The total partition function $\Omega_{\ell-1}$ is equal to $\sum_i \omega_{\ell-1}(i)$, where the sum runs over all lattice sites. The total number of chains of length ℓ that terminate on site i is clearly equal to the sum of the total number of chains of length $\ell-1$ that terminate on any of the neighbors of i, multiplied by the Boltzmann factor associated with site i. Using such a recursive scheme, we can compute *exactly* Ω_ℓ, the partition function of a single ideal polymer of arbitrary length ℓ on a lattice, in an arbitrary external potential. This scheme can be used as a starting point to study self-avoiding polymers [83, 84], but for the present purpose we limit ourselves to ideal polymers. Up to this point we have not specified the nature of the 'external' potential. We now assume that this potential is due to the presence of N hard, spherical colloidal particles each of which occupies many lattice sites. The polymer partition function clearly depends on the coordinates $\mathbf{r}^N$ of the colloidal particles: $\Omega_\ell(\mathbf{r}^N)$. The configurational part of the partition function of the system of N colloids plus one polymer of length ℓ in volume V is then given by:

$$Z(V, N, 1) = \int_V \mathrm{d}\mathbf{r}^N \mathrm{e}^{-\mathcal{V}_{\mathrm{hs}}(\mathbf{r}^N)} \Omega_\ell(\mathbf{r}^N), \tag{101}$$

where $\mathcal{V}_{\rm hs}(\mathbf{r}^N)$ denotes the hard-sphere interaction. Next, we make use of the fact that we are considering *ideal* polymers. In that case we can immediately write down the corresponding partition function for N colloids and M ideal polymers:

$$Z(V,N,M) = \int_V {\rm d}\mathbf{r}^N {\rm e}^{-\mathcal{V}_{\rm hs}(\mathbf{r}^N)} \left[\Omega_\ell(\mathbf{r}^N)\right]^M /M!, \tag{102}$$

where the factor $1/M!$ accounts for the fact that the polymers are indistinguishable. Using eqn (102) it is straightforward to transform to an ensemble where the polymer chemical potential (i.e. the osmotic pressure) is kept fixed. The corresponding grand-canonical partition function is given by:

$$\begin{aligned}\Xi(V,N,\mu) &= \sum_{M=0}^{\infty} {\rm e}^{M\mu_{\rm ch}}\, Z(V,N,M)/M! \\ &= \int_V {\rm d}\mathbf{r}^N {\rm e}^{-\mathcal{V}_{\rm hs}(\mathbf{r}^N)} \sum_{M=0}^{\infty} {\rm e}^{M\mu_{\rm ch}} \left[\Omega_\ell(\mathbf{r}^N)\right]^M /M! \\ &= \int_V {\rm d}\mathbf{r}^N {\rm e}^{-\mathcal{V}_{\rm hs}(\mathbf{r}^N)}\, {\rm e}^{z\Omega_\ell(\mathbf{r}^N)}\ . \end{aligned} \tag{103}$$

In the last line of eqn (103), we have introduced the polymer fugacity $z \equiv \exp(\mu_{\rm ch})$, where $\mu_{\rm ch}$ denotes the chemical potential of the chain molecules.

The important point to note is that eqn (103) allows us to evaluate the properties of the colloidal particles in osmotic equilibrium with a polymer reservoir. In particular, it shows that we can perform Monte Carlo sampling of the colloidal particles. The polymers only affect $\mathcal{V}_{\rm eff}(\mathbf{r}^N)$, the effective interaction between the colloidal particles:

$$\mathcal{V}_{\rm eff}(\mathbf{r}^N) \equiv \mathcal{V}_{\rm hs}(\mathbf{r}^N) - z\,\Omega_\ell(\mathbf{r}^N). \tag{104}$$

$z\Omega_\ell(\mathbf{r}^N)$ measures the entropic interaction between the colloids due to *all* possible polymer conformations. This entropic interaction is, in principle, not pairwise additive. In fact, it is shown in Ref. [82], that for all but the shortest chain molecules, this non-additivity of the polymer-induced interaction between the colloids, has a pronounced effect on the structure and stability of the mixture. For more details, we refer the reader to Ref. [82].

6.3. FROM SOAP TO SATURN?

Finally, I wish to discuss a more speculative, and somewhat exotic, example of an entropy-driven phase transition, namely the demixing of a system of ideal point fermions ("non-interacting electrons") and classical fluid particles. The analogy between this system and the previous ones rests on the fact that, in Feynman's path-integral formulation of quantum mechanics, there is an isomorphism between the partition-function of a d-dimensional quantum-mechanical system, and that of a $(d+1)$-dimensional 'polymer' system. Ideal quantum particles in d dimensions can be described by ideal, directed polymers in $d+1$ dimensions. This suggests that we may expect that mixtures of ideal quantum particles and atoms (effectively, hard spheres) behave much like a mixture of ideal polymers and colloids. In fact, the situation in the quantum case is complicated by the fact that one must take exchange into account. This is relatively straightforward in the case of numerical simulations of bosons [85], but it creates serious problems for fermions (the 'sign'-problem). However, in Ref. [86], it is shown how, at least for a mixture of atoms and ideal fermions, these problems may be overcome. Alavi and Frenkel [86] show that it is, in fact, possible to evaluate numerically (but exactly) the grand-canonical partition function of ideal fermions in the presence of arbitrary "obstacles" (in particular, atoms). This makes it possible to perform Monte Carlo sampling of the atomic positions in the presence of the fermion 'bath'. Here, I will not discuss the technical aspects of these simulations. Rather, I will summarize the main conclusion, which is that there is indeed strong evidence for a demixing transition as the chemical potential of the fermions is increased beyond a certain critical value. Again, there are no attractive forces in this system. Hence, this fermion-induced phase separation appears to be yet another example of "attraction through repulsion". For more details, the reader is referred to Ref. [86]. It is amusing to note that the phase-separation in a mixture of ideal fermions and hard-spherical "atoms" may have an intriguing counterpart in planetary astronomy, namely in the structure of the interior of Saturn. It is thought that the core of Saturn, which is a hydrogen-helium planet, consists of nearly pure helium [87], whilst at the conditions relevant to the interior of Saturn ($T \approx 10^4$K, $P \approx 1 - 10$Mbar) the hydrogen probably is metallic. If we view the helium as consisting of 'hard spheres', and the metallic hydrogen as a nearly free electron gas (in which the hydrogen nuclei play the role of a neutralising background), we might expect on the basis of the calculations in Ref. [86] that such a mixture would phase-separate. Clearly, this example is far removed from most other topics discussed during this school, but it does provide a nice illustration of the unity of physics.

Acknowledgements

The work of the FOM Institute is part of the research program of FOM and is supported by the 'Nederlandse Organisatie voor Wetenschappelijk Onderzoek'(NWO).

References

[1] N. Metropolis, A. W. Rosenbluth, M. N. Rosenbluth, A. H. Teller, and E. Teller, *J. Chem. Phys.* **21**, 1087 (1953).

[2] M. CREUTZ, *Phys. Rev. Lett.* **50**, 1411 (1983).

[3] K. BINDER, *The Monte Carlo Method in Condensed Matter Physics*, Springer, Berlin, 1992.

[4] W. W. WOOD, *J. Chem. Phys.* **48**, 415 (1968).

[5] I. R. MCDONALD, *Molec. Phys.* **23**, 41 (1972).

[6] Actually, there is no need to assume a *real* piston. The systems with volume V and $V_0 - V$ may both be isolated systems subject to their individual (periodic) boundary conditions. The only constraint that we impose is that the sum of the volumes of the two systems equals V_0.

[7] R. EPPENGA and D. FRENKEL, *Molec. Phys.* **52**, 1303 (1984).

[8] M. PARRINELLO and A. RAHMAN, *Phys. Rev. Lett.* **45**, 1196 (1980).

[9] M. PARRINELLO and A. RAHMAN, *J. Appl. Phys.* **52**, 7182 (1981).

[10] H. C. ANDERSEN, *J. Chem. Phys.* **72**, 2384 (1980).

[11] R. NAJAFABADI and S. YIP, *Scripta Metall.* **17**, 1199 (1983).

[12] G. E. NORMAN and V. S. FILINOV, *High Temp. Res. USSR* **7**, 216 (1969).

[13] D. J. ADAMS, *Molec. Phys.* **28**, 1241 (1974).

[14] D. J. ADAMS, *Molec. Phys.* **29**, 307 (1975).

[15] D. J. ADAMS, *Molec. Phys.* **32**, 647 (1976).

[16] D. J. ADAMS, *Molec. Phys.* **37**, 211 (1979).

[17] L. A. ROWLEY, D. NICHOLSON, and N. G. PARSONAGE, *J. Comput. Phys.* **17**, 401 (1975).

[18] J. YAO, R. A. GREENKORN, and K. C. CHAO, *Molec. Phys.* **46**, 587 (1982).

[19] M. MEZEI, *Molec. Phys.* **40**, 901 (1980).

[20] J. P. VALLEAU and K. L. COHEN, *J. Chem. Phys.* **72**, 3935 (1980).

[21] W. VAN MEEGEN and I. SNOOK, *J. Chem. Phys.* **73**, 4656 (1980).

[22] M. P. ALLEN and D. J. TILDESLEY, *Computer Simulation of Liquids*, Clarendon, Oxford, 1987.

[23] D. FRENKEL, Free-energy computation and first-order phase transitions, in *Molecular Dynamics Simulations of Statistical Mechanical Systems*, edited by G. CICCOTTI and W. G. HOOVER, pp. 151–188, North-Holland, Amsterdam, 1986.

[24] A. Z. PANAGIOTOPOULOS, *Molec. Phys.* **61**, 813 (1987).

[25] D. A. KOFKE and E. D. GLANDT, *Molec. Phys.* **64**, 1105 (1988).

[26] K. S. SHING, *Chem. Phys. Lett.* **119**, 149 (1985).

[27] K. S. SHING, *J. Chem. Phys.* **85**, 1105 (1986).

[28] P. SINDZINGRE, G. CICCOTTI, C. MASSOBRIO, and D. FRENKEL, *Chem. Phys. Lett.* **136**, 35 (1987).

[29] W. G. T. KRANENDONK and D. FRENKEL, *Molec. Phys.* **72**, 699 (1991).

[30] M. R. STAPLETON, D. J. TILDESLEY, and N. QUIRKE, *J. Chem. Phys.* **92**, 4456 (1990).

[31] B. WIDOM, *J. Chem. Phys.* **39**, 2808 (1963).

[32] D. J. ADAMS, *Molec. Phys.* **28**, 1252 (1974).

[33] U. HEINBUCH and J. FISCHER, *Molec. Simul.* **1**, 109 (1987).

[34] J. I. SIEPMANN, I. R. MCDONALD, and D. FRENKEL, *J. Phys. Cond. Mat.* **4**, 679 (1992).

[35] K. S. Shing and S. T. Chung, *J. Phys. Chem.* **91**, 1674 (1987).

[36] P. Sindzingre, C. Massobrio, G. Ciccotti, and D. Frenkel, *Chem. Phys.* **129**, 213 (1989).

[37] J. I. Siepmann, *Molec. Phys.* **70**, 1145 (1990).

[38] J. Harris and S. A. Rice, *J. Chem. Phys.* **88**, 1298 (1988).

[39] M. N. Rosenbluth and A. W. Rosenbluth, *J. Chem. Phys.* **23**, 356 (1955).

[40] D. Frenkel, *J. Phys. Cond. Mat.* **2(SA)**, 265 (1990).

[41] G. C. A. M. Mooij and D. Frenkel, *Molec. Phys.* **74**, 41 (1991).

[42] D. Frenkel and B. Smit, *Molec. Phys.* **75**, 983 (1992).

[43] J. J. de Pablo, M. Laso, and U. W. Suter, *J. Chem. Phys.* **96**, 6157 (1992).

[44] S. K. Kumar, I. Szleifer, and A. Z. Panagiotopoulos, *Phys. Rev. Lett.* **66**, 2935 (1991).

[45] B. Smit, G. C. A. M. Mooij, and D. Frenkel, *Phys. Rev. Lett.* **68**, 3657 (1992), Comment.

[46] K. S. Shing and K. E. Gubbins, *Molec. Phys.* **46**, 1109 (1982).

[47] K. S. Shing and K. E. Gubbins, *Molec. Phys.* **49**, 1121 (1983).

[48] C. H. Bennett, *J. Comput. Phys.* **22**, 245 (1976).

[49] G. M. Torrie and J. P. Valleau, *J. Comput. Phys.* **23**, 187 (1977).

[50] D. Chandler, *Introduction to Modern Statistical Mechanics*, Oxford University Press, New York, 1987.

[51] J. S. van Duijneveldt and D. Frenkel, *J. Chem. Phys.* **96**, 4655 (1992).

[52] K.-K. Han, *Phys. Lett. A* **165**, 28 (1992).

[53] P. J. Rossky, J. D. Doll, and H. L. Friedman, *J. Chem. Phys.* **69**, 4628 (1978).

[54] S. Duane, A. D. Stevenson, B. J. Pendleton, and D. Roweth, *Phys. Lett. B* **195**, 216 (1987).

[55] R. H. Swendsen and J.-S. Wang, *Phys. Rev. Lett.* **58**, 86 (1987).

[56] J. I. Siepmann and D. Frenkel, *Molec. Phys.* **75**, 59 (1992).

[57] J. J. de Pablo, M. Laso, and U. W. Suter, *J. Chem. Phys.* **96**, 2394 (1992).

[58] K. Kremer and K. Binder, *Comput. Phys. Rep.* **7**, 259 (1988).

[59] G. C. A. M. Mooij, D. Frenkel, and B. Smit, *J. Phys. Cond. Mat.* **4**, 3053 (1992).

[60] G. C. A. M. Mooij, D. Frenkel, and B. Smit, *J. Phys. Cond. Mat.* **4**, L255 (1992).

[61] M. Laso, J. J. de Pablo, and U. W. Suter, *J. Chem. Phys.* **97**, 2817 (1992).

[62] D. Wu, D. Chandler, and B. Smit, *J. Phys. Chem.* **96**, 4077 (1992).

[63] W. G. Hoover and F. H. Ree, *J. Chem. Phys.* **47**, 4873 (1967).

[64] J. P. Hansen and I. R. McDonald, *Theory of Simple Liquids*, Academic Press, London, 2nd edition, 1986.

[65] D. Frenkel and A. J. C. Ladd, *J. Chem. Phys.* **81**, 3188 (1984).

[66] D. Frenkel and B. M. Mulder, *Molec. Phys.* **55**, 1171 (1985).

[67] A. Stroobants, H. N. W. Lekkerkerker, and D. Frenkel, *Phys. Rev. A* **36**, 2929 (1987).

[68] D. Frenkel, H. N. W. Lekkerkerker, and A. Stroobants, *Nature* **332**, 822 (1988).

[69] J. L. LEBOWITZ and J. S. ROWLINSON, *J. Chem. Phys.* **41**, 133 (1964).

[70] T. BIBEN and J. P. HANSEN, *Phys. Rev. Lett.* **66**, 2215 (1991).

[71] D. FRENKEL and A. A. LOUIS, *Phys. Rev. Lett.* **68**, 3363 (1992).

[72] In fact, we can perform the same mapping for more extreme size-ratios, as long as the edges of the large particles exclude an integer number of small-particle sites.

[73] T. D. LEE and C. N. YANG, *Phys. Rev.* **87**, 410 (1952).

[74] L. ONSAGER, *Phys. Rev.* **65**, 117 (1944).

[75] L. E. REICHL, *A modern course in Statistical Physics*, University of Texas Press, Austin, 1980.

[76] P. J. FLORY, *Principles of Polymer Chemistry*, Cornell University Press, Ithaca, 1953.

[77] K. F. FREED and M. G. BAWENDI, *J. Chem. Phys.* **93**, 2194 (1989).

[78] A. BAUMGÄRTNER, in *Application of the Monte Carlo method in Statistical Physics*, edited by K. BINDER, pp. 145–180, Springer, Berlin, 1984.

[79] D. H. NAPPER, *Polymeric Stabilization of Colloidal Dispersions*, Academic Press, 1983.

[80] B. VINCENT, J. EDWARDS, S. EMMETT, and R. GROOT, *Coll. Surf.* **31**, 267 (1988).

[81] D. FRENKEL, *J. Phys. Cond. Mat.* **2**, SA265 (1990).

[82] E. J. MEIJER and D. FRENKEL, *Phys. Rev. Lett.* **67**, 1110 (1991).

[83] G. C. A. M. MOOIJ and D. FRENKEL, *Molec. Phys.* **74**, 41 (1991).

[84] D. FRENKEL, *Physica A* **176**, 54 (1991).

[85] D. M. CEPERLEY and E. L. POLLOCK, *Phys. Rev. Lett.* **56**, 351 (1986).

[86] A. ALAVI and D. FRENKEL, *J. Chem. Phys.* **97**, 9249 (1992).

[87] D. J. STEVENSON and E. E. SALPETER, *The Astrophysical Journal Supplement Series* **35**, 221 (1977).

Thermodynamic Constraints

M. Ferrario
Dipartimento di Fisica
Università di Messina
Sezione di Fisica Teorica
C.P. 50
98166 S. Agata di Messina
Italy

ABSTRACT. In theoretical developments, thermodynamic constraints are introduced by putting the system of interest in contact with some other virtually infinite system, the 'reservoir', with a coupling vanishingly small in the thermodynamic limit. Neither of these 'infinite' conditions can be reproduced in molecular dynamics simulations where the time evolution of an isolated system with a finite number of degrees of freedom is numerically integrated, producing trajectories representative of the microcanonical ensemble. Several ways have been proposed to overcome this limitation. Here the case of MD simulations at constant temperature and/or pressure will be treated within the extended system framework introduced by Andersen in his 1980 seminal paper, and later generalised mainly by Nosé. The N-particle physical system of interest is put in contact with external reservoirs, which are, in contrast to theoretical infinite ones, represented just by a few degrees of freedom. The equations of motion for the extended system are chosen in such a way that the dynamical trajectory in the phase space of the system of interest is representative of the desired ensemble. Moreover the coupling is non-linear yielding good ergodic properties and can be chosen to be weak enough to leave the dynamical properties of the system of interest unaltered.

1. Introduction

In standard molecular dynamics simulation, we usually consider the time evolution of a classical system composed of a fixed number N of particles in a periodically repeated volume V, whose shape and size is fixed. The state of the system is completely determined by the $3N$ coordinates $\{\mathbf{r}^N\} = \{\mathbf{r}_i, i = 1, \ldots, N\}$ and $3N$ momenta $\{\mathbf{p}^N\} = \{\mathbf{p}_i, i = 1, \ldots, N\}$, and its dynamics by the hamiltonian $\mathcal{H}$

$$\mathcal{H}(\mathbf{r}^N, \mathbf{p}^N) = \sum_{i=1}^{N} \frac{\mathbf{p}_i^2}{2m_i} + \mathcal{V}(\mathbf{r}^N) = \mathcal{K}(\mathbf{p}^N) + \mathcal{V}(\mathbf{r}^N) \qquad (1)$$

M.P. Allen and D.J. Tildesley (eds.), Computer Simulation in Chemical Physics, 153–171.

where m_i is the mass of the ith particle and $\mathcal{V}(\mathbf{r}^N)$ is the potential energy. In the absence of external perturbations, the total energy $E = \mathcal{H}(\mathbf{r}^N, \mathbf{p}^N)$ of the system is conserved. Therefore, averaging over the time and making use of the ergodic hypothesis, one can obtain macroscopic properties relative to constant N, V, E conditions, i.e. corresponding to the microcanonical ensemble of statistical mechanics. In particular, the temperature and pressure of the simulation cannot be fixed but can only be estimated by making use of the equipartition theorem

$$T = \frac{2}{3Nk_{\mathrm{B}}}\left\langle \sum_{i=1}^{N} \frac{\mathbf{p}_i^2}{2m_i} \right\rangle_t \tag{2}$$

and the virial theorem

$$P = \frac{1}{3V}\left\langle \sum_{i=1}^{N} \frac{\mathbf{p}_i^2}{m_i} + \mathbf{r}_i \cdot \mathbf{F}_i \right\rangle_t , \quad \mathbf{F}_i = -\frac{\partial \mathcal{V}(\mathbf{r}^N)}{\partial \mathbf{r}_i} \tag{3}$$

where $\langle \ldots \rangle_t$ stands for the average over the finite time trajectory generated by numerical integration of the equations of motion. Therefore only after the simulation is done can one know at what temperature and pressure it has been carried out. When studying the phase behaviour of a system this can be a major drawback of molecular dynamics with respect to the Monte Carlo method, which, however, does not yield any dynamical information.

From statistical mechanics we know that to produce a new ensemble one has simply to couple weakly the system of interest with a suitable reservoir. The way this is achieved theoretically, however, is not applicable to molecular dynamics simulation. The reservoir is usually a virtually infinite system with many more degrees of freedom than the system of interest, and the coupling is done through surfaces, becoming infinitely small in the thermodynamic limit. In simulation we really do not want to spend time integrating non-interesting degrees of freedom, and periodic boundary conditions are usually employed just to get rid of unwanted surfaces when studying bulk samples.

The key idea of Andersen's 1980 paper [1] is that MD reservoirs can be represented by only one or a few degrees of freedom, and the coupling can be applied uniformly to all particles. This is done by inventing new equations of motion for the extended system, composed of the system of interest plus the required reservoirs. These equations are derived by a Lagrangian, or in Hamilton's canonical form, only in a *virtual* system of coordinates and become non-canonical when expressed in term of the *real* variables of the system of interest. The particular form of these equations is chosen in such a way that the microcanonical equilibrium distribution produced by a constant-energy trajectory of the extended system reduces to the desired distribution when the coordinates and momenta of the system of interest are considered, that is after integration over the extra reservoir variables.

The extended system lagrangian, introduced by Andersen, was designed to produce a constant pressure, constant enthalpy simulation by adding one extra variable, representing the MD cell volume, which is therefore free to fluctuate, and the coupling is made through a dynamical equation of motion for the cell volume variable. When the cell volume changes there is a uniform spatial scaling of the particle positions.

In the same paper a constant-temperature method was also introduced. This, however, was based on a different mechanism in which the coupling between the heat reservoir and the system of interest is stochastic and is implemented by randomly sampling particle velocities from a maxwellian distribution. This method is of relevance when dealing with complex systems where there is a poor coupling between various degrees of freedom, for example molecular internal vibrational and translational or rotational degrees of freedom, and it is probably the best method of achieving adequate equilibrium sampling in these systems.

However we will not discuss here either the stochastic method or the constraint methods [2, 3], the latter having acquired their major application merits in non-equilibrium molecular dynamics simulations. We shall instead focus our attention on the method first proposed by Nosé [4], and later clarified by Hoover [5] and Nosé himself [6], referring the interested reader to his latest review article for a more detailed and fully comprehensive treatment of constant-temperature molecular dynamics methods [7]. Once again the reservoir is described by a single degree of freedom, which however does not have a simple physical meaning [8], and the coupling contains a scaling term which applies uniformly to all particle velocities and is in fact a scaling on simulation time. Hoover's approach simplifies matters as it does not require the use of *virtual* variables and scaling concepts and also makes a direct link to constraint methods.

Finally the so-called Parrinello-Rahman method [9] and the extension to molecular systems [10–12] deserve to be mentioned on their own. In the first method the pressure reservoir is represented by a set of nine variables which describe the complete geometry of the MD cell, thus allowing fluctuations of both its volume and shape. This is essential in the study of phase stability and phase transformations in solid systems. For molecular systems the key point lies in the application of the coupling of the volume fluctuations to the molecular centres of mass rather than to each individual atomic position [11].

All the methods mentioned so far share the property that it can be demonstrated that the correct thermodynamical ensemble averages are produced. Other methods have been proposed [13] for which a satisfactory demonstration from the point of view of statistical mechanics does not exist.

2. Andersen's demon

Rather than follow the original demonstration by Andersen for the NPH ensemble we shall combine it with the Nosé derivation and consider directly the case of the NPT ensemble. As a word of caution, it must be mentioned that different lagrangians can be set up, all leading to the same equilibrium distribution but with different definitions and dynamics for the reservoir variables.

2.1. VIRTUAL-VARIABLE REPRESENTATION

In the constant-pressure molecular dynamics method the volume V of the MD box is allowed to fluctuate in time. Andersen replaces the atomic coordinates $\{\mathbf{r}_i, i = 1, \ldots, N\}$ with *virtual* scaled coordinates $\{\boldsymbol{\rho}_i, i = 1, \ldots, N\}$ defined in the following way.

$$\boldsymbol{\rho}_i = V^{-1/3}\mathbf{r}_i. \tag{4}$$

Each component of $\{\boldsymbol{\rho}_i, i = 1, ..., 3N\}$ is therefore a dimensionless number between zero and one (taking care to consider always the image inside the MD box).

Similarly Nosé introduces a scaled time τ which is related to the *real* one through the reservoir coordinate s in the following way

$$\mathrm{d}\tau = s\mathrm{d}t. \tag{5}$$

This implies some caution is needed when considering relations between time derivatives. In fact

$$\frac{\mathrm{d}\boldsymbol{\rho}_i}{\mathrm{d}\tau} = \frac{\mathrm{d}\left(\mathbf{r}_i V^{-1/3}\right)}{s\mathrm{d}t} \Rightarrow \frac{\mathrm{d}\mathbf{r}_i}{\mathrm{d}t} = V^{1/3} s \frac{\mathrm{d}\boldsymbol{\rho}_i}{\mathrm{d}\tau} + \frac{\mathbf{r}_i}{3V}\frac{\mathrm{d}V}{\mathrm{d}t}. \tag{6}$$

We can now introduce the following Lagrangian for the extended system [14]

$$\begin{aligned}\mathcal{L} = {} & \sum_{i=1}^{N} \frac{1}{2} m_i Q^{2/3} S^2 \dot{\boldsymbol{\rho}}_i \cdot \dot{\boldsymbol{\rho}}_i - \mathcal{V}(\{\boldsymbol{\rho}_i Q^{1/3}, i = 1, \ldots, N\}) \\ & + \frac{1}{2} S^2 M_Q \dot{Q}^2 - P_{\mathrm{ext}} Q + \frac{1}{2} M_S \dot{S}^2 - g k_{\mathrm{B}} T \ln S \end{aligned} \tag{7}$$

in which the two new variables Q and S appear. If we interpret Q as the volume V then the two terms on the first line are just the lagrangian of the system of interest. The second line contains the new terms. First a kinetic energy term, followed by a potential energy term for the motion of Q in which two constants appear, the inertia factor M_Q and the external pressure P_{ext}. The third term is the kinetic term for the Nosé variable S, and the fourth the corresponding potential energy term. The constants appearing there are another inertia factor M_S and the temperature of the reservoir with Boltzmann constant $k_B T$, while g is the number of degrees of freedom which are in contact with the heat reservoir and, as shown later, it depends on the system studied.

Let us recall Andersen's physical interpretation of the Q terms. Imagine that the fluid to be simulated can be compressed in a container whose volume is changed by a piston. Thus M_Q represents the mass of the piston whose motion is described by the coordinate Q, and $P_{\text{ext}} Q$ is the potential derived from an external pressure P_{ext} acting on the piston. The piston is a *virtual* one and produces an isotropic expansion or contraction of the fluid. This interpretation is not completely consistent as, writing down the expected particle velocities

$$\dot{\mathbf{r}}_i = \dot{\boldsymbol{\rho}}_i Q^{-1/3} + \frac{1}{3} \boldsymbol{\rho}_i \dot{Q} Q^{-2/3} \tag{8}$$

one finds that the second term of this derivative does not appear in the kinetic terms of the lagrangian.

2.2. HAMILTONIAN EQUATIONS OF MOTION

However eqn (7) gives a well-defined Lagrangian for the extended system in terms of *virtual* coordinates. The next step is therefore to derive from it the conjugate momenta. First the momenta conjugate to the particle coordinates $\boldsymbol{\rho}_i$ will be denoted $\boldsymbol{\pi}_i$

$$\boldsymbol{\pi}_i = \frac{\partial \mathcal{L}}{\partial \dot{\boldsymbol{\rho}}_i} = m_i S^2 Q^{2/3} \dot{\boldsymbol{\rho}}_i . \tag{9}$$

The momentum conjugate to Q will be denoted Π_Q

$$\Pi_Q = \frac{\partial \mathcal{L}}{\partial \dot{Q}} = M_Q S^2 \dot{Q} \tag{10}$$

and the momentum conjugate to S will be denoted Π_S

$$\Pi_S = \frac{\partial \mathcal{L}}{\partial \dot{S}} = M_S \dot{S} . \tag{11}$$

The hamiltonian of the extended system can now be written down

$$\begin{aligned} \mathcal{H} &= \sum_{i=1}^{N} \dot{\boldsymbol{\rho}}_i \cdot \boldsymbol{\pi}_i + \dot{Q} \Pi_Q + \dot{S} \Pi_S - \mathcal{L} \\ &= \sum_{i=1}^{N} \frac{\boldsymbol{\pi}_i \cdot \boldsymbol{\pi}_i}{2 m_i S^2 Q^{2/3}} + \mathcal{V}(\{\boldsymbol{\rho}_i Q^{1/3}, i = 1, \ldots, N\}) \\ &\quad + \frac{\Pi_Q^2}{2 M_Q S^2} + P_{\text{ext}} Q + \frac{\Pi_S^2}{2 M_S} + g k_B T \ln S \end{aligned} \tag{12}$$

The equations of motion for the extended system in *virtual* variables can now be derived in the canonical way from the hamiltonian. First the equations of motion for the atomic *virtual* coordinates and momenta in terms of derivatives of the *virtual* time coordinate can be written

$$\begin{aligned}\frac{\mathrm{d}\boldsymbol{\pi}_i}{\mathrm{d}\tau} &= -\frac{\partial\mathcal{H}}{\partial\boldsymbol{\rho}_i} = -Q^{1/3}\frac{\partial\mathcal{V}(\{\boldsymbol{\rho}Q^{1/3}\})}{\partial(\boldsymbol{\rho}_i Q^{1/3})} = Q^{1/3}\mathbf{F}_i \\ \frac{\mathrm{d}\boldsymbol{\rho}_i}{\mathrm{d}\tau} &= \frac{\partial\mathcal{H}}{\partial\boldsymbol{\pi}_i} = \frac{\boldsymbol{\pi}_i}{m_i S^2 Q^{2/3}}\end{aligned} \tag{13}$$

where $\mathbf{F}_i$ is the total force on atom i

$$\mathbf{F}_i = -\frac{\partial\mathcal{V}(\{\mathbf{r}_1, \mathbf{r}_2, ..., \mathbf{r}_N\})}{\partial\mathbf{r}_i}. \tag{14}$$

The equations of motion for Q and its conjugate momentum are

$$\begin{aligned}\frac{\mathrm{d}\Pi_Q}{\mathrm{d}\tau} &= -\frac{\partial\mathcal{H}}{\partial Q} = \frac{1}{3Q}\sum_{i=1}^{N}\left[\frac{\boldsymbol{\pi}_i\cdot\boldsymbol{\pi}_i}{m_i S^2 Q^{2/3}} + Q^{1/3}\mathbf{F}_i\cdot\boldsymbol{\rho}_i\right] - P_{\mathrm{ext}} \\ \frac{\mathrm{d}Q}{\mathrm{d}\tau} &= \frac{\partial\mathcal{H}}{\partial\Pi_Q} = \frac{\Pi_Q}{M_Q S^2}\end{aligned} \tag{15}$$

and the equations of motion for S and its conjugate momentum are

$$\begin{aligned}\frac{\mathrm{d}\Pi_S}{\mathrm{d}\tau} &= -\frac{\partial\mathcal{H}}{\partial S} = \sum_{i=1}^{N}\frac{\boldsymbol{\pi}_i\cdot\boldsymbol{\pi}_i}{m_i S^3 Q^{2/3}} + \frac{\Pi_Q^2}{M_Q S^3} - \frac{g k_{\mathrm{B}} T}{S} \\ \frac{\mathrm{d}S}{\mathrm{d}\tau} &= \frac{\partial\mathcal{H}}{\partial\Pi_S} = \frac{\Pi_S}{M_S}.\end{aligned} \tag{16}$$

2.3. EQUATIONS OF MOTION IN REAL SPACE

These equations of motion could be solved numerically to produce a time trajectory in the *virtual* phase space of the extended system. However this turns out not to be the best choice as integrating with a constant time step $\Delta\tau$ over the *virtual* time implies producing a trajectory unevenly spaced in the *real* time of the system of interest. It is therefore more rewarding to go back from the coordinates and momenta in *virtual* space $\{\boldsymbol{\rho}^N, \boldsymbol{\pi}^N, Q, \Pi_Q, S, \Pi_S; \tau\}$ to the coordinates and momenta in *real* space $\{\mathbf{r}^N, \mathbf{p}^N, V, P_V, s, P_s; t\}$ by means of the following transform relations

$$\begin{aligned}&\mathbf{r}_i = \boldsymbol{\rho}_i Q^{-1/3} \qquad V = Q \qquad s = S \\ &\mathbf{p}_i = \boldsymbol{\pi}_i / S Q^{1/3} \qquad P_V = \Pi_Q / S \qquad P_s = \Pi_S\end{aligned} \qquad \mathrm{d}t = \mathrm{d}\tau / s. \tag{17}$$

The equations of motion become

$$
\begin{aligned}
\dot{\mathbf{r}}_i &= \frac{\mathbf{p}_i}{m_i} + \mathbf{r}_i \frac{\dot{V}}{3V} \\
\dot{\mathbf{p}}_i &= \mathbf{F}_i - \mathbf{p}_i \frac{\dot{s}}{s} - \mathbf{p}_i \frac{\dot{V}}{3V} \\
\dot{V} &= \frac{P_V}{M_Q} \\
\dot{P}_V &= \frac{1}{3V} \sum_{i=1}^{N} \left[\frac{\mathbf{p}_i^2}{m_i} + \mathbf{F}_i \cdot \mathbf{r}_i \right] - P_{\text{ext}} - P_V \frac{\dot{s}}{s} \\
\dot{s} &= s \frac{P_s}{M_S} \\
\dot{P}_s &= \sum_{i=1}^{N} \left[\frac{\mathbf{p}_i^2}{m_i} \right] + \frac{P_V^2}{M_Q} - g k_{\text{B}} T
\end{aligned} \tag{18}
$$

where the dot stands for the time derivative with respect to t.

The Hamiltonian $\mathcal{H}$ in eqn (12) can also be transformed, but it must be noticed that because the transformation is not canonical, the function

$$
\mathcal{H} = \mathcal{K}(\mathbf{p}^N) + \mathcal{V}(\mathbf{r}^N) + \frac{P_V^2}{2M_Q} + P_{\text{ext}} V + \frac{P_s^2}{2M_S} + g k_{\text{B}} T \ln s \tag{19}
$$

obtained cannot be used to derive hamiltonian equations of motion for the *real* coordinates and momenta $\{\mathbf{r}^N, \mathbf{p}^N, V, P_V, s, P_s; t\}$, while it still remains a constant of motion for the time evolution defined by eqns (18).

3. Time and ensemble averages

Solving numerically eqns (18), one can produce a time trajectory for the extended system and therefore calculate time averages as in the standard molecular dynamics to estimate the macroscopic properties of the system of interest. We must ask ourselves what kind of averages are we calculating with respect to statistical mechanics ensembles. We know that our time trajectory corresponds, by invoking the ergodic hypothesis, to a microcanonical ensemble distribution for the extended system in the *virtual phase space*. We are, however, interested in the properties of the system of interest which can be quite generally expressed as ensemble averages of an observable $\mathcal{F}(\mathbf{r}^N, \mathbf{p}^N, V; t)$ The corresponding observable $\mathcal{G}$ in the *virtual space* is defined as

$$
\begin{aligned}
\mathcal{F}(\mathbf{r}^N, \mathbf{p}^N, V; t) &\equiv \mathcal{F}(Q^{1/3}\boldsymbol{\rho}^N, S^{-1}Q^{-1/3}\boldsymbol{\pi}^N, S, Q; \tau), \quad \tau = \tau(t) \\
&= \mathcal{G}(\boldsymbol{\rho}^N, \boldsymbol{\pi}^N, Q, \Pi_Q, S, \Pi_S; \tau).
\end{aligned} \tag{20}
$$

What can be readily calculated from the dynamical trajectory is a time average of the form

$$\langle \mathcal{F} \rangle_t = \lim_{t_I \to \infty} \frac{\int_0^{t_I} \mathrm{d}t \; \mathcal{F}(\mathbf{r}^N(t), \mathbf{p}^N(t), V(t))}{\int_0^{t_I} \mathrm{d}t} \tag{21}$$

which then corresponds to an average over the *virtual time* τ

$$= \lim_{\tau_I(t_I) \to \infty} \frac{\int_0^{\tau_I} \mathrm{d}\tau \; \mathcal{G}(\boldsymbol{\rho}^N, \boldsymbol{\pi}^N, Q, \Pi_Q, S, \Pi_S; \tau)/S(\tau)}{\int_0^{\tau_I} \mathrm{d}\tau \; 1/S(\tau)} = \frac{\langle \mathcal{G}/S \rangle_\tau}{\langle 1/S \rangle_\tau}. \tag{22}$$

The above equation sets the equivalence between the dynamical averages in the two pictures of *virtual* and *real* time. But we know that, if the motion of the system is ergodic, time averages over τ are equivalent to microcanonical ensemble averages

$$\frac{\langle \mathcal{G}/S \rangle_\tau}{\langle 1/S \rangle_\tau} \Leftrightarrow \frac{\langle \mathcal{G}/S \rangle_{\mathrm{mc}}}{\langle 1/S \rangle_{\mathrm{mc}}} = \langle \mathcal{F} \rangle_{\mathrm{Ensemble?}} \tag{23}$$

which can be made more explicit

$$\begin{aligned} \frac{\langle \mathcal{G}/S \rangle_{\mathrm{mc}}}{\langle 1/S \rangle_{\mathrm{mc}}} &= \left\{ \frac{\int \mathrm{d}\boldsymbol{\Omega} \; [1/S]\delta(\mathcal{H} - E)}{\int \mathrm{d}\boldsymbol{\Omega} \; \delta(\mathcal{H} - E)} \right\}^{-1} \\ &\quad \times \frac{\int \mathrm{d}\boldsymbol{\Omega} \; [\mathcal{G}(\boldsymbol{\rho}^N, \boldsymbol{\pi}^N, Q, \Pi_Q, S, \Pi_S)/S)]\delta(\mathcal{H} - E)}{\int \mathrm{d}\boldsymbol{\Omega} \; \delta(\mathcal{H} - E)} \\ &= \frac{\int \mathrm{d}\boldsymbol{\Omega} \; [\mathcal{G}(\boldsymbol{\rho}^N, \boldsymbol{\pi}^N, Q, \Pi_Q, S, \Pi_S)/S]\delta(\mathcal{H} - E)}{\int \mathrm{d}\boldsymbol{\Omega} \; [1/S]\delta(\mathcal{H} - E)}. \end{aligned} \tag{24}$$

Here we introduced a short notation for the integration variables $\mathrm{d}\boldsymbol{\Omega}$ in the *virtual* phase space. The transformation relating it to the *real* phase space integration variables $\mathrm{d}\boldsymbol{\Gamma}$ is characterised by a jacobian which is readily calculated

$$\begin{aligned} \int \mathrm{d}\boldsymbol{\Omega} &= \int \mathrm{d}\boldsymbol{\rho}^N \mathrm{d}\boldsymbol{\pi}^N \mathrm{d}Q \mathrm{d}\Pi_Q \mathrm{d}S \mathrm{d}\Pi_S \\ &= \int \mathrm{d}\mathbf{r}^N \mathrm{d}\mathbf{p}^N \mathrm{d}V \mathrm{d}P_V \mathrm{d}s \mathrm{d}P_s \; s^{3N+1} = \int \mathrm{d}\boldsymbol{\Gamma} \; s^{3N+1}. \end{aligned} \tag{25}$$

What we have to do is now to find out to which ensemble the *real* time average

$$\langle \mathcal{F} \rangle_t \Leftrightarrow \langle \mathcal{F} \rangle_{\mathrm{Ensemble?}} = \frac{\int \mathrm{d}\boldsymbol{\Omega} \; [\mathcal{G}(\boldsymbol{\rho}^N, \boldsymbol{\pi}^N, Q, \Pi_Q, S, \Pi_S)/S]\delta(\mathcal{H} - E)}{\int \mathrm{d}\boldsymbol{\Omega} \; [1/S]\delta(\mathcal{H} - E)} \tag{26}$$

corresponds, in the ergodic limit. To do this we have to transform the integrals in the above equations into integrals over the space $\{\mathbf{r}^N, \mathbf{p}^N, V\}$ and perform the integral over the extra reservoir variables. Having already expressed the required jacobian we must make explicit the transformation of the microcanonical delta function distribution $\delta(\mathcal{H} - E)$. We first observe that

$$\delta(f(s)) = \delta(s - s_0)/f'(s_0) \qquad f'(s_0) = gk_BT/s_0. \tag{27}$$

Here we should really have a sum over all possible zeroes of f, but in this case we can state that there is only one zero at

$$s_0 = \exp\{-(\mathcal{H}_T - E)/gk_BT\} \tag{28}$$

because the dependence on s is logarithmic. It is useful at this time to introduce the function $\mathcal{H}_T$

$$\mathcal{H}_T = \sum_{i=1}^{N} \frac{\mathbf{p}_i^2}{2m_i} + \mathcal{V}(\mathbf{r}^N) + \frac{P_V^2}{2M_Q} + P_{\text{ext}}V + \frac{P_s^2}{2M_S} \tag{29}$$

in terms of which the microcanonical delta function

$$\delta(\mathcal{H}(\mathbf{r}^N, \mathbf{p}^N, V, P_V, s, P_s) - E) \tag{30}$$

can be reformulated in a compact way

$$\frac{1}{gk_BT} \exp\{-(\mathcal{H}_T - E)/gk_BT\}\delta(s - s_0). \tag{31}$$

We can now write for the denominator in eqn (24)

$$\int \mathrm{d}\mathbf{\Omega}\ [1/S]\delta(\mathcal{H} - E) = \int \mathrm{d}\mathbf{\Gamma}\ [s^{3N+1}]\ \frac{s_0}{gk_BT}(1/s)\delta(s - s_0) \tag{32}$$

which after the integration over the variable s has been performed becomes

$$= \frac{1}{gk_BT} \int \mathrm{d}\mathbf{r}^N \mathrm{d}\mathbf{p}^N \mathrm{d}V \mathrm{d}P_V \mathrm{d}P_s\ \exp\left\{-\left(\frac{3N+1}{g}\right)\frac{\mathcal{H}_T - E}{k_BT}\right\}. \tag{33}$$

Equivalently we can dispose of the numerator which becomes

$$\int \mathrm{d}\mathbf{\Omega}\ [\mathcal{G}/S]\delta(\mathcal{H} - E) =$$
$$\frac{1}{gk_BT} \int \mathrm{d}\mathbf{r}^N \mathrm{d}\mathbf{p}^N \mathrm{d}V \mathrm{d}P_V \mathrm{d}P_s\ \mathcal{F} \exp\left\{-\left(\frac{3N+1}{g}\right)\frac{\mathcal{H}_T - E}{k_BT}\right\}. \tag{34}$$

We can now fix the number of degrees of freedom g. By choosing $g = 3N + 1$ we can finally write

$$\langle \mathcal{F} \rangle_{\text{Ensemble?}} = \frac{\int \mathrm{d}\mathbf{r}^N \mathrm{d}\mathbf{p}^N \mathrm{d}V \mathrm{d}P_V \mathrm{d}P_s\ \mathcal{F} \exp\{-\mathcal{H}_T/k_BT\}}{\int \mathrm{d}\mathbf{r}^N \mathrm{d}\mathbf{p}^N \mathrm{d}V \mathrm{d}P_V \mathrm{d}P_s\ \exp\{-\mathcal{H}_T/k_BT\}} \tag{35}$$

which can be further simplified observing that the observable $\mathcal{F}$ does not depend on the momenta P_V, P_s. After integration over these variables the constant temperature - constant pressure NPT ensemble average is obtained

$$\langle \mathcal{F} \rangle_{NPT} = \frac{\int \mathrm{d}\mathbf{r}^N \mathrm{d}\mathbf{p}^N \mathrm{d}V \; \mathcal{F}(\mathbf{r}^N, \mathbf{p}^N, V) \exp\left\{-[\mathcal{K}(\mathbf{p}^N) + \mathcal{V}(\mathbf{r}^N) + P_{\mathrm{ext}} V]/k_{\mathrm{B}}T\right\}}{\int \mathrm{d}\mathbf{r}^N \mathrm{d}\mathbf{p}^N \mathrm{d}V \; \exp\{-[\mathcal{K}(\mathbf{p}^N) + \mathcal{V}(\mathbf{r}^N) + P_{\mathrm{ext}} V]/k_{\mathrm{B}}T\}}. \tag{36}$$

Therefore we arrive at the conclusion that the time averages computed over the dynamical trajectory obtained by integrating eqns (18) numerically are equivalent, if the motion is ergodic, to the NPT ensemble averages

$$\langle \mathcal{F} \rangle_t \Leftrightarrow \langle \mathcal{F} \rangle_{NPT}. \tag{37}$$

4. Nosé-Hoover thermostat

In 1985 Hoover [5] reformulated Nosé equations for NVT simulations in a way which is considered simpler and easier for making further generalizations, as it avoids the need for introducing *virtual* variables. The Hoover procedure follows the demonstration used for the constraint methods [3]. Consider the distribution function $\varrho(\mathbf{r}^N, \mathbf{p}^N, \ldots)$ defined in the extended phase space $\mathbf{\Gamma} = \{\mathbf{r}^N, \mathbf{p}^N, \ldots\}$, where we leave undefined the extra variables for the moment. A generalised Liouville equation expresses the conservation of probability in this phase space,

$$\frac{\partial \varrho}{\partial t} + \frac{\partial}{\partial \mathbf{\Gamma}} \cdot (\dot{\mathbf{\Gamma}} \varrho) = 0. \tag{38}$$

The change of ϱ along a phase space trajectory is given by the total time derivative

$$\frac{\mathrm{d}\varrho}{\mathrm{d}t} = \frac{\partial \varrho}{\partial t} + \dot{\mathbf{\Gamma}} \cdot \frac{\partial}{\partial \mathbf{\Gamma}} \varrho. \tag{39}$$

When combined one obtains

$$\frac{\mathrm{d}\varrho}{\mathrm{d}t} = -\varrho \left(\frac{\partial}{\partial \mathbf{\Gamma}} \cdot \dot{\mathbf{\Gamma}} \right) \tag{40}$$

which in the case of hamiltonian dynamics equals zero and expresses Liouville's theorem. As we already said the equations of motion in *real* variables for the extended systems cannot be derived directly from a hamiltonian, and therefore we have to use eqn (40) to find the evolution of the distribution function ϱ. So the idea is to write down a set of equations of motion and then find out what they predict for the time dependence of the distribution function ϱ, by putting them in eqn (40).

Rather then simply repeating the demonstration for the case considered in section 2 we shall write the equations for constant temperature - constant pressure simulation with a varying box shape, following the generalisation of the Andersen method due to Parrinello and Rahman[9]. In this case the volume variable is replaced by the 3×3 matrix variable $\overleftrightarrow{h}$, shorthand for the three lattice vectors $(\mathbf{a}, \mathbf{b}, \mathbf{c})$ which describe the shape and volume of the MD box and, as a consequence, the periodic boundary conditions in a Bravais lattice notation.

It has been pointed out [10] that only six of these nine variables have an interesting meaning, when the three degrees of freedom describing the overall rotation of the MD box are extracted out. The situation is very simple in atomic cases where one can show that the equations of motion conserve symmetric matrices $\overleftrightarrow{h}$. This is no longer true for the case of molecular systems, when intramolecular distances are kept fixed and the coupling is via the molecular centres of mass. In order to avoid overall rotation of the MD box, which, especially for solids, would simply confuse the analysis of the trajectories of the system, two options have been used both of which reduce the number of extra variables to six. In one case [10] the matrix $\overleftrightarrow{h}$ is kept symmetric by disregarding the asymmetric contributions in the equations of motion, and in the other one [12] the variables are reduced to six by fixing the MD box orientation with respect to a *laboratory* frame, and specifically by choosing $\mathbf{a}$ parallel to the x-axis and $\mathbf{b}$ to lie in the xy plane. This choice results in an upper triangular matrix $\overleftrightarrow{h}$. Also it is possible to make other choices (for example only allowing fluctuations of the box which maintain a parallelepiped shape, which makes 3 degrees of freedom and the matrix is diagonal), therefore we shall generally leave undefined the number of degrees of freedom of the box, and will indicate them by g_h. When using the Parrinello-Rahman technique, special care must be taken in the application of periodic boundary conditions as will be clarified later on.

Let's start by writing down the new equations of motions for the set of variables $\mathbf{\Gamma} = \{\mathbf{r}^N, \mathbf{p}^N, \overleftrightarrow{h}, \overleftrightarrow{P}_h, \zeta\}$

$$
\begin{aligned}
\dot{\mathbf{r}}_i &= \frac{\mathbf{p}_i}{m_i} + \left[\overleftrightarrow{h}^{-1}\dot{\overleftrightarrow{h}}\right]\mathbf{r}_i \\
\dot{\mathbf{p}}_i &= \mathbf{F}_i - \zeta\mathbf{p}_i - \left[\overleftrightarrow{h}^{-1}\dot{\overleftrightarrow{h}}\right]\mathbf{p}_i \\
\left[\dot{\overleftrightarrow{P}}_h\right]_{xy} &= \sum_{z=1}^{3}\left[\overleftrightarrow{h}^{-1}\right]_{zx}\left[\sum_{i=1}^{N}\frac{\mathbf{p}_i\mathbf{p}_i}{m_i} + \mathbf{F}_i\mathbf{r}_i\right]_{zy} - \left[\overleftrightarrow{h}^{-1}\right]_{yx} P_{\mathrm{ext}} \det\overleftrightarrow{h} - \zeta\left[\overleftrightarrow{P}_h\right]_{xy} \\
\dot{\zeta} &= \frac{1}{M_S}\left\{\sum_{i=1}^{N}\frac{\mathbf{p}_i^2}{m_i} + \sum_{x,y=1}^{3}\frac{1}{M_Q}\left[\overleftrightarrow{P}_h^2\right]_{xy} - g k_{\mathrm{B}} T\right\}
\end{aligned}
\tag{41}
$$

$$\left[\dot{\overleftrightarrow{h}}\right]_{xy} = \frac{1}{M_Q}\left[\overleftrightarrow{P}_h\right]_{xy}$$

where, following Hoover, the role of the Nosé coordinate s and the momentum P_s is played by the single coordinate ζ which is related to them by

$$\zeta = \frac{\mathrm{d}\ln s}{\mathrm{d}t} = \frac{\dot{s}}{s} = \frac{P_s}{M_S}. \tag{42}$$

The equations of motion do not depend on the variable s explicitly, but s appears when we consider the constant of the motion $\mathcal{H}$

$$\mathcal{H} = \mathcal{H}_T + g k_{\mathrm{B}} T \ln s \tag{43}$$

where $\mathcal{H}_T$ has been defined as

$$\mathcal{H}_T = \sum_{i=1}^{N} \frac{\mathbf{p}_i^2}{2m_i} + \mathcal{V}(\mathbf{r}^N) + \sum_{x,y=1}^{3} \frac{1}{2M_Q}\left[\overleftrightarrow{P}_h^2\right]_{xy} + P_{\mathrm{ext}} \det \overleftrightarrow{h} + \frac{M_S\zeta^2}{2}. \tag{44}$$

To demonstrate that $\mathcal{H}$ is indeed a conserved quantity let's first see what this requires for the time dependence of $\mathcal{H}_T$

$$\frac{\mathrm{d}\mathcal{H}}{\mathrm{d}t} = 0 \Leftrightarrow \frac{\mathrm{d}\mathcal{H}_T}{\mathrm{d}t} = -\frac{\mathrm{d}}{\mathrm{d}t}(g k_{\mathrm{B}} T \ln s) = -g k_{\mathrm{B}} T \zeta. \tag{45}$$

This can be readily verified

$$\begin{aligned}
\frac{\mathrm{d}\mathcal{H}_T}{\mathrm{d}t} &= \sum_{i=1}^{N}\left\{\dot{\mathbf{p}}_i \cdot \frac{\partial \mathcal{H}_T}{\partial \mathbf{p}_i} + \dot{\mathbf{r}}_i \cdot \frac{\partial \mathcal{H}_T}{\partial \mathbf{r}_i}\right\} + \sum_{x,y=1}^{3}\left[\dot{\overleftrightarrow{P}}_h\right]_{xy} \frac{\partial \mathcal{H}_T}{\partial \left[\overleftrightarrow{P}_h\right]_{xy}} \\
&\quad + \sum_{x,y=1}^{3}\left[\dot{\overleftrightarrow{h}}\right]_{xy} \frac{\partial \mathcal{H}_T}{\partial \left[\overleftrightarrow{h}\right]_{xy}} + \dot{\zeta}\frac{\partial \mathcal{H}_T}{\partial \zeta} \\
&= -g k_{\mathrm{B}} T \zeta\,.
\end{aligned} \tag{46}$$

The equation for the probability distribution can now be calculated

$$\begin{aligned}
\frac{\mathrm{d}\varrho}{\mathrm{d}t} &= -\varrho\left(\frac{\partial}{\partial \mathbf{\Gamma}} \cdot \dot{\mathbf{\Gamma}}\right) \\
&= -\varrho\left\{\left[\sum_{i=1}^{N} \frac{\partial \dot{\mathbf{p}}_i}{\partial \mathbf{p}_i} + \frac{\partial \dot{\mathbf{r}}_i}{\partial \mathbf{r}_i}\right] + \frac{\partial \dot{\overleftrightarrow{P}}_h}{\partial \overleftrightarrow{P}_h} + \frac{\partial \dot{\overleftrightarrow{h}}}{\partial \overleftrightarrow{h}} + \frac{\partial \dot{\zeta}}{\partial \zeta}\right\} \\
&= \varrho(3N + g_h)\zeta = -\frac{1}{k_{\mathrm{B}} T}\frac{(3N + g_h)}{g}\frac{\mathrm{d}\mathcal{H}_T}{\mathrm{d}t}\varrho
\end{aligned} \tag{47}$$

and this corresponds, by choosing $g = 3N + g_h$, to an equilibrium solution for ϱ of canonical form

$$\varrho \propto \exp\left\{-\frac{\mathcal{H}_T}{k_{\mathrm{B}} T}\right\}. \tag{48}$$

5. Molecular systems

Recently, Melchionna *et al.* [15] have proposed a set of equations in which they introduce a piston variable η for the MD box dynamics in the same spirit as Hoover's heat bath variable ζ. We shall only show their equations in the special case of systems composed of molecules with a description in terms of cartesian components and holonomic constraints.

In the presence of intramolecular geometrical constraints, the piston variable η will be applied only to the center of mass of the single molecules [11], while the heat bath variable ζ applies to all degrees of freedom [12].

Consider a system composed of N molecules, each one made of n atoms. The n atoms are bound between them with K rigid constraints

$$\sigma_l^\alpha(\{\mathbf{r}_{i\alpha}\}) = 0 \qquad l = 1, \ldots, K; \quad \alpha = 1, \ldots, N \tag{49}$$

where here and in the following the greek indices refer to the molecules. The constraining forces acting on the atoms can be written [11]

$$G_{i\alpha} = -\sum_{l=1}^{K} \lambda_l^\alpha \frac{\partial \sigma_l^\alpha}{\partial \mathbf{r}_{i\alpha}} \tag{50}$$

where $\{\lambda_l^\alpha\}$ are the associated Lagrange multipliers.

To deal with the possible overall rotations of the MD box induced by the fact that the pressure tensor is no longer symmetrical [10] for systems with fixed intramolecular distances, the equations of motion are written for an upper triangular matrix $\overleftrightarrow{\eta}$. By calling $\overleftrightarrow{P}_{\text{int}}$ the upper triangular part of the pressure tensor, one can write the following equations of motion which will evolve the matrix $\overleftrightarrow{\eta}$ in time keeping its upper triangular form exactly, thanks to the group properties of these matrices.

$$\begin{aligned}
\dot{\mathbf{r}}_{i\alpha} &= \frac{\mathbf{p}_{i\alpha}}{m_{i\alpha}} + \overleftrightarrow{\eta}(\mathbf{R}_\alpha - \mathbf{R}_0) \\
\dot{\mathbf{p}}_{i\alpha} &= \mathbf{F}_{i\alpha} + \mathbf{G}_{i\alpha} - \left(\zeta \mathbf{p}_{i\alpha} + \frac{m_{i\alpha}}{M_\alpha}\overleftrightarrow{\eta}\mathbf{P}_\alpha\right) \\
\dot{\zeta} &= \nu_T^2 \left(\frac{T_{\text{int}}(t)}{T_{\text{ext}}} - 1\right) \\
\dot{\overleftrightarrow{\eta}} &= \frac{\nu_P^2}{N k_B T} V(\overleftrightarrow{P}_{\text{int}}(t) - P_{\text{ext}}\overleftrightarrow{I}) \\
\dot{\overleftrightarrow{h}} &= \overleftrightarrow{\eta}\overleftrightarrow{h}
\end{aligned} \tag{51}$$

where

$$M_\alpha = \sum_i m_{i\alpha}$$

$$\begin{aligned}
\mathbf{R}_\alpha &= (M_\alpha)^{-1}\sum_i m_{i\alpha}\mathbf{r}_{i\alpha} \\
\mathbf{P}_\alpha &= \sum_i \mathbf{p}_{i\alpha} \\
\mathbf{R}_0 &= \left(\sum_\beta\sum_j m_{j\beta}\right)^{-1}\sum_\alpha\sum_i m_{i\alpha}\mathbf{r}_{i\alpha}.
\end{aligned} \tag{52}$$

The inertia factors M_S and M_Q have been rephrased in terms of a thermostating rate ν_T and a barostating rate ν_P respectively. The instantaneous *internal* pressure tensor and temperature are defined as

$$\begin{aligned}
\overleftrightarrow{P}_{\rm int}(t) &= \frac{1}{3V}\sum_\alpha\left[\frac{\mathbf{P}_\alpha\mathbf{P}_\alpha}{M_\alpha} + \mathbf{R}_\alpha\sum_i \mathbf{F}_{i\alpha}\right] \\
T_{\rm int}(t) &= \frac{1}{3nNk_{\rm B}}\sum_\alpha\sum_i\frac{p_{i\alpha}^2}{m_{i\alpha}}.
\end{aligned} \tag{53}$$

The first expression of eqn (53) depends only on the coordinates of the molecular centers of mass and contains only forces arising from intermolecular interactions. It is known as the *molecular* formulation of the pressure tensor in contrast with the corresponding *atomic* expression

$$\overleftrightarrow{P}_{\rm int}(t) = \frac{1}{3V}\sum_j\left[\frac{\mathbf{p}_j\mathbf{p}_j}{m_j} + \mathbf{r}_j\mathbf{F}_j\right] \qquad j = \{i\alpha\} = 1, nN \tag{54}$$

which has been shown [16] to be equivalent on average (and strictly identical for molecules with no internal degrees of freedom). The kinetic term in eqn (54) is different from the one in eqn (53) which represents only the translational part of the kinetic energy, and, correspondingly, the force term is different as it contains also all possible intramolecular contributions, including constraint forces. When using the constraint method, eqn (54) is attractive as it can be readily evaluated without any recourse to the concept of molecular assembly. In particular it would apply unchanged also when we replace some or all constraints by stiff springs or other kinds of intramolecular interactions.

The set of equations (51) produces a dynamics which is slightly different from what we would get by using box variables $\overleftrightarrow{h}$ and $\overleftrightarrow{P}_h$ as in the preceding section. Accordingly, when we write down the corresponding $\mathcal{H}_T$ function we notice that the kinetic term for the box variable is different and in particular the value of the constant g, i.e. the number of degrees of freedom thermostatted, is equal to $3nN$.

$$\mathcal{H}_T = \mathcal{K}(\mathbf{p}^N) + \mathcal{V}(\mathbf{r}^N) + P_{\rm ext}\det\overleftrightarrow{h} + \frac{3Nk_{\rm B}T}{2}\left(\frac{\zeta^2}{\nu_T^2} + \frac{\overleftrightarrow{\eta}^T : \overleftrightarrow{\eta}}{3\nu_P^2}\right) \tag{55}$$

where $\overleftrightarrow{A} : \overleftrightarrow{B}$ is the contraction of $\overleftrightarrow{A}$ and $\overleftrightarrow{B}$, $\overleftrightarrow{A} : \overleftrightarrow{B} = \mathrm{Tr}(\overleftrightarrow{A}\overleftrightarrow{B})$.

However it is possible to demonstrate that the equilibrium distribution function assumes the same canonical form in both cases.

$$\varrho(\mathbf{r}^{nN}, \mathbf{p}^{nN}, \overleftrightarrow{h}, \overleftrightarrow{\eta}, \zeta) \propto \exp\left\{-\frac{\mathcal{H}_T}{k_\mathrm{B}T}\right\}. \tag{56}$$

6. Numerical considerations

Different schemes can be used to integrate the equations of motion numerically for the various cases considered. A word of caution is needed when one chooses the Verlet central difference scheme. In the equations under consideration there is, in fact, a key difference from the usual hamiltonian (better say lagrangian) equations of motion used in the standard microcanonical molecular dynamics method. If we re-write eqns (18) in lagrangian form using coordinates and first time derivatives we obtain the set of second order differential equations

$$\begin{aligned}
\ddot{\mathbf{r}}_i &= \frac{1}{m_i}(\mathbf{F}_i - \zeta\mathbf{p}_i) + \mathbf{r}_i\left\{\frac{\ddot{V}}{3V} - 2\left(\frac{\dot{V}}{3V}\right)^2\right\} \\
\ddot{V} &= \frac{1}{3VM_Q}\sum_{i=1}^{N}\left[\frac{\mathbf{p}_i^2}{m_i} + \mathbf{F}_i\cdot\mathbf{r}_i\right] - \frac{P_\mathrm{ext}}{M_Q} - \zeta\dot{V} \\
\frac{\mathrm{d}^2\ln s}{\mathrm{d}t^2} &= \frac{1}{M_S}\left[\sum_{i=1}^{N}\frac{\mathbf{p}_i^2}{m_i} + M_Q\dot{V}^2 - gk_\mathrm{B}T\right]
\end{aligned} \tag{57}$$

where

$$\mathbf{p}_i = m_i\dot{\mathbf{r}}_i - m_i\mathbf{r}_i\frac{\dot{V}}{3V} \qquad \frac{\mathrm{d}\ln s}{\mathrm{d}t} = \zeta. \tag{58}$$

It is immediately evident that the accelerations contain velocity-dependent terms which make it impossible to use the standard Verlet algorithm where velocities and coordinates are not predicted at the same time. Let us define for the sake of compactness a unique set of $3N + 2$ variables

$$\{x_k\} = \{\mathbf{r}_i, V, \ln s\}. \tag{59}$$

If we know the set of initial conditions, that is coordinates and velocities at time t, and we use them to calculate the acceleration at the same time

$$\{x_k(t)\}, \qquad \{\dot{x}_k(t)\}, \qquad \ddot{x}_k(t) = \ddot{x}_k(\{x_k(t)\}, \{\dot{x}_k(t)\}) \tag{60}$$

we can write down the following 3-step integration scheme

(i) $x_k(t + \Delta t) = 2x_k(t) - x_k(t - \Delta t) + \ddot{x}_k(t)\Delta t^2$.

(ii) $\dot{x}_k(t) = [x_k(t + \Delta t) - x_k(t - \Delta t)]/2\Delta t$.

(iii) $\dot{x}_k(t+\Delta t) = \dot{x}_k(t-\Delta t) + 2\ddot{x}_k(t)\Delta t$.
As in the standard Verlet algorithm we also need to know the coordinates $\{x_k(t-\Delta t)\}$ at the previous time step, and in this case we need the velocities $\{\dot{x}_k(t-\Delta t)\}$ too. This simply means that the initial startup needs a little extra care. The scheme above predicts both coordinates and velocities at the same time and can be used as is. However differences between *Verlet* velocities (ii) and *predicted* ones (iii) can result in small drifts of the conserved quantities and in loss of time reversal accuracy with respect to the standard Verlet scheme. Such inconvenience is relevant only for very long simulations (more than hundred thousand steps) and when time reversal accuracy is needed for special purposes. This can be easily achieved iterating the first two expressions. The use of expression (iii) as a first guess ensures a quick convergence in few steps.

Andersen proposed an alternative scheme [13] which has been shown to be completely equivalent to the Verlet algorithm for accelerations with no velocity-dependent terms. This is known as the velocity version of the Verlet algorithm and solves the problem of the initial startup needing an initial condition written in terms of coordinates and velocities at the same time, rather than two sets of coordinates at subsequent times as with Verlet.

However it must be noted that when accelerations depend on velocity even this scheme cannot be applied without modification. The modifications can be expressed as follows. Let us start with the standard scheme which reads
(i) $x_k(t+\Delta t) = x_k(t) + \dot{x}_k(t)\Delta t + \ddot{x}_k(t)\Delta t^2/2$.
(ii) $\dot{x}_k(t+\Delta t) = \dot{x}_k(t) + [\ddot{x}_k(t+\Delta t) + \ddot{x}_k(t)]\Delta t/2$
where in our case the accelerations at time $t+\Delta t$ depend on the velocities at the same time

$$\ddot{x}_k(t+\Delta t) = \ddot{x}_k(\{x_k(t+\Delta t)\}, \{\dot{x}'_k(t+\Delta t)\}) \tag{61}$$

and one needs to predict these velocities, in the same spirit as above for Verlet's algorithm, with the extra formula
(iii) $\dot{x}'_k(t+\Delta t) = \dot{x}_k(t-\Delta t) + 2\ddot{x}_k(t)\Delta t$.
If necessary expression (ii) above can be readily iterated to get higher accuracy in the numerical integration.

Another numerical problem arises specifically in the Parrinello-Rahman method. In this case the MD box has no longer a cubic shape and the algorithm to compute the minimum image in the presence of periodic boundary conditions has to be modified. The easiest possibility is to use virtual coordinates $\boldsymbol{\rho}_i$. In this space, in fact, the MD box is again a cube of unit size. Given a particle k, one has that each component ξ of the distance $\mathbf{d}_j^{(\xi)}$ computed as follows with any other particle j

$$\mathbf{d}_j^{(\xi)} = (\rho_k^{(\xi)} - \rho_j^{(\xi)}) - \mathtt{NINT}(\rho_k^{(\xi)} - \rho_j^{(\xi)}) \tag{62}$$

Equation (65) can be reorganized after adding and subtracting terms $(x_{i\alpha} - x_{j\beta})$

$$T^{xz} = \sum_{\alpha,\beta>\alpha=1}^{N} \sum_{i,j=1}^{n} (x_{i\alpha} - x_{j\beta}) F^z_{i\alpha j\beta} + \sum_{\alpha,\beta>\alpha=1}^{N} \sum_{i,j=1}^{n} [(x_\alpha - x_{i\alpha}) - (x_\beta - x_{j\beta})] F^z_{i\alpha j\beta} \qquad (67)$$

and finally we obtain

$$T^{xz} = \sum_{\alpha,\beta>\alpha=1}^{N} \sum_{i,j=1}^{n} (x_{i\alpha} - x_{j\beta}) F^z_{i\alpha j\beta} - \sum_{\alpha=1}^{N} \sum_{i=1}^{n} (x_{i\alpha} - x_\alpha) \sum_{\beta\neq\alpha=1}^{N} \sum_{j=1}^{n} F^z_{i\alpha j\beta}. \qquad (68)$$

It is now possible to distinguish two terms which can be evaluated in a non-ambiguous way in the presence of periodic boundary conditions. The first term is simply the *atomic* term, where however only intermolecular interactions do appear, and the second term represents the correction needed to obtain the *molecular* property. It has to be stressed that in this representation intramolecular interactions, arising either from constraint forces or from potential terms, do not need to be considered explicitly for the calculation of the pressure, as their contribution to the total force on the molecule vanishes.

References

[1] H. C. ANDERSEN, *J. Chem. Phys.* **72**, 2384 (1980).

[2] W. G. HOOVER, A. J. C. LADD, and B. MORAN, *Phys. Rev. Lett.* **48**, 1818 (1982).

[3] D. J. EVANS, *J. Chem. Phys.* **78**, 3297 (1983).

[4] S. NOSÉ, *Molec. Phys.* **52**, 255 (1984).

[5] W. G. HOOVER, *Phys. Rev. A* **31**, 1695 (1985).

[6] S. NOSÉ, *Molec. Phys.* **57**, 187 (1986).

[7] S. NOSÉ, *Prog. Theor. Phys. Suppl.* **103**, 1 (1991).

[8] A. BRAŇKA and M. PARRINELLO, *Molec. Phys.* **58**, 989 (1986).

[9] M. PARRINELLO and A. RAHMAN, *J. Appl. Phys.* **52**, 7182 (1981).

[10] S. NOSÉ and M. L. KLEIN, *Molec. Phys.* **50**, 1055 (1983).

[11] J.-P. RYCKAERT and G. CICCOTTI, *J. Chem. Phys.* **78**, 7368 (1983).

[12] M. FERRARIO and J.-P.RYCKAERT, *Molec. Phys.* **54**, 587 (1985).

[13] H. C. ANDERSEN et al., New molecular dynamics methods for various ensembles, Report of CECAM Workshop, CECAM, Orsay, France, 1984.

[14] M. FERRARIO, in preparation.

[15] S. MELCHIONNA, G. CICCOTTI, and B. L. HOLIAN, *Molec. Phys.* , in press.

[16] G. CICCOTTI and J. P. RYCKAERT, *Comput. Phys. Rep.* **4**, 345 (1986).

Computer Simulations in the Gibbs Ensemble

B. Smit
Shell Research BV
Koninklijke/Shell-Laboratorium Amsterdam
Postbus 3003
1003 AA Amsterdam
The Netherlands

ABSTRACT. The Gibbs ensemble technique is an efficient method to study phase equilibria in a computer simulation. This Chapter gives an overview of this method. The focus is on the principles underlying the method, the practical aspects related to the implementation of the technique, and questions regarding the interpretation of the results.

The practical use of the method is illustrated with applications ranging from polar fluids to chain molecules. In particular, those systems are discussed which require special tricks and extend the range of applicability of the Gibbs method significantly.

1. Introduction

Conceptually, most computer simulations are simple: specify an intermolecular potential and press the 'enter' button. Typically, after a large amount of CPU time the computer reveals everything there is to know about the system. So why worry about techniques to calculate phase equilibria: when the particles phase separate, the interface can be located and the coexistence properties can be 'measured'! Indeed, this would be the case if both the computer and the computer budget were infinite. Unfortunately these limitations make it particularly difficult to study phase equilibria by computer simulations.

To see this, let us consider a system that phase separates into a liquid and a vapour. Assume that the system is equilibrated for sufficiently long so that two interfaces are formed. There are two points to note: firstly, the equilibration requires a very large amount of computer time, often considerably more than for the production run; secondly, since we assume that periodic boundary conditions are used, the formation of two interfaces can be expected. Since the properties of the fluid in the interfacial region differ substantially from the properties in the bulk, we have to discard all the particles in the interfacial region for a 'measurement' of a bulk property.

M.P. Allen and D.J. Tildesley (eds.), Computer Simulation in Chemical Physics, 173–209.

In Table I the percentage of the total number of particles in the interfacial region is given. These numbers show that for a system smaller than 1000 particles the properties are dominated by the interfaces. It is therefore essential to use relatively large systems to calculate reliable coexistence properties. Unfortunately for large systems equilibration requires not only much more CPU time because the systems are big, but also because the equilibration times are longer compared to those of small systems.

A direct calculation of phase equilibria turns out to be 'expensive'. An alternative would be to simulate the vapour and liquid separately and to determine the temperature at which the pressure and chemical potential are equal. Although 'thermal' quantities like the pressure or temperature can be calculated straightforwardly, a simulation does not yield quantities like the free energy or chemical potential directly. The problem of calculating the free energy can be compared with a survey. The percentage of statisticians who believe that statistics are lies, can be determined by a simple poll. The total number of statisticians who believe this, however, requires information on the total number of statisticians. Obtaining this number requires additional work. Similarly in a simulation, the pressure can be calculated from an (extremely) small sample of all possible configurations of the system. The free energy, which is related to the total number of configurations, requires additional techniques. Various methods have been developed to calculate the free energy or chemical potential (see [1] for a review).

TABLE I

The percentage of particles (P_{int}) in the interfacial region. The interfacial region is roughly defined as the volume within 2.5σ from the interface. N is the total number of particles.

N	P_{int}
100	95%
500	55%
1000	44%
10000	20%
50000	10%

One of these methods is to determine the equation-of-state of the system. From the equation-of-state the free energy can be calculated using standard thermodynamic relations [2]. This way of determining the free energy is identical to the experimental method. The method is simple and can be used for almost all systems. However, it requires a large number of equation-of-state points along a reversible path to a state with known free energy (for example the ideal gas limit). With this method the phase diagram of the Lennard-Jones fluid [3, 4] and the two-dimensional Lennard-Jones fluid [5] have been determined.

A method which yields the excess chemical potential in a single simulation is the Widom particle insertion method [6]. This method utilizes the fact that the change in the free energy in going from N to $N+1$ particles can be expressed as an ensemble average. This ensemble average is calculated by adding test particles to the system. The simulations can now be restricted to the coexistence region since both the pressure and the chemical potential are known. The phase diagram of a quadrupolar Lennard-Jones fluid [7] has been obtained with this method.

The methods described above require either long runs on large systems or many different simulations. Consequently, the calculation of a phase diagram is an elaborate task. A new technique proposed by Panagiotopoulos [8], the so called Gibbs ensemble method, reduces the computer time required for phase equilibrium calculations significantly. With this method, phase equilibria can be studied in a single simulation. The method is particularly useful for studying vapour-liquid and liquid-liquid equilibria. In its present form the method can not be used to study phase equilibria which involve a solid phase.

The Gibbs method utilizes two distinct simulation boxes, which are coupled using Monte Carlo rules. These rules are chosen to ensure that these subsystems are in equilibrium. When a conventional simulation is performed in the two-phase region, droplets of liquid or gas will be formed. In the Gibbs ensemble, the system can lower its free energy by filling one box with vapour and the other with liquid. In this way the formation of interfaces, which increase the free energy because of the interfacial tension between the liquid and the vapour, is avoided. The coexistence properties can be obtained directly from the two boxes. Since this method avoids the interface, it can be used with a relatively small number of particles.

In this Chapter, we study the Gibbs ensemble in detail. The introduction of a new ensemble brings up the question whether it is a 'proper ensemble', i.e. does it yield the same results as the conventional ensembles? To prove it does, we derive in section 2 the partition function and use this function to define a free energy. This free energy is used to show that, in the thermodynamic limit, the Gibbs ensemble and the canonical ensemble are equivalent. This proof gives considerable insight into the 'reason' why the method works.

Having established that the Gibbs ensemble is a sound ensemble, we turn in section 3 to a more practical issue, namely how to sample this ensemble using a Monte Carlo scheme. For this the appropriate acceptance rules are derived. In sections 4 and 5 we describe how the results can be analyzed and the kind of problems one can expect during the simulations. Some applications of the Gibbs ensemble technique are described in section 6. We focus on those applications that involve extensions of the general Gibbs ensemble method, such as phase equilibria involving dense liquids, mixtures, or chain molecules.

2. Theoretical aspects

In his original article [8], Panagiotopoulos introduced the Gibbs ensemble as a combination of the constant-NVT ensemble, the constant-NPT ensemble, and the constant-μVT ensemble. In this work we take a different point of view, and consider the Gibbs ensemble as a new ensemble [9]. For this ensemble we derive the partition function and define a proper free energy. By considering this free energy we demonstrate that, in the two-phase region, one of the boxes contains the liquid phase and the other the gas phase. Furthermore, we prove that, in the thermodynamic limit, the Gibbs ensemble is equivalent to the canonical ensemble.

2.1. THE PARTITION FUNCTION

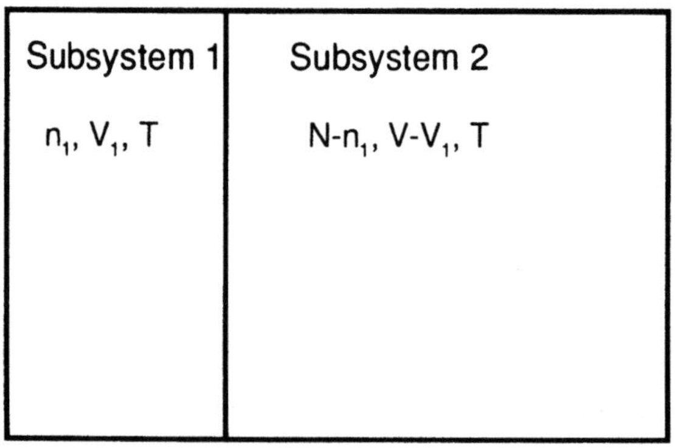

Fig. 1. A schematic picture of the Gibbs ensemble.

Consider a system at constant temperature (T), volume (V), and number of particles (N). This system is divided into two non-interacting subsystems 1 and 2 (see Fig. 1). In a simulation this implies that each box has periodic boundary conditions and particles in one box do *not* interact with the particles in the other box. The particles are distributed over the two subsystems keeping the total number of particles constant. The volume of each subsystem may vary in such a way that the total volume remains constant.

In the partition function of this ensemble, we have to take into account the number of possible distributions of N particles over the two subsystems, to allow for the subsystems to change volume between 0 and V, and to consider all possible configurations in each subsystem [10–12]

$$\bar{Q}_{NVT} \equiv \frac{1}{V\Lambda^{3N}N!}\sum_{n_1=0}^{N}\binom{N}{n_1}\int_0^V \mathrm{d}V_1 \int_{V_1}\mathrm{d}\mathbf{r}_1^{n_1}\exp\left[-\beta\mathcal{V}_1(n_1)\right]$$
$$\times \int_{V-V_1}\mathrm{d}\mathbf{r}_2^{N-n_1}\exp\left[-\beta\mathcal{V}_2(N-n_1)\right], \tag{1}$$

where n_1 denotes the number of particles in box 1, V_1 denotes the volume of box 1, $\mathbf{r}_1^{n_1} = \mathbf{r}_1, \ldots, \mathbf{r}_{n_1}$ and $\mathbf{r}_2^{N-n_1} = \mathbf{r}_{n_1+1}, \ldots, \mathbf{r}_N$ are the positions of the particles in subsystem 1 and subsystem 2 respectively, Λ is the thermal de Broglie wavelength, $\beta = 1/k_{\mathrm{B}}T$, and $\mathcal{V}(n_i)$ is the intermolecular potential.

This partition function will be used in the next section to define the free energy for the Gibbs ensemble, in section 2.3 to derive an expression for the chemical potential, and in section 3 to derive the acceptance rules for Monte Carlo simulations in this ensemble.

2.2. THE FREE ENERGY DENSITY

In this section we study the Gibbs ensemble, as introduced in the previous section in the thermodynamic limit. Before we proceed, we first list a few basic results for the free energy in the canonical ensemble.

2.2.1. Basic definitions and results for the canonical ensemble. Consider a system of N particles in a volume V and temperature T (the canonical ensemble). The partition function is defined as (see Ruelle [13])

$$Q_{NVT} \equiv \frac{1}{\Lambda^{3N}N!}\int_V \mathrm{d}\mathbf{r}^N \exp\left[-\beta\mathcal{V}(N)\right]. \tag{2}$$

The free energy density is defined in the thermodynamic limit by

$$f(\rho) \equiv \lim_{V\to\infty} f_V(\rho) \equiv \lim_{\substack{V\to\infty \\ N/V=\rho}} -\frac{1}{\beta V}\ln Q_{NVT}, \tag{3}$$

where $\rho = N/V$ is the density of the system. For a finite number of particles we can write

$$Q_{NVT} = \exp\left[-\beta V\left(f(\rho) + o(V)\right)\right], \tag{4}$$

where $g(V) = o(V)$ means: $g(V)/V$ approaches zero as $V \to \infty$. With this free energy we can derive some interesting properties of a canonical system in the thermodynamic limit.

For example, it can be shown that this free energy is a convex function of the density ρ [13]

$$f(x\rho_1 + (1-x)\rho_2) \leq x f(\rho_1) + (1-x) f(\rho_2), \tag{5}$$

for every ρ_1, ρ_2, and x where $0 \le x \le 1$. The equality holds in the case of a first-order transition, if $\rho_g \le \rho_1 \le \rho_2 \le \rho_l$, where ρ_g, ρ_l denote the density of the coexisting gas and liquid phases respectively (see Figs. 2 (a) and (b)).

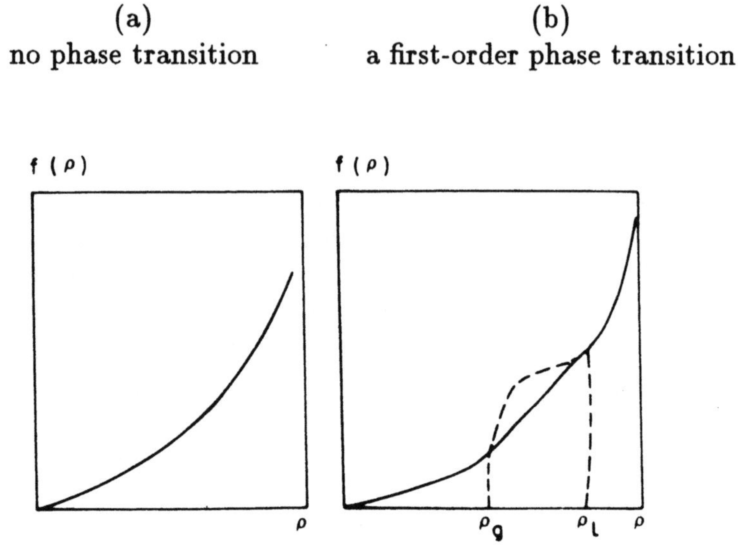

Fig. 2. The free energy density as a function of the density ρ.

Another interesting result, which plays a central rôle in what follows, is the well-known saddle-point theorem [14] (also called the steepest descent method and discussed in Chapter 3). This theorem is based on the observation that, for a macroscopic system (N very large) in equilibrium, the probability that the free energy density deviates from its minimum value is extremely small. Therefore, when we calculate for such a system an ensemble average, we have to take into account only those contributions where the free energy has its minimum value. Assume that Q_{NVT} can be written as

$$Q_{NVT} \equiv \int \mathrm{d}a_1, \ldots, \mathrm{d}a_m \exp\left[-\beta V \left(f_m(a_1, \ldots, a_m) + o(V)\right)\right], \tag{6}$$

where $a_1, \ldots, a_m$ are variables which characterize the thermodynamic state of the system. Furthermore, define

$$f(\rho) \equiv \min_{a_1, \ldots, a_m} f_m(a_1, \ldots, a_m) \tag{7}$$

and assume that $f_m(a_1, \ldots, a_m)$ and the term $o(V)$ satisfy a few technical conditions [14], which hold for most statistical mechanics systems. The saddle-point theorem states that in the thermodynamic limit the free energy of the system is equal to this minimum value $f(\rho)$ or

$$\lim_{\substack{V\to\infty \\ N/V=\rho}} -\frac{1}{\beta V}\ln Q_{NVT} = f(\rho). \tag{8}$$

Moreover, this saddle-point theorem can also be used to calculate the ensemble average of a quantity $\mathcal{A}$

$$\begin{aligned}\langle \mathcal{A}(a_1,\ldots,a_m)\rangle_V &\equiv \frac{1}{Q_{NVT}}\int \mathrm{d}a_1,\ldots,\mathrm{d}a_m \\ &\times \mathcal{A}(a_1,\ldots,a_m)\exp\left[-\beta V\left(f_m(a_1,\ldots,a_m)+o(V)\right)\right].\end{aligned} \tag{9}$$

In the thermodynamic limit, this ensemble average only has contributions from those configurations where $f_m(a_1,\ldots,a_m)$ has its minimum value. Let us define S as the collection of these minima

$$S = \left\{ y_1,\ldots,y_m \,\middle|\, f_m(y_1,\ldots,y_m) = \min_{a_1,\ldots,a_m} f_m(a_1,\ldots,a_m)\right\}. \tag{10}$$

We can now state the saddle-point theorem in a convenient form by introducing a function $G(a_1,\ldots,a_m) \geq 0$ with support on the surface S and normalization

$$\int_S \mathrm{d}a_1,\ldots,\mathrm{d}a_m\, G(a_1,\ldots,a_m) = 1 \tag{11}$$

such that, for an arbitrary function $\mathcal{A}$

$$\begin{aligned}\langle \mathcal{A}(a_1,\ldots,a_m)\rangle &\equiv \lim_{V\to\infty}\langle \mathcal{A}(a_1,\ldots,a_m)\rangle_V \\ &= \int_S \mathrm{d}a_1,\ldots,\mathrm{d}a_m\, G(a_1,\ldots,a_m)\mathcal{A}(a_1,\ldots,a_m).\end{aligned} \tag{12}$$

2.2.2. The free energy density in the Gibbs ensemble. The Gibbs ensemble is introduced in section 2.1 as a constant-NVT ensemble to which an additional degree of freedom is added: the system is divided into two subsystems which have *no* interactions with each other. We can rewrite the partition function of the canonical ensemble as (eqn (2))

$$\begin{aligned}Q_{NVT} = \frac{1}{\Lambda^{3N}N!}\sum_{n_1=0}^{N}\binom{N}{n_1}\int_0^V \mathrm{d}V_1\int \mathrm{d}\mathbf{r}_1^{n_1}\int \mathrm{d}\mathbf{r}_2^{N-n_1} \times \exp\left[-\beta\left(\mathcal{V}_1(n_1)\right.\right. \\ \left.\left. +\ \mathcal{V}_2(N-n_1) + \text{ interactions between the two volumes}\right)\right].\end{aligned} \tag{13}$$

The difference between this equation and the partition function of the Gibbs ensemble, eqn (1), is that in eqn (2) we do have interactions between the subsystems. In the case of short-range interactions, the last term in the exponent of eqn (13) is proportional to a surface term. This already suggests that both ensembles should behave similarly in many respects. We expound these ideas more rigorously in the following pages.

In the usual way, we define the free energy in the Gibbs ensemble as

$$\bar{f}(\rho) \equiv \lim_{\substack{V\to\infty \\ N/V=\rho}} -\frac{1}{\beta V} \ln \bar{Q}_{NVT}. \tag{14}$$

In the partition function of the Gibbs ensemble, eqn (1), we can substitute eqn (2)

$$\bar{Q}_{NVT} = \sum_{n_1=0}^{N} \int_0^V \mathrm{d}V_1 \; Q_{n_1,V_1,T} Q_{N-n_1,V-V_1,T}. \tag{15}$$

Introducing $x = N_1/N$ and $y = V_1/V$, and assuming that the number of particles is very large, we can write

$$\bar{Q}_{NVT} = NV \int_0^1 \mathrm{d}x \int_0^1 \mathrm{d}y \, \bar{Q}_N(x,y), \tag{16}$$

where

$$\begin{aligned} \bar{Q}_N(x,y) &= Q_{xN,yV,T}\, Q_{(1-x)N,(1-y)V,T} \\ &= \exp\left[-\beta V \left\{ y f\left(\frac{x}{y}\rho\right) + (1-y) f\left(\frac{1-x}{1-y}\rho\right) + o(V) \right\}\right]. \end{aligned} \tag{17}$$

Note that in this equation $f(\rho)$ is the free energy of a canonical system. So, we can apply the saddle-point theorem of the previous section (eqn (8)) to calculate the free energy density of the Gibbs ensemble $\bar{f}(\rho)$

$$\bar{f}(\rho) = \min_{\substack{0\lesssim x\lesssim 1 \\ 0\lesssim y\lesssim 1}} \left\{ y f\left(\frac{x}{y}\rho\right) + (1-y) f\left(\frac{1-x}{1-y}\rho\right) \right\} \equiv \min_{\substack{0\lesssim x\lesssim 1 \\ 0\lesssim y\lesssim 1}} \{\bar{f}(x,y)\}. \tag{18}$$

We now have to find the surface S on which the function $\bar{f}(x,y)$ reaches its minimum. For this we can use the fact that $f(\rho)$ is a convex function of the density (eqn (5)). This gives for $\bar{f}(x,y)$

$$\bar{f}(x,y) \geq f\left(y\frac{x}{y}\rho + (1-y)\frac{1-x}{1-y}\rho \right) = f(\rho). \tag{19}$$

We first consider the case where there is only one phase. For this case the free energy $f(\rho)$ is shown schematically as a function of the density in Fig. 2 (a). This figure demonstrates that any combination of x and y, which result in densities ρ_1 and ρ_2 in the subsystems different from ρ, will give a higher free energy. So, the equality in eqn (19) holds only if

$$\frac{x}{y}\rho = \frac{1-x}{1-y}\rho, \quad \text{or} \quad x = y. \tag{20}$$

Thus, in the case where there is only one phase, the free energy of the Gibbs ensemble has its minimum value (in the thermodynamic limit) when both boxes have a density equal to the equilibrium density of the canonical ensemble. Therefore, the surface S is given by

$$S = \{(x, y) \,|x = y\} . \tag{21}$$

Secondly, we consider the case of a first-order phase transition, for which the free energy as a function of the density is shown in Fig. 2 (b). Let ρ be such that $\rho_l \leq \rho \leq \rho_g$, and let us choose x and y such that

$$\rho_g \leq \frac{x}{y}\rho \equiv \rho_3 \leq \rho_l \quad \text{and} \quad \rho_g \leq \frac{1-x}{1-y}\rho \equiv \rho_4 \leq \rho_l. \tag{22}$$

For this case the equality in eqn (5) holds and we can write for $\bar{f}(x, y)$

$$\begin{aligned} \bar{f}(x, y) &= yf(\rho_3) + (1-y)f(\rho_4), \\ &= f(y\rho_3 + (1-y)\rho_4). \end{aligned} \tag{23}$$

Note that

$$(y\rho_3 + (1-y)\rho_4) = \rho, \tag{24}$$

which gives

$$\bar{f}(x, y) = f(\rho). \tag{25}$$

It can be shown that, if x, y do not satisfy eqn (22),

$$\bar{f}(x, y) > f(\rho). \tag{26}$$

Therefore, the surface S in the case of a first-order phase transition is given by

$$S = \left\{(x, y) \middle| \rho_g \leq \frac{x}{y}\rho \leq \rho_l, \quad \rho_g \leq \frac{1-x}{1-y}\rho \leq \rho_l \right\}. \tag{27}$$

This result shows that, in the case of a first-order transition, the (bulk) free energy of the Gibbs ensemble has its minimum value (in the thermodynamic limit) for all values of x, y where there is vapour-liquid coexistence in *both* boxes.

Equations (19) and (25) show that in the thermodynamic limit the free energy of the Gibbs ensemble is equal to the free energy of the canonical ensemble. In order to calculate an ensemble average, it simply remains to determine the function $G(x, y)$ (cf. eqn (12)).

In the case of a pure phase $G(x, y)$ needs to be of the form

$$G(x, y) = g(x)\,\delta(x-y). \tag{28}$$

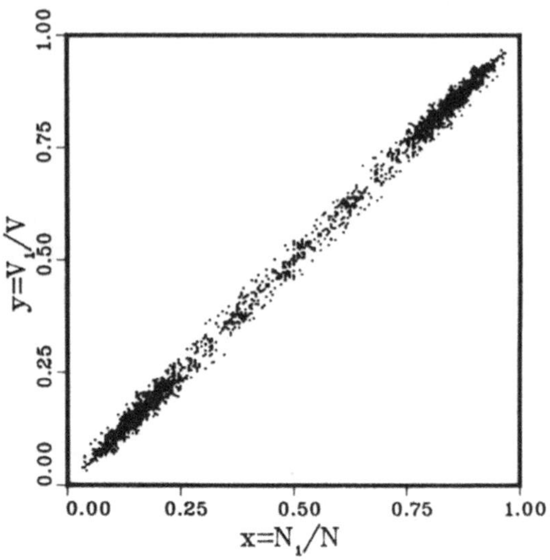

Fig. 3. A probability plot in the x, y plane ($x = n_1/N, y = V_1/V$) for a Lennard-Jones fluid at high temperature ($T = 5.21$, $N = 216$ and $V = 670$). The total number of cycles in this simulation was 6000 (see section 3 for details on the simulations). After every 6 cycles the points (x_i, y_i) and $(1 - x_i, 1 - y_i)$ were plotted in the x, y plane.

As shown the appendix of Ref. [9], $g(x) = 1$ for the ideal gas. We expect the same to be true for an interacting gas. Fig. 3 shows a probability plot in the x, y plane for a simulation of a finite system at high temperature. This figure shows that $x \approx y$. We have not attempted to estimate $g(x)$.

In the case of two phases we will show that the system will split into a liquid phase, with density ρ_l, in one box, and a vapour phase, with density ρ_g, in the other box.

Until now we have ignored surface effects, which arise from the presence of a liquid-vapour interface in the boxes. When the density in one of the boxes is between the vapour and liquid density, the system will form droplets of gas or liquid. The interfacial free energy associated with these droplets has (in the thermodynamic limit) a negligible contribution to the bulk free energy of the Gibbs ensemble. Nevertheless, it turns out that this surface free energy is the driving force that causes the system to separate into a homogeneous liquid in one box and a homogeneous vapour phase in the other.

These surface effects are taken into account in the next significant term in the expression for the free energy (eqn (4)), which is the term due to the surface tension. This gives, for the partition function,

$$Q_{NVT} = \exp\left[-\beta\left(Vf(\rho) + \gamma A + o(A)\right)\right], \tag{29}$$

where A denotes the area of the interface and γ denotes the interfacial tension. For three-dimensional systems this area will in general be proportional to $V^{2/3}$. using this form of the partition function for the Gibbs ensemble, eqn (9) can be written as

$$\langle \mathcal{A}(x,y)\rangle_V = \frac{\iint \mathrm{d}x\mathrm{d}y\, \mathcal{A}(x,y) \exp\left[-\beta\left(Vf(x,y) + \gamma V^{2/3}a(x,y) + o(V^{2/3})\right)\right]}{Q_{NVT}}, \tag{30}$$

where $a(x,y)$ is a function of the order of unity.

We know from the saddle-point theorem that the most important contribution to the integrals comes from the region S, defined by eqn (27). Thus

$$\begin{aligned}\langle \mathcal{A}(x,y)\rangle_V &\approx \frac{\iint_S \mathrm{d}x\mathrm{d}y\, \mathcal{A}(x,y) \exp\left[-\beta\left(Vf(x,y) + \gamma V^{2/3}a(x,y) + o(V^{2/3})\right)\right]}{\iint_S \mathrm{d}x\mathrm{d}y\, \exp\left[-\beta\left(Vf(x,y) + \gamma V^{2/3}a(x,y) + o(V^{2/3})\right)\right]} \\ &= \frac{\iint_S \mathrm{d}x\mathrm{d}y\, \mathcal{A}(x,y) \exp\left[-\beta\left(\gamma V^{2/3}a(x,y) + o(V^{2/3})\right)\right]}{\iint_S \mathrm{d}x\mathrm{d}y\, \exp\left[-\beta\left(\gamma V^{2/3}a(x,y) + o(V^{2/3})\right)\right]}\end{aligned} \tag{31}$$

and, applying the saddle-point theorem again,

$$\langle \mathcal{A}(x,y)\rangle_V \approx \frac{\iint_{S_A} \mathrm{d}x\mathrm{d}y\, \mathcal{A}(x,y) \exp\left[-\beta\gamma V^{2/3}a(x,y) + o(V^{2/3})\right]}{\iint_{S_A} \mathrm{d}x\mathrm{d}y\, \exp\left[-\beta\gamma V^{2/3}a(x,y) + o(V^{2/3})\right]} \tag{32}$$

and

$$\lim_{V\to\infty} \langle \mathcal{A}(x,y)\rangle_V = \iint_{S_A} \mathrm{d}x\mathrm{d}y\, G(x,y)\, \mathcal{A}(x,y), \tag{33}$$

where the surface S_A is now given by

$$S_A = \left\{(x,y)\,\middle|\, a(x,y) = \min_{\bar{x},\bar{y}} a(\bar{x},\bar{y})\right\}. \tag{34}$$

In the infinite system it is easily seen that the area of the interface is zero, if box 1 contains only gas (liquid) and box 2 only liquid (gas). Therefore, the surface S_A contains only two points, which corresponds to the vapour and liquid density

$$S_A = \left\{(x,y)\,\middle|\, \frac{x}{y} = \rho_\mathrm{l} \text{ and } \frac{1-x}{1-y} = \rho_\mathrm{g} \text{ or } \frac{x}{y} = \rho_\mathrm{g} \text{ and } \frac{1-x}{1-y} = \rho_\mathrm{l}\right\}. \tag{35}$$

It is straightforward to show that this surface gives for $G(x,y)$

$$G(x,y) = \frac{1}{2}\delta\left(x - \frac{\rho_{\rm g}}{\rho}\frac{\rho-\rho_{\rm g}}{\rho_{\rm l}-\rho_{\rm g}}\right)\delta\left(y - \frac{\rho-\rho_{\rm g}}{\rho_{\rm l}-\rho_{\rm g}}\right) + \frac{1}{2}\delta\left(x - \frac{\rho_{\rm l}}{\rho}\frac{\rho_{\rm l}-\rho}{\rho_{\rm l}-\rho_{\rm g}}\right)\delta\left(y - \frac{\rho_{\rm l}-\rho}{\rho_{\rm l}-\rho_{\rm g}}\right). \tag{36}$$

2.2.3. Summary. In this section, we have shown more formally that the free energy density for the Gibbs ensemble, as defined by eqn (14), becomes identical to the free energy density of the canonical ensemble.

Furthermore, it is shown that, at high temperatures, $x = y$, i.e. the densities in the two subsystems of the Gibbs ensemble are equal, and equal to the density in the canonical ensemble (see Fig. 3).

In the case of a first-order phase transition, if surface terms are unimportant, then x and y are restricted to the area defined by (cf. eqn (27))

$$\rho_{\rm g} \leq \frac{x}{y}\rho \equiv \rho_3 \leq \rho_{\rm l} \quad \text{and} \quad \rho_{\rm g} \leq \frac{1-x}{1-y}\rho \equiv \rho_4 \leq \rho_{\rm l}. \tag{37}$$

If we take surface effects into account, it is shown that this surface (eqn (37)) reduces to two points in the x, y plane. The densities of these points correspond to the density of the gas or liquid phase in the canonical ensemble.

It is interesting to compare this with the results of an actual simulation of a finite system. In Fig. 4 the results for a simulation at a temperature well below the critical point are given. This shows that the surface reduces to two points. This should be compared to the results of a simulation close to the critical point (Fig. 5). At those conditions the interfacial tension is very small and we see that the simulation samples the entire surface S. Note that, due to the finite size of this system, fluctuations are also possible in which the density of a subsystem becomes greater/smaller than the density of the liquid/gas phase.

2.3. THE CHEMICAL POTENTIAL

One of the steps in the Gibbs ensemble involves the insertion of a particle in one of the boxes. During this step, the energy of this particle has to be calculated (see section 3). Since this energy corresponds to the energy of a test particle, we can use the Widom insertion method [6] to calculate the chemical potential without additional costs [15]. At this point, it is important to note that the Gibbs method does not require computation of the chemical potentials. However, in order to test whether the system under consideration has reached equilibrium, or for comparison with other results, it is important to calculate the chemical potential of the individual phases correctly. The

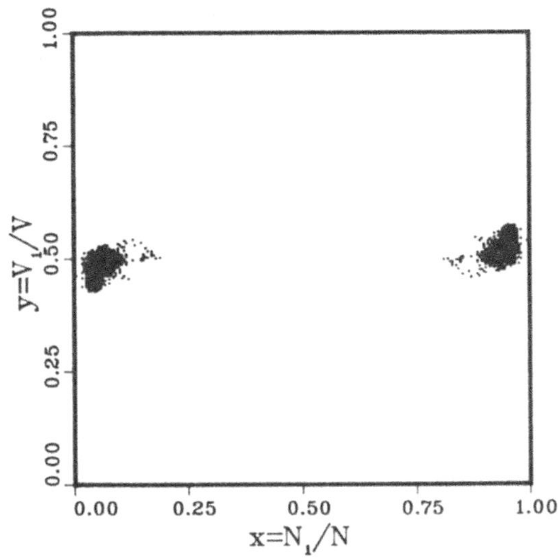

Fig. 4. A probability plot in the x, y plane for a Lennard-Jones fluid well below the critical temperature ($T = 1.15$, $N = 216$, $V = 670$); see also the caption to figure 3.

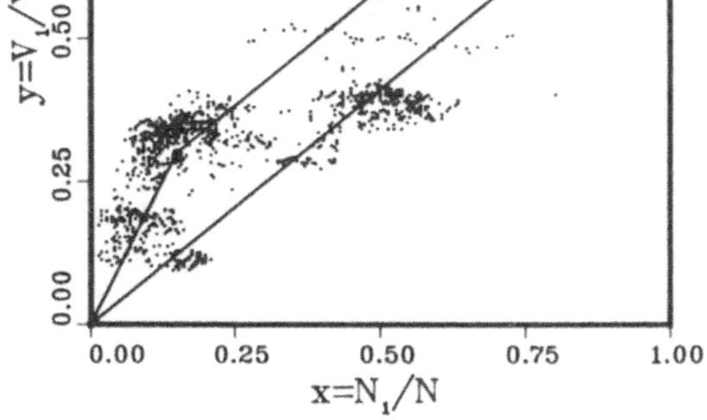

Fig. 5. A probability plot in the x, y plane for a Lennard-Jones fluid slightly below the critical temperature ($T = 1.30$, $N = 512$, $V = 1600$). The solid lines show the area S as defined in eqn (27); see also the caption to Fig. 3.

original Widom expression is only valid in the canonical ensemble. It can be modified for applications in the isothermal-isobaric ensemble [16, 17] or the microcanonical ensemble [1]. Here we derive an expression for the chemical potential for the Gibbs ensemble. We restrict ourselves to temperatures sufficiently far below the critical temperature such that the two boxes, after equilibration, do not change identity. For the more general case we refer to Ref. [11].

If we rescale the coordinates of the particles with the box length, the partition function for the Gibbs ensemble, eqn (1), becomes

$$\begin{aligned}\bar{Q}_{NVT} &\equiv \frac{1}{V\Lambda^{3N}N!}\sum_{n_1=0}^{N}\binom{N}{n_1}\int_0^V \mathrm{d}V_1\, V_1^{n_1}(V-V_1)^{N-n_1} \\ &\times \int \mathrm{d}\boldsymbol{\xi}_1^{n_1}\exp\left[-\beta\mathcal{V}_1(n_1)\right]\int \mathrm{d}\boldsymbol{\xi}_2^{N-n_1}\exp\left[-\beta\mathcal{V}_2(N-n_1)\right], \\ &= \sum_{n_1=0}^{N}\int_0^V \mathrm{d}V_1\, V_1^{n_1}(V-V_1)^{N-n_1}\,Q_1(n_1,V_1)\,Q_2(N-n_1,V-V_1),\end{aligned} \tag{38}$$

where $\boldsymbol{\xi} = \mathbf{r}/L$ is the scaled coordinate vector of a particle, L is the box length of the subsystem in which the particle is located, and $Q_i(n_i, V_i)$ is the partition function of the canonical ensemble (see also section 2.2.2).

The chemical potential of box 1 can be defined as

$$\begin{aligned}\mu_1 &\equiv -k_{\mathrm{B}}T\ln\sum_{n_1=0}^{N}\int_0^V \mathrm{d}V_1\, V_1^{n_1}(V-V_1)^{N-n_1} \\ &\times \left(\frac{Q_1(n_1+1,V_1)}{Q_1(n_1,V_1)}\right) Q_2(N-n_1,V-V_1).\end{aligned} \tag{39}$$

For the ratio of the partition functions of box 1 we can write

$$\begin{aligned}\frac{Q_1(n_1+1,V_1)}{Q_1(n_1,V_1)} &= \frac{V_1}{(n_1+1)\Lambda^3}\frac{\int \mathrm{d}\boldsymbol{\xi}_1^{n_1+1}\exp\left[-\beta\mathcal{V}_1(n_1+1)\right]}{\int \mathrm{d}\boldsymbol{\xi}_1^{n_1}\exp\left[-\beta\mathcal{V}_1(n_1)\right]} \\ &= \frac{V_1}{(n_1+1)\Lambda^3}\frac{\int \mathrm{d}\boldsymbol{\xi}_1^{n_1}\exp\left[-\beta\Delta v_1^+\right]\exp\left[-\beta\mathcal{V}_1(n_1)\right]}{\int \mathrm{d}\boldsymbol{\xi}_1^{n_1}\exp\left[-\beta\mathcal{V}_1(n_1)\right]},\end{aligned} \tag{40}$$

in which we have used the notation

$$\mathcal{V}_1(n_1+1) = \Delta v_1^+ + \mathcal{V}_1(n_1), \tag{41}$$

where Δv_1^+ is the test particle energy of a (ghost) particle in box 1. We can write eqn (39) as an ensemble average restricted to box 1

$$\mu_1 = -k_{\mathrm{B}}T\ln\frac{1}{\Lambda^3}\left\langle\frac{V_1}{n_1+1}\exp\left[-\beta\Delta v_1^+\right]\right\rangle_{\text{Gibbs, box 1}}, \tag{42}$$

where $\langle\ldots\rangle_{\text{Gibbs, box } i}$ denotes an ensemble average in the Gibbs ensemble restricted to box i (note that this ensemble average is well defined if the boxes do not change identity during a simulation).

3. Computational aspects

In section 2.1, the partition function of the Gibbs ensemble was derived and it was shown that in the two-phase region one of the sub-systems contains the vapour phase and the other subsystem the liquid phase. This property makes the Gibbs ensemble a convenient ensemble to study vapour-liquid (or liquid-liquid) phase equilibria. In this section we describe a Monte Carlo procedure which samples the Gibbs ensemble.

3.1. ACCEPTANCE RULES

The problem of devising a procedure which generates configurations with a specific probability distribution, can be formulated conveniently in terms of the theory of stochastic processes (van Kampen [18]). For example, one can generate a chain of configurations by specifying a Markov process, in which a new configuration $\mathbf{\Gamma}_m$ is obtained from the old configuration $\mathbf{\Gamma}_n$ with a transition probability $T_{(m|n)}$

$$\mathbf{\Gamma}_m = T_{(m|n)}\mathbf{\Gamma}_n \,. \tag{43}$$

Repetition of this procedure generates a chain of configurations (Markov chain). The conditions which need to be fulfilled by this transition probability to obtain configurations with the desired probability distribution are outlined by Feller [19, 20] or van Kampen [18]:

- the Markov chain needs to be *ergodic*, i.e. all possible configurations must be within reach from any other configuration in a finite number of steps;
- the transition probability must satisfy the condition of *detailed balance*, i.e. the number of transitions from configuration m to n must be equal to the number of transitions from n to m
  $$K_{(n|m)} = K_{(m|n)} \,. \tag{44}$$
  The number of transitions $K_{(n|m)}$ is the product of the statistical weight of configuration n (N_n) and the acceptance probability
  $$K_{(n|m)} = N_n \operatorname{acc}(n|m) \,. \tag{45}$$

The condition of detailed balance is stronger than strictly necessary to obtain an equilibrium distribution [18]. However, it has an important practical advantage. The condition of detailed balance guarantees that the equilibrium distribution will be reached in the course of a Markov process *irrespective* of the form of this distribution. Therefore, by imposing detailed balance we are guaranteed that the Markov chain will give the correct distribution of configurations for the Gibbs ensemble.

From the partition function of the Gibbs ensemble, it follows that we have to displace the particles in the two boxes, change the volume of the subsystems, and exchange particles between the boxes. Here we consider the version of the Gibbs ensemble where the total number of particles and the total volume of the two boxes remains constant, i.e. the total system is at constant-NVT conditions. The appropriate acceptance rules for the constant-NPT version can be found in Ref. [15].

In the Gibbs ensemble, the statistical weight of a configuration m with n_1 particles in box 1 with volume V_1, and $N - n_1$ particles in box 2 with volume $V - V_1$ is proportional to (cf. eqn (1))

$$N_m \propto \frac{V_1^{n_1}(V - V_1)^{N-n_1}}{n_1!(N - n_1)!} \exp(-\beta \mathcal{V}_m), \tag{46}$$

where $\mathcal{V}_m$ is the total potential energy (sum of the energy of box 1 and box 2).

The acceptance rules can be derived from fluctuation theory [8, 15] or directly from the partition function [10, 11]. Here we derive the acceptance rules using the condition of detailed balance.

For example, assume that state n is obtained from state m via the displacement of a particle in box 1. The ratio of the statistical weights of these two configurations is given by

$$\frac{N_n}{N_m} = \exp[-\beta(\mathcal{V}_n - \mathcal{V}_m)]. \tag{47}$$

Substituting eqn (47) into eqn (45), we find that the following acceptance rule for a particle displacement will satisfy the detailed balance condition, eqn (44)

$$\mathrm{acc}(m|n) = \min\{1, \exp[-\beta(\mathcal{V}_n - \mathcal{V}_m)]\}. \tag{48}$$

For a change of the volume of box 1 by an amount ΔV, the ratio of the statistical weights of the configurations after and before the move is given by

$$\frac{N_n}{N_m} = \frac{(V_1 + \Delta V)^{n_1}(V - (V_1 + \Delta V))^{N-n_1}}{V_1^{n_1}(V - V_1)^{N-n_1}} \exp[-\beta(\mathcal{V}_n - \mathcal{V}_m)]. \tag{49}$$

Imposing the condition of detailed balance gives as acceptance rule for the volume change

$$\mathrm{acc}(m|n) = \min\left\{1, \frac{(V_1 + \Delta V)^{n_1}(V - (V_1 + \Delta V))^{N-n_1}}{V_1^{n_1}(V - V_1)^{N-n_1}} \exp[-\beta(\mathcal{V}_n - \mathcal{V}_m)]\right\}. \tag{50}$$

If the configuration n is obtained from configuration m by removing a particle from box 1 and inserting this particle in box 2, the ratio of statistical weights is given by

$$\frac{\mathcal{N}_n}{\mathcal{N}_m} = \frac{n_1!(N-n_1)!V_1^{n_1-1}(V-V_1)^{N-(n_1-1)}}{(n_1-1)!(N-(n_1-1))!V_1^{n_1}(V-V_1)^{N-n_1}} \exp[-\beta(\mathcal{V}_n-\mathcal{V}_m)] \,. \quad (51)$$

Again, imposing detailed balance for a particle exchange move leads to the acceptance rule

$$\mathrm{acc}(m|n) = \min\left\{1, \frac{n_1(V-V_1)}{(N-n_1+1)V_1} \exp[-\beta(\mathcal{V}_n - \mathcal{V}_m)]\right\} . \quad (52)$$

The moves described above are identical to the ones proposed by Panagiotopoulos *et al.* in Refs. [8, 15]. A more natural choice for generating a new configuration in the volume-change step, is to make a random walk in $\ln(V_1/(V-V_1))$ instead of in V_1 (see also Ref. [21] for the NPT ensemble). This has the advantage that the domain of this random walk coincides with the possible values of V_1. Furthermore, the average step size turns out to be less sensitive to the density. In order to adopt this method for the Gibbs ensemble, the acceptance rule for the volume has to be modified. If we make a random walk in $\ln(V_1/(V-V_1))$, the ensemble averages that are calculated correspond to

$$\begin{aligned} \langle \mathcal{A} \rangle &= \frac{1}{Q_{NVT}} \frac{1}{\Lambda^{3N} N!} \sum_{n_1=0}^{N} \\ &\times \binom{N}{n_1} \int_{-\infty}^{\infty} \mathrm{d}\ln\left(\frac{V_1}{V-V_1}\right) \frac{V_1(V-V_1)}{V} V_1^{n_1}(V-V_1)^{N-n_1} \\ &\times \int \mathrm{d}\boldsymbol{\xi}_1^{n_1} \exp\left[-\beta\mathcal{V}_1(n_1)\right] \int \mathrm{d}\boldsymbol{\xi}_2^{N-n_1} \exp\left[-\beta\mathcal{V}_2(N-n_1)\right] \mathcal{A}(\xi^N) \quad (53) \end{aligned}$$

The statistical weight of a configuration m is now proportional to

$$\mathcal{N}_m \propto \frac{V_1^{n_1+1}(V-V_1)^{N-n_1+1}}{Vn_1!(N-n_1)!} \exp(-\beta\mathcal{V}_m) \,. \quad (54)$$

Imposing detailed balance for this move leads to the acceptance rule

$$\begin{aligned} &\mathrm{acc}(m|n) = \\ &\quad \min\left\{1, \left(\frac{V_1^n}{V_1^m}\right)^{n_1+1} \left(\frac{V-V_1^n}{V-V_1^m}\right)^{N-n_1+1} \exp[-\beta(\mathcal{V}_n - \mathcal{V}_m)]\right\} , \quad (55) \end{aligned}$$

in which V_1^m denotes the volume of box 1 of configuration m. The acceptance rules for the particle displacement or particle exchange are not affected.

3.2. COMPUTATIONAL ASPECTS

A convenient method to generate trial configurations is to perform a simulation in cycles. One cycle consists of (on average) N_{disp} attempts to displace a (random) particle in one of the (randomly chosen) boxes, N_{vol} attempts to change the volume of the subsystems, and N_{try} attempts to exchange particles between the boxes. It is important to ensure that at each step of the simulation the condition of microscopic reversibility is fulfilled (see for example the algorithm described in Fig. 6).

```
Npart1 : number of particles in box 1
Npart2 : number of particles in box 2
Nvol   : number of attempts to change the volume
Ntry   : number of attempts to exchange particles
Ncycle : number of Monte Carlo cycles
```

```
nvar = Npart1+Npart2+Nvol+Ntry
DO icl = 1,Ncycle
   DO ivar = 1,nvar
      ran = random_number
      IF (ran .LE. Npart1/nvar) THEN
         displace_particle_in_box_1
      ELSE IF (ran .LE. (Npart1+Npart2)/nvar) THEN
         displace_particle_in_box_2
      ELSE IF (ran .LE. (Npart1+Npart2+Nvol)/nvar) THEN
         change_volume
      ELSE IF (ran .LE. (Npart1+Npart2+Nvol+0.5*Ntry)/nvar) THEN
         exchange_particle_from_box_1_to_box_2
      ELSE
         exchange_particle_from_box_2_to_box_1
      ENDIF
   ENDDO
ENDDO
```

Fig. 6. The basic algorithm for a computer simulation in the Gibbs ensemble.

In most applications of the Gibbs ensemble, the simulations were performed slightly differently; instead of making a random choice at each Monte Carlo step which type of change (particle displacement, volume change, or particle exchange) will be made, the simulations were performed in a strict order. First, an attempt to displace each particle successively (the '*NVT* part'), then an attempt to change the volume (the 'NPT part'), and finally N_{try} attempts to exchange particles (the 'μVT part'). Although, with this scheme, microscopic reversibility is not fulfilled at every step in the algorithm, it can be expected that this method of generating new configurations will also lead to the correct probability distributions. However, the scheme described in the previous paragraph has some practical advantages. For example, there is no ambiguity when to sample a configuration, one does not have to make the choice of sampling after the NVT part, the NPT part, or the μVT part, but one can simply sample after each nth cycle.

A more serious disadvantage of the successive method is that in the latter scheme, when the probability of acceptance of an exchange becomes greater than 3%, erroneous can results occur. In Ref. [8] a bias in the liquid pressure was observed, which disappeared when the number of attempts to exchange particles was reduced. In section 4 it is shown that one should be extremely careful with this, since a simulation in the Gibbs ensemble can become easily trapped in such a region.

The details of the implementation of each step of the simulation depend on the intermolecular potential and on the conditions of the system. A general rule which should be considered in the implementation is that the simulation samples the configuration space efficiently. What efficient sampling means can be illustrated with a quotation from Ref. [10]:

> ...In very vague terms, sampling is efficient if it gives you good value for money. 'Good value' in a simulation corresponds to high statistical accuracy and 'money' is simply *money:* the money that buys your computer or even your own time. Let us assume that you are very poorly paid. In that case we only have to worry about your computer budget. Then we could use the following definition of an optimum sampling scheme: *A Monte Carlo sampling scheme can be considered optimum, if it yields the lowest statistical error in the quantity to be computed for a given expenditure of 'computer budget'.* ...it is reasonable to assume that the mean-square error is inverse proportional to the number of 'uncorrelated' configurations visited in a given amount of CPU time. ...

Below, some general points are made concerning the efficiency of the various steps in the Gibbs ensemble. However, one should be careful in applying these general rules for a particular system. Various examples exist where the optimal algorithm deviates significantly from these general rules. In section 6 some examples of these exceptions are described.

A particle displacement. In this step, a particle in one of the boxes is chosen at random and given a random displacement. The maximum displacement must be chosen in such a way that the sampling is efficient. For example, in the case of the Lennard-Jones potential the amount of CPU required to calculate the energy difference for a given trial move is independent of whether this move is rejected or accepted. Therefore, in this case the maximum displacement is set to give an acceptance ratio of approximately 50%. For other potentials, the optimum acceptance ratio can deviate significantly from this 50% (see section 6).

A volume change. In this step the volume of the boxes is changed in such a way that the total volume remains constant (in the case where the constant-NPT version of the Gibbs ensemble is used, the volumes of the two boxes are changed independently). For some systems, we can use the scaling properties of the intermolecular potential to calculate the energy of the new configuration efficiently (see Refs. [22, 23]).

Again, the maximum change of the volume must be chosen using the criterion of efficiency. In the case of the Lennard-Jones potential, there is no difference in required CPU time between an accepted and a rejected move. Therefore for this potential an acceptance ratio of approximately 50% is appropriate.

A particle exchange. In this step, a particle is exchanged between the two subsystems. Inspection of the main algorithm shows that a box, from which a random particle is to be deleted, is chosen at random. This particle will be placed at a random position in the other box.

The number of attempts to exchange a particle will depend on the conditions of the system. For example, it can be expected that close to the critical temperature the percentage of accepted exchanges will be higher than close to the triple point. A possible check whether the number of attempts is sufficient, is to calculate the chemical potential. Since the calculated energy of a particle which is to be inserted corresponds to just the test-particle energy, the chemical potential can be calculated without additional costs.

In Ref. [8] it is mentioned that close to the critical point some simulations failed because one of the boxes became empty. However, inspection of the partition function (eqn (1)) shows that one must allow for $n_1 = 0$ (box one empty) and $n_1 = N$ (box two empty) in order to calculate ensemble averages correctly. So, it is important to ensure that the program does not fail because of technical reasons when one of the boxes becomes empty. For example, when one of the boxes is empty the addition of yet another particle to the full box may cause a division by zero. This addition will lead to a situation with $N+1$ particles and inspection of the partition function shows that this configuration should not be sampled. However, if one also calculates the chemical potential during the exchange step one should be careful. In order to calculate the chemical potential correctly (see section 2.3) one should also add *test particles* when one of the boxes is full.

4. Finite-size effects

Computer simulations are usually performed on a relatively small number of particles. The results of these simulations will in general depend on the system size. Because we are usually interested in the infinite system, it is important to estimate the magnitude of the finite-size effects. For example, it is well known that the critical temperature of a finite system decreases when the system size is increased. For a finite system a true critical point does not exist. However, one can observe a pseudo-critical temperature when the fluctuations in the density are correlated over distances of the order of the system size. For lattice models, these finite-size effects have been studied in great detail using finite-size scaling techniques [24]. Here we analyze the finite-size effects that are specific to the Gibbs ensemble using the theoretical framework developed in section 2.

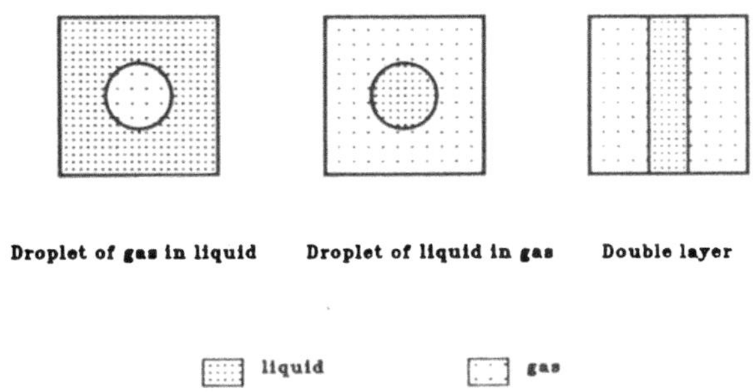

Fig. 7. A schematic representation of one of the boxes in a Monte Carlo simulation in the Gibbs ensemble slightly below the critical temperature.

For the infinite system we have shown that, in the case of a first-order phase transition, one of the subsystems will contain the gas (liquid) phase and the other subsystem will contain the liquid (gas) phase. For small systems we can expect to observe configurations that differ from the pure liquid or gas densities. The penalty of these deviations is related to the free energy increase caused by the formation of droplets of the other phase. Since the chemical potentials of the two phases are equal, this free energy increase arises from the surface between the two phases. For a finite system we have to calculate the surface area, $(A(x,y))$, for a given $x = n_1/N$ and $y = V_1/V$, The surface contribution of the free energy can be obtained from

$$F_{\text{surf}}(x,y) = \gamma A(x,y)\,, \tag{56}$$

where γ is the interfacial tension. In order to calculate $A(x,y)$ we assume that the surface area will have its minimum value for each x and y. If we perform a simulation with the normal periodic boundary conditions, we can assume that the surface energy is minimized either by a drop of liquid, a drop of gas, or a double layer (see Fig. 7). The volumes of the droplets or double layer can be expressed easily in terms of x and y (see Ref. [9] for details). In Fig. 8 we plotted the function $A(x, y)/V^{2/3}$ on the x, y plane. This plot shows that there are two minima, which correspond to the case where there is only vapour or liquid in the subsystems (no interface, so $A(x, y) = 0$). Furthermore, it shows that there are large parts of the x, y plane where the gradient towards these minima is very small. It is possible that a simulation becomes trapped in such a region. An example of such a simulation is described in section 5.

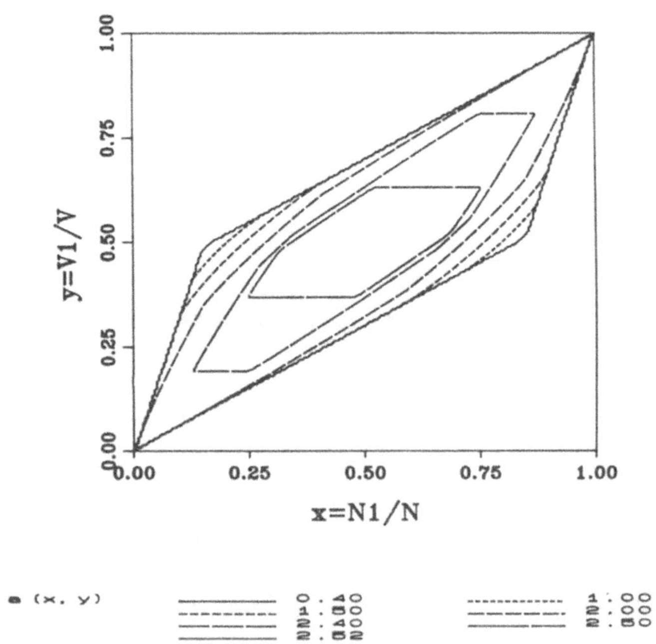

Fig. 8. A contour plot in the x, y plane ($x = n_1/N, y = V_1/V$) of the surface contribution ($a(x, y)$) to the free energy of the Gibbs ensemble. Note that the x, y values of the coexisting vapour and liquid phase are chosen arbitrarily.

It is interesting to calculate the probability of finding the density ρ in one of the sub-volumes for a finite system. This should be given approximately by

$$P(\rho) = \frac{\iint_S \mathrm{d}x\mathrm{d}y\, \delta(\frac{x}{y} - \rho) \exp\left[-\beta F_{\mathrm{surf}}(x, y)\right]}{\iint_S \mathrm{d}x\mathrm{d}y\, \exp\left[-\beta F_{\mathrm{surf}}(x, y)\right]}. \tag{57}$$

In order to study the temperature dependence, we have assumed that, close to the critical temperature, the interfacial tension depends on the temperature according to [25]

$$\gamma = \gamma_0 (1 - T/T_c)^{\mu} , \tag{58}$$

where μ is the critical exponent for the interfacial tension (for three dimensional systems $\mu = 1.26$ [25]). The area S is defined by the densities of the coexisting gas and liquid phases at the given temperature (cf. eqn (27)).

When the system is infinite, we have (combining eqns (36), (33), and (57))

$$P(\rho) = \frac{1}{2}\delta(\rho - \rho_g) + \frac{1}{2}\delta(\rho_l - \rho). \tag{59}$$

However, for a finite system, function (57) exhibits interesting behaviour.

In Figs. 9 (a) and (b) we have plotted $P(\rho)$ for a Lennard-Jones fluid ($\gamma_0 = 2.71$ [25] and $T_c = 1.32$ [8]) for 64 and 216 particles respectively. These figures demonstrate that, as the temperature is increased, the following transitions can be observed. Since at low temperature the surface tension is large, it is unlikely that vapour and liquid will coexist in the same sub-volume. Therefore, at low temperature there are two sharp peaks, which correspond to the two coexisting phases. As the temperature increases, the surface tension decreases and becomes comparable to the entropy associated with the formation of an interface (there are simply more overall densities where vapour and liquid can coexist). Note that around $x = y$ the interface will be a double layer, and therefore small deviations around this line do not change the surface free energy contribution (see Figs. 7 and 8). Furthermore, $x = y$ is the longest 'strip' in the x, y plane on the surface S. These entropic considerations predict the appearance of a third peak at the average density ρ. When the temperature is increased further this entropy effect will dominate the surface contribution and the peaks of the coexisting phases will disappear close to, but below, the critical temperature of the finite system.

Note that the temperature range over which these transitions occur depends on the number of particles. Comparison with the results for 64 particles (Fig. 9 (a)) shows that the third peak exists over a larger temperature range.

It is interesting to compare these results with probability distributions for the densities obtained from simulations for the Lennard-Jones fluid. In Fig. 10 the results are presented for simulations with 216 particles at $T = 1.30$ (a) and $T = 1.31$. Comparison with the theoretical results (Fig. 9 (b)) shows not only that the transitions from two peaks at low temperatures to three peaks and finally to one peak below the critical temperature are observed but also that the temperature range of these transitions is in good agreement with the theoretical predictions.

(a) $N = 64$

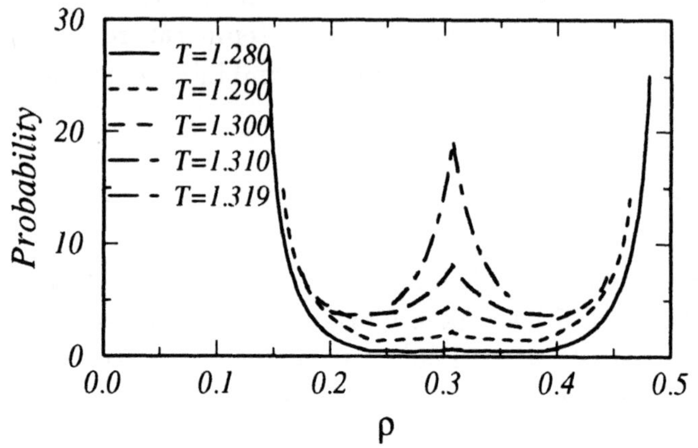

(b) $N = 216$

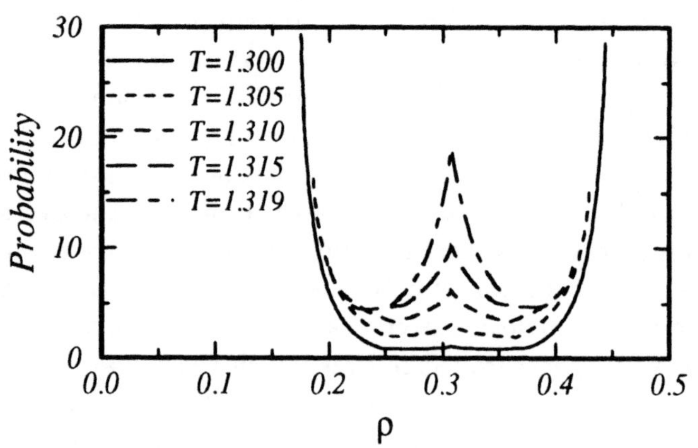

Fig. 9. The calculated density probability functions for a Lennard-Jones fluid (eqn (57)) at different temperatures for (a) $N = 64$ particles and (b) $N = 216$ particles. Note that in each figure the temperatures are different and that the curves are drawn for $\rho_g(T) < \rho < \rho_l(T)$.

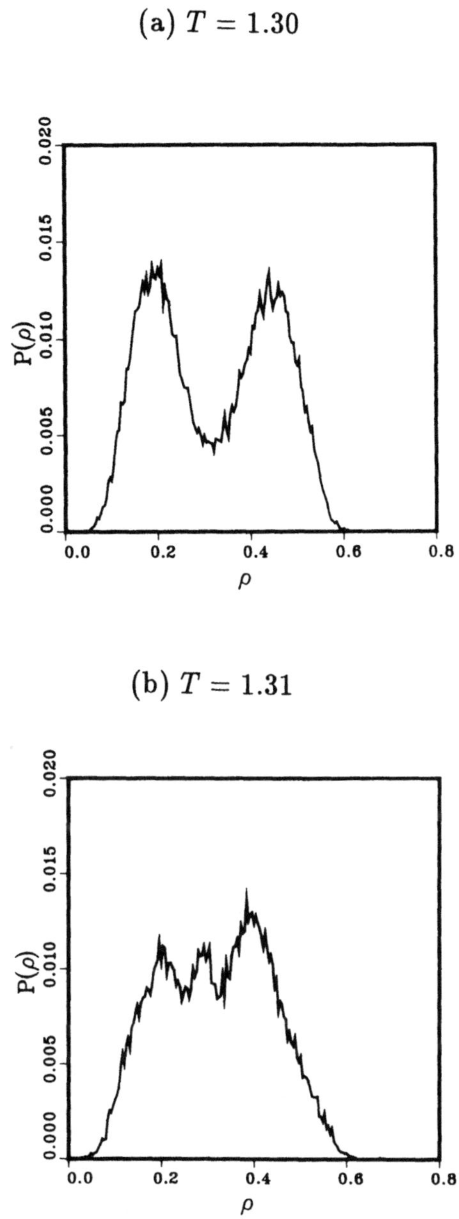

Fig. 10. The density probability functions for a Lennard-Jones fluid as obtained from simulations ($N = 216$) at two temperatures. These probability functions were obtained by calculating each cycle the volumes of the subsystems and densities in both subsystems and updating a histogram. At the end of the simulations these histograms were divided by the total number of cycles.

This theoretical analysis suggests that, in the Gibbs ensemble, finite-size effects can be expected which result in an underestimation of the critical temperature of the *finite* system. So, in the Gibbs ensemble, we are faced with finite-size effects of the opposite sign which will partly cancel. Therefore we can expect that the finite-size effects are smaller in this ensemble compared with conventional simulations. This would lead to a better estimate of the critical temperature of the infinite system. It would be interesting to see whether the arguments given above can be put on a more rigorous footing using finite-size scaling techniques.

In the above discussion it is assumed that eqn (58) is valid for a finite system. This assumption breaks down close to the critical temperature where we have to take into account finite-size rounding [26]. If we take these effects into account the interfacial tension is given by

$$\gamma \propto \gamma_0 (1 - T/T_c)^{2\nu} \tilde{f}(L/\xi)\,, \tag{60}$$

where $\tilde{f}(L/\xi)$ is a scaling function and ξ is the correlation length. For temperatures where $L \gg \xi$, eqn (60) reduces to eqn (58). However, close to T_c, if $L \ll \xi$ eqn (60) becomes

$$\gamma \propto \gamma_0 \frac{1}{\xi^2} \times \frac{\xi^2}{L^2} = \gamma_0 \frac{1}{L^2}\,. \tag{61}$$

As a result, the interfacial contribution to the free energy does not go to zero but to a constant value. If one takes this into account in the calculation of the density probability function, compare eqn (57), a third peak will only occur if γ_0 is sufficiently small. At present not much is known about the numerical value of γ_0, which will depend on the details of the intermolecular potential. Therefore it remains to be seen whether a third peak can be observed in a given system.

The finite-size effects for the simulations of phase coexistence in the Gibbs ensemble have been studied in detail by Mon and Binder [26]. In contrast to the theoretical arguments presented here, which predict that the Gibbs ensemble suffers less from finite-size effects than other ensembles, Mon and Binder conclude that the numerical magnitude of finite-size effects in the Gibbs ensemble and in the grand-canonical ensemble are very similar. In fact, detailed inspection of the results presented in Ref. [26] shows that the finite-size effects in the Gibbs ensemble are slightly larger. An important point is that Mon and Binder considered a two dimensional lattice gas. Furthermore, a "restricted" version of the Gibbs ensemble technique was used in which the volume fluctuations are suppressed. For this version of the Gibbs ensemble, we can also estimate the shape of the $P(\rho)$ curve using the same arguments as described above. The difference is that we now have to consider only the parameter x (no volume fluctuations, so y is constant).

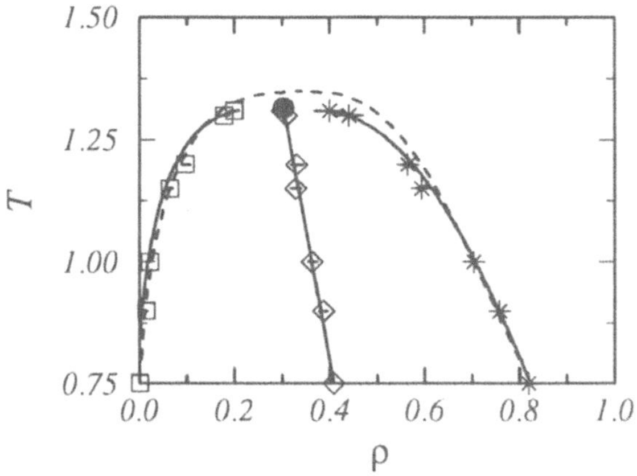

Fig. 11. The phase diagram of the Lennard-Jones fluid (with tail corrections, see [27] for details). The solid lines are the fits to the scaling law and the rectilinear law. The dashed line is the e.o.s. of Nicolas [3]. The points are the simulation results of [8, 9, 28], • is the estimate of the critical point. The simulation are performed with 108, 256, and 512 particles and show no systematic deviations.

As a result, the entropic contributions, which give rise to the third peak and result in an underestimation of the critical point, are completely suppressed. As a result, the theoretical arguments described here for the restricted form of the Gibbs ensemble would predict that the finite-size effects would be comparable to those in other ensembles.

An interesting point that follows from the work of Mon and Binder is the striking difference of the magnitude of the finite-size effects in lattices and continuum systems. For example, the results of Mon and Binder for the two-dimensional lattice gas show that a system of 100 spins overestimates the critical temperature of the infinite system by 20%. While in the results of Gibbs ensemble simulations of the two-dimensional Lennard-Jones fluid could not be detected [29] (see Fig. 11). More careful studies of the finite-size effects in continuum systems have been reported by Rovere *et al.* [30] and Wilding and Bruce [31]. finite-size scaling techniques, as developed for lattice models [24], are applied to the two-dimensional Lennard-Jones fluid. In lattice models the critical density is usually known from symmetry arguments and the critical point is can always be located by changing the temperature. In a continuum system the critical density has to be determined for each system size very accurately before one can apply the scaling techniques. Furthermore, these studies have been performed using many different cut-off values of the Lennard-Jones potential which has a significant influence on the phase diagram [29, 32]. As a result, the importance of finite-size effects in the simulation of a continuum model is still unclear. These studies do show, however, that in contrast to lattice simulations, finite-size effects in continuum systems are significantly smaller and harder to detect.

5. Analyzing the results

Assuming that we have a working algorithm for performing a Monte Carlo simulation in the Gibbs ensemble, we should address the question as to whether the numbers that are generated in a simulation are reliable. First of all, the equilibrium conditions should be fulfilled:

- the pressure in both subsystems must be equal;
- the chemical potential must be equal in both phases.

Unfortunately, the chemical potential and the pressure of the liquid phase are subject to relatively large fluctuations, and therefore the observation that the equilibrium conditions have been fulfilled (within the statistical error) is not always sufficient. In this section we introduce some methods to analyze the data and to judge whether a simulation has been successful.

5.1. IS THE SIMULATION RELIABLE?

An example of a simulation for which the chemical potential and the pressure did not show significant deviations is shown in Fig. 12. In this figure an (x, y) plot of the simulation is given. In section 2.2 it was shown that in the thermodynamic limit the two points in the (x, y) plane which correspond to the coexisting liquid and gas density are sampled. In Fig. 12 we see that, in this simulation, a drift away from these two points has occurred. The reason that these drifts can occur becomes clear if we consider Fig. 8. This figure shows a contour plot of the surface contributions to the free energy of each (x, y) point. Besides the two global minima, which correspond to the densities of the coexisting gas and liquid phase, this figure shows that there are large areas in this plane where the gradient towards the global minima is small. Comparison with Fig. 12 shows that the simulation accidentally became trapped in such an area.

These fluctuations are not immediately apparent when the density probability function is considered, but the average gas and liquid densities can be influenced substantially. Therefore it is important to check for the consistency of the results by making a probability plot in the (x, y) plane.

Another simple indication of a drift in the (x, y) plane can be obtained from a probability distribution of the volumes. At a given temperature the densities of the coexisting phases are determined (but unknown) and because the total number of particles and total volume remain fixed, the probability function of the volume must have two (sharp) peaks if the simulation samples

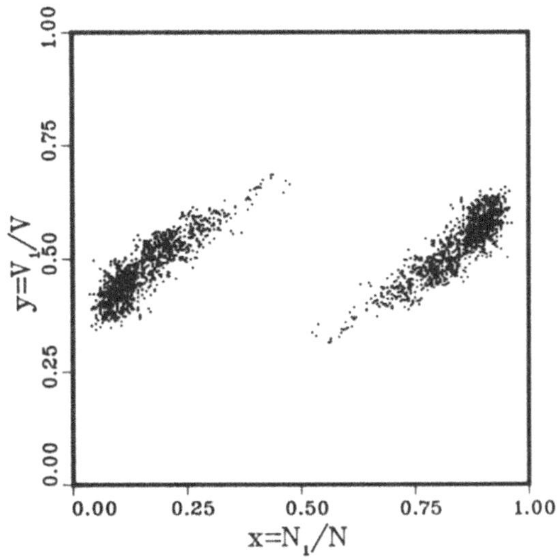

Fig. 12. The probability plot in the (x, y) plane for a Lennard-Jones fluid below the critical temperature ($T = 1.20$ and $N = 216$).

the (x, y) plane correctly. If this probability function does not give two sharp peaks, it is likely that the simulation is trapped in some metastable region.

Generally, if a simulation in the Gibbs ensemble is performed far below the critical temperature, it is not difficult to analyze the results. After the equilibration it becomes clear which of the boxes contains the vapour phase and which the liquid phase. The densities of the coexisting phases can simply be obtained by sampling the densities at regular intervals. For the estimation of the accuracy of the simulation one should be careful since these 'measured' densities are not sampled independently and in estimating the standard deviations of the results one should take this into account (this aspect is discussed in more detail in Appendix A of Ref. [33])

Close to the critical point, however, it is possible that the boxes continuously change 'identity' during a simulation. If, for such a system, the average density is calculated by calculating the average density in the two boxes, one would obtain the overall density. A more convenient method is to make a probability plot for an observed density in one of the boxes. In section 2.2 it was demonstrated that the two maxima of this curve correspond to coexisting vapour and liquid densities. Because this curve is obtained by sampling the density in both boxes, the results are not influenced when the boxes change identity. From this curve, an estimate of the densities can be obtained even very close to the critical temperature (see for example

Fig. 10). Furthermore, it is also clear from section 4 that the appearance of a third peak between the vapour and the liquid peak is an indication that the temperature is very close to the critical temperature (see for example Fig. 10(b)).

5.2. DETERMINING THE CRITICAL POINT

In section 4 it was shown that, due to finite-size effects which are specific to the Gibbs ensemble, the coexistence of vapour and liquid cannot be observed *just below* the critical temperature. Therefore, the highest temperature at which the coexistence can be observed is not a proper estimate of the critical temperature of the system. In order to estimate the critical temperature the results can be fitted to the law of rectilinear diameters [34]

$$\frac{\rho_l + \rho_g}{2} = \rho_c + A\,|T - T_c|\,, \tag{62}$$

where $\rho_l(\rho_g)$ is the density of the liquid (gas) phase, ρ_c the critical density and T_c the critical temperature. Furthermore, the results are also fitted to the scaling law for the density[25]

$$\rho_l - \rho_g = B\,|T - T_c|^\beta \tag{63}$$

where β is the critical exponent (for three dimensional systems $\beta = 0.32$ and for two dimensional systems $\beta = 0.125$ [25]). A and B depend on the system and are obtained from the fit. Note that this scaling law is strictly valid only close to the critical temperature and should be corrected at lower temperatures. In most practical applications the results of the Gibbs ensemble are not sufficiently accurate to justify these higher order terms.

Vega *et al.* [35] used Gibbs ensemble data to determine the effective exponent β (the exponent is not fixed *a priori* but is a result of the fit). Vega *et al.* observed that, within the limited accuracy, the data could be described with an effective exponent β very close to the critical value of 0.32.

6. Applications

The Gibbs ensemble has been used to study the phase behaviour of a various systems. The results of these simulations are reviewed in Ref. [12]. Here we discuss those applications of the Gibbs ensemble for which the algorithm that has been used differs significantly from the one described in section 3.

6.1. DENSE LIQUIDS

At high densities the number of exchange steps can become very large and requires a significant amount of CPU time. This problem occurs also in conventional grand canonical Monte Carlo simulations. Various methods, which are used to extend simulations in the grand canonical ensemble to higher densities, can also be used in the Gibbs ensemble. An example of such a technique is the so-called excluded volume map sampling. This technique, is based on the ideas of Deitrick *et al.* [36] and Mezei [37], and adopted by Stapleton and Panagiotopoulos [38] for the Gibbs ensemble. Before the energy of the particle which has to be inserted is calculated, a map of the receiving subsystem is made, by dividing this subsystem into small boxes which can contain at most one particle. These boxes are labelled when they contain a particle. This map can then be used as a look-up table to check whether there is 'space' for the particle to be inserted. If such space is not available the trial configuration can be rejected immediately.

6.2. POLAR AND IONIC FLUIDS

Because of the long range of the dipolar and Coulombic interactions, the dipolar and Coulombic potential can not simply be truncated. Special techniques, such as the Ewald summation or reaction field [39], have been developed to take into account the long-range nature of the potential in a simulation. It is a general (mis)belief that special techniques are only needed for quantities like the dielectric constant. For thermodynamic quantities it is commonly held that the differences between, for example, the Ewald summation and the minimum-image boundary conditions, are very small. It turns out that the phase diagram is very sensitive to the details of the way the long-range part of the potential is taken into account.

A study of the phase behaviour of the Stockmayer fluid (a Lennard-Jones potential plus a point dipole) [23, 40] using the Gibbs ensemble technique has shown that, without the Ewald summation, the N-dependence is much larger [41]. In fact, performing simulations in which the potential is simply truncated at half the box size can lead to wrong results. When the potential is long ranged, and simply truncated at half the box size, the effective potential is quite different for a large volume compared to a small volume. Since in the Gibbs ensemble the volume of a subsystem varies significantly, a situation can occur in which one volume is much larger than the other. The system then tries to find equilibrium between two boxes with *different* potentials.

Moreover, Panagiotopoulos [42] and Valleau [43, 44] have performed simulations of the restrictive primitive model (a hard-core potential with a point charge) to model an ionic solution. Panagiotopoulos used the Gibbs ensemble technique with Ewald summation, while Valleau used the density scaling technique [44] with the minimum image boundary condition. The results of the two studies turned out to be different. This suggests that the phase diagram is sensitive to the way the long-range interactions are taken into account. Also for the simple Lennard-Jones fluid it is observed that the phase diagram is very sensitive to the details of the truncation of the potential [27, 29] and that these differences are by no means small.

6.3. HARD-CORE POTENTIALS

Most of the rules of thumb as described in section 3.2 are not valid for systems in which particles interact with a hard-core potential. A configuration can immediately be rejected once *one* overlap between hard cores has been detected. In general, rejecting a configuration takes much less computer time than accepting of a configuration. For the hard-core fluids one can utilize some of the techniques developed by Wood [45] for constant pressure simulations of hard disks and which have been extended by Frenkel [46] for simulations of non-spherical convex bodies, to reduce the computer costs significantly. Below, some of these techniques as used by Smit and Frenkel for Gibbs-ensemble simulations of the hard-core Yukawa fluid [47] are described.

A particle displacement. During the simulations a neighbour list for each particle can be made without adding much additional computer time. This list contains all particles within a radius equal to the maximum displacement. This list can be used to test for possible overlaps before the energy of the new configuration is calculated. As a result the computer time required for a rejected move is much less than the time required for an accepted move. Therefore, the optimum maximum displacement results in an acceptance ratio which is far less than the normal 50%.

A volume change. It is straightforward to keep track of the minimum distance between a pair of particles in each box during the simulations. After a volume change, the attempt can be rejected immediately if one of these minima becomes smaller than the hard-core radius.

A particle exchange. Before the energy is calculated one can check whether the particle that is inserted overlaps with one of the other particles.

6.4. MIXTURES

The Gibbs ensemble technique can also be extended to study the behaviour of mixtures. The appropriate acceptance rules are given in Ref. [15]. Examples of mixtures that have been studied using the Gibbs ensemble technique are mixtures of Lennard-Jones particles [15, 48, 49].

One of the main problems in studying liquid-liquid systems is that both phases are relatively dense. It is therefore difficult to exchange particles between the two phases, in particular when the components differ significantly in size. For phase equilibria in mixtures it is sufficient that the chemical potential of one of the components, labelled i, is equal in both phases and for the other components only the difference between the chemical potential and the chemical potential of component i needs to be equal in the two phases. For a simulation this observation implies that it is sufficient to specify the chemical potential of one of the components and the differences for the others. Such simulations are called semi-grand canonical ensemble simulations [50]. In a simulation one can ensure that the difference in chemical potential between two components is constant by performing Monte Carlo moves in which the identity of the particles is changed.

The ideas of the semi-grand canonical ensemble simulations have been adopted by Panagiotopoulos [51] for the Gibbs ensemble. Instead of attempts to insert and remove particles for both components, only the smallest particles are inserted, while for the larger particles only moves involving a change of identity are used.

When mixtures are studied which are symmetric with respect to particle interchange, the density or compositions of the two phases must be equal by symmetry. As a result it is not necessary to perform volume changes [52, 53].

6.5. CHAIN MOLECULES

The applications of the Gibbs ensemble as described above are all limited to model systems containing atoms or small molecules. Using the conventional (random) sampling schemes, successful insertions can only be achieved for small molecules. At typical liquid densities, the probability of a successful insertion of a monomer is of the order of 0.5%; the probability of inserting a chain of eight of these monomers is less than 10^{-18}. Such a low probability of insertion would necessitate a prohibitive amount of computer time. Here we show that configurational-bias Monte Carlo techniques (see Frenkel's contribution, Chapter 4 in this volume) can be combined with the Gibbs ensemble method [54, 55].

We can use the configurational-bias scheme [56] to grow the chains in such a way that the "holes" in the system are found. However, such a scheme of insertion would bias the simulations if the ordinary acceptance rules were used. The correct distribution of configurations can be sampled if the acceptance rule for this step is modified.

Consider a trial move to remove a molecule from one box (say, 1) and insert it in the other box (2). To achieve this, we insert a chain molecule in box 2, using a stepwise method. First, we attempt to insert a single monomer in box 2. Next, k random trial segments are generated, such that the next monomeric unit is located somewhere on a spherical shell around the first monomer. For each of these trial segments, the potential energy due to interaction with the other particles in the system is calculated, and one of the directions, say direction i, is selected with a probability given by

$$P_2(n) = \frac{\exp[-\beta v_2(i)]}{\sum_{j=1}^{k} \exp[-\beta v_2(j)]}, \tag{64}$$

where $\beta = 1/k_{\mathrm{B}}T$ and $v(j)$ is the energy of a j-th trial direction and n labels the position of the segment in the chain. The subscript (in this case, 2) indicates the box in which these quantities are calculated. In addition, we compute a weight factor

$$W_2(n) = \left(\sum_{j=1}^{k} \exp[-\beta v_2(j)] \right) W_2(n-1). \tag{65}$$

Note that $W_2(0)$, the weight factor for the monomer, is simply the Boltzmann factor associated with the random insertion of a monomer. At the same time we calculate a corresponding weight factor $W_1(n)$ for the chain that we chose to remove from box 1

$$W_1(n) = \left(\sum_{j=1}^{k} \exp[-\beta v_1(j)] \right) W_1(n-1), \tag{66}$$

where the summation is over the same set of trial directions as in box 2, with the restriction that the orientation that is selected in box 2 is replaced by the actual orientation of the chain in box 1 of that particular segment. This procedure is repeated until the chain has the desired length ($n = l$).

To derive the appropriate acceptance rules, we impose detailed balance on our Monte Carlo procedure. This implies that, in equilibrium, the rate at which particles are moved from box 1 to 2 equals the reverse rate:

$$N(1)\, P(1|2)\, \mathrm{acc}(1|2) = N(2)\, P(2|1)\, \mathrm{acc}(2|1). \tag{67}$$

Note that we now have to take into account the probability that, starting from the current configuration, a configuration 2 is generated with $n_1 - 1$ particles in box 1 ($P(1|2)$) (compare section 3). The ratio of the statistical weights $N(1)$ and $N(2)$ follows from eqn (46). Substitution of this into eqn (67), using eqns (64)–(66) we find that the following acceptance rule for the exchange step will satisfy the detailed balance condition

$$\mathrm{acc}(1|2) = \min\left(1, \frac{n_1(V - V_1)}{(N - n_1 + 1)V_1} \frac{W_2}{W_1}\right). \tag{68}$$

This demonstrates that we can use the biased insertion and still sample the correct distribution of configurations, provided that we use the acceptance rules given by eqn (68). At this stage, it is worthwhile pointing out that, with trivial extensions, the above scheme can be applied to mixtures of chain molecules.

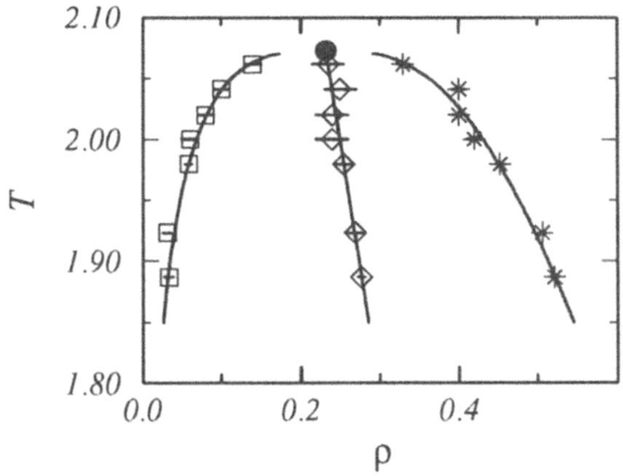

Fig. 13. Liquid-vapour coexistence curve of a system of 200 chains of 8 Lennard-Jones monomers. The monomers are connected by bonds of a fixed length σ. The bonds are allowed to rotate freely with respect to each other. The monomer-monomer interaction is modeled by a Lennard-Jones potential that is cut off at a radius $R_c = 2.5\sigma$ and shifted. The estimate of the critical point is indicated by a black dot. (Figure from Ref. [54])

An illustration of the scheme described above is shown in Fig. 13, This figure gives the phase diagram of a chain of eight Lennard-Jones 'monomers' [54]. This scheme has also been applied successfully to determine the phase diagram of alkanes by de Pablo *et al.* [55].

Acknowledgements

I would like to thank Daan Frenkel, Philip de Smedt, Kurt Binder, and Monica van Leeuwen for many useful discussions and suggestions.

References

[1] D. Frenkel, Free energy computations and first order phase transitions, in *Molecular Dynamics Simulations of Statistical Mechanics Systems*, edited by G. Ciccotti and W. G. Hoover, Proceedings of the 97th Int. Enrico Fermi School of Physics, 1986.

[2] J. M. Prausnitz, R. N. Lichtenthaler, and E. Gomes de Azevedo, *Molecular Thermodynamics of Fluid-Phase Equilibria*, Prentice-Hall, Englewood Cliffs N. J., second edition, 1986.

[3] J. J. Nicolas, K. E. Gubbins, W. B. Streett, and D. J. Tildesley, *Molec. Phys.* **37**, 1429 (1979).

[4] J. K. Johnson, J. A. Zollweg, and K. E. Gubbins, *Molec. Phys.* , in press.

[5] M. R. Reddy and S. F. O'Shea, *Can. J. Phys.* **64**, 677 (1986).

[6] B. Widom, *J. Chem. Phys.* **39**, 2802 (1963).

[7] K. S. Shing and K. E. Gubbins, *Molec. Phys.* **45**, 129 (1982).

[8] A. Z. Panagiotopoulos, *Molec. Phys.* **61**, 813 (1987).

[9] B. Smit, P. de Smedt, and D. Frenkel, *Molec. Phys.* **68**, 931 (1989).

[10] D. Frenkel, Monte Carlo simulations, in *Computer Modelling of Fluids, Polymers and Solids*, edited by C. R. A. Catlow, S. C. Parker, and M. P. Allen, Kluwer Academic Publishers, 1990.

[11] B. Smit and D. Frenkel, *Molec. Phys.* **68**, 951 (1989).

[12] A. Z. Panagiotopoulos, *Molec. Simul.* **9**, 1 (1992).

[13] D. Ruelle, *Statistical Mechanics: Rigorous Results*, Benjamin, Reading Massachusetts, 1969.

[14] R. B. Dingle, *Asymptotic Expansions, their Derivation and Interpretation*, Academic Press, 1973.

[15] A. Z. Panagiotopoulos, N. Quirke, M. Stapleton, and D. J. Tildesley, *Molec. Phys.* **63**, 527 (1988).

[16] P. Sindzingre, G. Ciccotti, C. Massobrio, and D. Frenkel, *Chem. Phys. Lett.* **136**, 35 (1987).

[17] K. Shing and S. T. Chung, *J. Phys. Chem.* **91**, 1674 (1987).

[18] N. G. Van Kampen, *Stochastic Processes in Physics and Chemistry*, North-Holland, Amsterdam, 1981.

[19] W. Feller, *An Introduction to Probability Theory and its Applications*, volume 1, Wiley, New York, 1957.

[20] W. Feller, *An Introduction to Probability Theory and its Applications*, volume 2, Wiley, New York, 1966.

[21] P. Eppenga and D. Frenkel, *Molec. Phys.* **52**, 1303 (1984).

[22] I. R. McDonald, *Molec. Phys.* **23**, 41 (1972).

[23] B. Smit, C. P. Williams, E. M. Hendriks, and S. W. de Leeuw, *Molec. Phys.* **68**, 765 (1989).

[24] K. Binder, *Z. Phys. B. Cond. Mat.* **43**, 119 (1981).

[25] J. S. Rowlinson and B. Widom, *Molecular Theory of Capillarity*, Clarendon Press, Oxford, 1982.

[26] K. K. Mon and K. Binder, *J. Chem. Phys.* **96**, 6989 (1992).

[27] B. Smit, *J. Chem. Phys.* **96**, 8639 (1992).

[28] B. Smit and C. P. Williams, *J. Phys. Cond. Mat.* **2**, 4281 (1990).

[29] B. Smit and D. Frenkel, *J. Chem. Phys.* **94**, 5663 (1991).

[30] M. Rovere, D. W. Hermann, and K. Binder, *J. Phys. Cond. Mat.* **2**, 7009 (1990).

[31] N. B. Wilding and A. D. Bruce, *J. Phys. Cond. Mat.* **4**, 3087 (1992).

[32] J. E. Finn and P. A. Monson, *Phys. Rev. A* **39**, 6402 (1989).

[33] B. Smit, *Computer simulation of phase coexistence: from atoms to surfactants*, PhD thesis, Rijksuniversiteit Utrecht, The Netherlands, 1990.

[34] J. S. Rowlinson and F. L. Swinton, *Liquids and Liquid Mixtures*, Butterworth, London, third edition, 1982.

[35] L. Vega, E. de Miguel, L. F. Rull, G. Jackson, and I. A. McLure, *J. Chem. Phys.* **96**, 2296 (1992).

[36] G. L. Deitrick, L. E. Scriven, and H. T. Davis, *J. Chem. Phys.* **90**, 2370 (1989).

[37] M. Mezei, *Molec. Phys.* **40**, 901 (1980).

[38] M. R. Stapleton and A. Z. Panagiotopoulos, *J. Chem. Phys.* **92**, 1285 (1990).

[39] M. P. Allen and D. J. Tildesley, *Computer Simulation of Liquids*, Clarendon, Oxford, 1987.

[40] M. E. van Leeuwen, B. Smit, and E. M. Hendriks, *Molec. Phys.* (1992), in press.

[41] D. J. Tildesley, private communication, 1989.

[42] A. Z. Panagiotopoulos, *Fluid Phase Equilibria* **76**, 97 (1992).

[43] J. P. Valleau, *J. Chem. Phys.* **95**, 584 (1991).

[44] J. P. Valleau, *J. Comput. Phys.* **96**, 193 (1991).

[45] W. W. Wood, *J. Chem. Phys.* **52**, 729 (1970).

[46] D. Frenkel, (private communications), 1990.

[47] B. Smit and D. Frenkel, *Molec. Phys.* **74**, 35 (1991).

[48] M. E. van Leeuwen, C. J. Peter, J. de Swaan Arons, and A. Z. Panagiotopoulos, *Fluid phase equilibria* **66**, 57 (1991).

[49] V. I. Harismiadis, N. K. Koutras, D. P. Tassios, and A. Z. Panagiotopoulos, *Fluid phase equilibria* **65**, 1 (1991).

[50] D. A. Kofke and E. D. Glandt, *Molec. Phys.* **64**, 1105 (1988).

[51] A. Z. Panagiotopoulos, *Int. J. Thermophys.* **10**, 447 (1989).

[52] J. G. Amar, *Molec. Phys.* **4**, 739 (1989).

[53] R. D. Mountain and A. H. Harvey, *J. Chem. Phys.* **94**, 2238 (1991).

[54] G. C. A. M. Mooij, D. Frenkel, and B. Smit, *J. Phys. Cond. Mat.* **4**, L255 (1992).

[55] M. Laso, J. J. de Pablo, and U. W. Suter, *J. Chem. Phys.* **97**, 2817 (1992).

[56] D. Frenkel, G. A. M. Mooij, and B. Smit, *J. Phys. Cond. Mat.* **4**, 3053 (1992).

Effective Pair Potentials and Beyond

M. Sprik
IBM Research Division
Zurich Research Laboratory
CH 8803 Rüschlikon
Switzerland

ABSTRACT. Interaction site models are the most popular type of potentials used in the simulation of molecular systems. This Chapter reviews the principles underlying this class of model, which is based on effective pair potentials supplemented, if necessary, with many-body terms. We will discuss these two elements and also give a few examples of the practice of model parametrization. Our point of view will be a rather pragmatic focus on potentials that have actually been applied in large-scale simulations.

1. Introduction

Computer simulation enables us to study the motion of the particles of model systems in microscopic detail. The method is now accepted as a direct approach to the understanding of structure and dynamics of condensed matter. Throughout its development in the past three decades, the emphasis has been on systems at finite temperature for which fluctuations are crucial. Thus, the first applications of both the Monte Carlo and molecular dynamics techniques were studies of simple atomic liquids, such as argon. In fact, the field of the statistical mechanics of liquids, with concepts and methods rather different from traditional theoretical solid state physics, has dominated, if not invented, computer simulation as a tool in condensed matter theory (for a review see Refs. [1–3]). Following the application to simple liquids, the method was extended to ionic and molecular liquids [1–3] and, most importantly, to polymers and biological systems [4]. More recently, covalent systems (liquids and solids) and metals have also been investigated (see Ref. [3] and two previous volumes [5, 6] in the NATO ASI series).

In the course of the development sketched above, simulation became increasingly dependent on the availability of realistic model potentials. The interactions in atomic liquids can still be described, to a fairly good approximation, by simple pair potentials (e.g. Lennard-Jones) which have been

M.P. Allen and D.J. Tildesley (eds.), Computer Simulation in Chemical Physics, 211–259.

fitted to properties of the dimer. When a quantitative comparison with experiment is required, the dimer model can be replaced by a similar, effective pair potential, which corrects for many-body effects in a mean-field approximation. Often a simple scale transformation is sufficient to relate the predictions of different models (correspondence of states, see e.g. [2]). For ionic systems, as well as the intermolecular interactions in hydrogen-bonded molecular liquids, many-body effects are more severe, and models based on atom-atom potentials derived from gas-phase (or *ab initio*) dimer surfaces can yield qualitatively wrong results. Fortunately, for several important systems of this class, e.g. water, it turned out that effective pair potentials acting between atomic centers are still adequate. For covalent materials and metals, on the other hand, the effective pair approximation fails, and many-body terms have to be included explicitly.

The study of complex molecular systems in physical chemistry and biology has become one of the major applications of computer simulations. The interaction models used in this field are essentially all of the same type. They are effective pair potentials supplemented, if necessary, with simple many-particle terms. The objective in this Chapter is to review the principles underlying this "standard" model. We start with a brief overview of the quantum theory of molecules and cohesion in condensed systems, which introduces the elements from which the "standard" model is built. The functional form of potentials is ultimately based on these first-principles considerations. However, the determination of the parameter set is a far less rigorous step in the development of a potential, and here the various implementations of the "standard" model may show considerable differences. Next, we will discuss two extensions that treat many-body effects explicitly. The first is an approach for dealing with electronic polarization in aqueous systems. As an application, we will present in some detail an example of the conversion of a non-polarizable effective pair potential for aqueous NaCl into a polarizable model, which also serves as an illustration of how effective pair potentials exploit the cancellation of errors. A polarizable model for a simple surfactant based on the same approach will be outlined. We conclude with an example of a three-body potential for the directional bonding in covalent systems and the generalization for the modelling of chemisorption of closed-shell molecules on metal substrates.

The emphasis of this Chapter is on the concepts which constitute the basis of the functional form of potentials and on the practice of the parametrization of models. Applications will be mentioned only as illustrations. For details, the reader is referred to the original literature. In the selection of examples, a certain bias towards the work of the author and his close colleagues is evident. We feel that yielding to our personal preferences can be excused in the present context, since the following is not intended as a review of any of the particular topics that serve as illustrations.

2. Where potentials come from

2.1. THE BORN-OPPENHEIMER PRINCIPLE

Perhaps the most fundamental property of materials in our everyday environment is their atomic structure, i.e. the fact that they can be characterized in terms of the position and motion of more or less precisely defined constituent particles, the "atoms". The question why this is at all possible is far from trivial. Even more amazing is the enormous complexity and variety of structures. Classical physics is unable to explain these basic observations and definite answers were only obtained with the development of quantum mechanics. According to quantum theory, the stability of atomic structure is a consequence of the Schrödinger equation, the Pauli principle and one of those fortunate "coincidences" in the constants of nature, namely the extremely small ratio of the electron mass m to the mass M of nuclei ($m/M \approx 10^{-4}$).

The implications of the small value of m/M is known as the Born-Oppenheimer principle (see e.g. Ref. [7]). In the formulation for zero temperature this principle states that the groundstate of a collection of n electrons and N nuclei can be found to a very good approximation by solving the Schrödinger equation for the electrons with fixed positions $\{\mathbf{R}_i, i = 1, N\}$ of the nuclei, and then minimizing the energy eigenvalue thus obtained with respect to the vectors $\mathbf{R}_i$. Hence, from the lowest eigenvalue solution $\mathcal{E}(\{\mathbf{R}_i\})$ of the electronic time-independent Schrödinger equation

$$\left(-\frac{\hbar^2}{2m}\sum_{\alpha=1}^{n}\frac{\partial^2}{\partial \mathbf{r}_\alpha^2} - e\sum_{\alpha=1}^{n}\sum_{i=1}^{N}\frac{q_i}{|\mathbf{r}_\alpha - \mathbf{R}_i|}\right)\Psi(\{\mathbf{r}_\alpha\}) = \mathcal{E}(\{\mathbf{R}_i\})\Psi(\{\mathbf{r}_\alpha\}) \quad (1)$$

where $\{\mathbf{r}_\alpha, \alpha = 1, n\}$ are the electronic coordinates and q_i is the charge of nucleus i, we derive a potential function for the nuclear coordinates by adding $\mathcal{E}(\{\mathbf{R}_i\})$ to the nuclear Coulomb energy as if both terms were classical potential energies

$$\mathcal{V}_{\mathrm{BO}}(\{\mathbf{R}_i\}) = \sum_{j<k}^{N}\frac{q_j q_k}{|\mathbf{R}_j - \mathbf{R}_k|} + \mathcal{E}(\{\mathbf{R}_i\}). \quad (2)$$

The function $\mathcal{V}_{\mathrm{BO}}$ is usually referred to as the Born-Oppenheimer function (or surface). The groundstate of the combined system is found by relaxing the nuclear configuration:

$$\frac{\partial}{\partial \mathbf{R}_i}\mathcal{V}_{\mathrm{BO}}(\{\mathbf{R}_i\}) = 0, \quad i = 1, N. \quad (3)$$

Only solutions of eqn (1) satisfying Pauli's antisymmetrization condition for fermions are permitted, which gives $\Psi(\{\mathbf{r}_\alpha\})$ an intricate shell structure with many anisotropic nodal surfaces. The complexity of the electronic states, in turn, implies that eqn (3) has numerous distinct locally-stable solutions even for small values of $N > 1$. Indeed, with only a few carbon, oxygen and hydrogen atoms, a chemist can make an astonishing variety of molecules with very different properties. Hence the basic answer to our very general question is that, given the fact that materials obey quantum laws, it is the Born-Oppenheimer principle that is the origin of the stability of atomic structure as well as its complexity.

It should always be kept in mind that eqns (1–3) are an approximation to the groundstate problem. The condition $m/M \ll 1$ imposes a separation of time scales on the zero-point motion of the two components, the nuclei moving much slower than the electrons. Violation of this condition is most likely to occur for lighter nuclei, such as hydrogen ($m/M_{\mathrm{H}} \approx 10^{-3}$). As a result of the high zero-point frequencies of light nuclei, the electronic state is unable to adjust instantaneously to the rapidly changing configuration $\{\mathbf{R}_i\}$ and the adiabatic (Born-Oppenheimer) approximation breaks down. Similarly, at finite temperature, the heavier nuclei may also acquire velocities high enough to induce electronic transitions, and the potential of eqn (2) is no longer a valid description of the interactions between the nuclei. Non-adiabatic effects can be very important in dynamics, e.g. in chemical reactions (see the contribution of Coker, Chapter 9). However, for finite-temperature equilibrium averages, the corrections to the Born-Oppenheimer approximation are minor. Therefore, we will assume that the nuclei can be treated as classical particles with a dynamics driven by the forces derived from the Born-Oppenheimer potential function $\mathcal{V}_{\mathrm{BO}}$

$$M_i \frac{\mathrm{d}^2 \mathbf{R}_i}{\mathrm{d}t^2} = -\frac{\partial}{\partial \mathbf{R}_i} \mathcal{V}_{\mathrm{BO}}(\{\mathbf{R}_j\}). \tag{4}$$

2.2. THE HYDROGEN AND HELIUM DIMER

The Born-Oppenheimer principle has restored only partly our intuitive picture of matter as consisting of classical particles with a structure determined by the positions. These particles hardly conform to our expectations of what atoms look like. The Born-Oppenheimer surface of eqn (2) is a global function and, as a result, the forces on a given particle (eqn (4)) depend on the coordinates of all other particles irrespective of whether they are near or far. The traditional concept of atoms, however, assumes that interactions involving real chemistry are local: atoms repel each other at short range and, in general, bonds are formed only with nearest neighbors.

A formal derivation of the locality of atomic interactions from quantum theory is not easy. In fact, such a derivation in all generality is not possible since some systems do exhibit highly non-local properties. Most textbooks (here we will follow Ref. [8], see also [7])) give a justification that proceeds in two steps. First the dimers of the two simplest atoms, hydrogen and helium, are considered. This introduces the two basic types of interactions, bonded and nonbonded. Let $\phi_a(\mathbf{r})$ be the electronic groundstate wavefunction of hydrogen atom a, and $\phi_b(\mathbf{r})$ the groundstate of hydrogen atom b. Expansion of the electron pair wavefunction $\Psi(\mathbf{r}_1, \mathbf{r}_2)$ of the hydrogen dimer in the two product wavefunctions $\phi_a(\mathbf{r}_1)\phi_b(\mathbf{r}_2)$ and $\phi_a(\mathbf{r}_2)\phi_b(\mathbf{r}_1)$ gives a 2×2 secular equation which is diagonalized by the symmetric and antisymmetric combinations

$$\Psi_\pm(\mathbf{r}_1, \mathbf{r}_2) = \frac{\phi_a(\mathbf{r}_1)\phi_b(\mathbf{r}_2) \pm \phi_a(\mathbf{r}_2)\phi_b(\mathbf{r}_1)}{\sqrt{2(1 \pm S^2)}} \tag{5}$$

with energy eigenvalues

$$E_\pm = 2E_{\mathrm{H}} + \frac{J \pm K}{1 \pm S^2}. \tag{6}$$

E_{H} is the energy of the electronic state ϕ of a single hydrogen atom and S is the overlap between the atomic orbitals (AO's)

$$S = \int \mathrm{d}\mathbf{r}\; \phi_a(\mathbf{r})\phi_b(\mathbf{r}). \tag{7}$$

The direct and exchange integrals J and K are defined as

$$J = \int \mathrm{d}\mathbf{r}_1 \mathrm{d}\mathbf{r}_2\; \phi_a(\mathbf{r}_1)\phi_b(\mathbf{r}_2) \left(\frac{1}{R_{ab}} + \frac{1}{r_{12}} - \frac{1}{r_{1b}} - \frac{1}{r_{2a}} \right) \phi_a(\mathbf{r}_1)\phi_b(\mathbf{r}_2) \tag{8}$$

$$K = \int \mathrm{d}\mathbf{r}_1 \mathrm{d}\mathbf{r}_2\; \phi_a(\mathbf{r}_1)\phi_b(\mathbf{r}_2) \left(\frac{1}{R_{ab}} + \frac{1}{r_{12}} - \frac{1}{r_{1b}} - \frac{1}{r_{2a}} \right) \phi_b(\mathbf{r}_1)\phi_a(\mathbf{r}_2), \tag{9}$$

where we have changed to atomic units for convenience ($e = 1$). Expanding J and using symmetry we obtain

$$J = \frac{1}{R_{ab}} + \int \mathrm{d}\mathbf{r}_1 \mathrm{d}\mathbf{r}_2\; \frac{\phi_a(\mathbf{r}_1)^2 \phi_b(\mathbf{r}_2)^2}{r_{12}} - 2 \int \mathrm{d}\mathbf{r}_1 \frac{\phi_a(\mathbf{r}_1)^2}{r_{1b}}. \tag{10}$$

This will be quite small, since it represents the Coulomb energy between two overlapping, classically distinguishable, spherically symmetric neutral atoms. If we write K in a similar way, we find

$$\begin{aligned} K = {} & \frac{S^2}{R_{ab}} + \int \mathrm{d}\mathbf{r}_1 \mathrm{d}\mathbf{r}_2\; \frac{\phi_a(\mathbf{r}_1)\phi_b(\mathbf{r}_2)\phi_b(\mathbf{r}_1)\phi_a(\mathbf{r}_2)}{r_{12}} \\ & -2S \int \mathrm{d}\mathbf{r}_1 \frac{\phi_a(\mathbf{r}_1)\phi_a(\mathbf{r}_1)}{r_{1b}}. \end{aligned} \tag{11}$$

The terms in eqn (11) depend on the exchange-charge density $\phi_a\phi_b$ located in the overlap region between the nuclei, and can be interpreted semiclassically as the combined Coulomb interaction of an overlap charge and two nuclei with reduced charge S. Provided R_{ab} is not too small, the last term, which is attractive, will dominate, and K is negative. Hence, according to eqn (6), the symmetric state has a lower energy than the antisymmetric state. Moreover in this range of nuclear separation we also have $|K| > |J|$. As a result, the E_+ level is pushed below, while the E_- state is raised above, the energy $E_{\rm H}$ of the free hydrogen atoms. Ψ_+ forms a bonding, and Ψ_- an antibonding, state.

Equations (5–11) summarize the celebrated Heitler-London treatment of the covalent bond of the H_2 molecule [7, 8]. We will continue now with a similar argument for why the interactions between a pair of helium atoms are nonbonding and repulsive. In He the nuclear charge is doubled and the atomic orbital ϕ is now occupied by two electrons. Therefore, to a first approximation, the groundstate of a helium dimer is the antisymmetric product (Slater determinant) $\phi_a(\mathbf{r}_1)\overline{\phi_a(\mathbf{r}_2)}\phi_b(\mathbf{r}_3)\overline{\phi_b(\mathbf{r}_4)}$, with the bar denoting the spin state. An alternative way of building the He_2 electron configuration is filling both the bonding and antibonding (Ψ_+ and Ψ_-) states of eqn (5). This is known as the MO (molecular orbital) picture. The expectation value of the total energy can be expressed in a form similar to eqn (6)

$$E = 2E_{\rm He} + \frac{4(J-K)}{1-S^2} + I. \tag{12}$$

J and K are the same direct and exchange integrals of eqns (8–11) containing the attractive interactions of the electrons with the nuclei. I is a sum of electron-electron Coulomb integrals. From expression (12) we see that the exchange integral K which, being negative, was responsible for the binding in the Ψ_+ state of H_2 (eqn (6)), now appears with the opposite sign and leads to the well known four-electron repulsion in a nonbonding state. In the MO picture, this is usually rationalized by the statement that the energies of bonding and antibonding states do not exactly cancel, but add up to a value larger than the sum of two isolated atoms, as is evident from eqn (5). The foremost important conclusion from our very brief excursion into quantum chemistry is that the interaction of the overlap charge with the nuclei is the origin of both bonding and nonbonding interactions. In bonded states, the effect is attractive and prevails over the nucleus-nucleus repulsion beyond a certain minimum separation. For nonbonded states, by contrast, it adds to the repulsion. The effective diameters for nonbonded hard cores are, as a result, considerably larger than covalent bond distances ($\approx$ a factor of two). Also, the corresponding overlap charge densities at equilibrium distances for nonbonded systems ($\rho < 0.001$ a.u.) are at least two orders of magnitude smaller than for covalent bonds ($\rho \approx 0.1$ a.u.).

2.3. DISPERSION FORCES

Since the hard-core forces are determined by the overlap, they are short range and very steep. At larger distances, however, the coupling between He atoms does not completely vanish. A very weak attraction remains. This attraction arises from electronic correlation effects and is called the dispersion or Van der Waals interaction. Atomic charge distributions, though on average spherical, are continuously distorted to some degree by zero-point motion. The instantaneous dipole moment $\boldsymbol{\mu}_1$ of such a fluctuation of atom 1 can induce a dipole moment $\boldsymbol{\mu}_2$ on atom 2 by means of the dipole-dipole coupling

$$E_{\mathrm{DD}} = \frac{\boldsymbol{\mu}_1 \cdot \boldsymbol{\mu}_2}{R_{12}^3} - \frac{3(\boldsymbol{\mu}_1 \cdot \mathbf{R}_{12})(\boldsymbol{\mu}_2 \cdot \mathbf{R}_{12})}{R_{12}^6}. \tag{13}$$

The magnitude of the induced moment is proportional to the electronic polarizability α and scales with distance as $1/R^3$. The effective strength of the interaction between a dipole fluctuation and its induction image on another atom is proportional to the square of these factors

$$E_{\mathrm{disp}} = -\frac{3\hbar\omega_0\alpha^2}{4R^6}, \tag{14}$$

where $\hbar\omega_0$ is an average electronic excitation energy. Expression (14) is due to London. Instead of repeating London's original derivation, we will give a simplified version based on the Drude model. We will return to this model later in our treatment of electronic polarization. An atom with induced dipole moment $\boldsymbol{\mu}$ is represented by a heavy charge q (the nucleus) and a light charge $-q$ (the electron) connected by a harmonic spring of length r; hence $\boldsymbol{\mu} = q\mathbf{r}$. The potential energy in an electric field $\mathbf{E}$ is the sum of the electrostatic interaction energy $-\boldsymbol{\mu} \cdot \mathbf{E}$ and a self-energy term

$$E_{\mathrm{pol}} = \frac{\boldsymbol{\mu}^2}{2\alpha}, \tag{15}$$

which can be interpreted as the work needed to polarize the atom. The Hamiltonian is obtained by associating a momentum $\mathbf{p}_\mu$ with the dipole moment $\boldsymbol{\mu}$, and a mass. For convenience, we will write the mass in terms of a spring constant (the inverse polarizability) and the oscillator frequency ω_0

$$\mathcal{H} = \alpha\omega_0^2 \frac{\mathbf{p}_\mu^2}{2} + \frac{\mu^2}{2\alpha} - \boldsymbol{\mu} \cdot \mathbf{E}. \tag{16}$$

Setting the forces to zero, we recover the linear polarizability approximation which underlies the Drude model

$$\frac{\partial}{\partial \boldsymbol{\mu}} \left(\frac{\mu^2}{2\alpha} - \boldsymbol{\mu} \cdot \mathbf{E} \right) = 0 \quad \Rightarrow \quad \boldsymbol{\mu} = \alpha \mathbf{E}. \tag{17}$$

Two Drude oscillators interact via the bilinear coupling of eqn (13). For the purpose of our qualitative discussion we can omit the second anisotropic term, but we will allow the two polarizabilities to be different. This reduces our problem to two coupled harmonic oscillators with a Hamiltonian of the form ($\mathbf{E} = 0$)

$$\mathcal{H}_{12} = \alpha_1 \omega_0^2 \frac{\mathbf{p}_{\mu 1}^2}{2} + \alpha_2 \omega_0^2 \frac{\mathbf{p}_{\mu 2}^2}{2} + \frac{\mu_1^2}{2\alpha_1} + \frac{\mu_2^2}{2\alpha_2} + \frac{\boldsymbol{\mu}_1 \cdot \boldsymbol{\mu}_2}{R_{12}^3} \tag{18}$$

or, in mass-scaled coordinates $\tilde{\boldsymbol{\mu}}_1 = \boldsymbol{\mu}_1 / \sqrt{\alpha_1}\omega_0$ and $\tilde{\boldsymbol{\mu}}_2 = \boldsymbol{\mu}_2 / \sqrt{\alpha_2}\omega_0$,

$$\mathcal{H}_{12} = \frac{\tilde{\mathbf{p}}_{\mu 1}^2}{2} + \frac{\tilde{\mathbf{p}}_{\mu 2}^2}{2} + \omega_0^2 \frac{\tilde{\mu}_1^2}{2} + \omega_0^2 \frac{\tilde{\mu}_2^2}{2} + \frac{\omega_0^2 \sqrt{\alpha_1 \alpha_2}}{R_{12}^3} \tilde{\boldsymbol{\mu}}_1 \cdot \tilde{\boldsymbol{\mu}}_2 \tag{19}$$

After decoupling by a normal mode transformation we obtain

$$\mathcal{H}_{12} = \frac{\mathbf{P}_+^2}{2} + \frac{\mathbf{P}_-^2}{2} + \omega_+^2 \frac{\mathbf{Q}_+^2}{2} + \omega_-^2 \frac{\mathbf{Q}_-^2}{2}, \tag{20}$$

$$\omega_\pm^2 = \omega_0^2 \pm \frac{\omega_0^2 \sqrt{\alpha_1 \alpha_2}}{R_{12}^3}. \tag{21}$$

Assuming the coupling is weak, the groundstate energy (zero-point motion) of two correlated quantum Drude oscillators can be expanded as

$$E_0 = \frac{3}{2}\hbar\omega_+ + \frac{3}{2}\hbar\omega_- \approx 3\hbar\omega_0 - \frac{3\hbar\omega_0 \alpha_1 \alpha_2}{4R_{12}^6}. \tag{22}$$

For completeness we give the general relation derived by London which also allows for different excitation frequencies

$$E_{\text{disp}} = -\frac{3\hbar\omega_1\omega_2\alpha_1\alpha_2}{2\left(\omega_1 + \omega_2\right) R^6}. \tag{23}$$

The example of the Drude atoms clearly demonstrates that the dispersion interaction is a quantum effect. In the classical limit, we recover equipartition of energy for the normal modes of the coupled oscillators and the non-interacting pair, with the result of zero net correlation energy.

2.4. BONDED AND NONBONDED COHESION

With our survey of the H_2 and He_2 dimer, we have introduced the two basic types of binding: bonded and nonbonded. The dimers of atoms with larger atomic numbers follow a similar pattern: the valence electrons play the role of the bonding or antibonding electron pairs, while the inner electron shells plus nucleus can be treated as effective ionic cores. Let us now return to our original question: why do molecules and even solids and liquids appear to consist of atoms? The answer is that binding in larger aggregates is essentially no different from the chemistry of small atomic clusters. It is this basic feature that ultimately justifies the use of potential functions to describe structure and dynamics of condensed phases.

In practice, of course, there are many complications, and some classes of materials are more amenable to modelling by potential functions than others. Let us examine, therefore, the various types of cohesion in condensed matter from this point of view. Again, the distinction between bonded and nonbonded interactions is crucial. The purest cases of nonbonded cohesion are the Van der Waals liquids and solids formed by the rare gases. The atomic closed shells of these elements hardly deform, and repel each other as though they were spheres. The solid structure is close packed and held together by the weak dispersive forces. Another class of nonbonded systems are the ionic compounds. Here, cohesion is provided by the long-range Coulombic forces, and the binding energy is several orders of magnitude larger than for Van der Waals solids. Ionic binding is the result of electron transfer between atoms with a large electronegativity difference, turning them into ions of opposite charge. To the extent that the electron transfer is complete, the outer electronic shells of ions are again closed with strong repulsive interaction at short range. In fact, the various structures of ionic solids (even the ones with low coordination numbers) can be understood surprisingly well in terms of electrostatic energy and packing of hard spheres with different radii.

The paradigm of bonded-cohesion is an sp^3 covalent solid, in particular the diamond phase of silicon. The chemist's view (see e.g. [8]) of binding in silicon is perhaps the most helpful in the present context. According to this picture, each of the electrons in the four singly occupied valence orbitals of a silicon atom is paired with a valence electron of only one of the four neighbors. The valence orbital is oriented and stretched towards its partner in order to maximize overlap. Two matching valence orbitals are superimposed to form a bonded orbital which is doubly occupied. This bonded orbital can be compared to the symmetric H_2 state of eqn (5). However, now there are several bonding orbitals emanating from an atomic center (four in the case of sp^3). Near the ionic core, the orbitals necessarily overlap and, as a result, they are not independent. Similar to the four-electron repulsion in filled $1s$ atomic orbitals of the helium dimer, two filled covalent bonding orbitals avoid each other in order to satisfy Pauli's principle. Unlike nonpenetrating closed shell orbitals of nonbonded systems however, the repulsive bonding orbitals are anchored at the same core. Hence, the only way to minimize the overlap is to optimize the orientation of the orbitals. This view of covalent bonding is called valence-shell electron pair repulsion (VSEPR) theory and is presented in most elementary text books on quantum chemistry [8]. It is very helpful for a qualitative understanding of the bond geometry of bonded interactions. It provides a simple explanation of the tetrahedral bonding structure in silicon which is also very suitable, as we will see, for the construction of many-body potentials.

The outer shells of elements that form metallic solids are also incomplete but now the valence electrons are not localized in well-defined bonds. The electron density is distributed more or less homogeneously over the region between the cores. Therefore, metallic bonding is considerably less directional than covalent bonding. Most metals have coordination numbers more characteristic of close-packed structures, suggesting that the cores behave like spherical atoms. The apparent neutrality of ionic cores is a result of screening by the conduction electrons. Nonetheless, the cohesion in metals cannot be classified as nonbonded, since the residual effect of the conduction electrons is not negligible. One manifestation is the well-known volume dependence of metal atom interactions, but local fluctuations in the ion density also affect the binding energy and therefore the forces. Using various approximate descriptions of metals, it is possible to design reasonably accurate models for metallic systems, in particular the transition metals of the d series. These potentials will not be discussed in this Chapter. For a review and several interesting applications we refer the reader to two previous volumes in the NATO-ASI series [5, 6].

2.5. POTENTIALS FOR CONDENSED PHASES

The conclusion of our discussion is that the quantum theory of condensed matter, in its real space version, seems to encourage us to resolve the Born-Oppenheimer surface (eqn (2)) into interactions between small groups of atoms. Such a partition has the form of a cluster expansion, i.e. with a pair interaction as the leading term followed by the interactions of triplets, quadruplets etc.

$$\mathcal{V}_{\mathrm{BO}}(\{\mathbf{R}_i\}) = \sum_{i<j} v_2\left(\mathbf{R}_i, \mathbf{R}_j\right) + \sum_{i<j<k} v_3\left(\mathbf{R}_i, \mathbf{R}_j, \mathbf{R}_k\right) + \dots . \tag{24}$$

The overview of the various forms of cohesion suggests that expansion of eqn (24) will be most successful for nonbonded systems. For these isotropic systems, we can hope that the higher-order (many-body) terms are small in comparison to the pair term. For systems with bonded cohesion, on the other hand, many-body contributions are crucial. For example, without a three-atom coupling, it will be very difficult to constrain the bond angles in the narrow range of values that are permitted in covalent solids.

Having justified expression (24) with the observation that atoms retain most of their identity in liquids and solids, it is tempting to go even further and interpret the first term in eqn (24) as the true interaction of a pair of atoms of a free (gas-phase) dimer and the next term as the correction required to reproduce the potential surface of the free trimer and so on. Taking this second step amounts to assuming that the (truncated) expansion (24) describes the interactions over the full phase diagram, from the low-density gas phase to the compact states of solids. Potentials having this desirable and ideal property are called phase transferable or *ab initio* when derived from accurate electronic structure calculations. However, it is not clear under which conditions this series converges and, if it does, how practical such a potential will be when implemented in a simulation. The alternative is the so-called effective pair approximation which treats the pair term as an effective two-atom interaction renormalized by the coupling to other particles in the condensed matter environment. Clearly, phase transferability is lost in this approximation, which is likely to fail outside the subset of thermodynamic states for which it has been designed. What is gained, on the other hand, is that in its region of validity, the effective pair approximation (possibly extended with many-body corrections) can be made more accurate and, at the same time, more convenient to use in simulation.

3. Simple atomic and ionic systems

3.1. SIMPLE ATOMIC LIQUIDS

The elementary pair potential that captures the essence of nonbonded interactions between two closed-shell atoms is the exponential-6 potential

$$v(\mathbf{r}) = A \exp(-Br) - \frac{C}{r^6}, \tag{25}$$

where $\mathbf{r}$ is used as a short-hand notation for vector $\mathbf{R}_j - \mathbf{R}_i$ specifying the separation of two atoms in eqn (24). The first term models the hard-core repulsion introduced in section 2.2. The functional form reflects the fact that the repulsion is largely determined by the overlap (see eqn (12)) of atomic wave functions which, at larger distances, have an exponential radial dependence. The second term is the attractive dispersion interaction of eqn (14). In the Lennard-Jones 12-6 interaction the exponent is replaced by an algebraic approximation which eliminates one parameter

$$v(\mathbf{r}) = \frac{A}{r^{12}} - \frac{C}{r^6}. \tag{26}$$

In its alternative form

$$v(\mathbf{r}) = 4\epsilon \left(\left(\frac{\sigma}{r}\right)^{12} - \left(\frac{\sigma}{r}\right)^{6} \right) \tag{27}$$

the Lennard-Jones potential has been most popular in studies of simple liquids [2]. Let us consider as an example the archetypal system of computer simulation, i.e. liquid argon. If we are free to choose the value of the parameters ϵ and σ to give an optimal fit to experimental data of the liquid, we find $\epsilon = 120\,\mathrm{K}$ and $\sigma = 3.4\,\text{Å}$. An *effective* Lennard-Jones potential with these parameters gives a surprisingly good description of the liquid indeed. Not only are the equilibrium averages, such as the radial distribution function, accurately reproduced, but dynamic properties such as diffusion are also not too far off. This, in fact, has been one of the early successes of computer simulation (see Refs. [1–3] and references quoted therein) and has contributed much to the acceptance of the method in theoretical statistical mechanics.

Argon is one of the few systems for which the ideal of a phase-transferable potential has been virtually realized [9]. We are, therefore, in the fortunate position that we can actually compare the effective 12-6 potential to the dimer term in eqn (24). The development of phase-transferable potentials for argon has a long history (see Ref. [9]). A more recent model that describes the full range of argon phases fairly accurately is the so-called BBMS (Barker-Bobetic-Maitland-Smith) potential. It was derived by considering a large quantity of experimental gas-phase data, including molecular beam scattering cross sections, spectroscopy of the dimer and the second virial coefficient. Also, solid-state data were used to determine the detailed shape of the repulsive hard core. To illustrate the level of sophistication of this class of models we give the BBMS functional form below:

$$v(r)/\epsilon = \exp\left[\alpha(1-\bar{r})\right]\sum_{i=0}^{n} A_i(\bar{r}-1)^i + \alpha' \exp\left[-50(\bar{r}-1.33)^2\right] +$$

$$\sum_{j=0}^{2} C^*_{2j+6}/\left(\delta + \bar{r}^{2j+6}\right), \tag{28}$$

where $\bar{r} = r/r_m$ is the nuclear separation scaled by the characteristic length r_m. In Fig. 1, the BBMS and the effective 12-6 potentials for argon are compared. Particularly in the region of the potential minimum, the two curves show significant differences. The origin of these discrepancies is twofold.

First of all, the functional form of eqn (26) is too simple to represent the fine details of the interaction of a pair of argon atoms. The repulsion at the smaller values of r ($\bar{r} < 1$) is considerably softer than $1/r^{12}$. A better approximation with a firm theoretical foundation (see section 2.2), is the exponent of eqn (25). However, at larger distances $\bar{r} > 1$, a simple exponent tends to overestimate the repulsion. Damping of the exponential tail is achieved in the BBMS potential by shifting the exponent and multiplication by polynomial factors (first term of eqn (28)). Another shortcoming of the 12-6 potential is the truncation of the dispersion interaction at the dipole-dipole term of eqn (14). Extended with the contributions of the higher-order induced multipole moments, the dispersion interaction can be expressed as

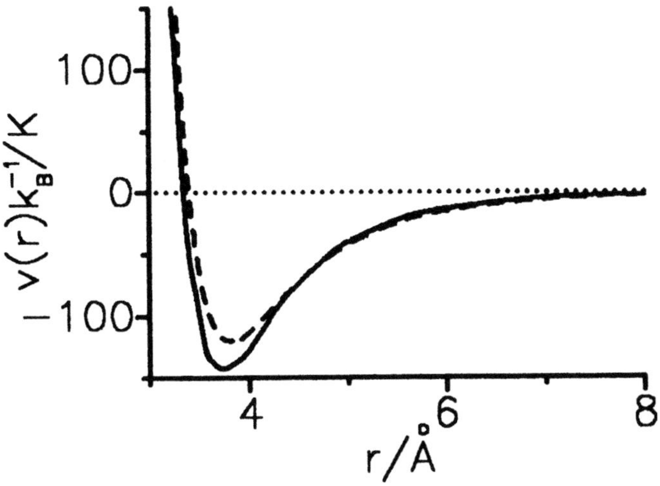

Fig. 1. The BBMS argon dimer potential (solid) compared to the effective Lennard-Jones potential (dashed) for the liquid (after Allen and Tildesley [1]).

a series of algebraic terms

$$E_{\text{disp}} = \frac{C_6}{r^6} + \frac{C_8}{r^8} + \frac{C_{10}}{r^{10}} + \dots \tag{29}$$

The series in eqn (29) is only valid at larger separation with no overlap of the atomic charge distributions. Under these conditions, the coefficient C_6 is rigorously equal to the numerical factor in eqn (14). Similar expressions can be derived for the higher-order dispersion coefficients. When evaluated for argon, at the distance of the equilibrium dimer bond length, it turns out that the contribution of the $1/r^8$ and $1/r^{10}$ term cannot be neglected. However, at this small separation, the atoms necessarily overlap, and the dispersion interaction is effectively reduced. The overestimation of the dispersion interaction at short range is progressively more serious for the higher-order terms. Hence, a potential model that includes these terms, must also introduce some kind of switching function which limits the effect again. This is the purpose of the offset δ in the denominators in the BBMS dispersion series (third term in eqn (28)).

In an L-J gas-phase potential, all the details of the two-body interaction not explicitly contained in the simple functional form can be taken into account in a good first approximation by adapting the values of σ and ϵ. As will be clear from the above discussion, the effective C coefficient obtained by such an optimization procedure can be substantially different from C_6 in eqn (29). However, this cannot be the complete story since, as a fit to the dimer surface, the *effective pair* potential of Fig. 1 is clearly not optimal. The reason is that the potential is intended for application in the liquid, and the discrepancy between the two curves in Fig. 1 is not due to neglected two-body terms but to renormalization of σ and ϵ by many-body interactions in the condensed phase. The leading non-additive term for argon is the Axilrod-Teller dispersion interaction which arises from a correlation between the dipole fluctuation of three atoms. Using the notation θ_{ijk} for the angle between the position vectors $\mathbf{r}_{ij} = \mathbf{r}_i - \mathbf{r}_j$ and $\mathbf{r}_{kj} = \mathbf{r}_k - \mathbf{r}_j$ of atoms i, j and k, the Axilrod-Teller term can be expressed as (see e.g. Ref. [9])

$$v_{\mathrm{DDD}}(\mathbf{r}_1, \mathbf{r}_2, \mathbf{r}_3) = \nu_3 \frac{3\cos(\theta_{123})\cos(\theta_{231})\cos(\theta_{312}) + 1}{(r_{12}\, r_{23}\, r_{13})^3}. \tag{30}$$

Employing again the Drude oscillator model of section 2.3, the coefficient ν_3 can be approximately related to the polarizability α and the excitation frequency ω_0, and we obtain (cf. eqn (14))

$$\nu_3 = 9/4\hbar\omega_0\alpha^3. \tag{31}$$

Contrary to the coefficient C_6 for pair correlations, ν_3 is positive. In dense phases, such as the liquid, the v_{DDD} term is on average repulsive, and the dispersive interaction is weakened. Indeed, the BBMS potential reproduces the structure and energetics of the liquid adequately only when applied in combination with the non-additive interaction of eqn (30). In the L-J pair potential for the liquid, the effect of the Axilrod-Teller term and further many-body interactions is taken into account by simply reducing ϵ.

3.2. SIMPLE IONIC SYSTEMS

The atomic pair potentials of eqns (25) and (26) can be generalized to potentials for simple ionic systems by attaching a point charge q in the atomic center. Ionic systems consist of at least two components, cations and anions. Labeling the species with the symbols a and b, the equivalent of eqn (25) is written as

$$v_{ab}(\mathbf{r}) = \frac{q_a q_b}{r} + A_{ab}\exp(-B_{ab} r) - \frac{C_{ab}}{r^6}. \tag{32}$$

Similar expressions hold for the interactions v_{aa} and v_{bb} between two ions of the same species. The equivalent of the Lennard-Jones interaction for ions is called the Born model:

$$v_{ab}(\mathbf{r}) = \frac{q_a q_b}{r} + \frac{A_{ab}}{r^n} - \frac{C_{ab}}{r^6}. \tag{33}$$

The Coulomb attraction is several orders of magnitude larger than dispersion interactions. Therefore, at the equilibrium bonding distance, the repulsion between two ions will also be stronger. As noted before, the $1/r^{12}$ dependence is too steep at the smaller radial separations. For ionic systems a smaller value of n is used. In most models n varies between 8 and 10.

As for atomic systems, the potentials of expressions (32) and (33) are effective pair potentials when applied in condensed phases. Because of the large ionic cohesive energies, non-additive effects can be quite severe. Unlike the three-body interaction of eqn (30), which correlates fluctuations, ion-ion interactions can lead to static distortions of the closed-shell groundstate wavefunctions. Hence, pair potentials, which ignore these effects, are sometimes referred to as the rigid ion approximation. The non-additive character of non-rigid ion effects is perhaps best illustrated by the electronic polarization. Assuming the linear approximation (see eqn (17)), it would seem at first glance that the dipoles induced on an ion i by the surrounding charges simply add up, since the contributions to the electric field at site $\mathbf{r}_i$ can be superimposed

$$\boldsymbol{\mu}_i^{(1)} = \alpha_i \sum_{j \neq i} \frac{q_j \mathbf{r}_{ij}}{r_{ij}^3}. \tag{34}$$

However, the first-order dipole moment $\boldsymbol{\mu}_i^{(1)}$ in eqn (34) itself creates an electric field which, in turn, will influence the induction on other sites. The actual induced dipole moments, therefore, are to be determined by solving the set of coupled equations

$$\boldsymbol{\mu}_i = \alpha_i \left[\sum_{j \neq i} \frac{q_j \mathbf{r}_{ij}}{r_{ij}^3} - \sum_{j \neq i} \left(\frac{\boldsymbol{\mu}_j}{r_{ij}^3} - \frac{3\mathbf{r}_{ij}\left(\mathbf{r}_{ij} \cdot \boldsymbol{\mu}_j\right)}{r_{ij}^5} \right) \right]. \tag{35}$$

From eqn (35) we see that it is the last term due to the dipole-dipole coupling of expression (13) that is responsible for the non-additive nature of electronic polarization. It should be kept in mind, that the first-order induction (eqn (34) and first term in eqn (35)) is the leading polarization effect. In the absence of permanent charges there would be no static induction at all. The average moments would vanish. Only fluctuations remain, either thermal or zero-point motion as discussed in section 2.3. A very rough estimate of the polarization can be obtained by ignoring the self-consistency problem (35) completely and evaluating the energy in the pair approximation. For a moment $\boldsymbol{\mu}$ induced by a single point charge q, the sum of polarization energy (see eqn (15)) and electrostatic interaction energy can be easily evaluated

$$v(r) = \frac{\boldsymbol{\mu}^2}{2\alpha} - \boldsymbol{\mu} \cdot \mathbf{E} = -\frac{\alpha q^2}{2r^4}. \tag{36}$$

For highly polarizable anions such as I^- this energy can be quite large. In spite of this, the effect of electronic polarization on the structure of ionic crystals is marginal. The solution to this paradox is that, as a result of the high symmetry of ionic crystals, the electric fields $\mathbf{E}_j$ in the vector sum (34) cancel. Of course, for the dynamics, the dipolar induction is important. In lattice dynamics calculations and molecular dynamics simulations ionic polarization is usually treated by so-called shell models. For a review the reader is referred to [10]. We will defer further discussion of electronic polarization to the section dealing with polar liquids for which the local symmetry is lower.

A uniform change of size is allowed by the high symmetry of a solid under equilibrium conditions, and can occur as a result of several mechanisms. Purely Coulombic interactions with the surrounding point charges may alter the effective radius of an ion, compressing anions and expanding cations. The opposite effect on anions and cations can be rationalized by considering the electrostatic potential profile which a charge cloud feels at the site of an ion. Anions sit in a potential well for negative charge, but the potential experienced at the site of a cation is a maximum. (A lucid discussion of non-rigid ion effects can be found in Ref. [11]; for molecular ions see e.g. Ref. [12]). Then there is the familiar hard-core repulsion due to overlap of the charge clouds. In ionic crystals the combined effect of the neighbors is very strong and non-additive. Again, for the purpose of qualitative understanding, one can imagine that the surrounding ionic shells create a cage which confines the electronic state of the central ion and compresses its volume.

Compression by overlap or electrostatic fields is compensated to some extent by increased three-shell overlap. When the ions are pressed together very closely by the enormous cohesive forces in the crystal, the integrals of products of orbitals belonging to three different cores can no longer be neglected. Similar to the overlap of a pair of closed shells (see section 2.2), the effect of triple overlap is also repulsive and approximately exponentially dependent on separation. Hence we can model this type of non-additive interaction by a product of exponents,

$$v_{aab}(\mathbf{r}_1, \mathbf{r}_2, \mathbf{r}_3) = A_{aab} \exp(-B_{aab} r_{13}) \exp(-B_{aab} r_{23}) \exp(-D_{aab} r_{12}), \quad (37)$$

where we have assumed that ion 3 of species b has contact with two ions 1 and 2 which are both of type a. The three-body repulsion of eqn (37) makes the core of the ions appear harder with increasing number of coordinated counter-ions.

The simple ion picture of eqns (32) and (33) is complicated even more by a chemical non-rigid ion effect. Only in the alkali halides is the electron transfer complete and the ions have unit charges. In the II-VI compounds and the post-transition metal halides (e.g. Ag and Zn salts) the bonds have partly covalent character and a fraction of the charge is shared between the ions. Ion softening by electron transfer can be modeled by reducing the charges q in the effective pair potentials of eqn (32) or (33). However, the effective charge is now a function of ion position. The dependence on configuration can be expected to be more complex than a sum of two-body potential functions. Pair potentials with effective ionic charges, when parametrized properly, can be highly successful in certain applications. The values of the parameters are usually determined empirically, i.e. by fitting them to experimental data. As an example of this approach, we will mention here the model for AgI of Ref. [13] This model has been developed to study the coupling between structural transitions and superionic conductivity. Upon heating the β phase of AgI, the iodine lattice undergoes a hcp-fcc transformation and silver ions become mobile. The model is of the type of expression (33) with power-law repulsion supplemented with a polarization term for the I^- ions in the pair approximation of eqn (36)

$$\begin{aligned} v_{\mathrm{AgAg}}(r) &= \frac{H_{\mathrm{AgAg}}}{r^{n\,(\mathrm{AgAg})}} + \frac{0.36}{r}, \\ v_{\mathrm{AgI}}(r) &= \frac{H_{\mathrm{AgI}}}{r^{n\,(\mathrm{AgI})}} - \frac{0.36}{r} - \frac{B_{\mathrm{I}}}{r^4}, \\ v_{\mathrm{II}}(r) &= \frac{H_{\mathrm{II}}}{r^{n\,(\mathrm{II})}} + \frac{0.36}{r} - \frac{2B_{\mathrm{I}}}{r^4} - \frac{C_{\mathrm{II}}}{r^6}. \end{aligned} \tag{38}$$

The coefficients H_{ab}, which define the magnitude of the short-range repulsion, were expressed in terms of hard-core radii σ_a for the two species $a = \mathrm{Ag}, \mathrm{I}$

$$H_{ab} = A\,(\sigma_a + \sigma_b)^{n\,(ab)}. \tag{39}$$

The parameters have been carefully optimized to reproduce both the structure of the low-temperature close-packed β phase and the high temperature α phase. This purely empirical procedure yielded a value of $q = 0.6e$ for the fractional charge and the exponents $n(\mathrm{AgAg}) = 11, n(\mathrm{AgI}) = 9$ and $n(\mathrm{II}) = 7$. With this parametrization the calculated $\alpha \rightleftharpoons \beta$ transition temperature and the structural and dynamic properties are in excellent agreement with experiment

3.3. PHASE AND ATOM TRANSFERABILITY

Summarizing our investigation of many-body effects in simple atomic and ionic systems we can conclude that the higher-order terms in expansion (24) of the Born-Oppenheimer surface are significant if we adopt "the quantum chemist's" approach, i.e. if we take the pair term to be the interaction of the free dimer. Also, it will be clear that many non-additive interactions are cooperative and involve more than just the three-body term. On the other hand, using effective pair potentials, the most important effects of the condensed-matter environment can be absorbed in the two-body term by adjusting the parameters, and explicit many-body interactions can be omitted. The price we pay is that the validity of the model is limited to the thermodynamic states that have been used to construct the potential. Global phase-transferability is lost. However, in a more restricted region of the phase diagram, effective pair potentials can be phase-transferable, as we have seen from the example of the AgI model.

The elementary effective pair potentials of eqns (32) and (33) have the major advantage of having a simple functional form. From the numerical point of view, they are convenient and inexpensive. However, the determination of the parameter set is a frustrating task involving many iteration cycles. In this respect, it is vital that the number of parameters is limited. For systems with one or two components, the parameter set is small enough, and a fairly accurate optimization is feasible. Again, the AgI model is a good illustration. In multicomponent systems the number of parameters proliferates rather quickly and the use of combination rules such as eqn (39) is essential. Combination rules are based on the assumption that the coefficients of pair interactions can be resolved in simple functions of single-particle characteristics. A trivial and exact realization of such a rule is the Coulomb pair parameter, which is a product of the individual charges. Similar relations for other properties such as diameters for hard-core repulsion are more approximate. A set of reliable combination rules is a very desirable feature of models, and represents a second type of transferability which is not only motivated by practical considerations but also by chemical intuition. This will be referred to as atomic transferability.

The coefficients A and B in the Lennard-Jones interaction in eqn (26) and the more general Born model (eqn (33)) are combined according to

$$A_{ab} = \sqrt{A_{aa}A_{bb}}, \qquad C_{ab} = \sqrt{C_{aa}C_{bb}}. \tag{40}$$

The rule of the geometrical mean for the A coefficient can be justified by the argument in section 2.2. The quantum mechanical origin of closed-shell repulsion is the exchange integral K (eqns (11) and (12)) which is determined by the integral of a product of two atomic orbitals. Similarly, according to the London relation (23), the dispersion coefficient C in lowest order is a product of atomic polarizabilities and excitation energies. Relations (40) imply that the ϵ and σ parameters in the Lennard-Jones potential eqn (27) also follow the rule of the geometric mean. An alternative is the so-called Lorentz-Berthelot set of rules, which take the Van der Waals diameter to be the arithmetic mean

$$\sigma_{ab} = (\sigma_{aa} + \sigma_{bb})/2, \qquad \epsilon_{ab} = \sqrt{\epsilon_{aa}\epsilon_{bb}}. \tag{41}$$

In the modelling of ionic liquids and crystals, combination rules for ionic radii are not only very helpful for the rationalization of the various structures (see any textbook on solid-state physics) but are also a practical necessity, since these systems consist of at least two components (cation and anion) and many interesting compounds are ternary or even more complex. An example of an additive definition of ionic radii for algebraic repulsion (Born model, eqn (33)) was already given in expression (39). A popular rule for exponential potentials (eqn (32)) is the Mayer-Huggins relation

$$\sigma_a + \sigma_b = B_{ab}^{-1} \ln A_{ab}. \tag{42}$$

Note that the ionic radii are dependent on the interaction strength A_{ab}.

4. Molecular liquids and solids

4.1. INTERACTION SITE MODELS

Using the terminology developed in this Chapter, molecules can be defined as clusters of atoms tightly held together by the bonded forces of predominantly covalent chemical bonds. The cohesive energy of molecular liquids and solids arises from nonbonded interactions similar to the interactions in simple atomic and ionic systems discussed in the previous section. Provided the distortion of molecular structure is not too large and dissociation is strictly forbidden, the bonding intra-molecular forces can be modelled by relatively simple three- and four-atom potentials. The smaller molecules, e.g. diatomics (N_2, CO), first row hydrides (CH_4, NH_3, H_2O) and small aromatic molecules (C_6H_6), can even be taken to be completely rigid in a first approximation. Flexibility is essential for the understanding of ordered and disordered condensed phases of the larger molecules, in particular chain-like molecules and polymers.

It is also convenient to resolve the intermolecular forces into atomic components. The atomic centers that specify the $\mathbf{r}$ vectors in the potential functions may coincide with the actual nuclear cores or may be fixed at special sites in the molecular structure, e.g. in regions of high electronic density. This type of intermolecular potential is referred to as an interaction site model. The concept of an atomic model for molecules immediately raises the question of pairwise additivity. In the previous section we argued that, in many mono-atomic and ionic systems, an effective pair potential is the only practical solution for the problem of many-body interactions. We will follow the same approach here for the design of interaction-site models for molecules. Only if absolutely necessary will we correct the pair approximation by explicit many-body terms.

The simple potentials of expression (32) or (33) will again be our first choice for the functional form of the site-site potentials. Hence, all electrostatic forces will be modeled by a distribution of point charges, and the short-range interactions either by power-law or exponential potentials. Since molecules consist, in general, of more than one type of site, and we want to restrict the number of parameters, the issue of atomic transferability is vital, even more so than for mono-atomic systems. Interaction site models have several features that make the design of potentials for complex molecular structures feasible in practice. The possibility of optimizing the location of the interaction sites gives these models a certain versatility, and modifications of the site distributions are easy to visualize. Two examples where this freedom is exploited are the potentials that have been developed for water and fullerenes. We will discuss these models below. Another consideration that favors the interaction site models is that they are very convenient for implementation in molecular dynamics code. Hence, the approach outlined here has found wide application and is the basis for most of the available software packages for the study of biological systems (see e.g. [4]).

The obvious disadvantage of the empirical approach sketched above is that any systematic procedure for setting up a model and optimizing the parameters is lacking. There have been several attempts to circumvent this serious problem. Most of these schemes are *ab initio* methods based on electronic structure calculations of molecular dimers. Here we want to mention an interesting alternative that is based on a multipole expansion of *ab initio* charge densities of monomers. This so-called intermolecular perturbation theory method (IMPT)[14] combined with a systematic method for generating a distribution of point multipoles for the monomer (DMA)[15] is capable of reproducing the electrostatic interactions between two (isolated) molecules essentially with the precision of the original monomer wavefunction. For a review of the IMPT/DMA method we refer the reader to Ref. [16]. An impressive example of an application (solid Cl_2) can be found in Ref. [17]. The drawback of this approach, apart from its computational complexity (see e.g. Ref. [17]), is that it is subject to the same uncertainties when applied in condensed phases as all other gas-phase models. If the high degree of precision is to be preserved in a liquid, a supplementary scheme is required that includes electronic polarization with comparable accuracy [16]. We will return to this issue at length in the following sections.

4.2. POTENTIALS FOR WATER

The simulation of liquid water has a long history, and numerous alternative models have been proposed with various degrees of sophistication. Regarding the question as to what to include in a model and what to leave out, we advocate drastic use of Ockham's razor to cut away all detail that is not strictly required to reproduce experiment. We will apply this pruning method even to elements of the molecular and electronic structure that have a firm basis in quantum chemistry, which is again very much in the spirit of the effective pair potential approach.

The key feature of water is the local tetrahedral structure of the liquid established by hydrogen bonding of a molecule to four nearest neighbors. It turns out that even a rather primitive interaction site model is able to give a realistic description of hydrogen bonding. These potentials consist of a Lennard-Jones sphere centered on the oxygen and fractional positive point charges on the hydrogen sites. The relative positions of the O and two H atoms are held fixed in the experimental geometry. The number and location of the negative point charges varies. To ensure tetrahedral coordination, the first successful model of this type, ST2 [18], placed two negative charges away from the (neutral) oxygen site in the direction of the lone pairs. Subsequently it was found that hydrogen bonding could be equally achieved by a single negative charge, either on the oxygen site (SPC model[19]) or at a small distance from the oxygen on the symmetry axis approaching the hydrogens (TIP4P [20]). As an illustration, we have drawn in Fig. 2 two TIP4P molecules in the equilibrium dimer configuration. Of these three roughly equivalent models, SPC contains the smallest number of interaction sites (only three) which is the main reason for its being the most popular model in computer simulation studies.

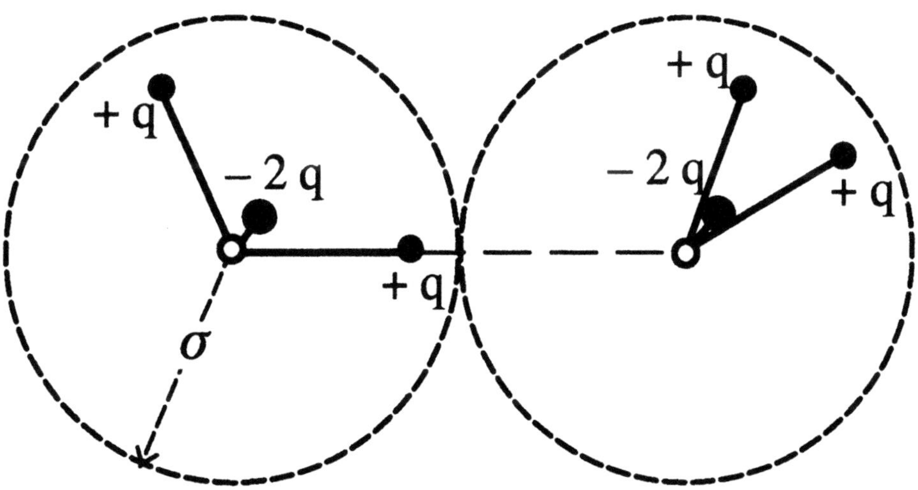

Fig. 2. Two TIP4P [20] water molecules in the equilibrium dimer configuration. The Van der Waals spheres (big circles) are centered on the oxygen atoms (small open circles). The solid circles are point charges on the hydrogen sites ($q_{\mathrm{H}} = 0.52e$) and a negative charge site ($q_{\mathrm{M}} = -1.04e$) at a distance of 0.15 Å from the oxygen.

A detailed comparison of the various models is beyond the scope of this Chapter (see e.g. [21]). What we are interested in is the role of effective potentials. One crucial aspect of water that requires the use of effective potentials is the electronic polarization. Water molecules have a polarizability of moderate magnitude (1.44Å^3). However, the polarization in the liquid is cooperative. The local order in the hydrogen-bond network imposes an average electric field that induces a finite dipole moment along the C_{2v} symmetry axis. This induced moment adds to the permanent (gas-phase) dipole moment of $\mu_0 = 1.85$D and increases the average moment in the liquid to $\mu_{\text{eff}} \approx 2.6$D. The dielectric properties are strongly dependent on the average dipole moment. Non-polarizable water models with the gas-phase dipole moment underestimate the static dielectric constant by as much as a factor of three (see e.g. [22]). The average polarization can be incorporated in a non-polarizable effective potential by enhancing the permanent dipole moment. In the interaction site models mentioned above (ST2, SPC, TIP4P) this is implemented by simply adjusting the value of the point charges. For example, the TIP4P and SPC model both have a dipole moment of about 2.2D.

The TIP4P and SPC models, despite their simplicity, reproduce the properties of liquid water surprisingly well. This holds for the thermodynamics and equilibrium structure as well as for the dynamics. Observables for which the agreement is less satisfactory are the self-diffusion coefficient (about 50% too large) and the dielectric constant (about 30% too small). Recently, the SPC model has been considerably improved in this respect. In its new version it is referred to as SPC/E [23]. It is very instructive to repeat the argument here, since the reparametrization was the result of recognizing an inconsistency in the design of effective pair potentials that had gone unnoticed until then. One of the quantities used for the parametrization of empirical water potentials is the heat of evaporation ($-41.5\,\text{kJmol}^{-1}$) which is identified with the total internal interaction energy U in the model. In a polarizable system such as water, U includes the average self-energy of eqn (15) of the induced moments $\boldsymbol{\mu}_i$ of the molecules i:

$$\langle E_{\text{pol}} \rangle = \sum_i \frac{\langle \boldsymbol{\mu}^2 \rangle}{2\alpha}. \tag{43}$$

In the effective pair potential approach, the fluctuating dipoles are replaced by a static average $\langle \boldsymbol{\mu} \rangle$. Hence we have in expression (43) $\langle \boldsymbol{\mu}^2 \rangle = \langle \boldsymbol{\mu} \rangle^2$. The permanent moment $\boldsymbol{\mu}_0$ is enhanced by this amount to give an effective total moment $\boldsymbol{\mu}_{\text{eff}} = \boldsymbol{\mu}_0 + \langle \boldsymbol{\mu} \rangle$ and a self-energy

$$E_{\text{pol}} = \sum_i \frac{(\boldsymbol{\mu}_{\text{eff}} - \boldsymbol{\mu}_0)^2}{2\alpha}. \tag{44}$$

Depending on the precise value for $\boldsymbol{\mu}_{\text{eff}}$, the value for water is $4\text{–}5\,\text{kJmol}^{-1}$. The observation made in Ref. [23] is that this term has been forgotten in the parametrization of effective pair potentials. Refitting the total SPC energy U, corrected with E_{pol}, to the heat of evaporation yields an average moment

$\mu_{\text{eff}} = 2.35\text{D}$ slightly higher than the $\mu_{\text{eff}} = 2.27\text{D}$ obtained previously with the inconsistent procedure. This small increase, as it turned out, was sufficient to improve the self diffusion coefficient [23] and dielectric constant [21] of the model significantly.

4.3. SOLID FULLERENES

The motivation for discussing the modelling of the intermolecular interactions of C_{60} is not so much that this is a good example of an effective potential. The reason is that the development of empirical interaction site models for C_{60} offers a good opportunity to show how experimental and theoretical input is used in practice to optimize a model. Interaction site models are particularly convenient for this type of trial and error process, because the distribution of sites can be easily visualized and modified. The C_{60} molecule is an almost perfectly spherical construction of carbon-carbon bonds arranged in 12 pentagons and 20 hexagons (of the numerous popular reviews that have appeared recently on "Bucky balls" we recommend Ref. [24], see also Fig. 3). The molecules retain their identity and shape in the solid and, at room temperature, form a textbook molecular crystal in a rotator phase. The centers of mass are ordered on a close-packed fcc lattice while the orientations are disordered. On cooling, a structural phase transition at $T_c = 250\text{K}$ orders the orientations in a simple cubic structure with four molecules per unit cell. The transition is first order, accompanied by a small discontinuity in the cubic lattice parameter. The space group was determined to be $Pa\bar{3}$ (for references to the experimental literature we refer again to Ref. [24] and the theoretical studies mentioned below).

All 60 C atoms in the molecule are equivalent. Similar to graphite, each atom is bonded to three neighbors. Hence, an obvious first guess for a model would be to represent the C atoms by Lennard-Jones interaction sites with ϵ and σ parameters borrowed from potentials for graphite [25, 26]:

$$\sigma = 3.4\,\text{Å}, \qquad \epsilon/k_B = 28\,\text{K}. \tag{45}$$

Indeed, this atom-atom model performs very well in the disordered phase and describes the structure and dynamics with satisfactory accuracy [26]. It came, therefore, as a surprise that the model fails to reproduce the correct structure of the ordered phase. Instead of the experimental structure with simple cubic symmetry, the graphite model predicts a tetragonal structure [25, 26]. Evidently, the intermolecular interactions in fullerenes are more complex than in graphite.

The element that could be missing was suggested by the imaginative analysis of diffraction data by a group of experimentalists [27]. These authors pointed out that the most revealing perspective of the intricate packing configuration of the C_{60} molecules is obtained by focusing on the way the molecules touch. All 12 nearest-neighbor contacts are similar. A bond connecting two pentagons (or equivalently a bond shared by two hexagons, see Fig. 3) faces the center of a pentagon of the other molecule. Unlike the atoms, the C – C bonds are not equivalent. The bonds shared by hexagons (6:6 bonds) are shorter than the bonds making up a pentagon (5:6 bonds, pentagons are surrounded on all five sides by hexagons, see Fig. 3). The difference in length has been confirmed by electronic structure calculations which also showed that the electron density on 6:6 bonds is larger than on 5:6 bonds [28]. The next step in the argument in Ref. [27] is the hypothesis that the stability of the $Pa\bar{3}$ structure is of electrostatic origin. The short electron-rich 6:6 bonds seek the electron-poor pentagons.

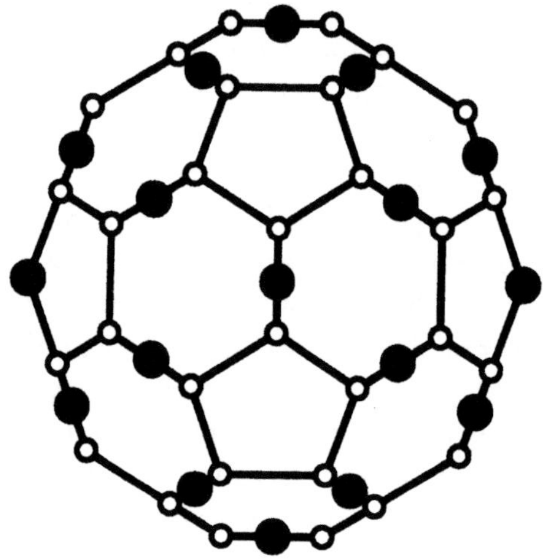

Fig. 3. Interaction site model of C_{60} for the electrostatic part of the intermolecular interaction [29]. Open circles indicate positive point charges $q_C = 0.175e$ on the carbon sites and the solid circles are the negative counter charges $-2q_C$ fixed in the centers of the short carbon-carbon bonds.

The suggestion of Ref. [27] can be straightforwardly implemented in an interaction site model. We can simply attach a negative charge q_D at the center D of the short ("double") bonds and place the positive counter charge q_C on the C atoms. There are 30 D sites, hence charge neutrality requires that $q_D = -2q_C$. In order to verify that the electrostatic hypothesis in Ref. [27] is basically sound we do the following test. All corrugation at the C_{60} surface is eliminated and the short-range carbon-carbon interactions, which caused so much trouble, are replaced by a single Lennard-Jones potential at the center of the C_{60} This caricature model works. With a typical value for a bond charge, $q_D = -0.35e$, we find a transition to the correct ordered state at about 110K. However, it is unrealistic to assume the C_{60} is completely smooth. Unfortunately, if the graphite C – C interactions are restored, keeping the charges, the system reverts to the tetragonal groundstate (i.e. for any reasonable value of q_D). The frustrating effect of the repulsive carbon atom profile prevails over the electrostatics, although the energy difference with respect to the target $Pa\bar{3}$ structure has been reduced.

How can we "engineer" our interaction site model to lower the energy of the $Pa\bar{3}$ state even further? There are two proposals in the literature. In Ref. [30], the electrostatics is drastically strengthened by placing the positive charges on the single bonds rather than on the C atoms and increasing the value of the bond charges to $q_D = -0.54e$. This operation proves to be very effective. The transition temperature obtained in a simulation of using this model is $T_c = 270$K [31]. Alternatively, the scheme proposed by us in Ref. [29] leaves the positive charges on the carbons. Instead, we assign a 12 – 6 interaction to the D sites (in addition to the L-J spheres at the C sites). The short bonds now acquire a more ellipsoid shape and this alteration of the repulsive profile turns out also to be sufficient to produce the $Pa\bar{3}$ order. The T_c of 220 K is somewhat lower than experiment. On the other hand, our model has the major advantage of reproducing the discontinuity of the lattice parameter at the transition in quantitative agreement with experiment, contrary to the first model which predicts a much more continuous phase transformation [31]. Clearly, the question of the nature of the intermolecular interactions in fullerenes has not yet been completely resolved. The point we want to demonstrate here is that interaction site models are a useful tool to quantify and test the various conjectures inferred from theory and experiment.

4.4. FLEXIBLE MOLECULES

The interaction sites in the models for water and C_{60} discussed in sections 4.2 and 4.3 were fixed in a rigid molecular frame. For some applications, in particular for the simulation of polymers, flexibility of the molecular structure is essential. The largely covalent intra-molecular bonded interactions can be modeled by relatively simple two-, three- and four-atom potentials, as long as the lengths of the bonds remain close to the equilibrium values. Thus, the distance between two directly bonded atom pairs (i, j) is usually taken to be either rigid or constrained by a harmonic potential (see e.g. Ref. [4])

$$\mathcal{V}_{\text{bond}}(\{\mathbf{r}_i\}) = \sum_{(i,j)} \frac{k_b}{2} (r_{ij} - b_0)^2 , \tag{46}$$

where k_b and b_0 are the bond-stretching force constant and equilibrium bond length respectively. The angles between covalent bonds have a well-defined geometry imposed by the chemistry of the bonding. For most stable molecules, the fluctuations of bond angles are small and approximately harmonic. Bond bending potentials can, therefore, be written as a quadratic function of the angle θ_{jik} between the bond vector $\mathbf{r}_{ji}$ of an atom j forming a bond with atom i and the vector $\mathbf{r}_{ki}$ of atom k bonded to the same atom i (see Fig. 4)

$$\mathcal{V}_{\text{bend}}(\{\mathbf{r}_i\}) = \sum_{(i,j,k)} \frac{k_\theta}{2} (\theta_{jik} - \theta_0)^2 , \tag{47}$$

where θ_0 is the equilibrium bond angle and k_θ is the bending force constant. Both k_b and k_θ can be readily determined from spectroscopic data or quantum chemistry calculations. For atoms coupled by bond stretching and bending potentials, short-range nonbonded interactions (Lennard-Jones) are usually omitted.

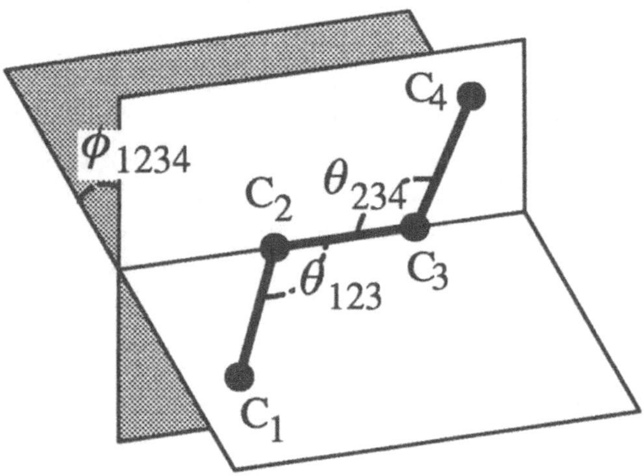

Fig. 4. Model for a flexible alkane chain [32]. Shown is a segment of four carbon atoms (solid circles). Bond bending (angles θ) is modeled by an interaction between three C atoms and torsion (angle ϕ) by a potential depending on the positions of four carbon atoms.

In addition to bond stretching and bending, long-chain alkanes and polymers exhibit a third type of conformational degree of freedom, namely torsion. Fig. 4 defines the torsion angle for a section of an alkane molecule as the dihedral angle ϕ between a pair of planes containing two sets of adjacent bonds with one bond in common [32]. Torsion about dihedral angles changes the relative positions of third neighbors. These chain units are far

TABLE I
Maxima and minima of the Ryckaert-Bellemans potential for butane.

ϕ	$v(\phi)$/kcalmol^{-1}	conformation
0	0	trans
$\pm\pi/3$	2.95	maximum
$\pm 2\pi/3$	0.70	gauche
$\pm\pi$	10.7	maximum

enough apart to be subjected to the regular nonbonded intermolecular interactions with short-range repulsion. However, the hindered rotation about bonds cannot be described in sufficient detail by an intermolecular potential optimized for the coupling between atoms of different molecules. Hence, the intermolecular interaction between third neighbors is replaced by an explicit torsional potential with the proper rotational symmetry and barriers. The simplest form such a potential might take is a periodic function in the dihedral angle ϕ associated with four atoms (i, j, k, l) in a butane-like bonding configuration.

$$\mathcal{V}_{\text{torsion}}(\{\mathbf{r}_i\}) = \sum_{(i,j,k,l)} k_\phi \left[1 + \cos\left(n\phi_{ijkl} - \phi_0\right)\right]. \tag{48}$$

Here k_ϕ, n and ϕ_0 are the torsional force constant, the number of periods and the phase, respectively. In the periodic potential of eqn (48), trans and gauche configurations ($n = 3, \Delta\phi = 2/3\pi$) are degenerate. For a more detailed understanding of the process of conformational defect formation in dense systems, it is essential to distinguish between the dihedral minima. In general, the barriers in between are also different. A torsional potential at this level of detail is the well-known Ryckaert-Bellemans potential for butane [32]

$$v_{\text{torsion}}(\phi)/k_{\text{B}} = \Big[1.116 + 1.462\cos\phi - 1.578\cos^2\phi \\ -0.368\cos^3\phi + 3.156\cos^4\phi - 3.788\cos^5\phi\Big]\,10^3\text{K} \tag{49}$$

Table I lists the energies of selected configurations. Note the appreciable difference between the trans-gauche barrier and the barrier between the right and left gauche conformation.

Bonded potentials of the type in eqns (48) and (49) have been applied in the study of conformational defects in partially disordered alkane crystals [33] and more recently also in substrate-supported and self-assembled monolayers of long-chain molecules (see e.g. Ref. [34–36]). Clearly, for these molecules, with sometimes large molecular weights, a simple interaction model would be an advantage. Major savings in computational expenses can be achieved by leaving out the hydrogen atoms of the non-polar C – H bonds. In these so-called united-atom models CH_2 links and terminal CH_3 groups are represented by a single enlarged Lennard-Jones sphere. The united-atom parameters in the Ryckaert-Bellemans potential, for example, have the values

$$\sigma = 3.92\,\text{Å}, \qquad \epsilon/k_B = 84\,\text{K} \tag{50}$$

which should be compared to the ϵ and σ of eqn (45) for C atoms in graphite. The H atoms in polar bonds, such as hydroxyl and amine groups, cannot be treated in the united-atom approximation. Similar to the OH bonds in the water models discussed above, these groups can participate in directional hydrogen bonds and the H atoms are usually represented by a fractional point charge.

The "pruning" of interaction site models for complex molecular systems is, in practice, a necessity since the total length of the run is one of the major limiting factors of simulation studies of these systems. United-atom models are useful in this respect but they are also just another implementation of the effective pair potential idea, and, as such, may lead to erroneous results when their use is extended beyond the original interval of thermodynamic states for which they were designed. The united-atom model for alkanes is another example of the accidents that might happen. An important property of ordered long-chain systems which is particularly sensitive to this aspect of the interaction potentials for flexible molecules is the collective tilt [34–37]. For self-assembled monolayers at moderate surface coverage, the collective tilt can be accounted for by the united atom model. The effective density in these systems is comparable to alkanes in the fluid phase, for which the united-atom model has been developed. However, for denser layers, there is convincing evidence that the tilt angle depends on the anisotropic shape of the monomeric units [35]. Similarly, the chain ordering in bulk crystals of alkanes below the rotator transition [33] senses the orientation of the CH bonds. Most likely, an accurate description of the packing in lipid bilayers in crystals also requires an all-atom model [37].

5. Introducing electronic polarization

5.1. DISTURBING THE CANCELLATION OF ERRORS

The effective pair potentials for water introduced in section 4.2 incorporate electronic polarization in a mean-field approximation. For pure water, these models are efficient and adequate. The second generation of so-called extended models such as SPC/E [23] hardly need further improvement. The main application of water potentials, however, is not in pure water but in the simulation of aqueous solutions. Being a mean-field theory, the effective pair approximation presupposes microscopic homogeneity of the system. The water in solvation shells of ions is in a highly inhomogeneous environment with strong electric field gradients. Under these conditions, the mean-field approximation might break down. A similar situation, perhaps less extreme, may arise at water-vapor and water-hydrocarbon interfaces. For these multicomponent or multiphase systems the polarization may have to be included explicitly.

It would be of great practical advantage if we could simply generalize the existing mean-field interaction site potentials. The first step in such a conversion procedure is to scale back the values of the point charges in the mean-field molecule to restore the gas-phase permanent dipole moment. This is easy. Unfortunately, the next step is not as straightforward as introducing variable dipole moments $\boldsymbol{\mu}_i$ at polarizable sites and then solving eqn (35) to determine the values for a given configuration of permanent charges. The complications that arise are best illustrated by an example, for which we take the TIP4P model of the $Na^+ - H_2O$ interaction [38]. The potential is of the form of expression (33) with exponent $n = 12$ and a parametrization that reproduces the experimental binding energy $E_1 = -100.4\,\mathrm{kJmol^{-1}}$ [39] and the Hartree Fock bond length $R_{\mathrm{NaO}} = 2.2\text{Å}$ of the Na^+H_2O dimer [38]. Removing the enhanced dipole moment (keeping the bond length fixed at $R_{\mathrm{NaO}} = 2.2\text{Å}$) gives

$$q = 0.52 \rightarrow 0.44e \quad \Rightarrow \quad E_1 = -80\,\mathrm{kJmol^{-1}}. \tag{51}$$

Assigning a variable dipole moment to the oxygen site with the experimental polarizability for water, $\alpha_{\mathrm{O}} = 1.44\text{Å}^3$, and substituting eqn (36) we regain an induction energy of

$$E_{\mathrm{pol}} = -\frac{\alpha_{\mathrm{O}} q_I^2}{2R_{\mathrm{NaO}}^4} = -45\,\mathrm{kJmol^{-1}} \quad \Rightarrow \quad E_1 = -125\,\mathrm{kJmol^{-1}}. \tag{52}$$

In the estimate (52) we have neglected the small polarizability $\alpha_{\mathrm{Na^+}} \approx 0.15\text{Å}^3$ of the cation. The overbinding is typical for the modeling of polarization and has a variety of origins. First, there is the hidden inconsistency of the effective pair potential we used as our starting point. As it turns out, not only the electrostatics, but also the parametrization of the short-range part, is modified by the mean-field approximation. Not all the relevant energies and bond distances can be simultaneously optimized by adjusting the water charges (the ion charges are fixed to unity). Hence, one is forced to use the A and C parameters to correct for all the remaining inaccuracies. In fact, all effective potentials rely ultimately on exploiting the cancellation of errors in some way or another.

In Fig. 5, we compare the TIP4P [38] binding energies E_n of $Na^+(H_2O)_n$ and $Cl^-(H_2O)_n$ clusters to experimental gas-phase hydration enthalpies [39, 40]. The Na^+H_2O and Cl^-H_2O dimer energies E_1 are in agreement with experiment by design. Adding a second water already gives a deviation which increases with n. For $n \geq 4$ both ions are overbound by at least 0.5 eV. The unacceptably large discrepancy is (partly) due to the fact that the short-range interaction contains some of the dimer polarization energy. Adding these energies to obtain the polarization energy for multiple coordination is clearly not allowed. A further indication that the short-range attraction in effective ion-water potentials includes energy contributions in excess of the dispersive interaction, is the inconsistency of the ratio of the C coefficients and the polarizabilities. Using again the TIP4P values we find

$$C_{\mathrm{Na^+O}}/C_{\mathrm{OO}} \approx 0.7 \gg \alpha_{\mathrm{Na^+}}/\alpha_{\mathrm{H_2O}} \approx 0.1. \tag{53}$$

According to eqn (23) these ratios should be roughly the same for purely dispersive interactions. The overbinding of hydrated ions in mean-field models has serious consequences for solvation in the bulk liquid, which leads to the well-known overcoordination in dilute solutions. For example, the number of waters in the first Cl^- solvation shell as predicted by TIP4P is 7.2 [41] compared to the experimental value of ≈ 5.5 [42].

The conclusion is that, with the introduction of polarizability in a previously unpolarizable model, readjustment of the A and C parameters is inevitable. In such a polarizable upgrade, the spurious effective polarization energy in the short-range interaction must be eliminated since it should now be accounted for by a proper induction. However, attempts to reproduce the experimental hydration curve of Fig. 5 by varying the A and C parameters are bound to end in disappointment and frustration. Even with a lot of patience, it seems impossible to find the right combination of A and C that compensates the 25kJmol^{-1} overbinding of eqn (52). What we encounter here is a phenomenon that, in unfavorable circumstances, may develop into a true instability and is then known as the polarization catastrophe. At small separations, the point-dipole approximation is no longer valid and leads to serious overestimation or even divergence of the induced polarization. In simulation, the problem is usually solved by multiplying the $1/r^4$ term by a switching function which damps the induction within a certain range.

In Ref. [43] we have proposed an alternative approach which is inspired by the actual quantum-mechanical root of the apparent damping with distance, namely the finite size of the molecular and ionic charge clouds. The induction center on the oxygen is represented by four closely spaced charges $q_i^{(\mu)}, i = 1 \ldots 4$, fixed in a tetrahedral configuration $\mathbf{r}_i^{(\mu)}$ in the water molecule (see Fig. 6). The values of the charges are variable. Hence, under the constraint of zero total charge, the cluster $q_i^{(\mu)}$ supplies the molecule with an induced

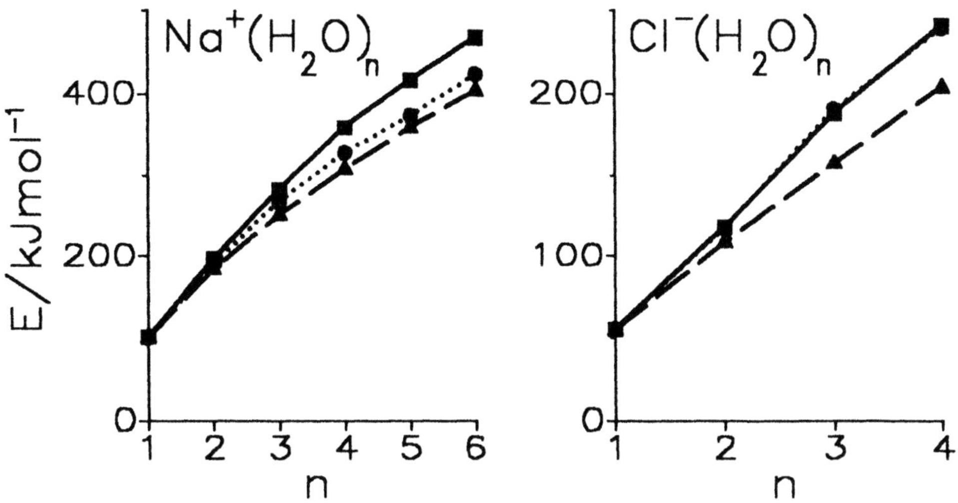

Fig. 5. Calculated hydration energies compared to experimental enthalpies (triangles) for the gas-phase hydration of a single Na^+ (left) and Cl^- ion (right). n is the number of water molecules. Squares give the values for TIP4P and circles for the polarizable model discussed in section 5.3.

dipole moment of arbitrary magnitude and direction while preserving the charge neutrality.

$$\boldsymbol{\mu} = \sum_{i=1}^{4} q_i^{(\mu)} \mathbf{r}_i^{(\mu)} , \qquad \sum_{i=1}^{4} q_i^{(\mu)} = 0. \tag{54}$$

Provided the dimensions $r_{ij}^{(\mu)}$ of the tetrahedron are small compared to the OH bond length, the higher moments can be neglected. In complete analogy with eqns (15), (16) and (17) for the case of an induced point dipole, the values of $q_i^{(\mu)}$ are determined from the local electric field and the polarization self-energy by a variational principle

$$\frac{\partial}{\partial q_i^{(\mu)}} \left(\frac{\boldsymbol{\mu}^2}{2\alpha} - \boldsymbol{\mu} \cdot \mathbf{E} \right) = \frac{\partial}{\partial q_i^{(\mu)}} \left(\sum_{i,j}^{4} \frac{q_i^{(\mu)} (\mathbf{r}_i^{(\mu)} \cdot \mathbf{r}_j^{(\mu)}) q_j^{(\mu)}}{2\alpha} - \sum_{i=1}^{4} q_i^{(\mu)} \mathbf{r}_i^{(\mu)} \cdot \mathbf{E} \right) = 0 \tag{55}$$

in conjunction with eqn (54). The polarization of ions can be treated in a similar approach. Since a simple ion has no anisotropic shape, we only need two variable charges $q_i^{(\mu)}, i = 1,2$ to represent the induced dipole. Again, the sum vanishes, hence $q_2^{(\mu)} = -q_1^{(\mu)}$. The charges are placed at a small distance on opposite sides of the permanent charge forming a little trimer (see Fig. 6). The induced moment changes direction by reorientation of the trimer.

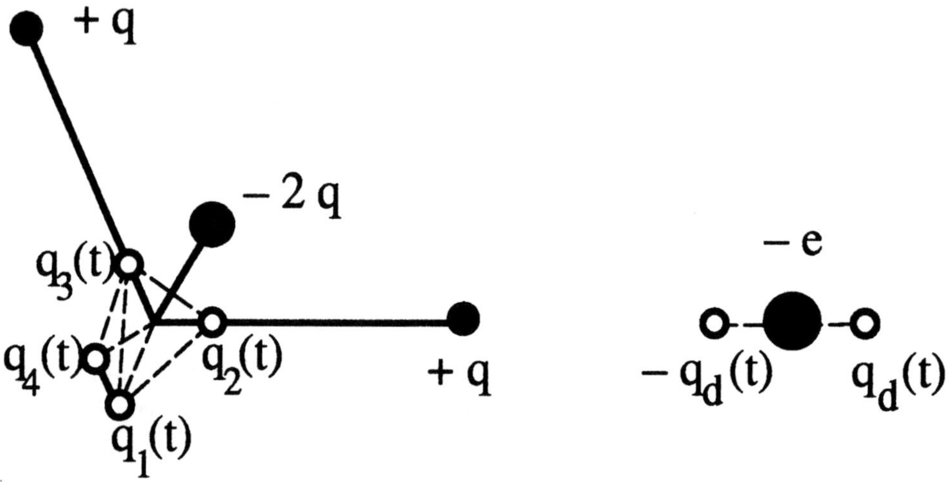

Fig. 6. Interaction site model for a polarizable water molecule [22] (left) and anion (right). Open circles indicate variable charges, solid circles are permanent charges. The induced moment of the molecule is defined by four variable charges with fixed positions in the molecular frame. The two variable charges representing the polarization of the anion form a trimer with the central permanent charge. The Van der Waals spheres have been omitted.

The charges $q_i^{(\mu)}$ can be interpreted as a very crude representation of the electronic charge density. This immediately suggests a model for the short-range damping of the induction which is intrinsically atom-transferable. All we have to do is to broaden the point charges $q_i^{(\mu)}$ into Gaussian charge distributions ρ_i with a width of ξ_a depending on the type a of the atom or molecule:

$$\rho_i^{(\mu)}(\mathbf{r}) = q_i^{(\mu)} \left(\frac{1}{2\pi\xi_a^2}\right)^{3/2} \exp\left[-\frac{(\mathbf{r}-\mathbf{r}_i^{(\mu)})^2}{2\xi_a^2}\right]. \tag{56}$$

Two Gaussian distributions i and j situated in different molecules of type a

and b interact through the error function potential

$$v\left(\mathbf{r}_{ij}^{(\mu)}\right) = \left(\frac{q_i^{(\mu)} q_j^{(\mu)}}{r_{ij}^{(\mu)}}\right) \operatorname{erf}\left[r_{ij}^{(\mu)} / \left(2(\xi_a^2 + \xi_b^2)\right)^{1/2}\right] \tag{57}$$

which gives us the combining rule for the effective radii ξ.

5.2. PURE WATER

The past decade has seen the appearance of a host of polarizable models for pure water [22, 43–46] (for a more complete list see e.g. Ref. [22]). The majority of these models are empirical. Experimental quantities commonly used for the parametrization are the heat of evaporation, the pressure and the structure of hydrogen bonding in the liquid, in particular the oxygen-oxygen pair correlation function. Most of the proposed models succeed in fitting these thermodynamic and structural properties with reasonable accuracy using the dipole approximation, i.e. without short-range damping of the polarization. Also dynamical properties, such as librational frequencies and diffusion rates, are in agreement with experiment. At least equally good results are obtained by the sophisticated "first principles" potential of Ref. [46], which has been derived solely from high-quality *ab-initio* calculations for gas-phase clusters. This so-called NCC model is the first successful example of an (almost, see below) completely transferable water model.

In Ref. [22] we have proposed an empirical model of the type outlined in section 5.1. The distinction with other models (including our own previous attempts [43]) is that now the static dielectric constant ϵ is also used as a target property in the optimization. The dielectric constant is extremely sensitive to the average molecular dipole moment $\langle\mu\rangle$ in the liquid. The freedom of fine-tuning $\langle\mu\rangle$ by adjusting the characteristic length ξ for the short-range damping (see eqns (56) and (57)) was found to be instrumental in obtaining the correct value of $\epsilon = 80$. The value $\langle\mu\rangle = 2.6$D, needed to reproduce the dielectric constant, confirmed the prediction of an earlier mean-field study [21], and is in agreement with the experimental estimate for ice. The values of $\langle\mu\rangle$ for alternative polarizable models (including the "first principles" potential [46]) tend to be at least 0.2D higher than this. The modest difference in $\langle\mu\rangle$ gives a disproportionate increase in ϵ. For our own now-superseded polarizable TIP4P potential [43] with $\langle\mu\rangle = 2.8$D we found $\epsilon = 140$!

The new model of Ref. [22] also performs well for other quantities. Our self diffusion constant of $2.6 \times 10^{-9}\text{m}^2\text{s}^{-1}$ is, within the statistical uncertainty, identical to the experimental value of $2.4 \times 10^{-9}\text{m}^2\text{s}^{-1}$. An important test for a polarizable potential is the binding energy of the dimer, since the electronic polarization in this smallest member of the gas-phase clusters is roughly 25 percent of the 0.8D in the liquid. Our result $E_{\text{dimer}} = -24.8\text{kJmol}^{-1}$ is slightly lower than the experimental energy of $E_{\text{dimer}} = -22.6\text{kJmol}^{-1}$ but still reasonably close. Hence, our model, which has been parametrized for the liquid, satisfies a "reverse" liquid-to-gas transferability.

Another result of Ref. [22] that is of interest here is the observation that the stability of the hydrogen-bonded network is less dependent on the average dipole moment. For a given $\langle \mu \rangle$, the value of the hydrogen charges q_H can be increased by a considerable amount by moving the location of the counter-charge away from the oxygen towards the hydrogens (see Fig. 2). It is mostly the precise value of q_H and the parameter A of the Van der Waals repulsion that determine the strength of the hydrogen bond and, therefore, the characteristic tetrahedral structure of water. Manipulating these parameters we were able to construct a "freak" water model with the gas phase dipole moment of 1.85D and a liquid structure comparable to that of traditional effective pair potentials such as SPC and TIP4P. The static dielectric constant, however, was found to be only $\epsilon = 25$. This model is no more than a curiosity, but it may explain why non-polarizable "first principles" potentials such as MCY [47] were apparently transferable to the liquid when judged by the $g(r)$'s, which were still more or less acceptable, although definitely inferior to the results of effective potentials (as mentioned above, the situation has drastically improved with the NCC model). The bulk property that non-polarizable "first principles potentials" fail to reproduce even approximately, is the dielectric constant, which may be a factor of three or more too small. ϵ, however, is a quantity that is notoriously hard to evaluate in a simulation (see e.g. Refs. [1] and [22]) and was for this reason largely ignored in the early stages of the long history of the modelling of water.

5.3. SIMPLE IONIC SOLUTIONS

Building on the potential for pure water, we will now outline the generalization to a fully polarizable model of aqueous NaCl. Following the effective pair potential program, we will fit the parameters of the ion-water potentials to known properties of single ions in hydrated cluster states and solvated in bulk water. The transferability rules constrain us to keeping all parameters for the water molecule the same as in the pure water potential. In particular, this means that we are not allowed to alter the carefully "calibrated" width ξ_O of the fluctuating charges representing the induction on the oxygen. Having determined the ion-water model, the ion-ion parameters will be derived from the transferability rules only, i.e. without any further adjustments. The comparison to experiment of the calculated ion-ion correlations in dilute solutions is left as an independent and, as we will see, severe test for the model, both for the parameter set and the assumptions regarding the relevant interactions which underlie the functional form of the potentials.

Similar to the pure water potential, the ion-water parameters are determined from a combination of binding energies and structural characteristics. We will use the experimental cluster hydration enthalpies of Fig. 5 [39, 40], the Hartree-Fock dimer bond distances [38] and the coordination numbers for bulk solvation, as measured by neutron diffraction. For the aqueous Cl^- ion, the experimental literature provides us even with a full H- (or rather D)-Cl^- radial distribution [42]. The reason that we refrain from fitting to bulk solvation enthalpies ΔH_s, which are known to high precision from experiment, and take cluster data instead, is that the current simulation techniques are unable to evaluate ΔH_s with sufficient accuracy. The long-range character of the ion-water interaction leads to major boundary effects which, for the commonly used sample sizes, may be of order of an eV (see e.g. [48]). As a technical side remark, we should perhaps mention here that, in our experience, the Ewald summation treatment of long-range forces brings no significant improvement over spherical cutoff. For the evaluation of absolute solvation enthalpies, both methods suffer from a substantial bias.

As pointed out in section 5.1, one of the most critical quantities for the optimization of polarizable ion-water models is the cluster hydration enthalpy. This is due to the very persistent tendency to overbind. Fig. 5 shows our final result for the hydration enthalpies after an extensive trial and error search for the best value for the A, C and ξ parameters. We were able to suppress the overbinding of Na^+ almost completely. However, agreement with experiment could not have been achieved (given the constraints of the functional form of our model) without broadening of the Na^+ permanent (unit) point charge. We settled for a width of $\xi_{Na} = 0.7$ Å which is comparable to ξ_O. With the introduction of polarization, the ratio of $C_{Na^+O}/C_{OO} = 0.06$ has been significantly reduced and is now more consistent with the low polarizabilities for cations (see eqn (53)). The Na^+ hydration number of 5.5 in bulk solution is also in good agreement with experiment.

The complications in the parametrization of models for the interaction of polarizable water with cations come to haunt us even more obnoxiously when we try anions, which themselves are highly polarizable. The gas-phase polarizability of Cl^- is $\alpha_{Cl^-} = 5.5\,\text{Å}^3$, almost four times larger than the α for water. The danger of overpolarization is amplified by the geometry of the dimer. One of the two positive H sites of the water molecule points directly towards the Cl^- (see Fig. 6) and closely approaches the ionic polarizable center. Resorting to our remedy and broadening the H-atom charge is not an option this time, since the water-water interactions would be also modified in the process. The restriction of not tampering with the water model is imposed by the transferability rules, which we have given priority, despite the implication that the control of the overbinding for Cl^- hydration becomes rather difficult. Indeed, all we could accomplish for chloride is a hydration energy curve (see Fig. 5) which virtually coincides with the mean-field result and is, therefore, still considerably overbound. The width of the fluctuating charges in Cl^- was set to 0.6 Å while the ξ of the permanent charge in the center was increased to 1.1 Å. The ratio $C_{Cl-O}/C_{OO} = 1.2$ was again chosen to limit the overbinding as much as possible and is not entirely consistent with the ratio of the polarizabilities.

Even though the hydration energies in Fig. 5 for $Cl^-(H_2O)_n$ clusters are almost identical, with and without explicit polarization, the shift of binding energy from the dispersion term to the polarization has non-trivial consequences. In bulk solution, more than one molecule is expelled from the first hydration shell, and the coordination number of $n_c = 5.5$ for the polarizable model is in excellent agreement with experiment. Furthermore, the larger clusters exhibit qualitatively different behavior as a result of many-body polarization effects. A striking example is described in Ref. [49] for the $n = 14$ cluster. In the mean-field approximation the Cl^- ion is residing in the center of the cluster solvated by the surrounding water molecules. However, when the polarization on the chloride and solvent is taken into account, a surface state with the anion on the outside of the cluster is the more stable mode of hydration. The model used in Ref. [49] is a polarizable extension of SPC developed in Ref. [45].

Overall, for single ions in solution, the polarizable model is definite progress with respect to the effective pair potential from which we started. Both the structure of the solvation shells and the diffusion coefficients of the ions are in better accord with experiment [50]. The real tests are now two-ion properties since the ion-ion potentials in our scheme are fully determined by combination rules. For the isolated NaCl dimer we obtain a binding energy of $E_{NaCl} = -570 \text{kJmol}^{-1}$ compared to the $E_{NaCl} = -560 \text{kJmol}^{-1}$ of experiment (the TIP4P NaCl dimer is, surprisingly enough, underbound, $E_{NaCl} = -515 \text{kJmol}^{-1}$). Another dimer property that we can use as a test is the dipole moment. The dipole moment in the unpolarizable model follows from the two permanent unit charges separated by the equilibrium bond distance, which gives 11.7D. The induction on the chloride screens the permanent dipole by about 2.2D, decreasing the total moment to 9.5D, which is in satisfactory agreement with estimates from quantum chemistry calculations ($\approx$ 9.2D). The screening is strongly dependent on the ξ parameters and the result for the dimer moment is, therefore, a clear validation of the short-range damping as implemented in our model.

When finally, after all this preparation, the ion-ion correlations at finite concentration are examined, it is disappointing to find that our polarizable model fails for this most important aspect of ionic solutions. In a more than 1 ns MD simulation of a 1M solution [50] we monitored the ion distribution and found that the fraction of fully dissociated ions is too small. This sad result is immediately evident when the Na^+, Cl^- radial distribution $g(r)$ is plotted on an energy scale using the relation to the reversible work function

$A(r)$ for bringing to ions together in solution (see e.g. Ref. [51] and also [2]):

$$A(r) = -k_B T \ln g(r). \tag{58}$$

Fig. 7 compares the $A(r)$ functions for a 1M solution of TIP4P and our fully polarizable model (details of the calculation will be published elsewhere [50]). In the TIP4P curve, a barrier of a few $k_B T$ separates the contact ion pair from a stable solvent-separated ion pair. This feature is characteristic for mean-field models and is found also in potential of mean force (PMF) calculations at infinite dilution (PMF studies for aqueous NaCl have been performed by numerous authors and all are in qualitative agreement on this point). The solvent-separated free energy minimum is the deeper of the two implying that the TIP4P solution behaves as a strong electrolyte. In our polarizable model, however, the barrier has disappeared. The much more pronounced contact ion pair minimum is responsible for the high degree of ion pair association. For verification, and in order to eliminate the possibility that our MD run was still not long enough, we also determined the corresponding PMF's for single ion pairs in the pure solvent [50] using the method of constraints [51]. These independent calculations confirmed the ion-ion correlation results of the MD simulation at finite concentration.

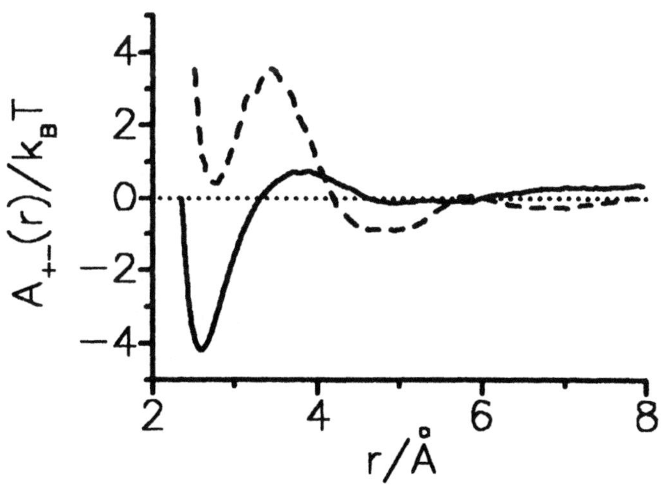

Fig. 7. Reversible work function for Na^+, Cl^- ion pairs as determined from the radial distribution functions in a 1M aqueous NaCl solution (see eqn (58)). The dashed curve gives the result for the TIP4P [38] model and the solid curve for the fully polarizable model discussed in the text.

In the 1M solution simulation, we did observe a clear preference for association in triplets, $(NaCl_2)^-$ and $(Na_2Cl)^+$, implying that the ionic character was retained, but our model has turned aqueous NaCl into a weak electrolyte. The distinctly unhappy ending of our salty water endeavor once again confirms that the modelling of ionic solutions is an extremely delicate affair, depending on the near cancellation of several large numbers, such as the binding energy of the ion, the solvent reconstruction energy and entropic contributions. At this stage we can only speculate about what went wrong. An indication, perhaps, is another deficiency of the model pointed out above, namely the persisting overbinding of the chloride ion in the cluster (the reason that TIP4P is not affected by its equally large energies of hydration is perhaps simply due to the fact that the NaCl dimer in this model is underbound, see above). The model in Ref. [45] (see also Ref. [49]) is considerably more successful in this respect. By including the three-body overlap term of expression (37) the gas-phase enthalpy curves could be fitted with satisfying accuracy. The implication suggested by this result is the perhaps unwelcome message that, in addition to electronic polarization, other non-rigid ion effects familiar from the modelling of ionic melts (section 3.2) also play a role in ionic solutions.

5.4. MACRO-IONS IN AQUEOUS SOLUTION

Larger molecules may contain several polarizable atoms. When not connected by a direct chemical bond, there is no reason to distinguish between mutual induction of a pair of atoms in the same and a different molecule. A special situation arises when a polyatomic molecule also contains several sites with permanent charges. Examples of such systems are polyelectrolytes and most biopolymers. Now the molecule will polarize itself, even in vacuum. Moreover, for flexible polymers, the self-polarization will be conformation-dependent and, hence, cannot be described by a distribution of static dipoles. The self-interaction may seem, at first, a major complication. However, in the framework of the model of polarization adopted here, with the width ξ of the fluctuating charges as the central parameter, we can also exploit the self-polarization effect. The ξ's can now be readily determined from the properties of a single molecule without having to consider the interaction with the solvent. As an illustration, we will treat here the modelling of a fully polarizable sodiumoctanoate molecule $CH_3(CH_2)_6COO^-Na^+$. In aqueous solution, this system forms micelles and we will also discuss, briefly, the effect of polarization on the counter-ion distribution.

Aqueous sodiumoctanoate is an elementary ionic surfactant and has been extensively studied by molecular dynamics simulations [52, 53]. The two main components of the interaction site model applied in these studies are the rigid hydrophilic head group and the flexible hydrophobic aliphatic tail.

The charge distribution of the non-polarizable carboxylate head group was taken from Ref. [54] and consists of permanent charges $q_O = -0.7$ on the two O sites and a charge $q_C = 0.4$ on the carboxyl C. The tail CH_2 and CH_3 are represented in the united-atom approach with neutral enlarged C atoms interacting via bond bending and torsion potentials of the type introduced in section 4.4. For water and the Na^+ ions, SPC type models were used [23, 48]. Similar to the aqueous NaCl of the previous section, we will start from this mean-field model and try to extend it in a stepwise procedure. We will place induction centers on all ten atoms, i.e. the COO^- head group as well as the tail C atoms. The oxygen sites will be assigned a polarizability of $\alpha_O = 1.5\text{Å}^3$ ($\approx \alpha_{H_2O}$). Effective polarizabilities for the carbon atoms are determined by interpolation based on counting CH bonds. This scheme gives us for the carboxyl C, CH_2 and CH_3 groups an $\alpha_C = 1.2, 1.9$ and 2.2D respectively.

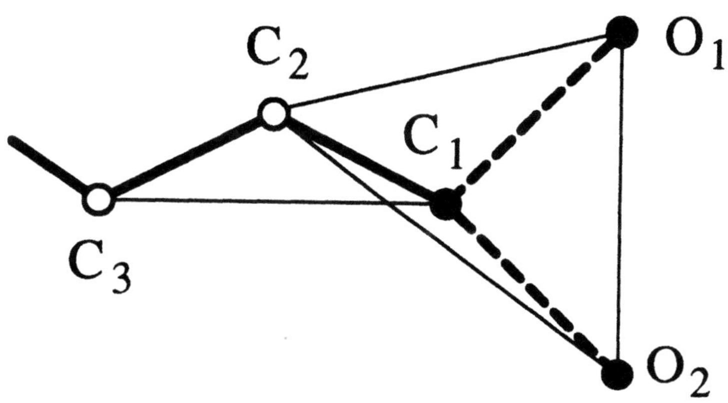

Fig. 8. Self-polarization in a model for the sodiumoctanoate ion [55]. All atomic sites (small circles) are polarizable. A solid circle indicates that the site also carries a permanent charge. Thin and solid lines indicate respectively nonbonded and bonded interactions which can induce moments. Coupling between charges and polarization of atoms joined by a polar bond (solid dashed lines) is excluded.

Before we can continue with the parametrization we have to decide which of the intramolecular induction interactions are allowed. We will use a coupling scheme where the charges and dipoles on all atoms interact except those directly joined by a polar bond. Hence, the permanent negative charge on an oxygen may polarize the opposite oxygen and all carbon atoms, but not the carboxyl C (see Fig. 8). The argument for excluding the induction across the CO bonds is that, since these bonds are polar, the polarization is already accounted for by the fractional permanent charges. The next question is which values to take for the permanent charges. The choice of the permanent charges is not as straightforward as it was for water. There, the charges are fixed by the gas-phase dipole moment. For self-interacting molecules, this is no longer possible since the gas-phase dipole moment also contains contributions of the self-polarization. The simplest ansatz we can make is not to modify the charges at all, and recover the gas-phase dipole from the induction screening, which can be adjusted by variation of the width parameter ξ (for simplicity we use the same ξ for all sites). Of course, this fitting procedure cannot be applied to the carboxylate ion since dipole moments are defined only for neutral molecules. We can circumvent this difficulty by using the equivalent atom-atom model of the acid or alcohol instead, and transferring the ξ needed to correct the mean-field dipole of the neutral molecule to the ion. For the 8 C atom system of interest here, the experimental estimate of the dipole moment for both $CH_3(CH_2)_6COOH$ and $CH_3(CH_2)_6CH_2OH$ is about 1.7D. The values for the non-polarizable model of Ref. [54] are 1.7 and 2.2D respectively. The induction with a width of $\xi = 0.7$Å gives total moments which are within 0.2D of the experimental value for both the acid and alcohol.

With the ξ of the surfactant sites fixed, we have reached the next stage, namely the renormalization of the A and C parameters of the short-range interaction with the Na^+ counter ion and solvent. This is an arduous task with an uncertain outcome, as we have seen from the example of the NaCl solution in section 5.3. For the purpose of a preliminary survey of the effect of polarization on the structure and dynamics of aqueous micelles we have reduced in Ref. [55] the complexity of the problem with a few drastic assumptions. The solvent polarizability was neglected altogether and, with no modification, the SPC model of the previous effective pair potential studies [52, 53] was used for the $H_2O - H_2O$, $H_2O - Na^+$ and $Na^+ - Na^+$ interaction. This simplified but somewhat inconsistent approach is, in fact, also the main motivation for not adapting the charges in the polarizable carboxylate head group. It also constrains us to keep a point charge on the Na^+ ion and not broaden it as we have done in the NaCl case. Furthermore, the parameters for the Na^+ interaction with the tail C sites can also be retained since the counter-ion is hardly ever in contact with these groups. That leaves us only with two parameters that need renormalization, namely A_{Na^+O} and C_{Na^+O} for the coupling to the carboxylate O. Since our main purpose was a comparison to the results of the unpolarizable model, the A_{Na^+O} and C_{Na^+O} parameters were fitted to the mean-field binding energy and equilibrium bond distance in vacuum.

As reported in Ref. [55] the polarization of surfactant ions had a dramatic effect on the distribution of counter-ions. In the effective pair approximation, about half of the Na^+ ions are directly bound to the COO^- head-groups forming stable contact ion pairs. Taking such a configuration as the initial state we switched to the polarizable model and watched the system evolve. All the contact ion pairs broke up in a relatively short period (10 ps). In the distribution established after equilibration, half of the Na^+ are now more loosely associated with the COO^- groups in solvent-separated ion pairs. The other half are scattered in the surrounding water as fully solvated free ions. Clearly, the inconsistency of treating only the polarization of the surfactant explicitly makes our results ambiguous, so verification using a completely polarizable model is imperative. Then the charge distribution of the carboxyl headgroup will also have to be reconsidered, which most likely will involve a more detailed analysis of the electrostatics of this complex functional group. Here, the *ab-initio* approach of Ref. [14–16] might be useful. The least we can conclude at this point, however, is that introduction of many-body polarization can alter the ion-ion correlations near macro-ions with possibly important consequences for the biological function of these molecules.

6. Bonded interactions

6.1. COVALENT SYSTEMS

Electronic polarization and other non-rigid ion effects can be treated, with varying success, in terms of a model because the underlying electronic processes still involve no real chemistry. In the present section we will consider true covalent bonding, with its strongly directional character and restrictions on coordination (valence). In section 4.4 we already encountered covalent bonds when we considered flexible molecules. However, the bonds were assumed to be stable and we could specify which atom is bound to which, since the bond connectivity remained unchanged during the evolution of the system. Hence, all difficulties related to bond breaking and saturation of valence could be avoided. Under the even stronger condition that the bond lengths are near their equilibrium value, we also resolved the interactions in harmonic radial bond stretching (eqn (46)) and angular bending potentials (eqn (47)).

The approach to modeling of covalent solids and liquids followed by Stillinger and Weber (SW) [56] is a generalization of the many-body potentials for flexible molecules. To allow for larger deformations and ultimately bond breaking, SW replaced the pair term by a short-range potential which decreases sufficiently fast beyond a certain separation. The three-body angular potential was extrapolated to the full range of angular degrees of freedom such that for, some configurations, the combination of bond angles can be nonbonding even though the atom-atom distances may be within bonding range. Hence, what limits in SW models the total number of bonds an atom can form, i.e. its valence, is the repulsive interaction between bond vectors established by the angular-dependent three-atom term. Depending on the thermodynamic state, the valence can fluctuate and, therefore, may be different in the liquid and the solid.

As an illustration, we will go through the famous SW model for solid and liquid silicon [56]. To make the potential resemble the familiar form of eqn (27) as closely as possible, SW introduce energy and length units ϵ and σ and scale the two-body and three-body potential functions v_2 and v_3, respectively, according to

$$\begin{aligned} v_2(\mathbf{r}_{ij}) &= \epsilon f_2(r_{ij}/\sigma) \\ v_3(\mathbf{r}_i, \mathbf{r}_j, \mathbf{r}_k) &= \epsilon f_3(\mathbf{r}_i/\sigma, \mathbf{r}_j/\sigma, \mathbf{r}_k/\sigma) \end{aligned} \tag{59}$$

where ϵ is chosen to give f_2 a depth of -1, and σ is chosen to make $f_2(2^{1/6})$ vanish. For f_2 SW used the functional form

$$f_2(r) = \begin{cases} A(Br^{-p} - r^{-q}) \exp\left[(r-a)^{-1}\right], & r < a \\ 0, & r \geq a \end{cases} \tag{60}$$

which automatically cuts off at $r = a$ without discontinuities in any r derivatives. For Si the cutoff was set to $a = 1.80$, while $p = 4$ and $q = 0$. The function f_3 is defined as

$$f_3(\mathbf{r}_i, \mathbf{r}_j, \mathbf{r}_k) = h(r_{ji}, r_{ki}, \theta_{jik}) + h(r_{ij}, r_{kj}, \theta_{ijk}) + h(r_{ik}, r_{jk}, \theta_{ikj}) \tag{61}$$

where θ_{jik} is the angle between the bond vectors $\mathbf{r}_{ji}$ and $\mathbf{r}_{ki}$ of atoms j and k which are in bonding distance of atom i. Extending the same perfectly smooth cutoff to the three-body term, SW write h for $r < a$ as the two parameter function

$$h(r_{ij}, r_{ik}, \theta_{jik}) = \lambda \exp\left[\gamma(r_{ij} - a)^{-1} + \gamma(r_{ik} - a)^{-1}\right] \left(\cos(\theta_{jik}) + \frac{1}{3}\right)^2 . \tag{62}$$

h vanishes for $r \geq a$. The "ideal" tetrahedral angle θ_t is such that $\cos(\theta_t) = -1/3$ and expression (62) clearly disfavors configurations violating the sp^3 geometry.

The two scale parameters ϵ and σ are fixed by the zero-temperature lattice spacing and cohesive energy. The optimization of the remaining six parameters A, B, p, q, λ and γ is a substantial task. In their "limited" search for a reasonable value for these parameters, SW first ensured that the diamond phase is the most stable at low pressure among the various alternative structures (some of these are actually observed at elevated pressure). In order to narrow down the interval of possible parameter values further, it was also necessary to use the melting temperature and the structure of the liquid as a selection criterion. Generating the correct liquid Si structure is particularly difficult, since the coordination number increases upon melting from four to about six. Again, we see that phase-transferability is a strong constraint for models. SW were able to find a parameter set that satisfies these requirements reasonably well, and their model was quite successful. The recent development of the *ab-initio* molecular dynamics method (see the contribution of Galli, Chapter 8) which is particularly suitable for Si and similar group III, IV, and V compounds, has superseded the application of SW models for these systems. However, SW models can still be very useful for more complex systems. We will see an example in the next section.

6.2. MODELLING CHEMISORPTION

Chemisorption of water molecules on transition metals such as Ni, Cu, Ag and Pt is a challenge for the construction of models since the water-metal interaction is clearly non-additive and qualitatively similar to chemical bonding. The observation that H_2O preferentially adsorbs on top sites with the oxygen down is convincing evidence for the chemical nature of the water metal substrate bond (for a comprehensive review we refer the reader to Ref. [57]). A pairwise additive potential must fail since it always attempts to maximize the number of nearest neighbors and, as a result, directs the adsorbates to the hollow or bridge sites with higher coordination numbers. On the other hand, typical adsorption energies are between $40 - 60 \mathrm{kJmol}^{-1}$ [57], which are values more characteristic of nonbonded interactions. Also the fact that H_2O is a closed-shell system suggests that the chemisorption is a "moderate" form of chemical bonding which, perhaps, permits modelling.

Water-metal adsorption has been the subject of numerous *ab-initio* studies (references relevant for our discussion are e.g. [58, 59]). The picture that emerged from electronic structure calculations is sketched in Fig. 9. The twofold-occupied lone-pair orbitals of the water donate into empty metal orbitals protruding from the surface. These orbitals are localized, oriented hybrids composed of s and p states of metal atoms with some admixture of d states. The adsorbate lone pair and the substrate orbital join to form a bonding state. Since the water-to-metal electron transfer is only minor, the binding energy is small in comparison to regular polar covalent bonds between open-shell atoms. The (for our purposes) attractive feature of this "molecular" view of chemisorption is that it explains, in a real space representation, why top sites are favored: this configuration optimizes the overlap between lone pair and metal orbitals. Also, it can be easily understood why the electron-rich oxygen end of the molecule attaches to the surface and not the H atoms.

The qualitative picture of chemisorption of Fig. 9 can be translated into a model using the general ideas about bonded and nonbonded cohesion developed in section 2. There, we argued that the outer doubly occupied single-electron orbitals can be considered independent entities with repulsive inter-

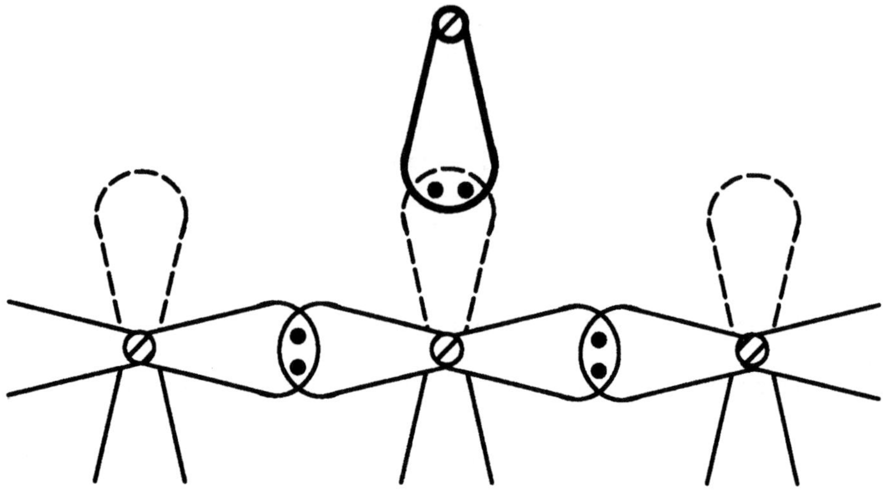

Fig. 9. Bonding pattern of chemisorption of water on metal substrates (after Ref. [58]). A doubly occupied lone pair orbital of the adsorbate (bold) donates into empty *spd*-like hybrid metal orbitals (dashed) protruding from the surface. Filled surface states are indicated by thin solid lines.

actions imposed by the exclusion principle. For nonbonded systems, these orbitals are the spherical atomic states centered on a single core. Bonded interactions, on the other hand, involve proper chemical bonds formed by hybrid bond orbitals linking two atomic centers. Hence, in the VSEPR view [8], we can think of the directional many-body interactions between covalently bonded atoms as arising from interactions between bonds. For example, in the SW model for Si, a repulsion between all four bond vectors emanating from an atomic vertex has the same effect of stabilizing the tetrahedral sp^3 geometry as the three-body potential function h of expression (62), which contains the tetrahedral angle as an explicit parameter.

The design of the SW-like model for chemisorption which we used in Ref. [60] was inspired by this philosophy. Similar to expression (60), we started by defining a pair function

$$v_2(r_{ij}) = Ar_{ij}^{-11} - Cr_{ij}^{-6} - Df_2(r_{ij})r_{ij}^{-3}\Phi_{ij}(\phi_{ij}). \tag{63}$$

The first two terms are the familiar isotropic soft-core repulsion and dispersive attraction for atoms. The purpose of the last term is to provide chemical binding. As in the SW expression (60), it is multiplied by a short-range cutoff function

$$f_2(r) = \exp\left[\frac{B}{(r-a)}\right]. \tag{64}$$

In order to ensure that the hydrogens point up, the strength of a bond is modulated by a factor

$$\Phi(\phi) = \exp\left[-8\left(\frac{\cos\phi - 1}{4}\right)^4\right] \tag{65}$$

which depends on the angle ϕ between the H_2O dipole vector and bond vector $\mathbf{r}_{ij}$ between the oxygen site i and metal atom j. Next we have to find a three-body function which will push the water molecule on top of a metal atom. After extensive trial and error we chose the function

$$v_3(\mathbf{r}_i, \mathbf{r}_j, \mathbf{r}_k) = E f_2(r_{ij}) r_{ij}^{-10} \exp\left[F\cos^2\left(\frac{\theta_{ijk}}{2}\right)\right] \tag{66}$$

where θ_{ijk} is the angle between the oxygen-metal bond $\mathbf{r}_{ij}$ and a metal-metal bond $\mathbf{r}_{kj}$ linking the absorbent atom j and its surface neighbors k. The function of expression (66) is entirely repulsive. It does the job we want it to do, in that it forces the oxygen-metal vector to avoid, as much as possible, all metal-metal bonds converging on the binding site. The most isolated site (on a flat surface), with all θ angles 90°, is the top of an atom. Hence the adsorbate will take refuge there. Second best is a bridge site, with some angles smaller than 90° and the environment of a hollow site is the most inhospitable.

Even for an H_2O on a top site, the v_3 term of expression (66) does not vanish. Each metal-metal bond adds a certain amount of positive energy and the sum total is proportional to the number of metal atoms coordinated to the binding site. As a result, our potential is also capable of accounting for a second important trend observed in experiment [57] and theory [58, 59], namely the sensitivity of adsorption to surface topology. Changing from a close-packed (111) surface to the more open (100) and then to the ridged (110) substrates, the desorption energy increases. The change in topology corresponds to a decrease of the number of nearest neighbors and hence our model predicts a more strongly bound adsorbate in accordance with experiment. In fact, for the parametrization used in Ref. [60] for Ni, the numbers are also in good quantitative agreement with quantum calculations [59].

In Ref. [60] we have applied the chemisorption model to study structure and dynamics of water on Ni(110) near and below monolayer coverage. The model was found to be able to give an explanation for the interesting phenomena related to the competition between hydrogen bonding in the water layer and the incommensurate surface corrugation. The crucial importance of representing the detailed atomic substrate structure when modelling water-metal interactions has been pointed out before [61, 62]. These earlier models, however, required the definition of a geometrically smooth boundary plane for the specification of a surface normal. The introduction of the three body SW term in our potential eliminates this artificial device. In fact, the validity of the model is not limited to corrugated flat surfaces. In principle, it can also be used for microscopic metal objects with non-planar shapes. However, it will be clear that a model of this complexity must be handled with care. There is always the possibility of encountering an unforeseen and unfortunate configuration where the model breaks down.

7. Conclusion

Potentials are powerful tools in simulation. If they work, they enable us to follow the evolution of a large and complex system over long periods of time. In particular for systems of biological interest, which require excessively long runs, there seems to be no alternative for the modelling of intermolecular interactions by potential functions. However, from the examples discussed in the previous sections, it will be clear that the condition that potentials "work" is at best locally satisfied. We discussed two different approaches to modelling, namely effective and first-principles potentials. Effective pair potentials, accounting for many-body effects in an average way, are likely to fail when applied outside the range of thermodynamic states for which they have been designed. First-principles potentials have been parametrized using gas-phase energy surfaces obtained from accurate electronic structure calculations of clusters. They may be superior under just those conditions but, when extrapolated to condensed phases, they suffer from exactly the same weakness as effective potentials. Being gas-phase models, there is no guarantee that they can be transferred to liquids and solids. In fact, in most dense systems, effective potentials are preferred since they have been developed for the situations in which they are applied.

The other aspect that distinguishes the first-principle from the effective-potential approach is that the first category of models uses only results of *ab-initio* calculations for parametrization while effective potentials are more of a pragmatic mix depending on input from experiment as well as theoretical data. We want to stress here that this distinction is separate from the issue of correcting pair potentials for many-body effects or not. Clearly, in complex systems with several components, it is an advantage to use the results of reliable quantum calculations, since the necessary experimental data are either not available or their interpretation in terms of elementary interactions is ambiguous (the notorious example is NMR data in biopolymers). This brings us to the general conclusion of this Chapter on potentials. What we want to advocate as the safest approach to modelling is to perform the electronic structure calculations in the condensed matter environment and use those results to construct effective potentials. If the complete system is too large to be treated fully on a quantum mechanical level, then part of it can be done *ab initio*. Classical modelling can still be used for the remainder, which surrounds the quantum cluster and imposes the thermodynamic state. This scheme has been pioneered in Ref. [11]. We hope that, combined with the Car-Parrinello technique (see Chapter 8), it can be developed into a reliable and systematic method for generating effective potentials.

Acknowledgements

The simulation studies which were used as illustrations in this Chapter and were taken from work by the author, were done in collaboration with Ilja Siepmann, John Shelley, Ailan Cheng, and Michael Klein. The author is particularly indebted to Michael Klein for guidance in the dangerous world of potential models and for continuing support. We also thank Mauro Ferrario for many helpful discussions on advanced molecular dynamics techniques.

References

[1] M. P. ALLEN and D. J. TILDESLEY, *Computer Simulation of Liquids*, Clarendon Press, Oxford, 1987.

[2] J.-P. HANSEN and I. R. McDONALD, *Theory of Simple Liquids*, Academic Press, London, 1986.

[3] G. CICCOTTI, D. FRENKEL, and I. R. McDONALD, editors, *Simulation of Liquids and Solids*, North Holland, Amsterdam, 1987.

[4] C. L. BROOKS III, M. KARPLUS, and B. M. PETTITT, Proteins: a theoretical perspective of dynamics, structure, and thermodynamics, in *Advances in Chemical Physics*, volume LXXI, John Wiley and sons, New York, 1988.

[5] C. R. A. CATLOW, S. C. PARKER, and M. P. ALLEN, editors, *Computer Modelling of Fluids, Polymers and Solids*, volume 293 of *NATO ASI series C*, Kluwer, Dordrecht, 1990.

[6] M. MEYER and V. PONTIKIS, editors, *Computer Simulation in Materials Science*, volume 205 of *NATO ASI series E*, Kluwer, Dordrecht, 1991.

[7] M. TINKHAM, *Group Theory and Quantum Mechanics*, McGraw-Hill, New York, 1964.

[8] R. McWEENY, *Coulson's Valence*, Oxford University Press, Oxford, 1979.

[9] G. C. MAITLAND, M. RIGBY, E. B. SMITH, and W. A. WAKEHAM, *Intermolecular Forces*, Clarendon Press, Oxford, 1981.

[10] M. J. L. SANGSTER and M. DIXON, *Adv. Phys.* **25**, 247 (1976).

[11] P. W. FOWLER and P. A. MADDEN, *Phys. Rev. B* **29**, 1035 (1984).

[12] M. FERRARIO, I. R. McDONALD, and M. L. KLEIN, *J. Chem. Phys.* **84**, 3975 (1986).

[13] M. PARRINELLO, A. RAHMAN, and P. VASHISTA, *Phys. Rev. Lett.* **50**, 1073 (1983).

[14] I. C. HAYES and A. J. STONE, *Molec. Phys.* **53**, 83 (1984).

[15] A. J. STONE and M. ALDERTON, *Molec. Phys.* **56**, 1047 (1985).

[16] S. L. PRICE, in *Computer Simulation in Materials Science*, edited by M. MEYER and V. PONTIKIS, volume 205 of *NATO ASI series E*, Kluwer, Dordrecht, 1991.

[17] R. J. WHEATLEY and S. L. PRICE, *Molec. Phys.* **71**, 1381 (1990).

[18] F. H. STILLINGER and A. RAHMAN, *J. Chem. Phys.* **60**, 1545 (1974).

[19] H. J. C. BERENDSEN, J. P. M. POSTMA, W. F. V. GUNSTEREN, and W. F. HERMANS, in *Intermolecular forces*, edited by B. PULLMANN, Reidel, Dordrecht, 1981.

[20] W. L. JORGENSEN, J. CHANDRASEKHAR, J. D. MADURA, R. W. IMPEY, and M. L. KLEIN, *J. Chem. Phys.* **79**, 926 (1983).

[21] K. WATANABE and M. L. KLEIN, *Chem. Phys.* **131**, 157 (1989).

[22] M. SPRIK, *J. Chem. Phys.* **95**, 6762 (1991).

[23] H. J. C. BERENDSEN, J. R. GRIGERA, and T. P. STRAATSMA, *J. Phys. Chem.* **91**, 6269 (1987).

[24] D. R. HUFFMAN, *Physics Today* , 22 (1991).

[25] Y. GUO, N. KARASAWA, and W. A. GODDARD III, *Nature* **351**, 464 (1991).

[26] A. CHENG and M. L. KLEIN, *J. Phys. Chem.* **95**, 6750 (1991).

[27] W. I. F. DAVID, R. M. IBBERSON, J. C. MATTHEWMAN, K. PRASSIDES, T. J. S. DENNIS, J. P. HARE, H. W. KROTO, R. TAYLOR, and D. R. M. WALTON, *Nature* **353**, 147 (1991).

[28] W. ANDREONI, F. GYGI, and M. PARRINELLO, *Chem. Phys. Lett.* **189**, 241 (1992).

[29] M. SPRIK, A. CHENG, and M. L. KLEIN, *J. Phys. Chem.* **95**, 2027 (1992).

[30] J. P. LU, X.-P. LI, and R. M. MARTIN, *Phys. Rev. Lett.* **68**, 1551 (1992).

[31] A. CHENG, M. L. KLEIN, M. PARRINELLO, and M. SPRIK, *Trans. Roy. Soc. A* (1992), in press.

[32] J.-P. RYCKAERT and A. BELLEMANS, *Chem. Phys. Lett.* **30**, 123 (1975).

[33] J.-P. RYCKAERT, M. L. KLEIN, and I. R. MCDONALD, *Phys. Rev. Lett.* **58**, 698 (1987).

[34] J. HAUTMAN and M. L. KLEIN, *J. Chem. Phys.* **91**, 4994 (1989).

[35] J. BAREMAN and M. L. KLEIN, *J. Phys. Chem.* **94**, 5202 (1990).

[36] J. I. SIEPMANN and I. R. MCDONALD, *Molec. Phys.* (1992), in press.

[37] I. CHANDRASEKHAR, preprint, 1992.

[38] J. CHANDRASEKHAR and D. C. S. W. L. JORGENSEN, *J. Am. Chem. Soc.* **106**, 903 (1984).

[39] I. DZIDIC and P. KEBARLE, *J. Phys. Chem.* **74**, 1466 (1970).

[40] M. ARSHADI, R. YAMDAGNI, and P. KEBARLE, *J. Phys. Chem.* **74**, 1475 (1970).

[41] M. SPRIK, M. L. KLEIN, and K. WATANABE, *J. Phys. Chem.* **94**, 6483 (1990).

[42] D. H. POWELL, A. C. BARNES, J. E. ENDERBY, G. W. NEILSON, and P. S. SALMON, *Faraday Disc. Chem. Soc.* **85**, 137 (1988).

[43] M. SPRIK and M. L. KLEIN, *J. Chem. Phys.* **89**, 7157 (1988).

[44] P. AHLSTROM, A. WALLQVIST, G. KARLSTROM, and B. JONSSON, *Molec. Phys.* **68**, 563 (1989).

[45] J. CALDWELL, L. X. DANG, and P. A. KOLLMAN, *J. Am. Chem. Soc.* **112**, 9145 (1990).

[46] U. NIESAR, G. CORIUNGIU, E. CLEMENTI, G. R. KNELLE, and D. K. BHATACHARYA, *J. Phys. Chem.* **94**, 7949 (1990).

[47] O. Matsuoka, E. Clementi, and M. Yoshimine, *J. Chem. Phys.* **64**, 1351 (1976).

[48] T. P. Straatsma and H. J. C. Berendsen, *J. Chem. Phys.* **89**, 5876 (1988).

[49] L. Perera and M. L. Berkowitz, *J. Chem. Phys.* **95**, 1954 (1991).

[50] M. Sprik, to be published.

[51] G. Ciccotti, M. Ferrario, J. T. Hynes, and R. Kapral, *Chem. Phys.* **129**, 241 (1989).

[52] K. Watanabe, M. Ferrario, and M. L. Klein, *J. Phys. Chem.* **92**, 819 (1988).

[53] J. Shelley, K. Watanabe, and M. L. Klein, *Electrochimica Acta* **36**, 1729 (1991).

[54] W. L. Jorgensen and J. Gao, *J. Phys. Chem.* **90**, 2174 (1986).

[55] J. Shelley, M. Sprik, and M. L. Klein, *Langmuir* (1992), in press.

[56] F. H. Stillinger and T. A. Weber, *Phys. Rev. B* **31**, 5262 (1985).

[57] P. A. Thiel and T. E. Madey, *Surf. Sci. Rep.* **7**, 211 (1987).

[58] M. W. Ribarsky, W. D. Luedtke, and U. Landman, *Phys. Rev. B* **32**, 1430 (1985).

[59] H. Yang and J. L. Whitten, *Surf. Sci.* **223**, 131 (1989).

[60] J. I. Siepmann and M. Sprik, *Surf. Sci. Lett.* **279**, L185 (1992).

[61] E. Spohr and K. Heinzinger, *Ber. Bunsenges. Phys. Chem.* **92**, 1358 (1988).

[62] K. Raghavan, K. Foster, K. Motakabbir, and M. Berkowitz, *J. Chem. Phys.* **94**, 2110 (1991).

First-principles Molecular Dynamics

G. Galli
IRRMA
PHB-Ecublens
CH-1015 Lausanne
Switzerland

and

A. Pasquarello
AT&T Bell Laboratories
600 Mountain Avenue
Murray Hill NJ 07974-2070
U.S.A.

ABSTRACT. We review the foundations and practical implementation of first-principles molecular dynamics and discuss some applications of the method to disordered systems and surfaces.

1. Introduction

Recent years have seen considerable progress in the scale and scope of molecular dynamics (MD) simulations with interatomic potentials derived from first principles (FP). Much of this progress is due to the novel approach to the study of electronic and thermodynamic properties of condensed matter systems, introduced in 1985 by Car and Parrinello [1–3]. This scheme combines the MD technique [4] for the computation of statistical properties of classical systems, with the FP treatment of interatomic forces due to the quantum electronic system, as provided by density functional theory (DFT) [5, 6].

Ab-initio MD simulations [1] involving a few hundred atoms are now performed in many laboratories, and there is a growing list of accomplishments in solving problems for real materials which were inaccessible until a few years ago. These include physical properties of disordered systems (such as carbon [7–10], silicon [11–14], silicon carbide [15, 16], gallium arsenide [17], arsenic [18, 19], I-IV alloys [20–22], selenium [23, 24] and copper [25]) in their

M.P. Allen and D.J. Tildesley (eds.), Computer Simulation in Chemical Physics, 261–313.

liquid and amorphous states, the investigation of processes relevant to semiconductor technology, e.g. the diffusion of hydrogen in Si [26], the study of phase diagrams [10, 27], surface reconstructions [28–36], and atomic clusters [37–44].

MD is a well-established methodology for the calculation of statistical properties of condensed matter systems [4]. The essence of the method is the numerical solution of Newton's equations of motion for an ensemble of atoms; these equations are solved with iterative techniques for appropriately long time intervals, and equilibrium statistical averages are computed as temporal averages over the observation time. In principle, explicit knowledge of the electronic ground state at each atomic position is required, in order to have a correct description of the interatomic forces. In conventional MD simulations, no attempt is made to solve this complex many-body problem, and interactions between atoms are modelled with empirical potentials. Although many systems have successfully been investigated with model potentials, it is difficult to find empirical interactions which work for different states of matter and for a wide class of materials. In complex cases, such as those involving covalent bonds, there is not general agreement about such basic questions as whether or not the interactions need to encompass 2-, 3-, 4- or higher n-body terms. Furthermore, the empirical MD approach suffers from an important conceptual limitation: the correlation between local atomic structure, e.g. bonding properties, and atomic dynamics is missed, as well as the effect of the atomic dynamics on the electronic properties.

DFT deals with the ground state of a quantum-mechanical many-body system of electrons. In principle, it can provide the *exact* charge density and thus, using the Hellman-Feynman theorem [45], exact forces on atoms, for a chosen ionic configuration. In practice, the local density approximation (LDA) [6, 46] to treat exchange and correlation has been used in most electronic structure calculations for solids and clusters, and has been shown to give very accurate energies and atomic forces for many systems of interest in material science. Thus, there is a well-established way to calculate forces from a fundamental electronic theory, in an *ab-initio* manner. However until recently, LDA energies and forces have been considered prohibitively complex, from a computational point of view, for direct applications to statistical mechanics simulations which usually require the study of a number of atoms ranging from several tens up to a few thousands, evolving through 10^3–10^6 configurations. Density functional theory within the local density approximation (LDF) has mainly been used to investigate zero-temperature properties of ordered systems, such as crystals and superlattices.

The unified approach to DFT and MD has made it possible to use forces derived within LDF in MD simulations. This has been accomplished by (i) devising a way to generate *at the same time* ionic trajectories *and* the corresponding electronic ground state; and (ii) developing efficient techniques for the solution of the Schrödinger equation for the single-particle electronic orbitals. The FP-MD technique has also been used, within a simulated-annealing (SA) approach [47, 48], to search for the global minimum of complex energy surfaces, depending upon a large number of degrees of freedom. This framework has been adopted, for example, to investigate the structural properties of different kinds of clusters [37–40] (S, Si, Na, etc.) with a number of atoms up to about 50.

The rest of the Chapter is organized as follows. In section 2, we briefly describe density functional theory and the local density approximation, and outline how an interatomic potential can be derived from FP within the LDF framework. In section 3, we present a scheme to perform MD simulations based on LDF (the Car-Parrinello (CP) method [1]), we analyze the reasons and conditions which make it a practical approach for MD simulations, and we discuss important differences of the CP method when applied to semiconductors, insulators and metals. In section 4, we describe the practical implementation of the DF-MD scheme within a pseudopotential-plane-waves framework, and we present the derivation of the total energy and forces for the case of norm-conserving pseudopotentials and for the case of ultrasoft (nonnorm-conserving) pseudopotentials, devised to treat efficiently first-row elements and transition metals. At the end of the section we briefly discuss recent developments of total energy minimization schemes which allow for an improvement of the scaling of the calculation as a function of the system size [49] and the efficiency of plane-waves basis sets [50]. In section 5, we present selected applications of the CP method to disordered systems and surfaces.

2. First-principles interatomic potential

2.1. CONDITIONS AND ASSUMPTIONS

The derivation of interatomic forces from FP presented here relies upon two basic assumptions. We assume that ions can be regarded as classical particles. Furthermore, we consider systems for which a separation between the classical motion of the ions and the quantum motion of the electrons can be achieved, i.e. systems satisfying the Born-Oppenheimer (BO) adiabatic approximation. This means that the electrons are assumed to be in a single energy eigenstate, and in particular in our derivation in their ground state which is required to be nondegenerate. We therefore exclude from our treatments those physical phenomena involving nonadiabatic transitions, which give rise to characteristic quantum effects on the *slow* nuclear dynamics, such as most of chemical reactions. The BO approximation is in general satisfied for semiconductors and insulators, where the electronic excitation spectrum has a gap between the ground and the first excited state much larger than the energy associated with the ionic motion. More subtle is the issue of the validity of the BO approximation for metals [51]. However it is generally believed that it holds also in this case, provided that the plasma frequency is much larger than the ionic thermal energies.

For systems for which the BO approximation holds, the interaction potential among the ions, $\mathcal{V}(\{\mathbf{R}_I\})$, can be derived from FP:

$$\mathcal{V}(\{\mathbf{R}_I\}) = \langle \Psi_0 | \mathcal{H} | \Psi_0 \rangle \, . \tag{1}$$

Here $\mathcal{H}$ is the hamiltonian of the system at fixed ionic positions $\mathbf{R}_I$ and Ψ_0 the corresponding instantaneous ground state. In order to use eqn (1) in realistic MD simulations, calculations of Ψ_0 for a number of configurations at least of the order of 10^4 are needed.

2.2. WHICH FIRST-PRINCIPLES DESCRIPTION?

There are two basic requirements to be fulfilled in order that the use of eqn (1) in MD simulations is meaningful and practical. An accurate theory for the many-body hamiltonian $\mathcal{H}$ is needed, capable of yielding with the *same* accuracy the ground state Ψ_0 for different ionic configurations. Indeed, rather different atomic arrangements are sampled during a MD simulation. Furthermore, the theory should lead to a computational scheme which is affordable in practice, with modern computer capabilities.

Computational considerations preclude the use of quantum chemical methods, such as the configuration-interaction method [52], which are often very accurate, but whose computational cost scales with a forbiddingly large power of the number (N) of electrons ($\mathcal{O}(N^m), m \geq 4$). A possible exception is the Hartree-Fock approximation [52] whose computational workload is of $\mathcal{O}(N^4)$. To date the only attempt to use Hartree-Fock for the construction of FP potentials relies, however, on an approximate treatment of the derivative of the total energy with respect to the ionic positions [53]. Furthermore, quantum chemical techniques are tailor-made for the study of molecules, but they are difficult to apply to bulk systems. Some interesting attempts in this direction can be found in Refs. [54–56].

A viable framework is offered by DF theory [6] which provides an accurate and practical means of calculating $\mathcal{V}(\{\mathbf{R}_I\})$. It scales acceptably as a function of system size ($\mathcal{O}(N^3)$ or less with newly developed techniques [49]) and has proved accurate for the description of many properties of both clusters and bulk systems. We note that approximate versions of LDF, such as those proposed by Harris [57] and more recently by Sankey [58], are computationally less expensive than the full LDF treatment, and in several cases have been shown to provide reasonably accurate descriptions.

2.3. DENSITY FUNCTIONAL THEORY IN THE LOCAL DENSITY APPROXIMATION

LDF electronic structure calculations are based on the Hohenberg-Kohn (HK) variational principle [5], which in principle allows the determination of the exact ground state of a system of interacting electrons in an external field; and on its formulation given by Kohn and Sham (KS) [46], who introduced a single-particle orbital picture in the theory.

According to the first statement of the HK theorem, for a system of fermions, the ground-state expectation value of any observable (e.g. the hamiltonian $\mathcal{H}_e$) is a *unique functional* of the exact ground-state density ρ_0:

$$\langle \Psi_0[\rho_0] \,|\mathcal{H}_e|\, \Psi_0[\rho_0]\rangle = F_{\rm HK}[\rho_0] \,. \tag{2}$$

The functional $F_{\rm HK}$ is unique in the sense that it does not depend on the external potential which acts on the system. This means, for example, that it is the same density functional for electrons in atoms, molecules and solids. Secondly, the HK theorem establishes the variational character of the energy functional $E_v[\rho] = F_{\rm HK}[\rho] + \int d\mathbf{r}\; v(\mathbf{r})\rho(\mathbf{r})$ of a system of fermions in an external potential v (e.g. of electrons in the potential of the nuclei). The functional attains a minimum with respect to variations of ρ when the density attains its ground-state value:

$$E_0 = \min_{\rho} E_v[\rho] \,. \tag{3}$$

ρ is meant to vary in the class of densities which are ground-state densities of a hamiltonian of the kind $\mathcal{H}_e + v'$, with some suitably chosen local external potential v'; such densities are called v-representable densities.

The central assertion used in establishing the Kohn-Sham single-particle orbital scheme is that, for any *interacting* system of electrons, we assume that there exists an auxiliary system of *noninteracting* particles (described by the hamiltonian $\mathcal{H}_s = T + V_s$, whose total energy is $E_s = T_s + \int d\mathbf{r}\, v_s\rho_s$ according to the HK theorem) such that the exact ground-state density of the interacting system equals the ground-state density of the auxiliary noninteracting system: $\rho_0 = \rho_s$. If the ground state of $\mathcal{H}_s = T + V_s$ (T is the kinetic energy of noninteracting particles) is nondegenerate, ρ_s, and then by assumption ρ_0, possess a unique representation,

$$\rho_0(\mathbf{r}) = \sum_i^{\rm occ} f_i |\psi_i(\mathbf{r})|^2, \tag{4}$$

in terms of the N single-particle orthonormal orbitals obtained from the Schrödinger equation:

$$\mathcal{H}_{\rm KS}\psi_i = (-\frac{1}{2}\nabla^2 + v_s)\psi_i = \epsilon_i\psi_i. \tag{5}$$

Atomic units $e = \hbar = m_e = 1$ are used throughout the Chapter; for simplicity the occupation numbers f_i are regarded all equal and will be omitted in the equations that follow. The total energy functional E_v of the interacting system in an external potential v can be written as

$$E_v[\rho_0] = T_s[\rho_0] + \int d\mathbf{r}\; v(\mathbf{r})\rho_0(\mathbf{r}) + \frac{1}{2}\int d\mathbf{r}\, d\mathbf{r}'\; \frac{\rho_0(\mathbf{r})\rho_0(\mathbf{r}')}{|\mathbf{r}-\mathbf{r}'|} + E_{xc}[\rho_0]. \quad (6)$$

Equation (6) formally defines the exchange and correlation energy functional E_{xc}; E has been split into four contributions: the kinetic energy of the corresponding noninteracting system; the interaction energy with the external potential; the classical electrostatic energy (E_H); and an exchange and correlation term which contains all the remaining quantum interactions between the electrons. It is then straightforward to show that the auxiliary potential which generates ρ_0 is given by:

$$v_s(\mathbf{r}) = v(\mathbf{r}) + \int d\mathbf{r}' \frac{\rho(\mathbf{r}')}{|\mathbf{r}-\mathbf{r}'|} + \frac{\delta E_{xc}[n]}{\delta\rho(\mathbf{r})}. \quad (7)$$

Equations (4–5) with the potential (7) represent the KS scheme: the application of the variational principle stated in the HK theorem has been recast into the solution of single-particle Schrödinger equations with an effective potential v_s. This is given by the sum of the external potential v, the Hartree potential $v_H = \int d\mathbf{r}'\, \rho(\mathbf{r}')/\,|\mathbf{r}-\mathbf{r}'|$, and the exchange and correlation potential $v_{xc} = \delta E_{xc}[n]\,/\delta\rho(\mathbf{r})$. Since v_s depends on the density, the set of eqns (4–5) has to be solved in a self-consistent fashion. The scheme contains, as particular approximations, the Hartree-Fock-Slater and the Xα methods. As it incorporates the description of correlation effects, it goes beyond the Hartree-Fock approximation. Moreover, if the central assumption of the Kohn-Sham derivation (the existence of v_s) can be justified, it constitutes an exact scheme.

However, in its present form, the KS formulation is not useful for practical computations, since the functional E_{xc}, which contains all the intricacies of the many-body electronic problem, is not known, and one must resort to approximations. An approximation to E_{xc} which has been widely used in the literature of the last decade is LDA [6,46]. The exchange-correlation energy density is approximated by the corresponding expression of the homogeneous electron gas ($e_{xc}(n)$) with the replacement of the constant density n by the local density $\rho(\mathbf{r})$ of the actual inhomogeneous system:

$$E_{xc} = \int d\mathbf{r}\; \rho(\mathbf{r}) e_{xc}[\rho(\mathbf{r})]. \quad (8)$$

In most applications of LDF, analytical parameterizations of the results of $e_{xc}[\rho(\mathbf{r})]$ for the homogeneous electron gas, as derived by Ceperley and Alder [59] with Monte Carlo calculations, are used. Application of the local density scheme should be restricted to situations with a slowly varying density. It is found, however, that the method works surprisingly well for many properties of systems beyond this limit (e.g. semiconductors with highly directional bonds). The LD scheme has some deficiencies such as the inadequate cancellation of self-interaction contributions. As a consequence, the potential v_{xc} does not exhibit the correct asymptotic behaviour proportional to $1/r$ for localized systems (atoms, molecules, etc.). A self-interaction corrected LDF scheme was proposed several years ago by Perdew and Zunger [60] and has recently been applied to the study of metal [61] and water [62] clusters using a FP-MD approach. The successes and failures of LDA have been analyzed in many papers: some of them are discussed, e.g. in Chapter 7 of Ref. [63]. We just mention that several gradient-correction schemes to improve over LDA have been suggested in the literature of the last decade. Recently the scheme proposed by Becke [64] has been extensively applied to the study of molecules [64–66] and bulk systems [67, 68] in an attempt to investigate whether a systematic improvement over LDA is possible.

2.4. SOLUTION OF THE KOHN-SHAM EQUATIONS

In conventional LDF electronic structure calculations for solids and molecules [69], the KS self-consistent equations are solved as follows. An initial value for $\rho(\mathbf{r})$ is guessed; for example $\rho(\mathbf{r})$ is taken to be a superposition of atomic charge densities. The potential of eqn (7) is then calculated, and eqn (5) solved by diagonalization of the KS Hamiltonian matrix: this is set up according to a chosen basis set for the expansion of ψ_i. From the eigenvectors of $\mathcal{H}_{KS}$ a new $\rho(\mathbf{r})$ is calculated, and the whole process is repeated till self consistency. Since the cost of a standard diagonalization grows as $\mathcal{O}(M^3)$, where M is the number of basis functions used to expand the Kohn-Sham orbitals, this procedure becomes very costly for large systems and one is forced to use iterative techniques [70, 71] such as Davidson's [72] or Lanckos' to find the lowest occupied levels, instead of a direct diagonalization of $\mathcal{H}_{KS}$.

The KS equations can be derived and solved in a different fashion if the minimization of the energy functional E with respect to the charge density is regarded as a nonlinear minimization of the functional with respect to the single-particle orbitals ψ_i. Minimization can then be achieved by introducing a fictitious dynamics in the space of electronic degrees of freedom ψ_i's. The simplest dynamics that one can conceive is provided by steepest descent [73] (SD), and can be formulated in terms of first-order equations in the fictitious time variable t:

$$\dot{\psi}_i(\mathbf{r},t) = -\frac{\delta E}{\delta \psi_i^*(\mathbf{r},t)}. \tag{9}$$

Here the dot indicates time derivative and the functional derivative gives $\delta E/\delta\psi_i^* = \mathcal{H}_{KS}\psi_i$. $\mathcal{H}_{KS}$ depends nonlinearly on the ψ_i. When solving eqn (9) numerically, orthonormality constraints must be explicitly imposed and the fictitious dynamics becomes:

$$\dot{\psi}_i(\mathbf{r},t) = -\frac{\delta E}{\delta\psi_i^*(\mathbf{r},t)} + \sum_j \Lambda_{ij}\psi_j \tag{10}$$

where the sum extends over the N occupied states. In practice, after an unconstrained step is performed, the predicted ψ_i are *corrected* by an appropriate orthonormalization procedure which gives the coefficients Λ_{ij}. Once the system has reached a minimum, $\dot{\psi}_i = 0$, i.e. $\mathcal{H}_{\rm KS}\psi_i = \sum_j \Lambda_{ij}\psi_j$, and by diagonalization of the constraint matrix $\mathbf{\Lambda}$ one recovers the KS equations. The unitary transformation which diagonalizes $\mathbf{\Lambda}$ can then be used to obtain the KS eigenstates and eigenvalues. Experience has shown that, at fixed nuclear configuration, E has only one minimum, and therefore the SD procedure leads to the absolute minimum. Note that in this dynamical way one achieves the objective of self-consistently minimizing E, without ever performing explicit diagonalizations.

The efficiency of the SD scheme is controlled by the number of steps that are necessary to achieve convergence. This can be quite large, especially in low-symmetry situations. Various attempts to improve the efficiency of the SD approach have been made, using either higher-order integration formulae or modified SD procedures [74]. Recently an efficient scheme has been obtained by adapting the well-known method of conjugate gradient (CG) minimization [75] to the problem of minimizing the HK energy functional. Also, preconditioning methods [76] and CG techniques with non-orthogonal orbitals [77] have been proposed, and in some cases shown to further speed up the rate of convergence of the iterative minimization.

As already mentioned, the basic step of iterative schemes such as SD and CG consists in evaluating the action of $\mathcal{H}_{\rm KS}$ on the ψ_i, followed by the orthonormalization of the single-particle orbitals. If the ψ_i are expanded in plane waves (with M plane waves for each of the N electronic states), the computation of $\mathcal{H}_{\rm KS}\psi_i$ requires, in principle, NM^2 floating point operations, whereas the orthonormalization requires N^2M operations. Such a dependence can be improved by using a splitting-operator technique to evaluate $\mathcal{H}_{\rm KS}\psi_i$. $\mathcal{H}_{\rm KS}$ consists of a kinetic $(-\frac{1}{2}\nabla^2)$ plus a potential energy $(v_{\rm s})$ term; the laplacian is diagonal in reciprocal space, so that $-\frac{1}{2}\nabla^2\psi_i$ requires $\mathcal{O}(NM)$ operations. Suppose that $v_{\rm s}$ is local ($v_{\rm s}$ can be non-local if a non-local pseudopotential is used to describe the core-valence interaction; this problem will be addressed in section 4.3); then its action on the Fourier coefficients of the wavefunctions is a convolution that can be conveniently calculated in real space by means of fast Fourier transform (FFT) techniques and this requires $\mathcal{O}(NM'\log_2 M')$, where $M' = xM$ is the number of points used to evaluate the FFT. The actual value of x depends upon the number of nonzero Fourier components of the potential, and on the accuracy required for the convolution $v_s\psi_i$. As apparent, in this iterative scheme the large number M (usually $M \gg N$) enters only linearly or via $M' \log_2 M'$ in the number of operations, and this gives a significant improvement over standard diagonalization techniques where M enters as M^3. Of course, for sufficiently large systems, the orthogonalization workload, growing like N^2M, dominates, and the numerical cost of the algorithm still grows as the cube of the system size. An additional advantage of SD and CG iterative techniques is that the $\mathcal{H}$ matrix is never explicitly required, thus permitting a significant saving in memory. The storage scales as N^2.

2.5. INTERATOMIC POTENTIAL CALCULATED WITHIN LDF

In principle, any of the methods described in the previous section can be used to solve the KS equations and then to derive *ab initio* the interaction potential amongst the ions $\mathcal{V}(\{\mathbf{R}_I\})$, at each atomic configuration $\{\mathbf{R}_I\}$. One could then perform *ab-initio* MD simulations by following three *separate* steps, at each MD move: (i) Solve self-consistently the KS equations for a given ionic configuration $\{\mathbf{R}_I\}$; (ii) compute forces acting on ions (F_I), according to the Hellman-Feynman theorem [45] ($\mathbf{F}_I = -\nabla_{\mathbf{R}_I}\mathcal{V}$); (iii) solve Newton equations of motions: $M_I\ddot{\mathbf{R}}_I = -\nabla_{\mathbf{R}_I}\mathcal{V}$. For some systems, this is computationally feasible and some authors are pursuing this approach [77, 78].

An alternative methodology was proposed some years ago by Car and Parrinello [1], in which electrons and ions are treated on the same footing, i.e. ionic trajectories and the corresponding electronic ground state are generated at the same time, without repeating the full self-consistent procedure to solve the KS equations at each MD step. This approach, together with global optimization techniques, will be discussed in detail in the next section.

Before doing so, it is useful to see how to generalize the minimization procedure of the energy functional E with respect to the electronic degrees of freedom to include the ionic degrees of freedom. The idea of simultaneously relaxing both ions and electrons was first discussed by Bendt and Zunger [70]. Within SD, this is easily done by considering the following two sets of equations:

$$\bar{\mu}_i\dot{\psi}_i = -\frac{\delta E}{\delta\psi_i^*} + \sum_j \Lambda_{ij}\psi_j \tag{11}$$

$$\overline{M}_I\dot{\mathbf{R}}_I = -\frac{\partial E}{\partial\mathbf{R}_I} \tag{12}$$

where the "masses" $\bar{\mu}_i$ and $\overline{M}_I$ have been introduced in order to account for the difference in time scales associated with the two sets of parameters $\{\psi_i\}$

and $\{\mathbf{R}_I\}$. Equations (11–12) define a mass-weighted SD trajectory on the E surface that moves downhill from the starting point to a nearby local minimum. By varying the ψ_i at fixed ionic positions, a single minimum on the E surface is found, which coincides with a point on the BO potential energy surface $\mathcal{V}$ defined by $\mathcal{V}(\{\mathbf{R}_I\}) = \min_{\{\psi_i\}} E\left[\{\psi_i\}, \{\mathbf{R}_I\}\right]$. As a consequence, the local minima of the E surface are also local minima of the $\mathcal{V}$ surface. Equations (11–12) can be used to optimize complex atomic structures, by relaxing simultaneously electronic and nuclear degrees of freedom. In conventional gradient methods this is accomplished by following a trajectory on the $\mathcal{V}$ surface, which means that perfect electronic self-consistency is required for any nuclear configuration. By using eqns (11–12) one follows instead a trajectory on the fictitious E surface, and the optimal relative relaxation rate for electrons and ions can be achieved by adjusting the respective mass parameters. The scheme may be generalized to include additional degrees of freedom like the volume or the shape of the unit cell. Simultaneous relaxation of electrons and ions substantially equivalent to that defined in eqns (11–12) has been used successfully in several problems involving local optimization of complex structures [28–41].

The scheme just presented has a severe limitation, allowing only for local and not global optimizations. Indeed, it is well known that the potential energy surface $\mathcal{V}$ can have several minima. To deal with this problem one needs a global optimization method. We show in the next section that a satisfactory solution to the global optimization problem is based on statistical mechanics, and it is intimately connected to the calculation of finite-temperature properties and the possibility of performing LDA-based MD simulations.

3. Molecular dynamics based on density functional theory

3.1. THE CAR-PARRINELLO METHOD

The key-point in the FP-MD method proposed in Ref. [1] is the definition of a *fictitious* dynamical system, associated with the *physical* system, whose potential energy surface E is an appropriate functional of both ionic and electronic degrees of freedom; the fictitious system is devised in such a way that the trajectories generated by its dynamics reproduce very closely those of the physical system with potential energy surface $\mathcal{V}$. The classical lagrangian of the physical system is given by the sum of the ionic kinetic energy and the ionic potential energy with reversed sign:

$$\mathcal{L}^{\mathrm{cl}} = \frac{1}{2} \sum_I M_I \dot{\mathbf{R}}_I^2 - \mathcal{V}\left[\{\mathbf{R}_I\}\right], \tag{13}$$

where M_I are the physical masses of the ions. The generalized classical lagrangian [1, 79] of the fictitious system is defined as:

$$\mathcal{L} = \sum_i^{\mathrm{occ}} \int \mathrm{d}\mathbf{r}\ \mu_i \left|\dot{\psi}_i(\mathbf{r})\right|^2 + \frac{1}{2}\sum_I M_I \dot{\mathbf{R}}_I^2 - E\left[\{\psi_i\}, \mathbf{R}_I\right]$$
$$+\sum_{ij} \Lambda_{ij} \left(\int \mathrm{d}\mathbf{r}\ \psi_i^*(\mathbf{r})\psi_j(\mathbf{r}) - \delta_{ij} \right). \tag{14}$$

$\mathcal{L}$ does not depend explicitly on time, and is a functional of two sets of classical degrees of freedom, the ψ_i and the $\mathbf{R}_I$, which depend on time. The μ_i are arbitrary parameters of units (mass)×(length)^2 which play the role of generalized masses for the electronic degrees of freedom; how to choose these parameters remains to be specified. For simplicity we will consider a unique μ for the ψ_i , independent from the electronic state, even though this is by no means necessary. The first and second term in lagrangian (14) are the kinetic energy of the electronic and ionic degrees of freedom, K_e and K_{I}, respectively. E is the potential energy of the coupled electron-ion fictitious system. The Lagrangian multipliers Λ_{ij} are used to impose orthonormality conditions on the ψ_i; in the language of classical mechanics, they are just simple holonomic constraints. The Euler equations associated with the lagrangian (14) are:

$$\mu\ddot{\psi}_i = -\frac{\delta E}{\delta\psi_i^*} + \sum_j \Lambda_{ij}\psi_j, \tag{15}$$

$$M_I\ddot{\mathbf{R}}_I = -\frac{\partial E}{\partial \mathbf{R}_I}. \tag{16}$$

At first sight there is no relation between the ionic dynamics generated from eqn (16) and that obtained from the correct equation of motion (EOM) for the nuclei, derived from the lagrangian (13):

$$M_I\ddot{\mathbf{R}}_I = -\frac{\partial \mathcal{V}[\{\mathbf{R}_I\}]}{\partial \mathbf{R}_I}. \tag{17}$$

Indeed, in general the nuclear trajectories generated by eqn (16) and those obtained from eqn (17) do not coincide, unless $E\left[\{\psi_i\}, \{\mathbf{R}_I\}\right]$ is at the instantaneous minimum. However, the parameter μ and the initial conditions $\{\psi_i\}_0$, $\{\dot{\psi}_i\}_0$ can be chosen in such a way that the time-scale for the electronic degrees of freedom is much shorter than that of the nuclei. In this case nuclear trajectories, initially lying on the BO surface, will deviate from it only after times that are significantly longer than the MD time step and, in some cases, even longer than the MD observation time. In other words, if μ and $\{\psi_i\}_0$, $\{\dot{\psi}_i\}_0$ are chosen so that the two sets of classical degrees of freedom, ions and electrons, are only weakly coupled, the transfer of energy between them is small enough, to allow the electrons to follow adiabatically the ionic motion, remaining close to the BO surface. In such a metastable situation, meaningful temporal averages can be computed. This dynamics, which we call classical adiabatic dynamics, is meant to reproduce in a computationally effective way what indeed occurs in real life, that is electrons following adiabatically the motion of the ions. The role of eqn (15) is that of keeping the electrons on the BO surface, without solving self-consistently the KS equations at each step, but for the initial configuration.

We note that by variation of the velocities $\dot{\psi}_i$ and $\dot{\mathbf{R}}_I$ one can vary the temperature of the system, thus making the fictitious dynamical system undergo various thermal treatments, such as annealing and quenching. During such processes, all the relevant degrees of freedom, both nuclear and electronic, are relaxed simultaneously. In particular, one can set up a dynamical process that brings the system to its equilibrium state at $T = 0$ by slowly reducing the ionic temperature. We call such a process *dynamical simulated annealing*. It is just a specific application of the concept of SA, introduced by Kirkpatrick, Gelatt and Vecchi [47, 48] to solve complex optimization problems. The difference from the original methodology in that here a MD rather than a Monte Carlo technique is used to explore configuration space. At variance with SD, the SA method offers a means of exploring globally the potential energy surface of a given system and, possibly, of locating the absolute minimum. This is important if the ions are allowed to move: in such a case E can have several minima, even for systems with a small number of ionic degrees of freedom, and local optimization methods can easily miss some of the relevant ones.

3.2. WHY DOES IT WORK?

In order to understand why the CP method works we consider the dynamics generated by eqn (15) when the ions are not allowed to move. Although entirely classical, this dynamics has some remarkable features. The most noticeable one is that it accounts for the properties of the quantum eigenvalue spectrum of the electrons. In particular, it is easy to show that for small deviations from the ground state, the dynamics of the KS orbitals can be well described as a superposition of oscillations whose frequency is given by:

$$\omega_{ij} = \left(\frac{2(\epsilon_j - \epsilon_i)}{\mu} \right)^{1/2} \tag{18}$$

where ϵ_j (ϵ_i) is the KS eigenvalue of an empty (occupied) state. Thus, the lowest frequency that in a regime of small deviations from the ground state can appear in the system is $\omega_{\min} = (2E_g/\mu)^{1/2}$, where E_g is the energy gap. This point is illustrated in Fig. 1 (from Ref. [80]), which shows the frequency spectrum of the velocity-velocity autocorrelation

$$\gamma(\omega) = \int_0^\infty \mathrm{d}t \cos(\omega t) \sum_i \langle \dot{\psi}_i(t) \dot{\psi}_i(t) \rangle \tag{19}$$

for the occupied KS orbitals of a model system of 8 Si atoms in the diamond structure.

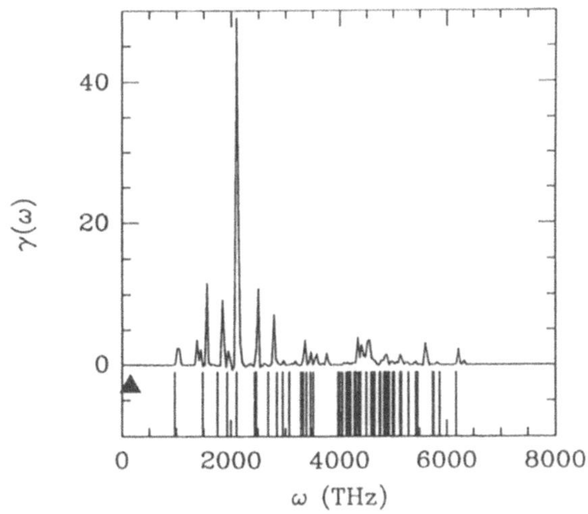

Fig. 1. The vibrational spectrum of the normal modes of the electronic degrees of freedom (full line) for a system with a large gap (8 Si atoms in the diamond structure; see text). The solid triangle indicates the position of the highest ionic frequency. The vertical bars below the spectrum represent the frequencies obtained from eqn (18) (from Ref. [80]).

The details of the system are not very realistic, but the example is instructive. It is seen that there is a lower bound for the electronic frequency spectrum at about 1010 THz. This value is in good agreement with the expression for $\omega_{\rm min}$ given above, and evaluated with $\mu = 300$ a.u. (the value used in the simulation) and the energy gap of the model $E_{\rm g} = 2.24$ eV. This value is to be compared with the highest ionic frequency for this model $\Omega = 140$ THz, indicated by an arrow in Fig. 1. This large difference in frequencies, e.g. the *clear separation between electronic and ionic characteristic frequencies*, is the clue to understanding how the CP method works. If the system is initially prepared in the instantaneous ground state with zero velocity, then the high-frequency motion of the electrons will ensure that they follow the motion of the atoms adiabatically, while the irreversible energy exchange between the fast and slow degrees of freedom is kept very small. An important point to underscore is that, if the conditions for the adiabatic motion of the electrons are satisfied, the instantaneous value of the forces does not coincide with the Hellman-Feynman forces but their *average* value does to a very high degree of accuracy. This is illustrated in Fig. 2 where the difference between the instantaneous forces and the BO ones is plotted. Two different oscillations can be seen: one very fast and averaging to zero, which corresponds to the electronic motion, and one much slower, which is due to the dragging of the electrons by the ions. The latter effect is small and can be cast in the form of a renormalization of the ionic mass M by a few percent.

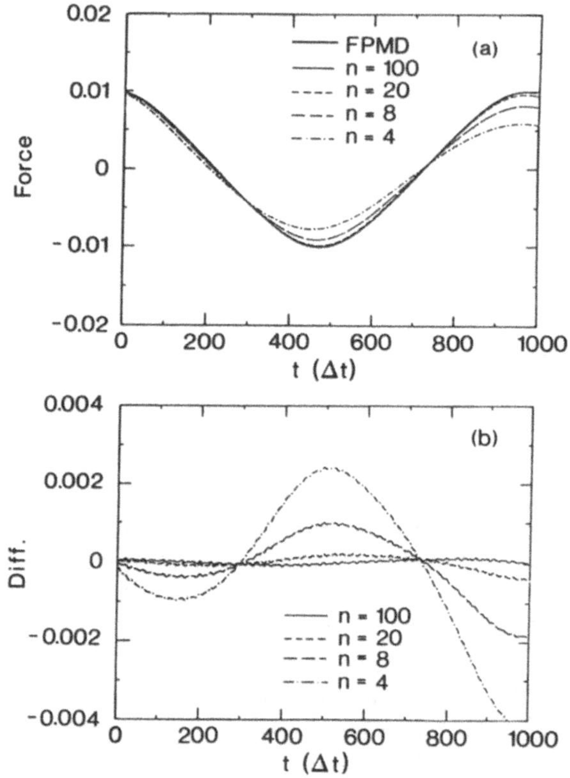

Fig. 2. The time dependence of the x-component of the force on the ions in a two-atom Si cell, as computed by solving the coupled eqns (15–16) (FP-MD) and by minimizing the electronic states at each step by steepest descent (SD). (a) n indicates the number of SD iterations per step. In (b) the differences between the forces plotted in (a) and the FP-MD force are shown (from Ref. [80]).

3.3. DEALING WITH SEMICONDUCTORS AND INSULATORS

In order to illustrate how the CP method works for system with a clearly defined single-particle gap we will consider an example [81]: a simulation for a system consisting of 16 silicon atoms in the diamond structure in a fcc unit cell. Initially the system has been heated to a temperature of $\simeq$ 400 K. The EOM have been integrated with a Verlet algorithm, with a time-step of $\delta t = 7$ a.u. The mass μ has been chosen to be 300 a.u.. The smaller the value of μ the better the adiabaticity is preserved. However small values of μ result in small values for the integration time step. The choice $\mu = 300$ a.u. represents a satisfactory compromise between these two opposite requirements in this case .

The calculation has been started with the electrons exactly in the ground state. Fig. 3(a) shows the ionic kinetic energy $K_{\rm I}$, the potential energy $E\left[\{\psi_i\},\{\mathbf{R}_I\}\right]$ and their sum $U_{\rm I}'$. If the electrons were exactly in their ground state, $U_{\rm I}'$ would be the total physical energy of the isolated system and therefore a constant as a function of time. It is seen from the figure that this is true to a large extent: on the scale of the figure, $U_{\rm I}' \approx$ constant. $U_{\rm I}'$ is conserved only approximately. This is not due to numerical errors but to the fact that the constant of motion of the coupled system described by eqns (15–16) is $U_{\rm eI}' = K_{\rm e} + U_{\rm I}'$, where $K_{\rm e}$ is the kinetic energy of the electronic degrees of freedom. However, within the adiabatic classical dynamics framework, one can choose μ and the initial conditions for ψ_i so that $K_{\rm e}$ can be made very small and kept as such as a function of time; therefore $U_{\rm I}' \approx$ constant. We note that one has to enlarge the scale of Fig. 3(a) by two orders of magnitudes in order to see the variations in $U_{\rm I}'$ (Fig. 3(b)). In Fig. 3(b) $K_{\rm e}$ is seen to be small and not to show any appreciable drift as the simulation proceeds. Since $K_{\rm e}$ measures the deviation from the BO surface, we see that, in this case, the CP dynamics ensures that at all times $E\left[\{\psi_i\},\{\mathbf{R}_I\}\right] \approx \mathcal{V}(\{\mathbf{R}_I\})$, and that the ion dynamics is close to the correct BO dynamics. In Fig. 3(b) it is also seen that the integration algorithm preserves the constant of motion $U_{\rm eI}'$, as it must.

3.4. DEALING WITH METALS

It is clear from the discussion of paragraph 3.2 that if $E_{\rm g}$ is small one needs a very small value of μ to obtain an adiabatic separation of ionic and electronic frequencies in eqns (15–16). As discussed earlier, the use of small electronic masses has the serious drawback that a small δt has to be used in the numerical solution of the EOM. In the case of metals, $E_{\rm g}$ is vanishing in an infinite system and very small in a finite one. Therefore in practice, for reasonable values of δt, an irreversible transfer of energy between ions and electrons is observed. Then electrons quickly slide away from the BO surface, as opposed to the case of semiconductors and insulators. At the same time a drift in the temperature of the ions is observed, due to the exchange of energy between cold electrons and hot ions. This leads to a spontaneous cooling of the ions. These two effects, the drift of the electrons from the BO

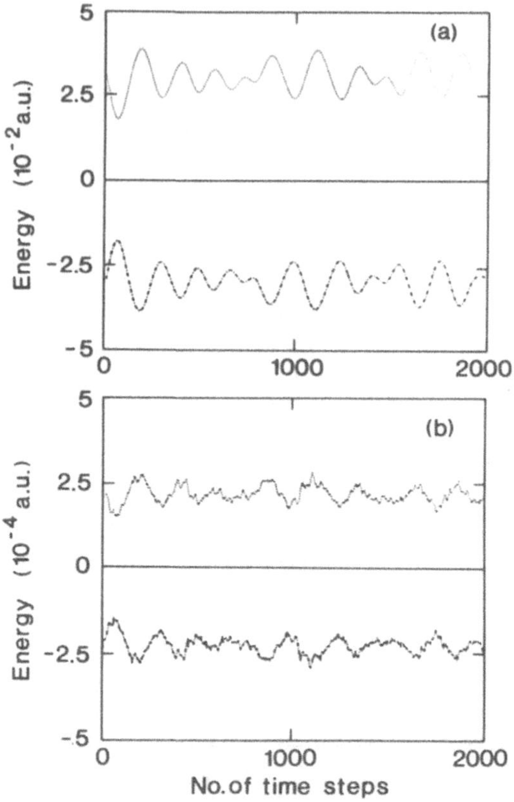

Fig. 3. The time variation of the electronic and ionic energies on a segment of 2000 steps for a run of 16 Si atoms in the diamond structure (see text). K_{I} (dotted line), $E[\{\psi_i\},\{R_I\}]$ (dashed-dotted line) and $U_{\mathrm{I}}' = K_{\mathrm{I}} + E$ (full line) are shown in the upper part of the figure, (a), on an energy scale of 10^{-2} a.u. K_{e} (dotted line), U_{I}' (dash-dotted line) and $U_{\mathrm{eI}} = U_{\mathrm{I}}' + K_{\mathrm{e}}$ (full line) are shown in the lower part, (b), on an energy scale of 10^{-4} a.u. E, U_{I}' and U_{eI} are measured relative to the initial value $U_{\mathrm{eI}} = -61.62887$ (from Ref. [3]).

surface and simultaneous cooling of the ions, are illustrated in Fig. 4.

A practical, although not entirely satisfactory, way forward has been to control the temperature of the ions using, for instance, the Nosé thermostat [82], and to perform periodic minimizations of the electrons so that the deviations from the BO surface remain within a preassigned value [9]. The quality of the results obtained in this way has been good. However the ad-hoc nature of the procedure and the additional computational burden due to the extra minimizations calls for a different approach.

One alternative has been proposed by Sprik [83] and implemented in the context of FP-MD by Blöchl and Parrinello [84] and Fois *et al.* [85]: two

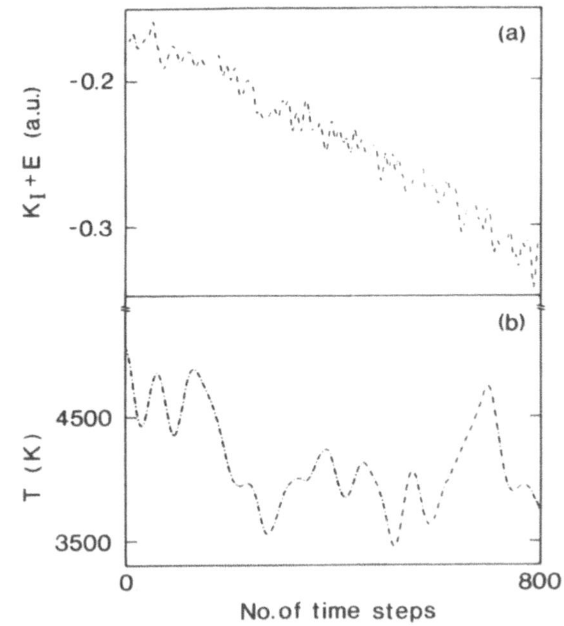

Fig. 4. The time variation of (a) $U_I' = K_I + E$ and (b) the ionic temperature T, on a segment of 800 steps for a run of 54 C atoms in a liquid-like (diffusive) state. The system is metallic (from Ref. [3]).

different thermostats are applied, one to the ionic and the other to the electronic degrees of freedom. The role of the two thermostats is to maintain the ions at the desired temperature while at the same time keeping the electrons "cold", namely very close to the ground state. Applications of this method to liquid aluminum (l-Al) [84] and to the study of dissociation/recombination of Na_2 and isomerisation of Na_3 [85] are very satisfactory. A practical difficulty associated with using a Nosé thermostat to maintain the electrons close to their ground state is that one introduces velocity-dependent forces in the EOM for the electronic degrees of freedom. In the presence of such forces, ordinary integration schemes require the Lagrange multipliers Λ_{ij} to be computed several times until self consistency between predicted positions and velocities is achieved. In FP-MD simulations, this would result in a catastrophic increase in the computational time, since the calculation of the constraints is a substantial part of the computation. Solutions to this problem have been proposed in Ref. [84, 85] and we refer the reader to the original papers for details of the algorithms.

One problem that should also be considered in the context of simulations of metals is the introduction of fractional occupation numbers for the electronic states. For example, this problem has been addressed by Fernando *et al.* [86], and Pederson and Jackson [87].

4. Practical implementation of a DF-MD scheme

4.1. BASIS SET FOR THE ELECTRONIC ORBITALS AND MD SUPERCELLS

The FP-MD scheme just described was originally based on a pseudopotential description of the interaction between ionic cores and valence electrons, in conjunction with plane-wave (PW) basis sets [1]. To date, the method has been mostly applied in its original version, for reasons of computational efficiency, although all-electron formulations of the CP method within an augmented plane-waves scheme have been proposed in the literature [88–90]. The choice of PW expansions for the single-particle orbitals has several advantages. It permits the use of FFT techniques which are computationally very efficient; PW's do not depend on atomic positions and therefore forces acting on atoms can be easily computed via a straightforward application of the Hellman-Feynman theorem [45]. PW's are free of basis-set superposition errors and allow computation of the total energy of different atomic arrangements with the same accuracy. This is particularly relevant in MD simulations, where rather different atomic configurations are explored. Finally the convergence of PW calculations can be controlled in a very simple manner, since it depends only upon the *number* of Fourier components included in the expansion of the ψ_i. The disadvantage of PW expansions is the very large number (M) of basis functions needed to represent the electronic orbitals, compared, for example, to basis sets of gaussian or atomic-like orbitals. PW basis sets call, of course, for the use of pseudopotentials, since the number of PW's needed to describe localized core states is too large and cannot be afforded from a computational point of view. A nontrivial problem is posed by first-row and transition metal elements which have very localized charge densities also in a pseudopotential description. This problem will be addressed in 4.4.

In simulations of bulk systems, the MD box is periodically repeated to infinity. This is a natural choice for perfect crystals, whose hamiltonian has translational symmetry. For disordered systems, such as amorphous solids or liquids, periodic boundary conditions (PBC) can still be adopted, provided the MD box is chosen to be large enough so that the imposed periodicity does not affect the dynamical properties of the system. Therefore the techniques [69] developed in electronic structure calculations within LDF, which take advantage of the translational symmetry of the hamiltonian, can also be used for LDF-MD simulations of both ordered and disordered systems.

As a consequence of the periodicity introduced by PBC, the single-particle orbitals entering the lagrangian (14) satisfy the Bloch theorem and can be expanded in PW's:

$$\psi_i^{\mathbf{k}}(\mathbf{r}) = e^{i\mathbf{k}\cdot\mathbf{r}} \sum_{\mathbf{g}} c_i^{\mathbf{k}}(\mathbf{g}) e^{i\mathbf{g}\cdot\mathbf{r}}, \tag{20}$$

where $\mathbf{g}$ is a reciprocal lattice vector of the MD supercell and the wave vector $\mathbf{k}$ lies within the Brillouin Zone (BZ) of the reciprocal lattice of the supercell. $c_i^{\mathbf{k}}(\mathbf{g})$ are the Fourier components (FC) of the single-particle wavefunction, which are treated as time-dependent degrees of freedom in the calculation. The basis set specified by eqn (20) is reduced to a finite set by truncating the sum over $\mathbf{g}$ to include only those plane waves with a kinetic energy $E_{\mathbf{k}} = \frac{1}{2}(\mathbf{k}+\mathbf{g})^2$ less than a given energy E_{cut}. The value of E_{cut} depends on the specific system and in particular upon the choice of the pseudopotential for the description of the core-valence interaction. For a given pseudopotential the choice of E_{cut} determines the accuracy of the calculation.

The computation of the electronic density ($\rho_e(\mathbf{r})$) and then of the total energy requires an integral over the BZ:

$$\rho_e(\mathbf{r}) = \sum_{\mathbf{k}} w_{\mathbf{k}} \sum_i |\psi_i^{\mathbf{k}}(\mathbf{r})|^2, \tag{21}$$

where $w_{\mathbf{k}}$ is the $\mathbf{k}$-point weight. In principle, an infinite number of points should be included in the summation (21). In practice, an appropriate small set chosen according to the symmetry properties of $\rho_e(\mathbf{r})$ leads to a rapid convergence of the summation. Sets of special $\mathbf{k}$ points have been computed [91] for supercells with different symmetries, such as simple cubic, face centred and body centred cubic. The source of computational saving in BZ sampling with special points comes from the symmetry of the system. For example, for diamond Si, a calculation with the Baldereschi point [92] ($\mathbf{k} = (1/4, 1/4, 1/4)$) provides a sampling of the 8 $\mathbf{k}$ points contained in the corresponding star and leads to an accurate description of the properties of the crystal. However, a general lattice distortion breaks the point symmetry of the lattice and the sole symmetry point for $\rho_e(\mathbf{r})$ is $\mathbf{k} = (0,0,0)$ (Γ point). In this case BZ sampling with special points is not much help. Therefore the computation of finite-temperature properties of crystals and the simulation of disordered systems is most appropriately performed with large supercells and the use of the Γ point for BZ sampling.

There is a further computational advantage in the use of the Γ point: at $\mathbf{k} = 0$, one can choose the single-particle orbitals $\psi_i(\mathbf{r})$ to be real, since the phase factor of the wave function is arbitrary. Thus, for each reciprocal lattice vector $\mathbf{g}$, the Fourier components $c_i(\mathbf{g})$ of the orbitals satisfy the symmetry relation: $c_i(-\mathbf{g}) = c_i^*(\mathbf{g})$. One can take advantage of this property to reduce the independent FC (e.g. the basis set) by a factor of two, substantially reducing the computational cost of a MD simulation.

4.2. PSEUDOPOTENTIALS

In most molecular or solid-state systems, it is possible to separate atomic states into valence states, which determine the bonding properties, and core states which are much deeper in energy and can therefore be considered as chemically inert. In the pseudopotential scheme the electronic problem is simplified by considering only the valence states. Note that the separation between valence and core states is not strict, and it is even possible that for the same kind of atom the number of valence states changes according to the environment or to the degree of approximation allowed. The core states, which can be considered to be essentially unchanged with respect to the free atom, affect the valence wave functions only because of the orthogonality condition. These inert states can be eliminated by replacing the true atomic Coulomb potential by pseudopotentials, designed in such a way that the effect of the core states on the valence states is effectively reproduced. We note that the pseudopotential needs to give an accurate description of the core-valence interaction only in a finite energy range, in order to correctly reproduce the bonding properties of a given system.

In a pioneering paper, Hamann, Schlüter and Chiang (HSC) [93] suggested a scheme for constructing pseudopotentials within the framework of LDA. Since, for most elements, a fully local pseudopotential cannot reproduce accurately the properties of the all-electron wave functions, a semi-local form is proposed in which, in the core region, a different pseudopotential v_l is used for each angular-momentum channel l:

$$v_{\mathrm{ps}}(\mathbf{r}) = \sum_{l=0}^{\infty} v_l(r)\hat{P}_l, \tag{22}$$

where $\hat{P}_l$ is a projector operator onto the lth angular momentum. Since for high l (e.g. $l \geq 2$ for most elements) the l-component of the wave functions is very small in the core region, v_l can eventually be taken to be independent of l, and is usually taken to be the same as for the highest l which needs to be considered separately. The pseudopotentials are derived in the case of the free atom and required to satisfy the following conditions. For every angular momentum channel, the pseudo-eigenstates are required to be at the same energy as the all-electron eigenstates. A core radius cutoff is defined beyond which the bound pseudo-wave-function matches the all-electron wave-function. In the core region, the pseudo-wavefunction is nodeless and carries the same charge as the all-electron wavefunction (the norm-conserving condition). The latter condition implies that the logarithmic derivatives of the pseudo-eigenstate match the corresponding all-electron ones at any given point beyond the cutoff radius. This ensures that the pseudopotential gives an accurate description of the wave functions at energies close to the eigenvalues. This kind of pseudopotential has been derived for all the atoms of the periodic table [94].

An additional requirement for pseudopotentials to be used in conjunction with PW basis sets is to be as smooth as possible, e.g. to require the lowest possible E_{cut}. Since the original formulation [93], different schemes have been suggested for constructing norm-conserving pseudopotentials in order to reduce as much as possible the number of required plane waves [95–98]. Recently, improvements which contribute to the flexibility of the pseudopotential scheme have been obtained. A generalized scheme has been proposed which relaxes the condition that the reference energy be an eigenvalue [99]. This is particularly useful when, for example, the atomic eigenvalue is not close to the energy region relevant to the bulk calculation. It has also been shown that the energy range of validity of a pseudopotential can significantly be increased by extending the correctness of the logarithmic derivative one order further in energy [100].

In a recent development in the theory of pseudopotentials, Vanderbilt has been able to relax the norm-conserving condition on the pseudo-wave-functions [101]. The associated pseudopotentials are termed *ultrasoft*. In this new scheme, the pseudo-wave-functions are not required to have the same charge as the all-electron wave functions, and can therefore be designed to be as smooth as possible in the core region. In Fig. 5, we illustrate the result of this procedure in the case of the $3d$ orbital of Cu. Note that in this case the radial all-electron wavefunction has no nodes, since it is the lowest state of d symmetry, and it is therefore particularly difficult to pseudize it effectively in norm-conserving pseudopotential schemes. The implementation of norm-conserving and ultrasoft pseudopotentials in a LDF-MD scheme, as well as the differences between the two cases, are discussed in the next two sections.

4.3. ENERGY AND FORCES WITH NORM-CONSERVING PSEUDOPOTENTIALS

Fig. 6 outlines the basic steps of an *ab-initio* MD simulation. At $t = 0$ the ground-state single-particle orbitals ψ_i associated with the electronic states of the system in a chosen initial ionic configuration have to be provided. The total energy and its derivatives with respect to the ionic positions can then be computed. The action of the hamiltonian on each FC is evaluated in order to solve the equation

$$\mu\ddot{\psi}_i = -\mathcal{H}_{\text{KS}}\psi_i \tag{23}$$

for each wave vector $\mathbf{k}$ and electronic state i. Equation (23) does not contain the forces coming from the orthogonality constraints that the wavefunctions must satisfy. These are accounted for in a subsequent step where the electronic states are orthogonalized. When both the electronic and ionic degrees

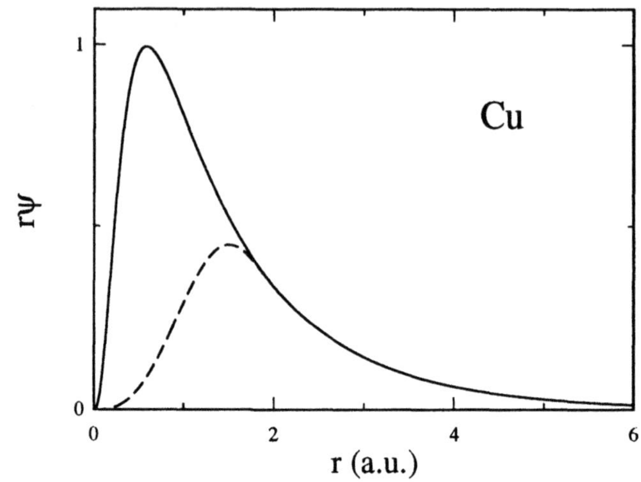

Fig. 5. The all-electron (solid) and pseudized (dashed) radial wave functions of the 3*d* orbital of Cu. A cutoff radius of 2 a.u. has been used in the pseudization.

of freedom are evolved in time according to second-order EOM and the MD run is performed, the choice of the orthogonalization procedure is crucial and has to be appropriate for the treatment of holonomic constraints. This point will be discussed in detail in section 4.5. After solving the EOM for the electronic degrees of freedom, one solves those for the ionic degrees of freedom. Finally the time step is incremented and the *new* wavefunctions obtained from the integration of eqn (15) and new positions from eqn (16) are used as input for the next step.

A similar scheme can be used for total energy minimizations at fixed ionic positions. Instead of solving first-order equations with, e.g. SD or CG techniques (see section 2.4), one can choose to solve second-order ones where the velocities of the electronic degrees of freedom are slowly rescaled towards zero. In this case the number of δt corresponds to the number of iterations and $\delta t/\mu$ is just a parameter governing the convergence rate. At $t = 0$ an initial guess for the wavefunctions is provided and the same procedure outlined above can be followed, where the solution of the EOM for ions is skipped. The basic steps involved in the computation of the total energy E and $\mathcal{H}_{\rm KS}\psi_i$ are illustrated in Fig. 7, which shows the most convenient space (real or Fourier space) for the evaluation of the different terms and the number of FFT needed in the calculation.

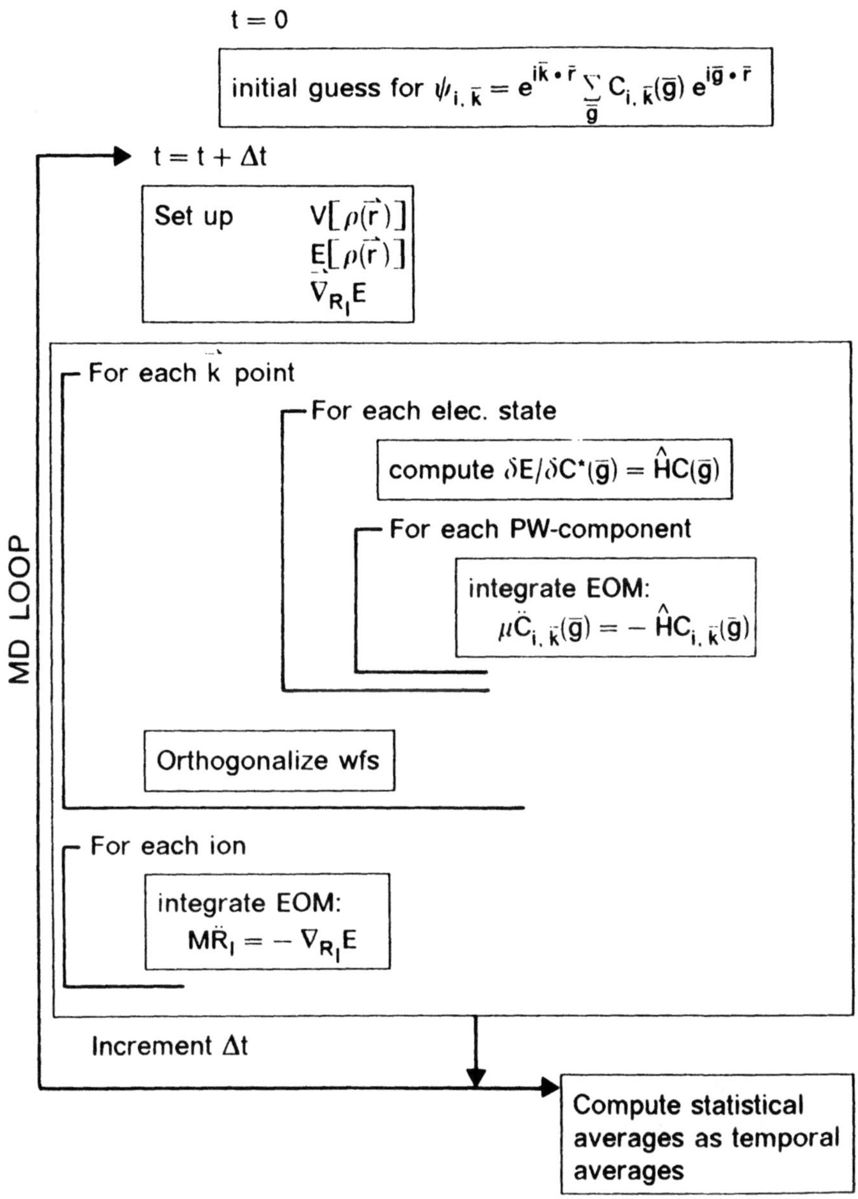

Fig. 6. The flow chart of an *ab-initio* molecular dynamics run (from Ref. [3]).

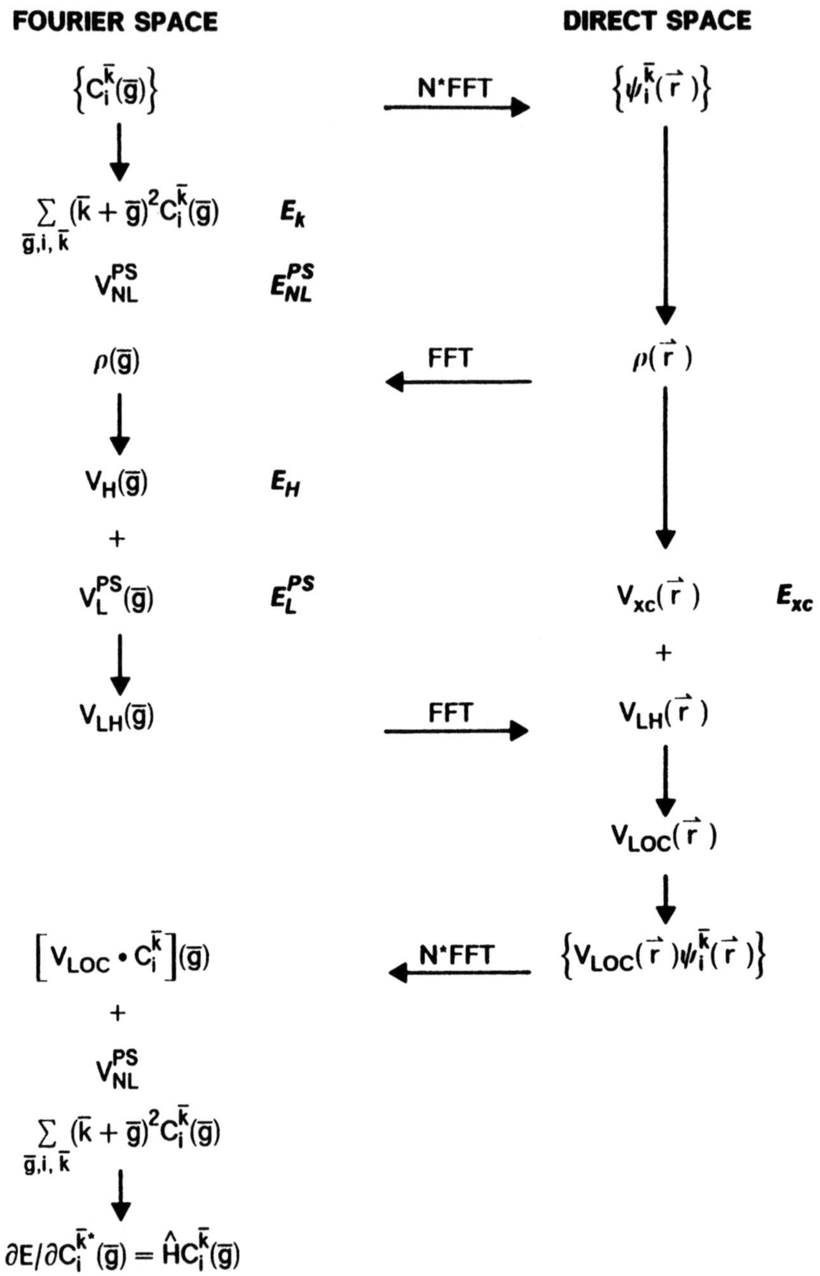

Fig. 7. Block diagram illustrating the basic steps involved in the calculation of the total energy. FFT indicates a direct (→) and inverse (←) Fast Fourier Transform and N the number of electronic states (from Ref. [3]).

The total energy E is given by the sum of the HK energy functional and of an ion-ion interaction energy $E_{\rm M}$:

$$E = E_{\rm k} + E_{\rm xc} + E_{\rm H} + E_{\rm ps} + E_{\rm M}, \tag{24}$$

where

$$E_{\rm M} = \frac{1}{2} \sum_{I \neq J} \frac{Z_I Z_J}{|\mathbf{R}_I - \mathbf{R}_J|}. \tag{25}$$

The kinetic energy of the electrons, $E_{\rm k}$, is easily evaluated in Fourier space:

$$E_{\rm k} = \frac{1}{2} \sum_{i,\mathbf{g}} \mathbf{g}^2 c_i^*(\mathbf{g}) c_i(\mathbf{g}). \tag{26}$$

For simplicity of notation, here we assume that only one $\mathbf{k}$-point is included in the BZ sampling and drop the summations over $\mathbf{k}$. The exchange-correlation energy functional expressed within LDA (eqn (8)) is computed in real space: once the energy density $e_{\rm xc}$ is tabulated on a real space grid, the integral in eqn (8) is performed in a straightforward way. $E_{\rm H}$ and $E_{\rm ps}$ denote the Hartree and pseudopotential energy, respectively:

$$E_{\rm ps} = \int \mathrm{d}\mathbf{r}\, v(\mathbf{r}) \rho_{\rm e}(\mathbf{r}), \tag{27}$$

where v is the total ionic potential acting on the electrons, expressed as a sum of ionic pseudopotentials $v_{\rm ps}$:

$$v(\mathbf{r}) = \sum_I v_{\rm ps}(\mathbf{r} - \mathbf{R}_I). \tag{28}$$

In the formulation proposed by HSC and discussed in the previous section, $v_{\rm ps}$ is a norm-conserving angular momentum dependent pseudopotential given by eqn (22). For most elements of the periodic table, it is a good approximation to assume $v_l(r) = v_{\bar{l}}$ for $l \geq \bar{l}$, where $\bar{l}$ is taken to be either 1 (for first-row elements) or 2. Then eqn (22) becomes:

$$v_{\rm ps}(\mathbf{r}) = v_{\rm loc}(r) + \sum_{l=0}^{\bar{l}-1} \Delta v_l(r) \hat{P}_l, \tag{29}$$

where $\Delta v_l = v_l(r) - v_{\bar{l}}(r)$ is a short-ranged function. The first term on the right-hand side is called the local part of the pseudopotential, whereas the second term is the non-local part: indeed, it acts in different ways on different angular components of the wavefunction. Accordingly, the energy $E_{\rm ps}$ is split into two contributions, a local ($E_{\rm ps}^{\rm L}$) and a non-local ($E_{\rm ps}^{\rm NL}$) one:

$$E_{\rm ps}^{\rm L} = \int \mathrm{d}\mathbf{r}\, \rho_{\rm e}(\mathbf{r}) \sum_I v_{\rm loc}(|\mathbf{r} - \mathbf{R}_I|), \tag{30}$$

$$E_{\rm ps}^{\rm NL} = \sum_i \langle \psi_i | v_{\rm NL} | \psi_i \rangle, \tag{31}$$

where

$$v_{\rm NL} = \sum_I \sum_{l=0}^{\bar{l}-1} \Delta v_l(|\mathbf{r} - \mathbf{R}_I|)\hat{P}_l(\mathbf{R}_I). \tag{32}$$

In eqn (32), the operator $\hat{P}_l(\mathbf{R}_I)$ is the projector on the $l^{\rm th}$ angular momentum with respect to the point $\mathbf{R}_I$. The evaluation of the terms $E_{\rm H}$, $E_{\rm ps}^{\rm L}$ and $E_{\rm M}$ requires some care due to the presence of slowly decaying Coulomb forces. One can proceed as follows. As far as the term $E_{\rm M}$ is concerned, it is convenient to replace the ionic core point charges by smeared charge distributions ρ_I, e.g. gaussians centred at the ionic sites:

$$\rho_I(\mathbf{r} - \mathbf{R}_I) = -\frac{Z_I}{(R_I^c)^3}(\pi)^{-3/2} \exp\left(-\frac{|\mathbf{r} - \mathbf{R}_I|^2}{(R_I^c)^2}\right). \tag{33}$$

R_I^c determines the width of the gaussian distribution associated with site I. The term $E_{\rm M}$ can now be expressed as follows:

$$E_{\rm M} = \frac{1}{2}\sum_{I,J} \int d\mathbf{r}d\mathbf{r}' \, \frac{\rho_I(\mathbf{r} - \mathbf{R}_I)\rho_J(\mathbf{r}' - \mathbf{R}_J)}{|\mathbf{r} - \mathbf{r}'|} - E_{\rm self} - E_{\rm ovrl}, \tag{34}$$

where the sum runs over all I and J. The term $E_{\rm self}$ is the self-interaction of the smeared charge (33) and must be subtracted out in eqn (34) because it does not appear in eqn (25) (where the sum runs over I different from J):

$$E_{\rm self} = \frac{1}{2}\sum_I \int d\mathbf{r}d\mathbf{r}' \, \frac{\rho_I(\mathbf{r} - \mathbf{R}_I)\rho_I(\mathbf{r}' - \mathbf{R}_I)}{|\mathbf{r} - \mathbf{r}'|} = \frac{1}{\sqrt{2\pi}}\sum_I \frac{Z_I^2}{(R_I^c)^2}. \tag{35}$$

The term $E_{\rm ovrl}$ accounts for the difference between the original point-charge interactions and the interactions between smeared charges:

$$E_{\rm ovrl} = \frac{1}{2}\sum_{I \neq J} \left[\frac{Z_I Z_J}{|\mathbf{R}_I - \mathbf{R}_J|} - \int d\mathbf{r}d\mathbf{r}' \, \frac{\rho_I(\mathbf{r} - \mathbf{R}_I)\rho_J(\mathbf{r}' - \mathbf{R}_J)}{|\mathbf{r} - \mathbf{r}'|}\right], \tag{36}$$

which can be evaluated using eqn (33):

$$E_{\rm ovrl} = \frac{Z_I Z_J}{R_{IJ}} {\rm erfc}\left(\frac{R_{IJ}}{\sqrt{(R_I^c)^2 + (R_J^c)^2}}\right), \tag{37}$$

where $R_{IJ} = |\mathbf{R}_I - \mathbf{R}_J|$.

On performing the summation of the three terms $E_{\rm M}$, $E_{\rm H}$ and $E_{\rm ps}^{\rm L}$: according to eqns (30) and (34) we obtain:

$$E_{\rm M} + E_{\rm H} + E_{\rm ps}^{\rm L} = \frac{1}{2}\int \mathbf{dr}\mathbf{dr}' \, \frac{\rho(\mathbf{r})\rho(\mathbf{r}')}{|\mathbf{r}-\mathbf{r}'|} + \int \mathbf{dr}\, \rho_e(\mathbf{r}) \sum_I v_{\rm loc}^I(|\mathbf{r}-\mathbf{R}_I|) \\ -E_{\rm self} - E_{\rm ovrl}, \tag{38}$$

where $\rho = \rho_e + \rho_I$, and

$$v_{\rm loc}^I(r) = \left[v_{\rm loc}(r) - \frac{Z_I}{r} \mathrm{erf}\left(\frac{r}{R_I^c}\right)\right], \tag{39}$$

with $-(Z_I/r)\mathrm{erf}(r/R_I^c)$ resulting from $\int \mathbf{dr}'\, \rho_I(\mathbf{r}')/\, |\mathbf{r}-\mathbf{r}'|$. The smeared core charge which replaces the ionic core point charges can be treated in Fourier space, on the same footing as the electronic charge. The sum of electronic and smeared charges ρ corresponds to the charge density of a *neutral* system, and thus explicit evaluation of Madelung sums is avoided. As far as the last two terms of eqn (38) are concerned, $E_{\rm self}$ is a constant contribution given by eqn (35), whereas $E_{\rm ovrl}$ can easily be evaluated using eqn (37).

The only term which is now left out is $E_{\rm ps}^{\rm NL}$; according to eqns (20), (29) and (31) we can write it as:

$$E_{\rm ps}^{\rm NL} = \sum_i \sum_I \sum_{\mathbf{g},\mathbf{g}'} \mathrm{e}^{-i\mathbf{g}\cdot\mathbf{R}_I} c_i^*(\mathbf{g}) v_{\rm NL}(\mathbf{g},\mathbf{g}') \mathrm{e}^{i\mathbf{g}'\cdot\mathbf{R}_I} c_i(\mathbf{g}'), \tag{40}$$

where

$$v_{\rm NL}(\mathbf{g},\mathbf{g}') = 4\pi \sum_{l=0}^{\bar{l}-1} (2l+1) P_l[\cos(\theta_{\mathbf{g}\mathbf{g}'})] \int \mathrm{d}r\, r^2 j_l(gr) j_l(g'r) \Delta v_l(r). \tag{41}$$

P_l is the Legendre polynomial of order l; j_l indicates a spherical Bessel function of the first kind of order l, and $\theta_{\mathbf{g}\mathbf{g}'}$ is the angle between $\mathbf{g}$ and $\mathbf{g}'$. In principle, $N_k \cdot N \cdot N_{\rm at} \cdot M(M+1)/2$ integrals of the kind entering eqn (40) are needed in order to compute $E_{\rm ps}^{\rm NL}$, where N_k is the number of points used to sample the BZ and, as usual, M, N and $N_{\rm at}$ the number of basis functions, electronic states and atoms, respectively. This usually amounts to a very large number of integrals making the evaluation of $E_{\rm ps}^{\rm NL}$ computationally very demanding. The evaluation of $E_{\rm ps}^{\rm NL}$ can be considerably simplified if the matrix elements can be cast into a separable form:

$$v_{\rm NL}(\mathbf{g},\mathbf{g}') = \sum_{l=0}^{\bar{l}-1} F_l(\mathbf{g}) F_l'(\mathbf{g}'), \tag{42}$$

where F_l and F_l' are appropriate functions, each depending upon only one wave vector component. An effective way to accomplish this separability has been proposed by Kleinman and Bylander [102]. In their formulation, the second term in eqn (29) (which is non-local in the angular part but acts as a multiplication operator on the radial part of the wavefunction) is replaced by a fully non-local operator given by:

$$v_{\mathrm{NL}}^{\mathrm{KB}} = \sum_{l,m} \frac{|\Delta v_l(r)\Phi_{l,m}\rangle\langle\Phi_{l,m}\Delta v_l(r)|}{\langle\Phi_{l,m}|\Delta v_l|\Phi_{l,m}\rangle}, \tag{43}$$

where $\Phi_{l,m} = \phi_l(r)Y_{l,m}$ is an eigenfunction of the atomic hamiltonian in which the core valence interaction is given by the pseudopotential v_{ps} of eqn (29). The form of $v_{\mathrm{NL}}^{\mathrm{KB}}$ allows a reduction of the number of integrals that need to be evaluated to $N_k \cdot N \cdot N_{\mathrm{at}} \cdot M$. The number of operations can be further decreased to $\mathcal{O}(N_k \cdot N \cdot M)$ if the matrix elements v_{NL} are evaluated in real space, taking advantage of the fact that Δv_l are functions localized around the atomic sites. Details of this formulation can be found in Ref. [103].

Recently, a new scheme has been suggested which allows the construction of separable pseudopotentials directly from all-electron calculations with an arbitrary number of reference energies [101, 104]. We note that separable pseudopotentials have to be handled with some caution, since in some cases they can give rise to undesired spurious solutions (ghost states) at energies lower than the ground-state energy. The appearance of these ghost states has been traced back to the choice of the local pseudopotential in eqn (29) [105, 106].

4.4. ENERGY AND FORCES WITH ULTRASOFT PSEUDOPOTENTIALS

The ultrasoft pseudopotentials proposed by Vanderbilt [101] can efficiently be implemented in a FP-MD scheme [25, 107, 108]. However, the relaxation of the norm-conservation condition adds some complications to the formulation. In this section, we describe how ultrasoft pseudopotentials have been implemented and point out the differences between the norm-conserving and the ultrasoft schemes.

In order to account for that part of the charge which is not described by the pseudowave functions ψ_i, the electron density ρ_e has to be augmented with appropriate functions (which we call $Q_{nm}^I(\mathbf{r})$) localized in the core region and defined as follows:

$$\rho_e(\mathbf{r}) = \sum_i \left[\psi_i^*(\mathbf{r})\psi_i(\mathbf{r}) + \sum_{nm,I} Q_{nm}^I(\mathbf{r})\langle\psi_i|\beta_n^I\rangle\langle\beta_m^I|\psi_i\rangle \right], \tag{44}$$

where the functions β_n^I are strictly localized in the core region and are also used to define the non-local pseudopotential:

$$v_{\rm NL} = \sum_{nm,I} D^0_{nm} |\beta^I_n\rangle\langle\beta^I_m|. \tag{45}$$

The functions β^I_n and Q^I_{nm} are related to the atomic functions β_n and Q_{nm} by

$$\beta^I_n(\mathbf{r}) = \beta_n(\mathbf{r} - \mathbf{R}_I), \tag{46}$$

$$Q^I_{nm}(\mathbf{r}) = Q_{nm}(\mathbf{r} - \mathbf{R}_I). \tag{47}$$

The $Q_{nm}(\mathbf{r})$ are constructed in the atomic pseudization procedure in such a way that, at the reference energies, the electron density of the pseudo-wave-functions as defined by eqn (44) be the same as the all-electron density. The functions β_n, Q_{nm} and the parameters D^0_{nm} are obtained from first principles in the ultrasoft pseudopotential scheme, and characterize the atomic species. Here, for simplicity, we consider a single atomic species.

Technically, the possibility of separating the description of the wave functions from that of the electron density is acquired by making use of a generalized orthonormality condition for the wave functions:

$$\langle\psi_i|S|\psi_j\rangle = \delta_{ij}, \tag{48}$$

where the operator S is defined by

$$S = 1 + \sum_{nm,I} q_{nm} |\beta^I_n\rangle\langle\beta^I_m|. \tag{49}$$

Here $q_{nm} = \int d\mathbf{r}\, Q_{nm}(\mathbf{r})$. It can easily be checked that condition (48) is consistent with the definition of the electron density (44) giving

$$\int d\mathbf{r}\ \rho_e(\mathbf{r}) = N_{\rm elec}, \tag{50}$$

where $N_{\rm elec}$ is the number of valence electrons.

In Vanderbilt's ultrasoft pseudopotential scheme the total energy is given by the same expression as in the norm-conserving case, eqn (24), where the non-local contribution to the pseudopotential is now $v_{\rm NL}$ of eqn (45). However, the functional derivative entering eqn (15) is modified, with respect to the norm-conserving scheme, since $\delta\rho_e(\mathbf{r})/\delta\psi^*_i(\mathbf{r}')$ is not simply given by $\psi_i(\mathbf{r})\delta(\mathbf{r} - \mathbf{r}')$, but is:

$$\frac{\delta\rho_e(\mathbf{r})}{\delta\psi^*_i(\mathbf{r}')} = \psi_i(\mathbf{r})\delta(\mathbf{r} - \mathbf{r}') + \sum_{nm,I} Q^I_{nm}(\mathbf{r})\langle\beta^I_m|\psi_i\rangle\beta^I_n(\mathbf{r}'), \tag{51}$$

This leads to

$$\frac{\delta E}{\delta\psi^*_i} = \left[-\nabla^2 + \sum_{nm,I} D^I_{nm}|\beta_n\rangle\langle\beta_m| + v_s\right]\ \psi_i = \epsilon_i S\psi_i, \tag{52}$$

where ϵ_i are the eigenvalues of the hamiltonian defined by the terms between squared brackets, v_s is given by eqn (7) in which v is the local part of the pseudopotential, and

$$D^I_{nm} = D^0_{nm} + \int \mathrm{d}\mathbf{r}\, v_s(\mathbf{r}) Q^I_{nm}(\mathbf{r}). \tag{53}$$

Note, that D^0_{nm} are just parameters, whereas the D^I_{nm} depend through v_s on the wave functions and have to be updated in the self-consistent procedure.

In norm-conserving pseudopotential schemes the energy cutoff E_{cn} required to describe the electron density is four times the energy cutoff E_{cut} of the wave functions. The simple relationship $E_{\mathrm{cn}} = 4 \cdot E_{\mathrm{cut}}$ does not always hold in the ultrasoft pseudopotential scheme, because of the augmentation functions Q_{nm} entering the expression of the electron density (eqn 44). It is therefore appropriate to introduce two independent energy cutoffs. The *soft* part of the electron density, i.e. the first term on the right hand side of (44), can again be described with a cutoff which is four times E_{cut}. In contrast, the part of eqn (44) which depends on the augmentation functions Q_{nm} requires a much higher cutoff, E_{cn}. The cutoff E_{cn} can be reduced by smoothing the Q_{nm}. An inner core region (determined by an inner radius which is smaller than the cutoff radius of the core region) is defined in which the Q_{nm} are modified in such a way that all the electrostatic properties outside the inner core region are preserved. In general, the smoothing of the Q_{nm} allows the use of a cutoff E_{cn} which is approximately equal to $4 \cdot E_{\mathrm{cut}}$. Hence, the calculation can be carried out similarly to the norm-conserving case. For some elements, such as transition metals, the Q_{nm} require $E_{\mathrm{cn}} > 4 \cdot E_{\mathrm{cut}}$, even after smoothing. In this case the problem of having a large number of plane waves can still be overcome by recognising that the Q_{nm} are localized in real space.

In the ultrasoft pseudopotential scheme, the inclusion of orthonormality constraints in the lagrangian (14) has to be generalized and the last term entering eqn (14) becomes:

$$\sum_{ij} \Lambda_{ij} (\langle \psi_i | S | \psi_j \rangle - \delta_{ij}). \tag{54}$$

The Euler equations associated with the lagrangian eqn (14) including the constraints of eqn (54) are:

$$\mu \ddot{\psi}_i = -\frac{\delta E}{\delta \psi_i^*} + \sum_j \Lambda_{ij} S | \psi_j \rangle, \tag{55}$$

$$M_I \ddot{\mathbf{R}}_I = -\frac{\partial E}{\partial \mathbf{R}_I} + \sum_{ij} \Lambda_{ij} \langle \psi_i | \frac{\partial S}{\partial \mathbf{R}_I} | \psi_j \rangle. \tag{56}$$

The expression for the ionic forces can easily be obtained by recalling that, in the present scheme, the electron density also depends on $\mathbf{R}_I$ [108]:

$$\frac{\partial \rho_e}{\partial \mathbf{R}_I} = \sum_i \sum_{nm} \left[Q^I_{nm}(\mathbf{r})(\langle\psi_i|\frac{\mathrm{d}\beta^I_n}{\mathrm{d}\mathbf{R}_I}\rangle\langle\beta^I_m|\psi_i\rangle + \langle\psi_i|\beta^I_n\rangle\langle\frac{\mathrm{d}\beta^I_m}{\mathrm{d}\mathbf{R}_I}|\psi_i\rangle) + \frac{\mathrm{d}Q^I_{nm}}{\mathrm{d}\mathbf{R}_I}\langle\psi_i|\beta^I_n\rangle\langle\beta^I_m|\psi_i\rangle \right]. \tag{57}$$

The contribution to the ionic forces from the orthonormality constraints (48) appears because the operator S depends on the ionic positions through the functions β^I_n (see eqn 46). Note that the contribution of orthogonality constraints to the ionic forces is absent in the case of norm-conserving schemes. In order to illustrate the complication produced by the dependence of the orthonormality constraints on the ionic coordinates, we discretize eqns (55) and (56). When the Verlet algorithm is used, $\psi_i(t+\delta t)$ and $\mathbf{R}_I(t+\delta t)$ can be expressed in terms of the wave functions and the ionic coordinates at the previous two time-steps; *both* ψ_i and $\mathbf{R}_I$ depend on the Lagrange multipliers Λ_{ij}. These Λ_{ij} are determined by the condition [109]:

$$\langle\psi_i(t+\delta t)|S(t+\delta t)|\psi_j(t+\delta t)\rangle = \delta_{ij}, \tag{58}$$

where $S(t+\delta t)$ means that S (eqn (49)) is evaluated for the ionic coordinates $\mathbf{R}_I(t+\delta t)$. In norm-conserving schemes the identity operator used instead of S is found in eqn (58); in this case substitution into eqn (58) of $\psi_i(t+\delta t)$ and $\mathbf{R}_I(t+\delta t)$ in terms of their known values at (t) and $(t-\delta t)$ leads to a quadratic matrix equation for Λ_{ij} which can easily be solved iteratively (details are given in the next section [79]). In the ultrasoft pseudopotential case, an additional dependence on Λ_{ij} occurs in eqn (58), because $S(t+\delta t)$ depends upon the $\mathbf{R}_I(t+\delta t)$, which depend on Λ_{ij} (see eqn (56)). This problem is solved by making a good estimate of $\mathbf{R}_I(t+\delta t)$ assuming a continuous evolution for the Λ_{ij}. Then, eqn (58) is solved for Λ_{ij} in a similar way to the norm-conserving case [79,108]. In most cases the estimated $\mathbf{R}_I(t+\delta t)$ are found to be very accurate, when compared to those obtained with a converged iterative procedure. Thus the orthonormality problem is solved with the same numerical effort as in the norm-conserving scheme.

As long as the electron density (thus the Q_{nm}) can be described with a cutoff E_{cn} given by $4 \cdot E_{\mathrm{cut}}$, MD simulations with ultrasoft pseudopotentials can be performed without other modifications with respect to norm-conserving schemes. As already mentioned, sometimes the condition $E_{\mathrm{cn}} = 4 \cdot E_{\mathrm{cut}}$ cannot be satisfied in the ultrasoft pseudopotential scheme. For example this is the case for Cu, in which a large part of the charge of the tightly bound *d*-orbitals is incorporated in the Q_{nm} [25]. Another example is the first-row element F, where the *p* electronic states are extremely localized [110]. In order to permit an optimal choice for E_{cn} and E_{cut}, it is convenient to implement computer codes in which these two parameters can be chosen independently of each other.

FFT's are used to calculate the convolutions $\psi_i^*(\mathbf{r})\psi_i(\mathbf{r})$, which enter the expression of the electron density, and $v_s(\mathbf{r})\psi_i(\mathbf{r})$ which appears in eqn (52). The FFT-grid must contain all plane waves determined by an energy cutoff of $4 \cdot E_{\text{cut}}$ in order to fully describe the results of those convolutions. On the other hand, in order to describe the Q_{nm} a grid determined by the higher cutoff E_{cn} is required. A possible solution is to use a single grid determined by E_{cn}. However, the number of FFT's needed for $\psi_i^*(\mathbf{r})\psi_i(\mathbf{r})$ and $v_s(\mathbf{r})\psi_i(\mathbf{r})$ is proportional to the number of electronic states (N), whereas the number needed for the evaluation of ρ_e and of the Hartree potential is limited and is independent of N. It is therefore more convenient to introduce two different FFT-grids [25] in the calculation, a coarse and a dense one, corresponding to $4 \cdot E_{\text{cut}}$ and E_{cn}, respectively. In this way, the FFT's for the convolutions $\psi_i^*(\mathbf{r})\psi_i(\mathbf{r})$ and $v_s(\mathbf{r})\psi_i(\mathbf{r})$ are calculated on the coarse grid with the same numerical effort required in norm-conserving schemes when using the same wave-function cutoff E_{cut}. The numerical cost of the few FFT's on the dense grid is negligible in simulations of large systems. The connection between the two different grids is established in Fourier space.

We now focus on the integrals entering eqn (53) which give the D^I_{nm}'s as well as on other integrals involving $\mathrm{d}Q^I_{nm}/\mathrm{d}\mathbf{R}_I$, which appear in the expression of the ionic forces. These integrals are conveniently calculated in real space because the Q_{nm} are localized functions. In order to perform the integrals the functions must be known on the grid points while the ions are moving, requiring an accurate interpolation from tabulated values. This interpolation is performed by introducing little boxes, one for every ion, in which FFT's are carried out [25]. The boxes are large enough to contain the core region, and follow the motion of the ions. In the boxes an FFT-grid is defined, which coincides locally with the dense grid. Because of the small size of the box, FFT's on the box-grid can be carried out with a negligible computational cost. Note that the cutoff associated with the box-grid is E_{cn}. In the Fourier space of the box-grid, the change of the ionic positions is accounted for by a simple phase factor. The use of this technique is essential to make very high charge density cutoffs, E_{cn}, computationally affordable.

The main feature of Vanderbilt's ultrasoft pseudopotential scheme is to reduce the required plane-wave cutoff for the valence wave functions. This is achieved by allowing the pseudo-wave-function to be as soft as possible outside the core region, which is determined by a given cutoff radius (r_c). The larger the cutoff radius, the lower the energy cutoff for the plane waves. The upper physical limit for the cutoff radius is given by half a characteristic bond length (r_b). It turns out that for large cutoff radii, $r_c \simeq r_b$, it is necessary to use more than one reference energy for every angular momentum channel in order to have pseudopotentials which are as accurate as conventional norm-conserving pseudopotentials [94]. Therefore, besides the complication of having a generalized eigenvalue problem (eqn (52)) and wave function dependent D^I_{nm} (eqn (53)), the number of β_n functions has generally to be doubled, increasing the related computational cost correspondingly. In spite of these complications, the number of required plane waves is so reduced with respect to the norm-conserving case that the present scheme is still competitive compared to conventional methods for a variety of systems. For example, E_{cut} needed for Cu in the HSC norm-conserving scheme is about 300 Ry [111], whereas $E_{\text{cut}} \simeq 20$ Ry can be used in the Vanderbilt pseudopotential scheme. Presently, for first-row elements and transition metals, the Vanderbilt pseudopotential scheme seems to be the only accurate method which can be used for relatively large systems. We note that an additional advantage, coming from the reduction of the number of PW's, is the possibility of choosing a larger timestep. So far, the present method has been successfully applied to study the dynamical and structural properties of small water clusters [112], of ice under high pressure [27], and of liquid Cu [25]. The last example will be discussed in some detail in section 5.3.

4.5. ORTHOGONALITY CONSTRAINTS

We now discuss how to impose orthogonality constraints during FP-MD simulations.

If the orthogonality constraints are fulfilled exactly at some initial time $t = t_0$, then the *analytical* solution of eqn (12) ensures that they are satisfied for each $t \geq t_0$. However the *numerical* integration of the large set of coupled differential eqn (12) leads to a rapid deterioration in the constraints for $t \geq t_0$. In order to avoid these errors in the conservation of constraints, which accumulate with time, an algorithm has to be used that allows them to be satisfied at each step, despite the approximate character of the numerical integration [4, 109]. It has to be stressed that the orthonormality requirements of eqn (5) are holonomic constraints, that is they are expressible as $f(\psi_i(r)) = 0$ where the function f does not depend explicitly on time and therefore the constraints do not do any work. Within the lagrangian framework, the parameters Λ_{ij} can easily be calculated (the constraint equation $\int d\mathbf{r}\psi_i^*(\mathbf{r})\psi_j(\mathbf{r}) = \delta_{ij}$ is differentiated twice with respect to time and ψ_i obtained from the EOM, $\psi_i = -\mathcal{H}_{\text{KS}}\psi_i + \Sigma_{ij}\Lambda_{ij}\psi_j$, is substituted into the resulting expression):

$$\Lambda_{ij} = \mathcal{H}_{ji} - \mu \int d\mathbf{r}\dot{\psi}_j^*(\mathbf{r},t)\dot{\psi}_i(\mathbf{r},t) \tag{59}$$

where $\mathcal{H}_{ji} = \int d\mathbf{r}\psi_j^*(\mathbf{r},t)\mathcal{H}_{\text{KS}}\psi_i(\mathbf{r},t)$. The matrix $\mathbf{\Lambda} = [\Lambda_{ij}]$ is Hermitian. Equation (59) is the only orthonormalization procedure consistent with the lagrangian (14). Indeed, whereas E is invariant under unitary transformations in the occupied orbital subspace, K_e and then $\mathcal{L}$ are not [79]. The

Lagrangian multipliers entering eqn (15) depend upon the trajectories of the electronic degrees of freedom. Therefore the explicit implementation of eqn (59) depends on the specific algorithm, e.g. Verlet algorithm, used to integrate the EOM. We note that orthonormalization schemes different from that of eqn (59), such as the Gram Schmidt (GS), lead to incorrect trajectories and to nonconservation of the total energy U_{eI}. An example is presented in Fig. 8. This shows U_{eI} and K_e as a function of simulation time for the example discussed in section 3.3, in two cases: one in which orthogonalization is performed according to eqn (59) and the other in which a GS procedure is used.

As illustrated in Fig. 6, the solution of EOM consists of two steps: an unconstrained one which, e.g. according to the Verlet algorithm, is

$$\bar{\psi}_i(t+\Delta t) = -\psi_i(t-\Delta t) + 2\psi_i(t) + \frac{\Delta t^2}{\mu}[-\mathcal{H}_{KS}\psi_i(t)] \tag{60}$$

and a subsequent step where the newly-predicted wavefunctions $\bar{\psi}_i$ are "corrected" to add the constraint forces. This is accomplished as follows. By imposing the orthonormality condition on the "corrected" wavefunctions

$$\psi(t+\Delta t) = \bar{\psi}_i(t+\Delta t) + \sum_j x_{ij}\psi_j(t) \tag{61}$$

one is led to an equation for the unknown $x_{ij} = (\Delta t^2/\mu)\Lambda_{ij}$, which, in matrix form, reads as:

$$\mathbf{X}\mathbf{X}^+ + \mathbf{X}\mathbf{B} + \mathbf{B}^+\mathbf{X}^+ = \mathbf{1} - \mathbf{A} \tag{62}$$

where $A_{ij} = \langle\bar{\psi}_i(t+\Delta t)|\bar{\psi}_j(t+\Delta t)\rangle$ and $B_{ij} = \langle\psi_i(t)|\bar{\psi}_j(t+\Delta t)\rangle$. We note that $\mathbf{A} = \mathbf{1} + \mathcal{O}(\Delta t^2)$ and $\mathbf{B} = \mathbf{1} + \mathcal{O}(\Delta t)$. Equation (62) can be solved iteratively:

$$\mathbf{X}^{(0)} = \frac{1}{2}(\mathbf{1} - \mathbf{A}) \tag{63}$$

$$\mathbf{X}^{(n+1)} = \frac{1}{2}\left[\mathbf{1} - \mathbf{A} + \mathbf{X}^{(n)}(\mathbf{1} - \mathbf{B}) + (\mathbf{1} + \mathbf{B}^+)\mathbf{X}^{(n)} - (\mathbf{X}^{(n)})^2\right] \tag{64}$$

Here n denotes the nth iteration. Typically a few iterations (2–5) are necessary to solve eqn (62). Finally, we stress that the choice of the orthonormalization procedure is only crucial in MD simulations: in contrast, for SD dynamics or other minimization procedures, any efficient algorithm can be conveniently adopted.

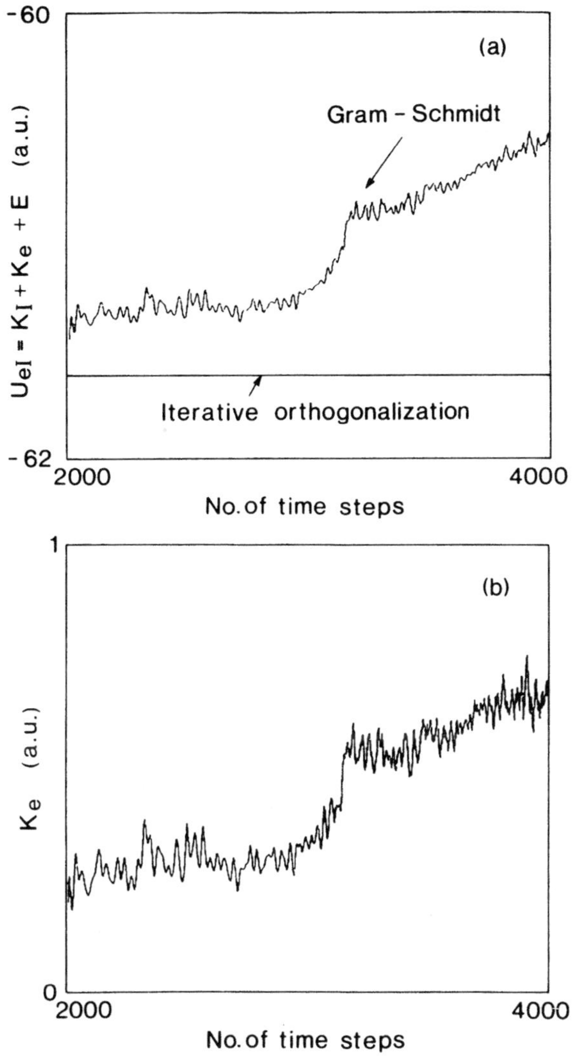

Fig. 8. The time variation of (a) $U_{\mathrm{eI}} = K_{\mathrm{e}} + K_{\mathrm{I}} + E$, and (b) K_{e}, on a segment of 2000 steps for the same system as in Fig. 3. Two different orthogonalization procedures have been used: Gram Schmidt (GS) and the iterative procedure described in section 4.5. The variations of K_{e} corresponds to the a run performed with GS orthogonalization (from Ref. [3]).

4.6. RECENT DEVELOPMENTS OF TOTAL ENERGY MINIMIZATION TECHNIQUES AND FP-MD SCHEMES

There are two major problems related to the total energy minimization techniques and FP-MD schemes that have been discussed in this Chapter. One problem concerns the scaling of the calculations as a functions of system size, which is of $\mathcal{O}(N^3)$ and therefore has so far prevented the treatment of systems with more than 400–500 electrons. (The use of parallel machines has recently allowed the study of systems as big as 1000 electrons [113, 114]) and the study of phenomena which involve simulation times longer than a few ps). A second problem concerns the efficiency of the plane-waves basis set, which, in principle, is not well suited to the distribution of charge densities localized in given regions of space (for example, the charge density of molecules and clusters or of an impurity in a bulk system).

The problem of the computational workload of typical total energy minimizations has been recently addressed in Ref. [49], where a novel formulation of the KS equations in terms of nonorthogonal localized orbitals is presented. The orbitals, which are expanded in a PW basis set, are localized using a simple method. Ref. [49] demonstrates that a scaling of total energy functional minimizations of order $NM\log(M)$ can be obtained with straightforward modifications of existing iterative procedures: no orthogonalization is needed and advantage is taken of the rapid decay of the overlap and Hamiltonian matrix elements as a function of distance, due to the localized character of the orbitals. Ref. [49] also shows how the goal of a PW-based method of order N can be achieved, by only using the volume where the localized wave functions are different from zero to expand the KS orbitals and to express the total energy and self-consistent potential.

A way of improving the efficiency of PW basis sets has recently been proposed in Ref. [50]. The plane-wave method is reformulated using adaptive curvilinear coordinates. In this approach, the search for the solutions of the Schrödinger equation is cast into an optimization problem in which both the plane-wave expansion coefficients and the coordinates are treated as variational parameters. This allows the effective plane-wave energy cutoff to vary in space so as to adapt optimally to the physics of the problem. In spite of the overhead involved in the calculation of coordinates and their optimization, this approach retains much of the simplicity of the plane-wave formulation, and allows the use of fast Fourier transform algorithms to calculate real-space representations of wavefunctions and densities.

5. Selected applications of first-principles molecular dynamics

Since its formulation seven years ago, the CP method has been used to study a variety of systems, including amorphous materials and liquids, surfaces and atomic clusters. In this Chapter we discuss three types of applications which have recently appeared in the literature. The first one concerns the study of amorphous silicon carbide (a-SiC) and of liquid potassium silicide (l-KSi): these investigations illustrate the crucial role of computer simulations with realistic descriptions of chemical bonds in order to probe the microscopic structure of amorphous and liquid semiconducting alloys. The second application regards the determination of Si and C (2×1) (111) surface reconstructions, which have been long-standing problems, recently solved by using FP-SA techniques. Finally, we discuss the study of liquid Cu, which shows that FP-MD simulations for transition metals are feasible, when using PW basis sets in conjunctions with ultrasoft pseudopotentials.

5.1. SHORT RANGE ORDER IN AMORPHOUS AND LIQUID SEMICONDUCTING ALLOYS: ILLUSTRATIVE EXAMPLES.

Amorphous semiconducting alloys (ASA) are potentially useful as high temperature semiconductors in or close to engines, turbines and reactors, and in the technology of solar cells and photovoltaic devices. This widespread technological interest has motivated many experimental investigations and several theoretical studies in the last decade. Of particular interest and complexity are the materials like amorphous silicon carbide (a-SiC), composed of species whose chemical behaviour is radically different. It is well known that C can have different bonding configurations, such as sp, sp^2 and sp^3, giving rise to crystals and disordered solids with a great diversity of properties, whereas Si usually prefers fourfold sp^3 coordination at low T and P. Different interpretations of the many experiments devised to study a-SiC have generated two views. Some authors describe the material as chemically disordered, with both homonuclear and heteronuclear bonds, whereas others characterize the system as highly ordered, with a predominance of Si-C bonds, as in the corresponding crystal.

The study of an equimolar SiC alloy [15, 16, 115] with FP-MD has allowed the characterization of its microscopic structure and given information about the material which are complementary to experiment. The short-range order of the a-SiC networks generated by FP computer simulations has been characterized through computed *partial* and *total* distribution functions; the latter can be directly related to diffraction experiments. Fig. 9 shows the radial distribution functions (RDF) $g_{ij}(r)$ $(i,j=\text{C,Si})$. It is immediately clear from the intense first peak of $g_{\text{CC}}(r)$ and the first maximum present in $g_{\text{SiSi}}(r)$ that there exists a considerable proportion of homonuclear bonds in the amorphous system, quite unlike c-SiC, where only heteronuclear bonds are present. The first maxima of $g_{\text{SiSi}}(r)$ and $g_{\text{CSi}}(r)$ are located at the first-neighbour distances computed for the corresponding crystals; the first peak of $g_{\text{CC}}(r)$ is at a distance intermediate between the C-C nearest-neighbour

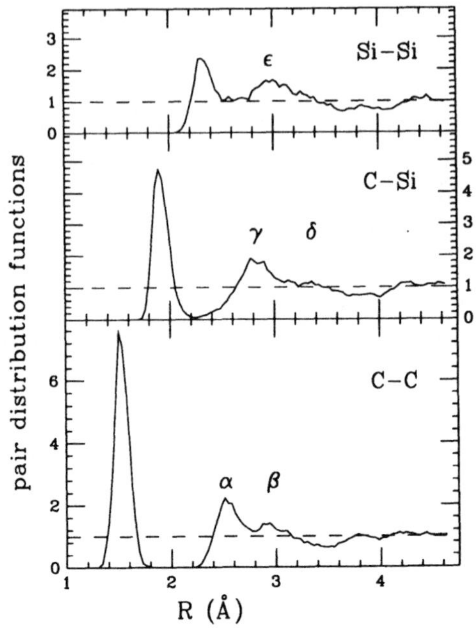

Fig. 9. The radial distribution functions g_{ij}, i, j=C,Si, of an equimolar a-SiC network as obtained in FP-MD simulations (see text). The peaks labelled with greek letters refer to $A_1 - \mathbf{B} - A_2$ configurations, where the second-neighbour atoms A_1, A_2 are both bonded to a **B**-type atom: α) C-**C**-C; β) C-**Si**-C; γ) C-**C**-Si; δ) C-**Si**-Si; ϵ) Si-**C**-Si.

bond lengths in graphite and diamond. This is consistent with the presence of both three-fold and four-fold C sites in the computer generated samples. Further analyses of the ionic trajectories obtained in the simulation show that C atoms tend to form chain-like structures, surrounded by distorted Si-C and Si-Si tetrahedral networks, where Si gives rise to strongly distorted diamond-like sites.

The picture of a-SiC which emerges from the calculation is consistent with many experimental studies; in particular the presence of sp^2 carbon constituting a kind of polymeric network with C=C bonds, has recently been detected in ESR experiments [116]. Nevertheless, the results of the simulation are in disagreement with several interpretations of experiments appearing in the recent literature. In order to understand the reasons of this discrepancy, it is useful to compare directly the data of the simulation to X-ray diffraction measurements.

From the computed partial structure factors (SF) S_{ij}, one can construct the total SF

$$S(q) = \frac{\sum_{i,j} f_i(q) f_j(q) S_{i,j}(q)}{\sum_i f_i(q)^2} \tag{65}$$

where $f_i(q)$ are X-ray atomic scattering factors. The total RDF, $j_{XR}(r)$, can then be obtained as Fourier Transform (FT) of $S(q)$, and compared directly to the FT of measured scattered intensities. $j_{XR}(r)$, reported in the lower panel of Fig. 10, has a first intense peak at $\simeq 1.9$ Å preceded by a shoulder at 1.5 Å, consistent with the experimental data reported in Refs. [117, 118]. Note that in the expression of $S(q)$ Si is weighted approximately twice as much as C by its atomic scattering factor, and therefore the intense peak of g_{CC} (Fig. 9) is reduced to just a weak shoulder in j_{XR}. The peak of j_{XR} at $\simeq$ 1.9 Å, close to the Si-C distance in *c*-SiC (1.89 Å), could be interpreted as a clear signature of chemical order in a-SiC, thus leading to a structural model rather different from that found in the simulation, where approximately half of C bonds are found to be homonuclear.

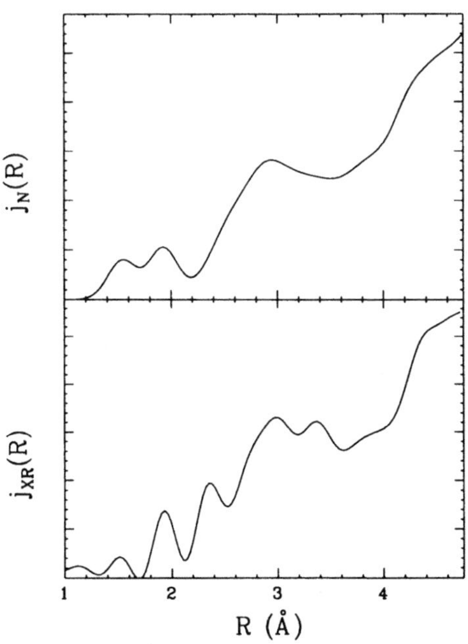

Fig. 10. The total distribution functions of an equimolar a-SiC network as obtained in FP-MD simulations (see text), to be compared with neutron (upper panel) and X-ray (lower panel) diffraction experiments, respectively.

Technical difficulties have so far prevented the analysis of a-SiC films by neutron diffraction. This method could be very useful in further clarifying the structure of the material, since it is complementary to X-ray diffraction. The upper panel of Fig. 10 shows the RDF $j_N(r)$ computed from

$$j_N(r) = \frac{\sum_{i,j} f_i^N f_j^N g_{i,j}(r)}{\sum_i (f_i^N)^2}, \tag{66}$$

where f_i^N are neutron scattering lengths. It is seen that $j_N(r)$ is quite different from $j_{XR}(r)$, with its double peak at 1.5 Å and 1.9 Å, and indicates the presence of different kinds of bonds in the amorphous materials with respect to the crystal.

In summary, the results of the FP computer simulation show that a-SiC has a more complex structure than previously thought: it cannot be classified as either chemically ordered or as a random network. These results and their comparison with experimental findings indicate that a detailed analysis of *each* atomic species is crucial to the understanding of the material properties. Different structural models implying different kinds of chemical bonds and coordination disorder can be compatible with the same set of measurements. This general conclusion is expected to hold for the whole family of ASA.

We now turn to the discussion of potassium silicide (KSi). This compound is a member of a large class of alloys of the form AB, where A is an alkali metal and B one of the group IV elements Si, Ge, Sn or Pb, which have been intensively studied in recent years. Experiments performed mainly on Sn and Pb liquid compounds reveal several anomalies. Their thermodynamic properties show deviations from the ideal solution behaviour [119–122]. All the pure components, including Si and Ge, are metallic in the molten state, yet their alloys have an electrical resistivity which exceeds that of the metallic regime [120]. The structure factors of Sn and Pb alloys, the only ones obtained so far by neutron scattering experiments [123–127], exhibit a first sharp diffraction peak (FPDP) at $\approx 1\,\text{Å}^{-1}$ [128].

In order to rationalize these findings, models [129–131] have been proposed that are based on the structure that AB compounds have in the solid state. These alloys do not obey the Hume-Rothery rule [132] about the relationship between the electron-to-atom ratio and the crystal structure, but are instead an example of the so-called Zintl phases [132, 133]. According to the Zintl picture, they can be nominally described by the formula $(A^+)_4$ $(B_4)^{4-}$, with the $(B_4)^{4-}$ anion (Zintl ion) being isoelectronic with the tetrahedral molecules P_4 and As_4. If it is assumed that Zintl ions are also present in sufficiently large numbers in the melt, this would provide a natural explanation for the high resistivity of the liquid. FSDP can also be related to the ordering of negatively charged Zintl ions and positive cations.

As in the case of amorphous SiC, FP simulations of liquid KSi [20, 21] have provided valuable information about the structure, vibrational and electronic properties of the binary alloy. Here we will briefly discuss only the structure of the system. Fig. 11 shows the partial distribution functions obtained in the simulation as compared to the corresponding bond length distributions in the solid.

The computed total structure factor $S(q)$

$$S(q) = \frac{\Sigma_{ij} f_i^{\mathrm{N}} f_j^{\mathrm{N}} S_{ij}}{\Sigma_i f_i^2} \tag{67}$$

(where the $S_{ij}(q)$ are obtained from their definition $S_{ij}(q) = \Sigma_{k,l} e^{i\mathbf{q}\cdot\mathbf{R}_i^k} e^{-i\mathbf{q}\cdot\mathbf{R}_j^l}$) is reported in Fig. 12, where it is seen that it has a FSDP at $q \simeq 1\,\text{Å}^{-1}$. This was found to be the case for most APb alloys [125].

The partial radial distribution functions with their similarities to the crystal, the coordination number of Si (from the integral of $g_{\mathrm{SiSi}}(r)$ up to its first minimum) which is found to be very close to 3, and the prepeak found in $S(q)$, all are features that appear to be consistent with the picture of a liquid composed of tetrahedral $(Si_4)^{4-}$ units embedded in a matrix of K^+ ions, as suggested by several authors [125, 130]. However, a closer examination of the simulation data reveals that the structure of molten KSi is much more complex than that of an ionic liquid modelled by $(A^+)_4$ and $(B_4)^{4-}$ units. This is borne out by detailed analyses of the ionic trajectories. The coordination of Si atoms as a function of simulation time shows that, on average, only 50% of these atoms are 3-fold coordinated, the remaining ones being 2-fold (30%) and 4-fold (20%) coordinated. These findings are clearly incompatible with the picture that most of the Si atoms are to be found in a Zintl form. In fact, the number of isolated $(Si_4)^{4-}$ anions is found to be rather low in the calculation. It varies in time from 0 to 2, in the small sample (64 atoms) considered in the simulation. Rather, Si atoms are arranged in large three-dimensional networks, whose number of constituents varies in time, and individual Zintl ions are seen to leave or join the clusters. These networks are composed of a variety of distorted tetrahedral units, which are connected by those atoms whose coordination is different from 3. The partial persistence of Zintl-like structures in going from the crystal to the melt, as well as the formation of new kinds of bonds, are also reflected in the bond-angle distributions ($A_{i,j,k}$, where j is the corner atom). As seen in Fig. 13 they are a broadened version of their solid counterparts, with the important exception of $A_{\mathrm{Si,Si,Si}}$. This has a sharp maximum at 60 °, as in the solid, plus a broader secondary peak centred at about 110–120°, which is absent in the crystal. Again, the presence of this secondary peak contradicts a simplified Zintl picture and clearly reflects the presence of several different bonding configurations in the liquid.

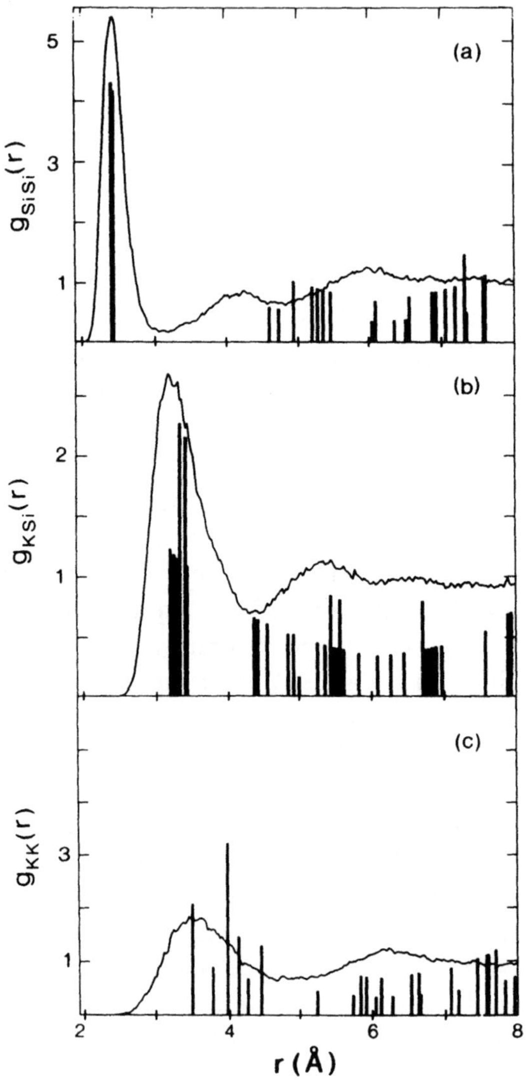

Fig. 11. The radial distribution functions g_{ij} $(i, j = \mathrm{K, Si})$ of liquid KSi as obtained in FP-MD simulations (see text) are compared to the corresponding bond-length distributions in the solid (from Ref. [21]).

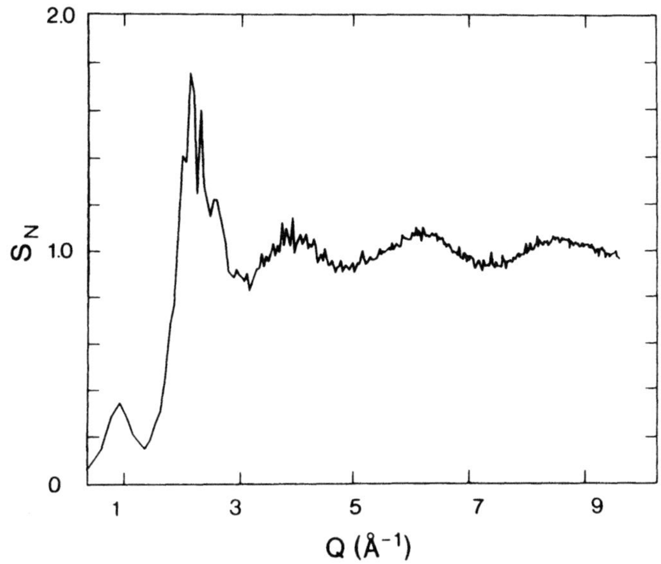

Fig. 12. The total structure factor of l-KSi as obtained in FP-MD simulations (see text). (from Ref. [21]).

5.2. RECONSTRUCTION OF SILICON AND CARBON SURFACES

A central issue of semiconductor surface physics is the change of geometrical arrangement that the surface layers often undergo with respect to the bulk structure (surface reconstruction). In this context, important questions concern the details of the resulting surface structure, its formation process and the induced changes on its electronic and vibrational properties.

C and Si(111) are unstable against reconstruction; their reconstructed (2×1) surfaces have been the subject of intense theoretical and experimental research in the last twenty years [134, 135]. The key to the understanding of these reconstructions was provided about ten years ago by Pandey's suggestion of a π-bonded chain model [136, 137]. In this model, by a change in the topology of the ideal structure, the dangling bonds formed upon truncation of the solid along the (111) face are brought to nearest-neighbour atoms leading to a much stronger interaction. This results in a substantial total energy gain and gives rise to highest occupied and lowest unoccupied surface energy bands which come from bonding and antibonding combinations of π orbitals, respectively.

In spite of the success of Pandey's model, several issues concerning the static and dynamic structure of C and Si (111) 2×1 surfaces, and their similarities and differences have long been unclear.

The reconstruction of the bulk-terminated C and Si(111) surfaces have

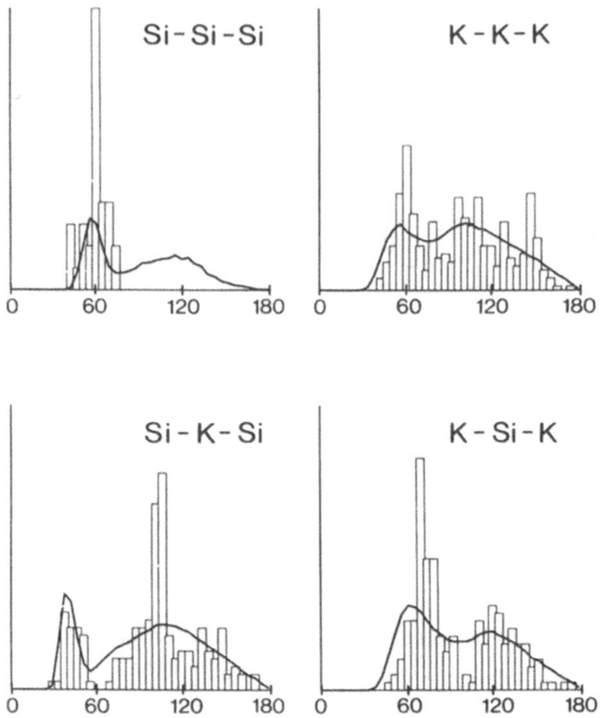

Fig. 13. The bond-angle distributions in solid (histogram) and molten KSi (from Ref. [21]).

been studied with FP-MD [28, 35, 36] by modelling the surfaces with periodically repeated slabs of C or Si layers plus a vacuum region to decouple each slab from its repeated images. The lower (ideal) surface of each slab has been stabilized by a monolayer of hydrogen.

Fig. 14 shows the transformation process leading to the (2×1) reconstructed Si(111) surface, starting from the ideal surface, as obtained in Ref. [28]. The characteristic features are an initial 2×1 buckling distortion of the topmost layer (see structure (b)), and the bond-switching process between (b) and (c), which leads to the formation of 5-fold and 7-fold rings of atoms near the surface. The intermediate structure ((c), "chain high") is already chain-like, but the direction of the tilt angle of the topmost chains relative to the substrate is reversed with respect to the ground-state geometry. In the final phase (d), which corresponds to the lowest energy, the sign of the tilt is changed in agreement with experiment [138]. The "chain high" arrangement turns out to be a shallow local minimum only a few meV per surface atom higher in energy than the (d) structure.

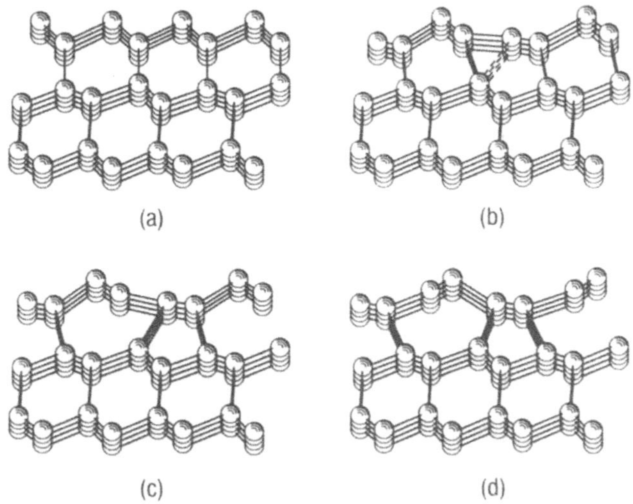

Fig. 14. The atomic structures of the Si (2×1) surface in the $(\bar{1}10)$ plane, corresponding to selected points in the reconstruction path found in FP-MD simulations: (a) ideal (initial), (b) intermediate, (c) chain "high", (d) chain "low" (final). (from Ref. [28]).

These results show that the ideal surface is unstable and that at least two relevant minima exist with $2{\times}1$ periodicity and a chain-like structure. Coexistence of the (c) and (d) structures may result in domains, which could be detected by large-scan scanning tunnelling microscopy imaging. At variance with Pandey's original suggestion, the Si reconstructed surface is characterized by a substantial buckling of the surface chains ($b \sim 0.49$ Å), i.e. the chains run along alternating "up" and "down" atoms. A nonzero buckling ($b = 0.3$–0.4 Å) is essential for a quantitative interpretation of the LEED data [138, 139]. This buckling is found to be accompanied by a net charge transfer from the "down" to the "up" atoms of the surface chains, with respect to the case $b = 0$. This is likely to have a further stabilizing effect on the surface.

The reconstructed (2×1) C(111) surface has a topology similar to the Si corresponding surface, but there are remarkable differences between the two geometrical arrangements. Fig. 15 shows the C(111) surface geometry as obtained after a series of FP calculations [35, 36]. It consists of dimerized π-bonded chains with no buckling. The optimal dimerization parameter $d = (l_1 - l_2)/(l_1 + l_2)$ (l_1 and l_2 are the two non-equivalent nearest-neighbour distances in the top chain) is found to be 1.4%.

Although comparison with experimental findings is not yet possible, this amount of dimerization is compatible with current interpretations of both LEED [135, 140] and ion scattering structural data [141]. It is also interesting

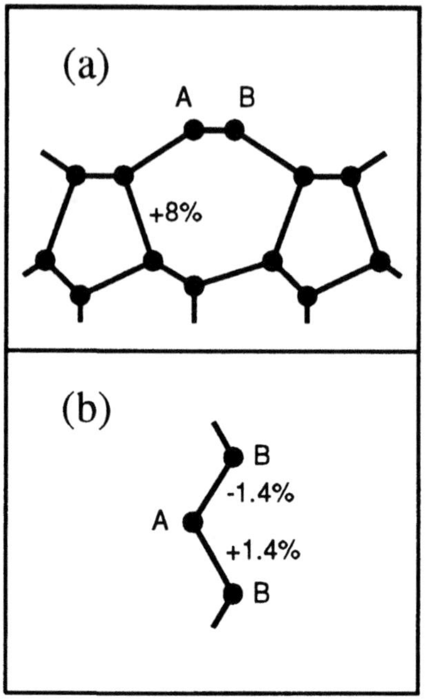

Fig. 15. The C(111) (2×1) reconstructed surface as obtained from FP calculations (see text): (a) side view; (b) top view (see text).

to note that in polyacetylene, which is another example of a π-bonded chain carbon system, the measured value of d ranges from 1.6 to 2%, quantitatively close to that found for the surface chains. The unbuckled dimerization of chains on (2×1) C(111) is in sharp contrast with the undimerized buckling of the corresponding Si(111) and Ge(111) surfaces. This can be explained as follows. The buckled state is partly ionic and implies a charge transfer between the raised and the lowered atoms. This is more costly in C, than in either Si or Ge, due to a stronger intra-atomic Coulomb repulsion, thus favouring dimerization in C. This distinctive feature of C is also reflected in the fact that stable adatom-type reconstructions, which involve a sizeable charge transfer from adatom to rest atom, are absent in C(111) but present and stable in Si (7×7) and in Ge (2×8).

Additional features of the computed 2×1 C(111) optimal surface structure are (i) an average π bond length $l_0 = (l_1 + l_2)/2$ of 1.44 Å; (ii) an expansion $\Delta d_3/d_3 = 8\%$ of the spacing between the first and second layer; and (iii) all other bond lengths within 1–2% of that of bulk diamond. These results are in good agreement with the corresponding experimental numbers ($l_0 = 1.45$ Å and $\Delta d_3/d_3 = 8.1\%$) extracted from LEED data [140].

In the FP simulations, the C(111) reconstructed surface geometry is very stable against thermal fluctuations, up to 2500 K. At this temperature the distance between the first and second bilayer grows, becoming considerably larger than at $T = 0$. The value of d correspondingly decreases by roughly 50%. The topology of the π-bonded chain reconstruction also starts being modified. This might indicate a tendency towards a structural transition, such as graphitization of the surface, found to occur experimentally [142] after ion bombardment or after annealing at very high temperature. However, the shape and size of the MD cell used in the calculation prevents a complete characterization of this transition. This would require MD simulations of several hundreds of atoms.

5.3. LIQUID COPPER

In transition metals the $3d$ orbital plays an important role in the bonding properties and must be included among the valence wave functions in an accurate description. The localized nature of the $3d$ orbitals (see Fig. 5) in transition and noble metals, has until recently prevented large scale FP-MD simulations of such materials. The development of the ultrasoft pseudopotential scheme [101] and its implementation in a MD framework [25, 107], have now made it possible to simulate a d-bonded liquid metal, such as l-Cu [25].

An ultrasoft pseudopotential has been constructed for Cu using two reference energies for each angular momentum ($l = 0, 1, 2$), which gives a total of 18 different β_i functions to be used in the non-local part of the pseudopotential. The chosen cutoff radii are $r_{cs} = 2.1$, $r_{cp} = 2.4$, $r_{cd} = 2.0$ a.u.. With this choice and the condition that the pseudo-wave-function matches the all-electron wavefunction smoothly, the two wavefunctions do not deviate significantly beyond a distance of approximately 2 a.u., which corresponds to about half the equilibrium distance in the copper dimer. A calculation for the dimer showed that a cutoff $E_{\text{cut}} = 18$ Ry for the wavefunctions and $E_{\text{cn}} = 200$ Ry for the electron density gave results for binding energy, equilibrium distance, and vibrational frequency, which differed only by a few percent from the fully converged results ($E_{\text{cut}} = 300$ Ry) obtained with a norm-conserving scheme [111].

In spite of the large number of β_i functions, $E_{\text{cut}} = 18$ Ry is low enough to allow a study of a larger number of Cu atoms. Liquid Cu has been simulated by a system of 50 Cu atoms in a periodically repeated simple cubic box. The density of Cu atoms has been taken to be the experimental one at the melting temperature ($\rho = 7.97$ g/cm^3 at $T_{\text{M}} = 1356$ K [143]). In order to simulate a constant-temperature evolution at 1500 K the equations of motion of the ionic coordinates has been coupled to a Nosé-Hoover thermostat [82, 144], with a mass of $Q_p = 5 \times 10^6$ a.u.

The electronic wave functions have been described by a Γ sampling of the Brillouin zone. Since liquid Cu is metallic, excited energy levels are found very close to the Fermi energy. In order to prevent the ionic energy flowing to rapidly into the electronic degrees of freedom, the system can be coupled to a second thermostat as proposed by Blöchl and Parrinello [84] and discussed in section 3.4. This second thermostat couples to the electronic variables and is used to keep their energy close to a constant E_e during the simulation.

Stable MD can now be carried out. In the simulation an electronic mass of $\mu = 1500$ a.u. has been used which allowed a choice of $\delta t = 10$ a.u. as time-step. During a run of 2 ps various quantities could be measured, such as the pair correlation function, the self-diffusion coefficient, and the electronic density of states. The overall shape of the pair correlation is found to be in very good agreement with experiment [143] (see Fig. 16). The first two peaks

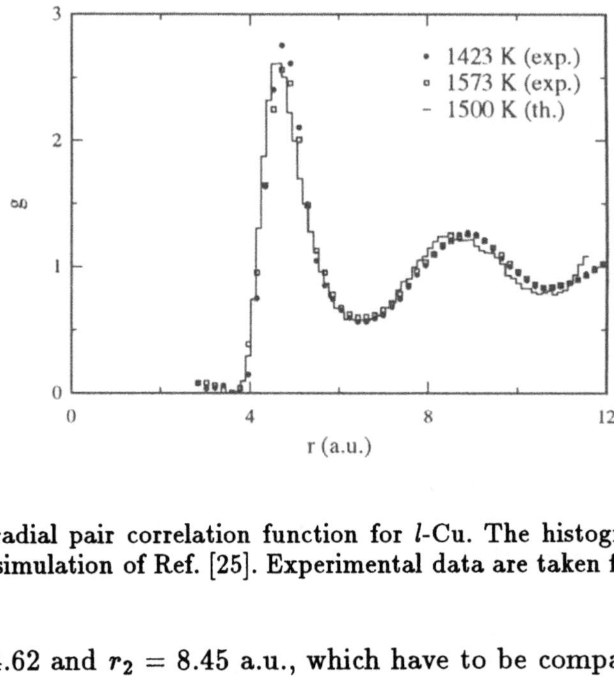

Fig. 16. The radial pair correlation function for l-Cu. The histogram is obtained from the MD simulation of Ref. [25]. Experimental data are taken from [143] (from Ref. [25]).

are at $r_1 = 4.62$ and $r_2 = 8.45$ a.u., which have to be compared to the experimental values of $r_1 = 4.86$ and $r_2 = 9.07$ a.u. respectively ($T = 1423$ K). The peaks from the calculation are found at slightly lower values compared to experiment, in agreement with a general trend of LDA. The self-diffusion coefficient extracted from the simulation is $D = 2.8 \times 10^{-9}\,\mathrm{m^2s^{-1}}$. The experimental value at 1500 K is $D = 5.6 \times 10^{-9}\,\mathrm{m^2s^{-1}}$ [145]. The agreement is satisfactory, considered the strong temperature dependence [145], and the fact that the melting temperature for the system studied is unknown. It has also been possible to acquire information about the electronic density of states during the motion. An average of many configurations taken at regular intervals shows that the density of states of liquid Cu does not differ much from crystalline Cu.

This study of liquid Cu shows that nowadays it is possible to perform FP-MD of extended *d*-bonded systems, and opens a way to the study of systems, such as transition metal clusters and surfaces, amorphous transition metals, as well as all possible combinations with other elements, including first-row elements such as oxygen.

Acknowledgement

One of us (GG) acknowledges support by the Swiss National Science Foundation under grant No. 21-31144.91.

References

[1] R. CAR and M. PARRINELLO, *Phys. Rev. Lett.* **55**, 2471 (1985).

[2] D. K. REMLER and P. A. MADDEN, *Molec. Phys.* **70**, 921 (1990).

[3] G. GALLI and M. PARRINELLO, in *Computer Simulation in Material Science*, edited by M. MEYER and V. PONTIKIS, Kluwer, Dordrecht, 1991.

[4] G. CICCOTTI and W. HOOVER, editors, *Molecular Dynamics Simulation of Statistical Mechanical Systems*, Plenum Press, New York, 1986.

[5] P. HOHENBERG and W. KOHN, *Phys. Rev. B* **136**, 864 (1964).

[6] W. KOHN and P. VASHISHTA, in *Theory of the inhomogeneous electron gas*, edited by S. LUNDQVIST and N. H. MARCH, p. 79, Plenum Press, New York, 1983.

[7] G. GALLI, R. M. MARTIN, R. CAR, and M. PARRINELLO, *Phys. Rev. Lett.* **62**, 555 (1989).

[8] G. GALLI, R. M. MARTIN, R. CAR, and M. PARRINELLO, *Phys. Rev. Lett.* **63**, 988 (1989).

[9] G. GALLI, R. M. MARTIN, R. CAR, and M. PARRINELLO, *Phys. Rev. B* **42**, 7470 (1990).

[10] G. GALLI, R. MARTIN, R.M. CAR, and M. PARRINELLO, *Science* **250**, 1547 (1990).

[11] R. CAR and M. PARRINELLO, *Phys. Rev. Lett.* **60**, 204 (1988).

[12] I. STICH, R. CAR, and M. PARRINELLO, *Phys. Rev. Lett.* **63**, 2240 (1989).

[13] I. STICH, R. CAR, and M. PARRINELLO, *Phys. Rev. B* **44**, 11092 (1991).

[14] F. BUDA, G. CHIAROTTI, R. CAR, and M. PARRINELLO, *Phys. Rev. B* **44**, 5908 (1991).

[15] F. FINOCCHI, G. GALLI, M. PARRINELLO, and C. M. BERTONI, *J. Non Cryst. Solids* **137**, 153 (1991).

[16] F. FINOCCHI, G. GALLI, M. PARRINELLO, and C. M. BERTONI, *Phys. Rev. Lett.* **68**, 3044 (1992).

[17] Q. M. ZHANG, G. CHIAROTTI, A. SELLONI, R. CAR, and M. PARRINELLO, *Phys. Rev. B* **42**, 5071 (1990).

[18] X. P. LI, P. B. ALLEN, J. Q. BROUGHTON, R. CAR, and M. PARRINELLO, *Phys. Rev. B* **41**, 3260 (1990).

[19] E. FOIS, A. SELLONI, G. PASTORE, Q. M. ZHANG, and R. CAR, *Phys. Rev. B* **45**, 13378 (1992).

[20] G. GALLI and M. PARRINELLO, *J. Phys. Cond. Mat.* **3**, SA227 (1991).

[21] G. GALLI and M. PARRINELLO, *J. Chem. Phys.* **95**, 7504 (1991).

[22] G. SEIFERT, G. PASTORE, and R. CAR, *J. Phys. Cond. Mat.* **4**, L179 (1992).

[23] D. HOHL, R. O. JONES, R. CAR, and M. PARRINELLO, *J. Am. Chem. Soc.* **111**, 826 (1989).

[24] D. HOHL and R. O. JONES, *Phys. Rev. B* **43**, 3856 (1991).

[25] A. PASQUARELLO, K. LAASONEN, R. CAR, C. LEE, and D. VANDERBILT, *Phys. Rev. Lett.* **69**, 1982 (1992).

[26] F. BUDA, G. CHIAROTTI, R. CAR, and M. PARRINELLO, *Phys. Rev. Lett.* **63**, 294 (1989).

[27] C. LEE, D. VANDERBILT, K. LAASONEN, R. CAR, and M. PARRINELLO, *Phys. Rev. Lett.* **69**, 462 (1992).

[28] F. ANCILOTTO, W. ANDREONI, A. SELLONI, R. CAR, and M. PARRINELLO, *Phys. Rev. Lett.* **65**, 3148 (1990).

[29] F. ANCILOTTO, A. SELLONI, W. ANDREONI, S. BARONI, R. CAR, and M. PARRINELLO, *Phys. Rev. B* **43**, 8930 (1991).

[30] F. ANCILOTTO, A. SELLONI, and E. TOSATTI, *Phys. Rev. B* **43**, 5180 (1991).

[31] F. ANCILOTTO, A. SELLONI, and E. TOSATTI, *Phys. Rev. B* **43**, 14726 (1991).

[32] F. ANCILOTTO and A. SELLONI, *Phys. Rev. Lett.* **68**, 2640 (1991).

[33] I. MOULLET, W. ANDREONI, and M. PARRINELLO, *Phys. Rev. B* **46**, 1842 (1992).

[34] N. TAKEUCHI, A. SELLONI, and E. TOSATTI, *Phys. Rev. Lett.* **69**, 648 (1992).

[35] S. IARLORI, G. GALLI, F. GYGI, M. PARRINELLO, and E. TOSATTI, *Physica B* (1992), in press.

[36] S. IARLORI, G. GALLI, F. GYGI, M. PARRINELLO, and E. TOSATTI, *Phys. Rev. Lett.* **69**, 2947 (1992).

[37] P. BALLONE, W. ANDREONI, R. CAR, and M. PARRINELLO, *Phys. Rev. Lett.* **60**, 271 (1988).

[38] D. HOHL, R. O. JONES, R. CAR, and M. PARRINELLO, *Chem. Phys. Lett.* **139**, 540 (1987).

[39] D. HOHL, R. O. JONES, R. CAR, and M. PARRINELLO, *J. Chem. Phys.* **89**, 6823 (1988).

[40] P. BALLONE, W. ANDREONI, R. CAR, and M. PARRINELLO, *Europhys. Lett.* **8**, 73 (1989).

[41] U. ROETHLISBERGER and W. ANDREONI, *J. Chem. Phys.* **94**, 8129 (1991).

[42] B. FUESTON, W. ANDREONI, E. CLEMENTI, and M. PARRINELLO, *Phys. Rev. B* **44**, 4056 (1991).

[43] W. ANDREONI, F. GYGI, and M. PARRINELLO, *Chem. Phys. Lett.* **189**, 241 (1992).

[44] K. LAASONEN, W. ANDREONI, and M. PARRINELLO, *Science* (in press).

[45] R. FEYNMAN, *Phys. Rev.* **56**, 340 (1939).

[46] W. KOHN and L. J. SHAM, *Phys. Rev. A* **140**, 1133 (1965).

[47] S. KIRKPATRICK, C. D. GELATT, and M. P. VECCHI, *Science* **220**, 671 (1983).

[48] S. KIRKPATRICK, *J. Stat. Phys.* **34**, 975 (1984).

[49] G. GALLI and M. PARRINELLO, *Phys. Rev. Lett.* **69**, 3547 (1992).

[50] F. GYGI, *Europhys. Lett.* **19**, 617 (1992).

[51] G. CHESTER, *Adv. Phys.* **10**, 357 (1961).

[52] H. F.SCHAEFER, editor, *The Electronic Structure of Atoms and Molecules*, Addison-Wesley, Reading, Massachusetts, 1972.

[53] S. M. YOUNGER, A. K. HARRISON, and G. SUGIYAMA, *Phys. Rev. A* **40**, 5256 (1989).

[54] F. GYGI and A. BALDERESCHI, *Phys. Rev. B* **34**, 4405 (1986).

[55] S. MASSIDDA, M. POSTERNAK, and A. BALDERESCHI, *Phys. Rev. B* **in press** (1992).

[56] R. ORLANDO, C. PISANI, C. ROETTI, and E. STEFANOVICH, *Phys. Rev. B* **45**, 592 (1992), and references therein.

[57] J. HARRIS, *Phys. Rev. B* **31**, 1770 (1985).

[58] O. F. SANKEY and D. J. MIKLWESKI, *Phys. Rev. B* **40**, 3979 (1989).

[59] D. M. CEPERLEY and B. J. ALDER, *Phys. Rev. Lett.* **45**, 566 (1980).

[60] J. P. PERDEW and A. ZUNGER, *Phys. Rev. B* **23**, 5048 (1981).

[61] E. S. FOIS, J. I. PENMAN, and P. A. MADDEN, *J. Chem. Phys.* (1992), submitted for publication.

[62] A. FISHER and P. BLOECHL, private communication.

[63] R. M. DREIZLER and E. K. U. GROSS, *Density Functional Theory,*, Springer-Verlag, Berlin, 1990.

[64] A. D. BECKE, *Phys. Rev. A* **38**, 3098 (1988).

[65] G. ORTIZ and P. BALLONE, *Phys. Rev. B* **44**, 5881 (1991).

[66] G. ORTIZ and P. BALLONE, *Phys. Rev. B* **44**, 6376 (1991).

[67] G. ORTIZ, *Phys. Rev. B* **45**, 11328 (1992).

[68] A. GARCIA, C. ELSAESSER, J. ZHU, S. G. LOUIE, and M. L. COHEN, *Phys. Rev. B* **(in press)** (1992).

[69] W. E. PICKETT, *Comput. Phys. Rep.* **9**, 115 (1989).

[70] P. BENDT and A. ZUNGER, *Phys. Rev. Lett.* **50**, 1688 (1983).

[71] D. M. WOOD and A. ZUNGER, *J. Phys. A Math. Gen.* **18**, 1343 (1985).

[72] E. R. DAVIDSON, *J. Comput. Phys.* **17**, 87 (1975).

[73] H. W. PRESS, P. B. FLANNERY, S. A. TEUKOLSKY, and W. T. VETTERLING, *Numerical Recipes*, Cambridge Univeristy Press, 1986.

[74] M. C. PAYNE, J. D. JOANNOPOULOS, D. C. ALLAN, M. P. TETER, and D. H. VANDERBILT, *Phys. Rev. Lett.* **56**, 2656 (1986).

[75] I. STICH, R. CAR, M. PARRINELLO, and S. BARONI, *Phys. Rev. B* **39**, 4997 (1989).

[76] M. P. TETER, M. C. PAYNE, and D. C. ALLAN, *Phys. Rev. B* **40**, 12255 (1989).

[77] T. A. ARIAS, M. C. PAYNE, and J. D. JOANNOPOULOS, *Phys. Rev. Lett.* **69**, 1077 (1992).

[78] J. L. MARTINS and R. WENTZCOVITCH, *Sol. St. Commun.* **78**, 831 (1991).

[79] R. CAR and P. PARRINELLO, in *Simple Molecular Systems and Very High Density*, edited by A. POLIAN, P. LOUBEYRE, and N. BOCCARA, p. 455, Plenum Press, New York, 1988.

[80] G. PASTORE, E. SMARGIASSI, and F. BUDA, *Phys. Rev. A* **44**, 6334 (1991).

[81] F. BUDA, PhD thesis, SISSA Trieste, 1989.

[82] S. NOSE, *Molec. Phys.* **52**, 255 (1984).

[83] M. SPRIK, *J. Phys. Chem.* **95**, 2283 (1991).

[84] P. BLOECHL and M. PARRINELLO, *Phys. Rev. B* **45**, 9413 (1992).

[85] E. S. FOIS, J. I. PENMAN, and P. A. MADDEN, *J. Chem. Phys.* (1992), submitted for publication.

[86] G. W. FERNANDO, G. X. QIAN, M. WEINERT, and J. W. DAVENPORT, *Phys. Rev. B* **40**, 7985 (1989).

[87] M. R. PEDERSON and K. A. JACKSON, *Phys. Rev. B* **43**, 7312 (1991).

[88] J. M. SOLER and A. R. WILLIAMS, *Phys. Rev. B* **40**, 5414 (1989).

[89] J. M. SOLER and A. R. WILLIAMS, *Phys. Rev. B* **42**, 9728 (1990).

[90] P. A. SERENA, A. BARATOFF, and J. M. SOLER, *J. Chem. Phys.* (1992), in press.

[91] K. J. CHADI and M. L. COHEN, *Phys. Rev. B* **8**, 5747 (1973).

[92] A. BALDERESCHI, *Phys. Rev. B* **7**, 5212 (1973).

[93] D. R. HAMANN, M. SCHLUETER, and C. CHIANG, *Phys. Rev. Lett.* **43**, 1494 (1979).

[94] G. B. BACHELET, D. R. HAMANN, and M. SCHLUETER, *Phys. Rev. B* **26**, 4199 (1982).

[95] G. KERKER, *J. Phys. C* **13**, L189 (1980).

[96] D. VANDERBILT, *Phys. Rev. B* **32**, 8412 (1985).

[97] A. M. RAPPE, K. M. RABE, E. KAXIRAS, and J. D. JOANNOPOULOS, *Phys. Rev. B* **41**, 1227 (1990).

[98] N. TROULLIER and J. L. MARTINS, *Phys. Rev. B* **43**, 1993 (1991).

[99] D. R. HAMANN, *Phys. Rev. B* **40**, 2980 (1989).

[100] E. L. SHIRLEY, D. C. ALLAN, R. M. MARTIN, and J. D. JOANNOPOULOS, *Phys. Rev. B* **40**, 3652 (1989).

[101] D. VANDERBILT, *Phys. Rev. B* **41**, 7892 (1990).

[102] L. KLEINMAN and D. M. BYLANDER, *Phys. Rev. Lett.* **48**, 1425 (1982).

[103] R. D. KING-SMITH, M. C. PAYNE, and J.-S. LIN, *Phys. Rev. B* **44**, 13063 (1991).

[104] M. Y. CHOU, *Phys. Rev. B* **45**, 11465 (1992).

[105] X. GONZE, P. KAECKELL, and M. SCHEFFLER, *Phys. Rev. B* **41**, 12264 (1990).

[106] X. GONZE, R. STUMPF, and M. SCHEFFLER, *Phys. Rev. B* **44**, 8503 (1991).

[107] K. LAASONEN, R. CAR, C. LEE, and D. VANDERBILT, *Phys. Rev. B* **43**, 6796 (1991).

[108] K. LAASONEN, A. PASQUARELLO, R. CAR, C. LEE, and D. VANDERBILT, to be published.

[109] J.-P. RYCKAERT, G. CICCOTTI, and H. J. C. BERENDSEN, *J. Comput. Phys.* **23**, 327 (1977).

[110] A. SATTA, A. PASQUARELLO, and K. LAASONEN, unpublished.

[111] P. BALLONE and G. GALLI, *Phys. Rev. B* **42**, 1112 (1990).

[112] K. LAASONEN, R. CAR, M. PARRINELLO, C. LEE, D. VANDERBILT, and G. ORTIZ, to be published.

[113] I. STICH, M. C. PAYNE, R. D. KINGSMITH, J. S. LIN, and L. J. CLARKE, *Phys. Rev. Lett.* **68**, 1351 (1992).

[114] K. D. BROMMER, M. NEEDELS, B. E. LARSON, and J. D. JOANNOPOULOS, *Phys. Rev. Lett.* **68**, 1355 (1992).

[115] F. FINOCCHI, G. GALLI, M. PARRINELLO, and C. M. BERTONI, *Physica B* (1992), in press.

[116] O. Chauvet et al., *Materials Science Forum* **83**, 1201 (1992).

[117] C. Meneghini et al., *J. Non Cryst. Solids* **75**, 137 (1991).

[118] A. Sproul, D. R. McKenzie, and D. J. H. Cockayne, *Phil. Mag. B* **54**, 113 (1986).

[119] G. K. Johnson and M.-L. Saboungi, *J. Chem. Phys.* **86**, 6376 (1987).

[120] M.-L. Saboungi, H. T. J. Reijers, M. Blander, and G. K. Johnson, *J. Chem. Phys.* **89**, 5869 (1988).

[121] C. van der Marel, W. Geertsma, , and W. van der Lugt, *J. Phys. F Met. Phys.* **12**, 2349 (1982).

[122] P. J. Tumidajski, A. Petric, T. Takenaka, A. D. Pelton, and M.-L. Saboungi, *J. Phys. Cond. Mat.* **2**, 209 (1990).

[123] M.-L. Saboungi, R. Blomquist, K. J. Volin, and D. L. Price, *J. Chem. Phys.* **87**, 2278 (1987).

[124] H. T. J. Reijers, W. van der Lugt, and C. van Dijk, *Physica B* **144**, 404 (1987).

[125] H. T. J. Reijers, M.-L. Saboungi, D. L. Price, J. W. Richardson Jr., K. Volin, and W. van der Lugt, *Phys. Rev. B* **40**, 6018 (1989).

[126] B. P. Alblas, W. van der Lugt, W. Dijkstra, W. Geertsma, and C. van Dijk, *J. Phys. F Met. Phys.* **16**, 845 (1983).

[127] D. L. Price, M.-L. Saboungi, R. Reijers, G. Kearley, and R. White, *Phys. Rev. Lett.* **66**, 1894 (1991).

[128] S. Eliot, *Phys. Rev. Lett.* **67**, 711 (1991).

[129] W. Geertsma, *J. Phys. Cond. Mat.* **2**, 8517 (1990).

[130] K. Toukan, H. T. J. Reijers, D. L. Price, and M.-L. Saboungi, *Phys. Rev. B* **41**, 11739 (1990).

[131] H. T. J. Reijers, W. van der Lugt, and M.-L. Saboungi, *Phys. Rev. B* **42**, 3395 (1990).

[132] R. Nesper, *Prog. Sol. State Chem.* **20**, 1 (1990).

[133] E. Zintl, *Angew. Chem.* **52**, 1 (1939).

[134] D. Hanemann, *Rep. Prog. Phys.* **50**, 1045 (1987).

[135] B. B. Pate, *Surf. Sci.* **165**, 83 (1986).

[136] K. Pandey, *Phys. Rev. Lett.* **47**, 1913 (1981).

[137] K. Pandey, *Phys. Rev. B* **25**, 4338 (1982).

[138] F. J. Himpsel, P. M. Marcus, R. Tromp, I. P. Batra, M. R. Cook, F. Jona, and H. Liu, *Phys. Rev. B* **30**, 2257 (1984).

[139] H. Sakama, A. Kawazu, and K. Ueda, *Phys. Rev. B* **34**, 1367 (1986).

[140] E. C. Sowa, G. D. Kubiak, R. H. Stulen, and M. A. van Hove, *J. Vac. Sci. Technol. A* **6**, 832 (1988).

[141] T. E. Derry, L. Smit, and J. Van der Veen, *Surf. Sci.* **167**, 502 (1986).

[142] G. D. Kubiak and K. W. Kolasinski, *Phys. Rev. B* **39**, 1381 (1989).

[143] Y. Waseda, *The Structure of Non-Crystalline Materials*, McGraw-Hill, New York, 1980.

[144] W. G. Hoover, *Phys. Rev. A* **31**, 1695 (1985).

[145] T. Iida and R. I. L. Guthrie, *The Physical Properties of Liquid Metals*, Clarendon Press, Oxford, 1988.

Computer Simulation Methods for Nonadiabatic Dynamics in Condensed Systems

D. F. Coker
Department of Chemistry
Boston University
590 Commonwealth Avenue
Boston MA 02215
U.S.A.

ABSTRACT. As nuclear rearrangements take place during chemical reactions, the reacting molecules can undergo nonradiative electronic excitation or relaxation. These radiationless changes in electronic state or nonadiabatic transitions are particularly important when the electronic states of the reacting system are close in energy, as is often the case for reactions of molecules in excited electronic states. When the energy difference between these close-lying electronic states is comparable to the energy of the nuclear motion, transfer of energy between nuclear and electronic degrees of freedom can have a profound effect on the outcome of these types of reactions. Electronically nonadiabatic effects are thus important in many fundamental molecular processes including: excited-state electronic energy transfer, charge transfer reactions, and quenching of electronic excitation. These types of electronic transitions can also play an important role in chemical reactions of ground-state species at room temperature.

In the last few years there have been significant theoretical and computational advances which have enabled the development of new molecular dynamics methods which incorporate nonadiabatic electronic transitions explicitly. In this Chapter the fundamental theories underlying these advances are presented, and a number of approximate semiclassical surface-hopping trajectory algorithms implementing these theories are detailed. Various one-dimensional examples are studied with these approximate semiclassical methods, as well as full quantum calculations to explore the effectiveness of these algorithms. The methods are used to study the nonadiabatic dynamics of excited excess electrons in various fluids as examples of applications to realistic model systems.

1. Introduction

The aim of this Chapter is to explore approaches for generalizing the molecular dynamics (MD) methods used widely to study the dynamics of condensed molecular systems, so as to include the possibility of transitions between different electronic states driven by the nuclear motion. The fundamental assumptions underlying traditional MD methods are that the nuclear dynamics is governed by classical equations of motion, and that the forces on

M.P. Allen and D.J. Tildesley (eds.), Computer Simulation in Chemical Physics, 315–377.

the nuclei are determined by the ground-state Born-Oppenheimer electronic energy surface. The methods which we shall explore in this Chapter involve relaxing the second of these assumptions, and considering classical nuclear dynamics with forces determined by several coupled electronic energy surfaces, between which we allow for the possibility of electronic transitions.

There is a long history of development of theories and methods for treating these nonadiabatic transitions, and various levels of sophistication ranging from classical, through semiclassical to fully quantum mechanical techniques have been developed. Due to our interest here in condensed-phase applications [1–9] where large numbers of degrees of freedom can be treated classically whilst relatively few require full quantum description, our focus in this Chapter is on the semiclassical techniques known as surface-hopping trajectory methods [3, 4, 10–27]. Other general semiclassical and quantum methods for treating molecular dynamics in the presence of electronic transitions are detailed in various articles [28–36].

If the nuclei move infinitely slowly, the electronic charge distribution has an infinitely long time to adjust to a change in nuclear geometry; thus the adiabatic theorem implies that the electronic distribution will remain in its ground state throughout such an infinitely slow, or adiabatic nuclear rearrangement. In reality, however, when electronic states get close in energy, the finite velocity of nuclear motion through these regions where the states are closely spaced can result in the electronic distribution becoming a superposition of the close-lying electronic states, rather than staying purely in the ground electronic state distribution. This mixing of the electronic states can be described in terms of the quantum mechanical electronic subsystem making nonadiabatic transitions between these closely spaced or strongly coupled electronic energy levels. Changes in the electronic charge distribution associated with these nonadiabatic transitions often lead to dramatic differences in the forces experienced by the nuclei compared with when motion over a single surface is considered. These different forces have a profound influence on the dynamics of nuclear rearrangement.

From the preceding discussion we have seen that properties of the nuclear trajectory, such as its velocity through the region of strong electronic coupling, and the proximity with which it approaches these coupled regions, will have considerable influence on the mixing of the states, i.e. the probabilities of transitions between them. As argued above, these transitions will influence the nuclear forces which ultimately determine the trajectory. Thus we see that within this semiclassical formulation of classical nuclear dynamics in the presence of quantal electronic transitions, the dynamical evolution of the coupled classical and quantal subsystems must be determined self-consistently.

With the various considerations discussed above in mind, the Chapter is organized into four sections. In section 2 we develop the basic concepts central to a semiclassical formulation of coupled quantum and classical subsystems. We first motivate the basic ideas by considering a simple two-state molecular example, then we derive the Landau-Zener [37, 38] formula and explore the properties of the coupled classical and quantum subsystems which influence the branching of the classical subsystem trajectory into the different quantum subsystem states. Next, these ideas are generalized by studying the formally exact separation of nuclear and electronic coordinates and considering the semiclassical approximation to this exact separation. Within this semiclassical approximation the classical and quantum subsystems become nonlocally coupled in time, and a discussion of the nature of the nonlocal dynamical coupling concludes section 2.

Section 3 details the surface-hopping trajectory methods for treating the nonadiabatic dynamics of mixed quantum-classical systems. The important issues such as electronic phase coherence [12, 39–41] and energy conservation [10, 12, 13], which useful algorithms must address, are detailed. A stochastic model in which an ensemble of hopping classical trajectories is propagated over the different quantum subsystem energy surfaces is derived so that the ensemble dynamics gives the correct statistical occupation probabilities for the quantum states [12]. *Ad hoc* velocity rescaling methods for energy redistribution after a trajectory hop are outlined [10, 15]. These *ad hoc* procedures become necessary because these approximate surface-hopping methods localize the coupling between the quantum and classical subsystems in time.

The self-consistent nature of the nonlocal dynamical coupling between the subsystems is formalized in section 4, where we outline the nonadiabatic dynamical theory of Pechukas [35, 36]. Next we describe an implementation of Pechukas theory in a time-stepping algorithm developed by Webster *et al.* [6]. With this approach the full nonlocal dynamical coupling is approximated so that only points along the classical trajectories that are nearest neighbors in time have self-consistent coupling between the classical and quantum subsystems. With this level of nonlocality the approach rigorously conserves energy and no *ad hoc* velocity rescaling is needed. The problem with Webster's implementation is that the quantum subsystem is only propagated coherently over a single classical time step. Outside this time scale Webster's approach assumes that the quantal dynamics is completely incoherent. The validity of this approximation for condensed systems has been discussed [5] and many nonadiabatic dynamics methods in the literature make this assumption [20–24]. We conclude section 4 with the description of a new self-consistent, energy-conserving nonadiabatic dynamics algorithm which integrates the quantal subsystem dynamics coherently on all time scales. The coupling of the classical and quantum subsystems should provide a natural damping of the quantum coherence with this approach.

The Chapter is concluded in section 5 where we test the algorithms by comparing exact results for several 1D two-state problems with surface-hopping trajectory results. Finally we outline the results of various applications of these methods to studying electron solvation dynamics.

2. Formal background and basic principles

2.1. TWO-STATE MOLECULAR EXAMPLE TO MOTIVATE CONCEPTS

To motivate many of the concepts which are central to the discussion of nonadiabatic dynamics, we will consider a simple one-dimensional molecular example. Consider a diatomic molecule PQ and suppose that at large internuclear separation R, the lowest energy electronic state is a pair of neutral atoms P + Q. We will call this state the atomic covalent state, covalent because the valence electron density is shared between the two atoms. Let us assume further that around the minimum energy geometry the most stable electronic configuration is ionic in nature, and that the valence electrons become localized around one of the two atoms forming a $P^+ + Q^-$ ionic molecule, for example. Suppose further that the excited electronic state undergoes exactly the opposite variation in charge distribution with nuclear separation i.e. at large internuclear separations the first excited state is ionic $P^+ + Q^-$, whilst in the region of the equilibrium excited-state geometry, a stable covalent electronically excited-state molecule is formed. This situation is illustrated in Fig. 1. Also in this figure we present a schematic representation of how the charge distribution for the valence electrons may change for these two states as a function of internuclear separation.

Working within the Born-Oppenheimer approximation, the adiabatic eigenstates $\psi_n(\mathbf{r}, \mathbf{R})$ are functions of electronic coordinates $\mathbf{r}$ and are parameterized by the nuclear configuration $\mathbf{R}$. These functions are the solutions of the eigenvalue problem $\hat{\mathcal{H}}_{el}(\mathbf{r}, \mathbf{R})\psi_n(\mathbf{r}, \mathbf{R}) = E_n(\mathbf{R})\psi_n(\mathbf{r}, \mathbf{R})$ where $\hat{\mathcal{H}}_{el}(\mathbf{r}, \mathbf{R})$ is the electronic hamiltonian which will be described in more detail in section 3 and $E_n(\mathbf{R})$ are the Born-Oppenheimer energy surfaces. If we suppose that for our model problem there are two valence electrons, the atomic covalent ground state at large separations may be approximated by an antisymmetrized product of atomic orbitals with terms of the form $1s_P(1)1s_Q(2)$, while the ionic excited state will have terms like $1s_Q(1)1s_Q(2)$ involving both electrons localized on a single atom. For our model system we suppose that the roles of these two states interchange so that for internuclear separations around the equilibrium bond length, R_{eq}, the equilibrium ground state has the ionic form $1s_Q(1)1s_Q(2)$ and the excited state is well approximated by a function involving covalent terms like $1s_P(1)1s_Q(2)$. For our example the adiabatic eigenstates ψ_1 and ψ_2 (labelled according to their relative energies) interchange their character (e.g. covalent $\rightarrow$ ionic for ψ_1) as we pass through the transition geometry R_0. On the other hand the so called diabatic states $\phi_1 = \phi_c$ and $\phi_2 = \phi_i$ are labelled according to the character of their

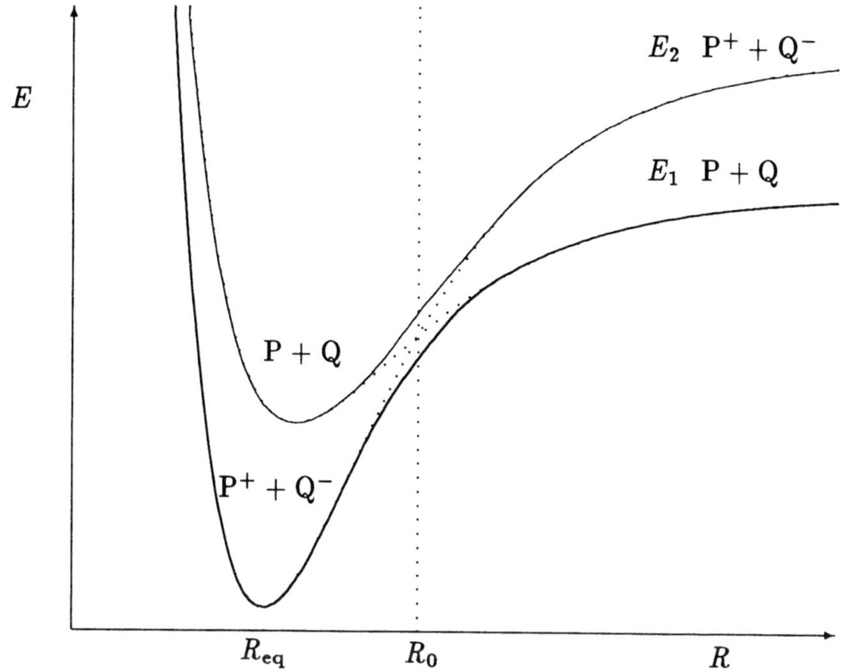

$R \approx R_{eq}$		$R \gg R_0$	
$\psi_1 \equiv$ ionic $\phi_2 = 1s_Q(1)1s_Q(2)$	P^+ Q^-	$\psi_1 \equiv$ atomic covalent $\phi_1 = 1s_P(1)1s_Q(2)$	P Q
$\psi_2 \equiv$ covalent $\phi_1 = 1s_P(1)1s_Q(2)$	P Q	$\psi_2 \equiv$ ionic $\phi_2 = 1s_Q(1)1s_Q(2)$	P^+ Q^-

Fig. 1. Adiabatic (solid lines) and diabatic (dotted lines) potential energy curves and schematic wavefunctions for two-state model molecular problem. R_{eq} is in the region of the equilibrium geometry.

charge distributions, $\phi_i \sim 1s_Q(1)1s_Q(2)$ and $\phi_c \sim 1s_P(1)1s_Q(2)$. These diabatic states approximate the adiabatic eigenstates in the asymptotic regions ($R \sim R_{eq}$ and $R \to \infty$) and interchange their energy ordering in the region of the transition geometry R_0. The adiabatic functions are well defined, being the eigenstates of $\hat{\mathcal{H}}_{el}$. The diabatic functions, on the other hand, are not unique and merely provide a convenient basis set in terms of which we can represent the changing charge distribution in different dynamical situations.

If we take our molecule initially to be in its ground state at large separations $R \gg R_0$ and we bring the atoms together infinitely slowly so that at each new value of R the electronic distribution has time to equilibrate with the new nuclear geometry, the charge distribution will stay in the ground state, and in our example it will change from covalent to ionic in character as R is decreased. The instantaneous adiabatic states ψ_n will provide an excellent representation of the changing charge distribution under these circumstances. On the other hand, if the nuclei come together very rapidly, the electronic distribution will have no time to relax to each nuclear arrangement so the shape or character of the wave function will not have time to change during such a rapid collision. Under these conditions, the diabatic states ϕ_n provide a good representation of the changing charge distribution in that the distribution may remain purely covalent in nature, for example, throughout the motion. In the intermediate situation, where R changes with finite velocity, the electronic distribution may become a mixture of states which could be equally well represented as a linear combination of adiabatic or diabatic functions.

2.2. THE LANDAU-ZENER THEORY

With the surface-hopping trajectory ideas which we will develop in this Chapter, the rates of change of the mixing coefficients are related to the transition probabilities between the states of the quantum subsystem. We need to understand the factors which influence the linear combination coefficients. The square moduli of these coefficients give the occupation probabilities of the different states. Within the Born-Oppenheimer picture we suppose that the nuclei experience forces associated with the charge distribution of the occupied instantaneous adiabatic eigenstate. Thus the charge distribution determining the nuclear forces will be governed by the coefficients $A_1(\mathbf{R})$ and $A_2(\mathbf{R})$ in the expansion $\psi(\mathbf{r},\mathbf{R}) = A_1(\mathbf{R})\psi_1(\mathbf{r},\mathbf{R}) + A_2(\mathbf{R})\psi_2(\mathbf{r},\mathbf{R})$. We will assume, for the time being, that we can ignore the need for self-consistency between the classical and quantum subsystems and develop an understanding of what properties of a fixed classical trajectory influence the probability of transition between quantum states.

For convenience we will work in a diabatic basis [38] rather than the adiabatic basis discussed above. A complete orthonormal diabatic basis can be constructed, say $|\phi_1\rangle$ and $|\phi_2\rangle$ for our two-state example. These diabatic functions are not eigenstates of $\hat{\mathcal{H}}_{\text{el}}$ so

$$\begin{aligned}\hat{\mathcal{H}}_{\text{el}}|\phi_1\rangle &= \epsilon_{11}|\phi_1\rangle + \epsilon_{12}|\phi_2\rangle \\ \hat{\mathcal{H}}_{\text{el}}|\phi_2\rangle &= \epsilon_{21}|\phi_1\rangle + \epsilon_{22}|\phi_2\rangle\end{aligned} \tag{1}$$

or, representing the wave function of our two-state system as a vector, we can write the above as

$$\underline{\underline{\mathcal{H}}}^{\text{d}}\underline{\Phi} = \begin{pmatrix}\epsilon_{11} & \epsilon_{12} \\ \epsilon_{21} & \epsilon_{22}\end{pmatrix}\begin{pmatrix}|\phi_1\rangle \\ |\phi_2\rangle\end{pmatrix} \tag{2}$$

where $\epsilon_{ij}(\mathbf{R}) = \langle\phi_j|\hat{\mathcal{H}}_{\text{el}}(\mathbf{R})|\phi_i\rangle$ and $\underline{\underline{\mathcal{H}}}^{\text{d}}$ is the Hamiltonian matrix in the diabatic representation. The matrix $\underline{\underline{\mathcal{H}}}^{\text{d}}$ is diagonalized by the matrix $\underline{\underline{P}}$ thus

$$\underline{\underline{\mathcal{H}}}^{\text{a}} = \underline{\underline{P}}^{\text{T}}\underline{\underline{\mathcal{H}}}^{\text{d}}\underline{\underline{P}} \tag{3}$$

where $\underline{\underline{\mathcal{H}}}^{\text{a}}$ is the diagonal Hamiltonian matrix in the adiabatic representation and the vector of adiabatic eigenstates is obtained from the similarity transform matrix thus: $\underline{\Psi} = \underline{\underline{P}}\,\underline{\Phi}$. The diagonal Hamiltonian matrix in the adiabatic representation has diagonal elements

$$E_{\substack{2\\1}} = \frac{1}{2}\left[(\epsilon_1 + \epsilon_2) \pm \sqrt{(\epsilon_1 - \epsilon_2)^2 + \epsilon_{12}^2}\,\right] \tag{4}$$

and the energies of the two adiabatic curves are closest when the diabatic curves cross at R_0 and $\epsilon_1(R_0) = \epsilon_2(R_0)$ where $E_2(R_0) - E_1(R_0) = \epsilon_{12}(R_0)$. We proceed by assuming that the nuclear motion is a known function of time $\mathbf{R} = \mathbf{R}(t)$ which gives rise to a time-dependent electronic hamiltonian $\hat{\mathcal{H}}_{\text{el}}(\mathbf{R}(t))$ and a time-dependent electronic wavefunction $\psi(\mathbf{r}, \mathbf{R}(t))$ which satisfies

$$i\hbar\frac{\partial\psi}{\partial t} = \hat{\mathcal{H}}_{\text{el}}\psi\,. \tag{5}$$

It will prove more convenient to express the model hamiltonians of interest in the diabatic basis so we choose to expand the dynamical wavefunction in terms of the functions $\{\phi_j(\mathbf{r}, \mathbf{R})\}$ writing

$$\psi(r,t) = c_1(t)\exp\left[-\frac{i}{\hbar}\int\epsilon_1\ \text{d}t\right]\phi_1(r) + c_2(t)\exp\left[-\frac{i}{\hbar}\int\epsilon_2\ \text{d}t\right]\phi_2(r). \tag{6}$$

Here we have defined the time-dependent coefficients $c_i(t)$ by pulling out an oscillatory phase factor

$$\exp\left[-\frac{i}{\hbar}\int\epsilon_i\ \text{d}t\right]$$

which would determine the dynamics of the coefficient of the function $\phi_i(r)$ if the system were to stay on the diabatic curve ϵ_i throughout its dynamics. Substituting into the time-dependent Schrödinger equation and eliminating $c_1(t)$ we obtain the following general equation of motion for the expansion coefficient $c_2(t)$

$$\frac{\mathrm{d}^2 c_2}{\mathrm{d}t^2} - \left[\frac{i}{\hbar}(\epsilon_2 - \epsilon_1) + \frac{\dot{\epsilon}_{12}}{\epsilon_{12}}\right] \frac{\mathrm{d}c_2}{\mathrm{d}t} + \left(\frac{\epsilon_{12}}{\hbar}\right)^2 c_2 = 0 \,. \tag{7}$$

Normalization requires $|c_1|^2 + |c_2|^2 = 1$ for our two-state system. The approach outlined above was first written down by Zener [38] and in a different but equivalent form by Landau [37]. Zener studied the crossing of two diabatic curves which changed linearly with time so $(\epsilon_2 - \epsilon_1)/\hbar = \alpha t$ and for which the diabatic basis functions $\{\phi_k(r, R)\}$ and off-diagonal hamiltonian coupling matrix elements ϵ_{12} were assumed to be independent of time so $\dot{\epsilon}_{12} = \dot{\phi}_i = 0$. The energy curves for this model are presented in Fig. 2 for the case of constant relative nuclear velocity $\mathrm{d}R/\mathrm{d}t = v$.

The conditions outlined above would approximate a realistic molecular situation, such as the model system described in section 2.1 and depicted in Fig. 1, if the transition region where the two electronic states were coupled was small i.e. the states were only coupled for a small time, or over a small range of internuclear separations. We suppose that this linear model represents the coupling region for the model diatomic curves presented in Fig. 1. To determine the probability of transition from one molecular surface to the other within the approximations of this linear model, we need only concern ourselves with the properties of the system at the crossing point of the diabatic curves R_0. If we start our atoms out at large separations $R \gg R_0$ at $t = -\infty$ in the atomic covalent ground state $\psi_1 = \phi_1$ (see Fig. 1) then the initial conditions for our diabatic expansion coefficients are $|c_1(-\infty)| = 1$ and $c_2(-\infty) = 0$. Making use of the model assumptions, eqn (7) can be transformed into the Weber equation [38] whose asymptotic solution gives $|c_2(\infty)|^2 = 1 - \exp[-2\pi\gamma] = 1 - |c_1(\infty)|^2$ where

$$\gamma = \epsilon_{12}^2 \Bigg/ \hbar \left|\frac{\mathrm{d}(\epsilon_2 - \epsilon_1)}{\mathrm{d}t}\right| \,.$$

We are interested in the probability of transition from the incoming adiabatic ground state $\psi_1(R \gg R_0)$ to the adiabatic excited state ψ_2 in the asymptotic region $R \sim R_{eq}$ which, if the coupling region is small, we can regard as occurring at $t \to \infty$. We see that the probability of transition between adiabatic state ψ_1 and ψ_2 is the probability that we end up in state ψ_2 i.e. $|A_2(\infty)|^2$, which is the same as the probability that we stay in diabatic state ϕ_1; thus the adiabatic transition probability is $P = |A_2(\infty)|^2 = |c_1(\infty)|^2 =$

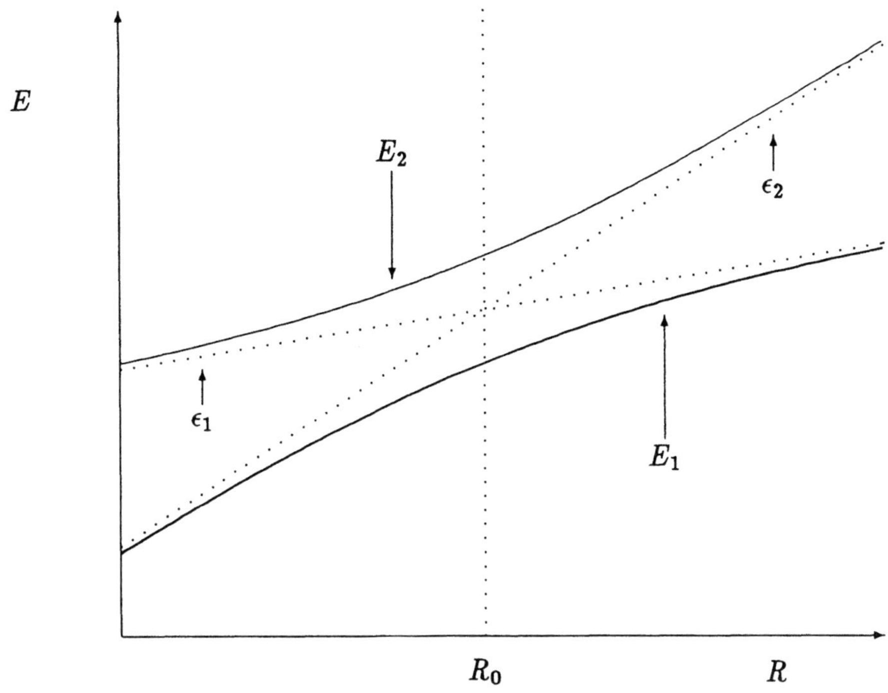

Fig. 2. Linear diabatic curve crossing. Diabatic (dotted) curves are denoted by ϵ_i, adiabatic (solid) curves are E_i. R_0 is the position of the crossing.

$\exp[-2\pi\gamma]$. Finally since

$$\frac{\mathrm{d}(\epsilon_2 - \epsilon_1)}{\mathrm{d}t} = \left(\frac{\mathrm{d}\epsilon_2}{\mathrm{d}R} - \frac{\mathrm{d}\epsilon_1}{\mathrm{d}R}\right)\frac{\mathrm{d}R}{\mathrm{d}t}$$

we obtain the celebrated Landau-Zener formula

$$P = \exp\left[-\frac{2\pi\epsilon_{12}^2}{\hbar v|F_1 - F_2|}\right] \tag{8}$$

where $F_i = -\mathrm{d}\epsilon_i/\mathrm{d}R$ are the forces arising from the diabatic energy curves, and all quantities in this expression are evaluated at the crossing point R_0. We see that in the limit of large velocity ($v \to \infty$), or as the gap between the states becomes small ($\epsilon_{12} \to 0$) the probability of transition between adiabatic states $P \to 1$. When the velocity is small, or the states are widely separated, $P \to 0$ and the motion is adiabatic.

In these first two subsections we have put together much of the conceptual framework of nonadiabatic dynamics in terms of some simple approximate examples. In the next subsection we consider the formal generalization of these ideas to an arbitrary system of electrons and nuclei. We first consider the exact separation of electronic and nuclear coordinates in the full quantum description and then formulate the semiclassical approximation.

2.3. FORMAL SEPARATION OF NUCLEAR AND ELECTRONIC COORDINATES

The hamiltonian for a system of N nuclei and n electrons in the absence of spin orbit interactions is $\hat{\mathcal{H}} = \hat{\mathcal{K}}_R + \hat{\mathcal{H}}_{\text{el}}$ where the nuclear kinetic energy is

$$\hat{\mathcal{K}}_R = \sum_{I=1}^{N} -\frac{\hbar^2}{2M_I}\nabla^2_{R_I} \tag{9}$$

and the electronic hamiltonian has the form [11,42]

$$\hat{\mathcal{H}}_{\text{el}} = \sum_{i=1}^{n} -\frac{\hbar^2}{2m_e}\nabla^2_{r_i} + \sum_{i<j}^{n}\frac{1}{r_{ij}} - \sum_{i=1}^{n}\sum_{J=1}^{N}\frac{Z_J}{|\mathbf{r}_i - \mathbf{R}_J|} + \sum_{I<J}^{N}\frac{Z_I Z_J}{|\mathbf{R}_I - \mathbf{R}_J|}\,. \tag{10}$$

The total wavefunction for the electronic and nuclear degrees of freedom satisfies the time-independent Schrödinger equation $\hat{\mathcal{H}}\Psi(\mathbf{r},\mathbf{R}) = \mathcal{E}\Psi(\mathbf{r},\mathbf{R})$. Next we choose an orthonormal basis set of electronic wavefunctions $\{\phi_k(\mathbf{r},\mathbf{R})\}$ parameterized by nuclear configuration $\mathbf{R}$ and $\langle\phi_k|\phi_{k'}\rangle = \delta_{kk'}$ (the bracket $\langle\,|\,\rangle$ indicates integration over only electronic degrees of freedom). The stationary state of the full system is expanded in terms of the electronic basis functions with coefficients $\chi_k(\mathbf{R})$ which depend on the nuclear coordinates so

$$\Psi(\mathbf{r},\mathbf{R}) = \sum_k \chi_k(\mathbf{R})\phi_k(\mathbf{r},\mathbf{R})\,. \tag{11}$$

Substituting this general form into the Schrödinger equation, using the fact that

$$\hat{\mathcal{K}}_R\chi(\mathbf{R})\phi_k(\mathbf{r},\mathbf{R}) = \sum_{I=1}^{N} -\frac{\hbar^2}{2M_I}\left[\chi_k\nabla^2_{R_I}\phi_k + 2\nabla_{R_I}\chi_k\nabla_{R_I}\phi_k + \phi_k\nabla^2_{R_I}\chi_k\right] \tag{12}$$

multiplying by $\langle\phi_{k'}(\mathbf{r},\mathbf{R})|$ and integrating over electronic coordinates gives an infinite system of coupled equations for the nuclear functions

$$\left[\mathcal{E} - \hat{\mathcal{K}}_R\right]\chi_k = \sum_{k'}\left[E_{kk'} + \hat{C}_{kk'}\right]\chi_{k'} \tag{13}$$

where $E_{kk'} = \langle\phi_k|\hat{\mathcal{H}}_{\text{el}}|\phi_{k'}\rangle$ and

$$\hat{C}_{kk'} = \sum_{I=1}^{N}\left\{-\frac{\hbar^2}{2M_I}\langle\phi_k|\nabla^2_{R_I}|\phi_{k'}\rangle - \frac{\hbar^2}{M_I}\langle\phi_k|\nabla_{R_I}|\phi_{k'}\rangle\nabla_{R_I}\right\} \tag{14}$$

is the nonadiabatic operator. With the adiabatic basis, $E_{kk'} = E_{kk}(\mathbf{R})\delta_{kk'}$ and $E_{kk}(\mathbf{R})$ are the Born-Oppenheimer potential surfaces. We see that in eqn (13) we have formally decoupled the nuclear and electronic degrees of freedom. If we set $\hat{C}_{kk'} = 0$ and use the adiabatic basis, the system of equations (13) further decouples and we obtain a set of Schrödinger equations for the nuclear wavefunctions describing the different states of nuclear motion over the uncoupled Born-Oppenheimer surfaces.

Solving the full coupled system in eqn (13) for many vibrational, rotational, and translational states is computationally intractable for even moderate-sized systems, so the approach which the methods described in this Chapter adopt involves a semiclassical treatment in which the fast electronic subsystem is treated quantum mechanically whilst the slow nuclear subsystem is described by classical equations of motion [11, 42].

Proceeding in much the same way as in the two-state example presented in section 2.2, we first assume that the nuclei follow some known trajectory $\mathbf{R}(t)$; thus $\hat{\mathcal{H}}_{\text{el}}(\mathbf{r}, \mathbf{R}(t))$ becomes time dependent through its parameterization by $\mathbf{R}(t)$. The wavefunction $\psi(\mathbf{r}, t)$ describing the quantum mechanical electronic distribution is nonstationary due to the time dependence of $\hat{\mathcal{H}}_{\text{el}}$ and it satisfies the time-dependent Schrödinger equation

$$i\hbar \frac{\partial \psi(\mathbf{r}, t)}{\partial t} = \hat{\mathcal{H}}_{\text{el}}(\mathbf{r}, \mathbf{R}(t))\psi(\mathbf{r}, t) \,. \tag{15}$$

We expand the time-dependent electronic wavefunction in terms of a set of instantaneous electronic basis functions which are parameterized by the nuclear coordinates as in eqn (11). Now, however, the expansion coefficients are not functions of the nuclear variables, but rather, they are simple functions of time to be determined, thus [42]

$$\psi(\mathbf{r}, t) = \sum_k a_k(t)\phi_k(\mathbf{r}, \mathbf{R}(t)) \,. \tag{16}$$

Substituting into the time-dependent Schrödinger equation, we obtain the following set of coupled differential equations for these coefficients

$$i\hbar \frac{\partial a_k}{\partial t} = \sum_{k'} \left[E_{kk'} + C_{kk'}^{\text{SC}} \right] a_{k'} \tag{17}$$

where

$$C_{kk'}^{\text{SC}} = \langle \phi_k | - i\hbar \frac{\partial}{\partial t} | \phi_{k'} \rangle = -i\hbar \sum_{I=1}^{N} \langle \phi_k | \nabla_{R_I} | \phi_{k'} \rangle \cdot \frac{\mathrm{d}\mathbf{R}_I}{\mathrm{d}t} \,. \tag{18}$$

If we make use of the classical-quantum correspondence and replace the classical nuclear velocity by the quantum momentum operator, i.e. replace

$$-i\hbar \frac{\mathrm{d}\mathbf{R}_I}{\mathrm{d}t} \to -\frac{\hbar^2}{M_I} \nabla_{R_I}$$

we see that

$$C_{kk'}^{\mathrm{SC}} \rightarrow -\sum_{I=1}^{N} \frac{\hbar^2}{M_I} \langle \phi_k | \nabla_{R_I} | \phi_{k'} \rangle \nabla_{R_I} \tag{19}$$

which is just the second term in the full quantum expression for the nonadiabatic operator $\hat{C}_{kk'}$ in eqn (14). This is consistent with the formulation of the semiclassical approximation in which we suppose that the electronic wavefunctions vary slowly with nuclear coordinates, and thus neglect the second order term $\langle \phi_k | \nabla_{R_I}^2 | \phi_{k'} \rangle$ compared with the term which is first order with respect to variations of the electronic wavefunctions with nuclear coordinates i.e. $\langle \phi_k | \nabla_{R_I} | \phi_{k'} \rangle$ in eqn (14).

We can get a clearer idea of the physical behaviour of the expansion coefficients if we choose the adiabatic basis set and factor out the simple oscillatory solution as if the motion were over the uncoupled Born-Oppenheimer surfaces. Thus we write the time-dependent wavefunction as [10, 42–44]

$$\psi(\mathbf{r}, t) = \sum_k \tilde{a}_k(t) \psi_k(\mathbf{r}, \mathbf{R}) \exp\left[-\frac{i}{\hbar} \int_0^t E_k(\mathbf{R}(t')) \mathrm{d}t' \right] \tag{20}$$

and in place of eqn (17) we obtain the following equations of motion for the adiabatic expansion coefficients

$$\frac{\mathrm{d}\tilde{a}_k}{\mathrm{d}t} = -\sum_{k'} \langle \psi_k | \frac{\partial}{\partial t} | \psi_{k'} \rangle \exp\left[-\frac{i}{\hbar} \int_0^t (E_{k'}(t') - E_k(t'))\, \mathrm{d}t' \right] \tilde{a}_{k'} \,. \tag{21}$$

If the basis set is orthonormal we find that the matrix elements satisfy the following relationship $\langle \psi_k | \partial/\partial t | \psi_{k'} \rangle = -\langle \psi_{k'} | \partial/\partial t | \psi_k \rangle^*$ and that the diagonal elements are zero. Suppose that at some time, $\tilde{a}_{k'}$ has some finite amplitude and $\tilde{a}_k$ is zero. If the energy gap $E_{k'} - E_k$ is large and approximately independent of time, the factor

$$\exp\left[-\frac{i}{\hbar} \int_0^t (E_{k'} - E_k)\, \mathrm{d}t' \right]$$

will quickly oscillate in time, changing sign very rapidly. The average effect of these rapid oscillations will be that $\mathrm{d}\tilde{a}_k/\mathrm{d}t \sim 0$. If the energy gap closes and $E_{k'} - E_k \sim 0$ for a time, $\mathrm{d}\tilde{a}_k/\mathrm{d}t$ may become nonzero; thus $\tilde{a}_k$ will "steal" amplitude from $\tilde{a}_{k'}$. The rate at which the amplitude is exchanged between these coefficients will depend on how long E_k and $E_{k'}$ remain close together and also on the size of the coupling matrix element $\langle \psi_k | \partial/\partial t | \psi_{k'} \rangle$. Thus we see that the expansion coefficients only change in regions of stationary phase where the adiabatic state energies remain close to one another for a time.

2.4. DYNAMICAL COUPLING OF THE CLASSICAL AND QUANTUM SUBSYSTEMS

All of the semiclassical results obtained above have been derived assuming a given nuclear trajectory $\mathbf{R}(t)$. We have not as yet discussed in any detail how this trajectory is to be determined, other than hinting in the introduction that within a semiclassical formulation there is an apparent need to determine self-consistently the trajectory followed by the classical subsystem. The need for a self-consistent iteration process for determining $\mathbf{R}(t)$ is outlined schematically in Fig. 3. First we choose an initial guess for the nuclear trajectory $\mathbf{R}^{(0)}(t)$. This initial guess is used to determine the time dependence of the electronic hamiltonian, and the time-dependent Schrödinger equation is solved to give an electronic distribution whose time variation is consistent with the time dependence of the initial guessed trajectory. Now, however, the changing electronic distribution will provide a time-varying force on the nuclei through an effective potential derived from this dynamical charge distribution. Newton's equations are solved in the presence of these forces to give a new trajectory $\mathbf{R}^{(1)}(t)$ which is used as input for the next iteration. This cyclic iteration process, which is formalized in section 4, is continued until the trajectory no longer changes from one iteration to the next.

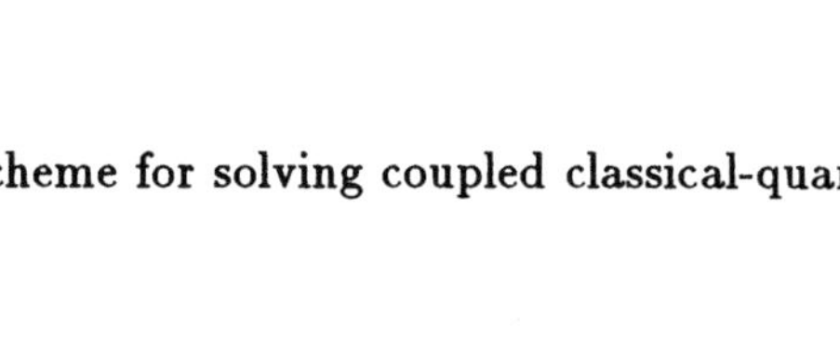

Fig. 3. An iterative scheme for solving coupled classical-quantum equations of motion.

The central quantity which remains to be specified in this self-consistent semiclassical theory is the effective potential $\mathcal{V}_{\text{eff}}$ which describes how the quantum mechanical charge distribution interacts with the nuclei. Within the Born-Oppenheimer framework we usually suppose that the nuclei move over the adiabatic electronic energy surfaces and the forces they experience during this motion derive from the charge distributions associated with the adiabatic eigenstates of $\hat{\mathcal{H}}_{\text{el}}$. If the electronic distribution is in the adiabatic eigenstate $\psi_k(\mathbf{r},\mathbf{R}(t))$ at time t then the force on the nuclei from this adiabatic state is $\mathbf{F}_k^{\text{eN}} = -\langle\psi_k|\nabla_{\mathbf{R}}\mathcal{V}_{\text{eN}}|\psi_k\rangle$ where $\mathcal{V}_{\text{eN}}$ is the model potential describing the interactions between the electronic distribution and the nuclear degrees of freedom. When these forces propel the nuclear trajectory into regions where the dynamical electronic wave function becomes a strong mixture of the adiabatic eigenstates (e.g. passing through a region of strong electronic coupling) the system can be viewed as making nonadiabatic transitions between the adiabatic component states making up the mixture. When such transitions are taking place the specification of which electronic distribution to use to determine the forces on the nuclei becomes ambiguous.

The so called "best trajectory" methods such as the "self-consistent eikonal" approach [45, 46], Miller's classical electron model [47–50], or the "time-dependent Hartree" [51–62] method have been widely used in the literature in many different applications. The effective potential employed with these methods is simply the expectation of the electronic Hamiltonian for the instantaneous dynamical mixed-state charge distribution i.e. $\mathcal{V}_{\text{eff}}^{\text{BT}}(\mathbf{R}) = \langle\psi(\mathbf{r},\mathbf{R})|\hat{\mathcal{H}}_{\text{el}}(\mathbf{r},\mathbf{R})|\psi(\mathbf{r},\mathbf{R})\rangle$. If ψ becomes a mixture of two adiabatic eigenstates as a result of passage through a region of strong coupling so that $\psi = a_1\psi_1 + a_2\psi_2$ then we see that the "best trajectory" effective potential will be a linear combination of the adiabatic energy surfaces $\mathcal{V}_{\text{eff}}^{\text{BT}}(\mathbf{R}) = |a_1|^2 E_1(\mathbf{R}) + |a_2|^2 E_2(\mathbf{R})$. An example of the potential surface over which the nuclear trajectory would move in this "best trajectory" formulation is presented in Fig. 4 for our two-state model. Here we suppose that initially the system is in electronic state ψ_1, so $|a_1|^2 \sim 1$, and that the nuclei start out widely separated and heading towards one another. As the nuclei move through the coupling region the dynamical electronic state ψ becomes a mixture: suppose for example that $|a_1|^2 \sim |a_2|^2 \sim 0.5$ after emerging from the coupling region. The actual relative proportions of the mixture will depend on how long the trajectory spends in the coupling region, and the coupling strength. After the trajectory has moved through the coupling region the expansion coefficients stop changing with time until the next coupling region is encountered, as discussed in section 2.3. As we see in the figure, the effective potential predicted by the "best trajectory" model will lie somewhere in between the two adiabatic curves (half way in this example) and the subsequent nuclear motion after traversing the coupling region in this model will be determined by a weighted-average force derived from the two adiabatic curves.

This idea of nuclear motion over a weighted average of adiabatic potential surfaces is inconsistent with the intuition underlying the Born-Oppenheimer approximation. After the trajectory leaves the region of strong coupling where electronic transitions can occur, the nuclear motion should proceed adiabatically over a single Born-Oppenheimer energy surface, not some mixture of surfaces. The probability that subsequent motion occurs over a particular surface associated with adiabatic state ψ_k once the trajectory leaves

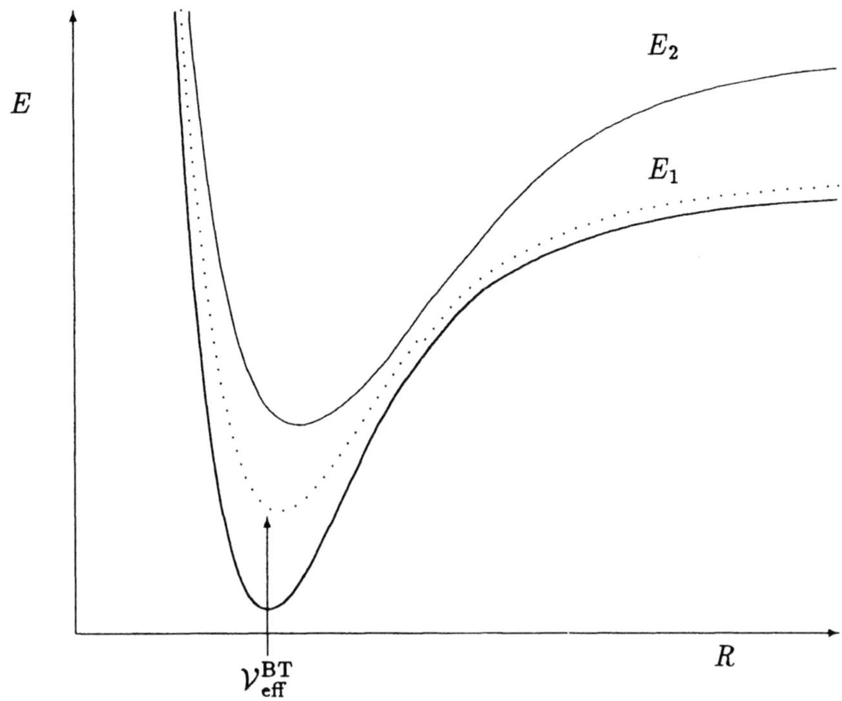

Fig. 4. A best trajectory effective potential. Solid lines are adiabatic curves. Dotted line is "best trajectory" effective potential assuming equal mixing of states after incoming trajectory passes through the strong-coupling region.

the coupling region is determined by the magnitude of the expansion coefficient, $|a_k|^2$, asymptotically after leaving the coupling region. Best trajectory methods propagate the nuclear trajectory over a meaningless mixture of the adiabatic potential curves. The surface-hopping trajectory methods which we will describe in detail later in this Chapter propagate an ensemble of trajectories over the adiabatic surfaces, and allow for transitions between electronic surfaces in the strong-coupling regions. Outside these regions, motion of each trajectory in the ensemble occurs adiabatically with appropriate numbers of ensemble members on each of the adiabatic surfaces to be consistent with the occupation probabilities $|a_k|^2$.

To illustrate the fundamental problems with best trajectory methods, consider the atom-surface scattering experiment schematically depicted in Fig.5 [12, 63, 64]. A variety of different exit channels exist for this atom-surface scattering problem. The atom can simply scatter elastically on a repulsive atom-surface interaction potential (trajectory I, traversing points $\mathbf{A} \to \mathbf{B} \to \mathbf{A}$). Some of the atom's translational energy can be deposited into surface electronic excitations such as electron-hole pairs, in which case the atom may either scatter inelastically (trajectory II, $\mathbf{A} \to \mathbf{B} \to \mathbf{C}$), or it can lose sufficient energy to these surface excitations to become trapped on the surface (trajectory III, $\mathbf{A} \to \mathbf{B} \to \mathbf{D}$). When these surface electronic excitations occur, the atomic trajectory can hop from, say, the repulsive electronic energy curve to the electron-hole excited potential curve which may support nuclear states in which the atom is bound to the surface. Model potential curves for this atom-surface scattering problem indicating various trajectories leading to these different outcomes are presented in Fig. 5. The branching probability for these different outcomes will vary with collision energy. The important point is that a given collision energy can have a number of different possible channels open simultaneously, the probabilities of each being determined by the strength of the electronic coupling between the states.

A physically realistic description of a competitive multichannel process as outlined above is not possible within the "best trajectory" formulation. If we fix the collision energy there is only one possible best trajectory over the $\mathcal{V}_{\text{eff}}^{\text{BT}}$ surface. Thus there is only one possible outcome for a given collision energy within this "best trajectory" model, i.e. the best trajectory at a given energy will exclusively either scatter elastically, inelastically, or trap. The "best trajectory" approach cannot give disproportionation into the different open exit channels for a given collision energy which is observed experimentally. With the surface-hopping trajectory models discussed in the next section, on the other hand, the ensemble of trajectories will bifurcate into the different possible channels with correct electronic proportioning. Thus this picture is physically more realistic as well as being consistent with our Born-Oppenheimer intuition.

3. Surface-hopping trajectory methods

3.1. DESIRABLE FEATURES OF A NONADIABATIC DYNAMICS ALGORITHM

The most popular approaches for studying electronically nonadiabatic processes are the stochastic surface-hopping trajectory methods [3, 4, 10–27] which are based on the pioneering work of Tully and Preston [10, 11]. The principal idea behind these techniques is to propagate an ensemble or swarm of classical nuclear trajectories over the coupled electronic energy surfaces. When a trajectory enters a region of strong coupling between various alternative energy surfaces we use a stochastic method to choose one of the electronic states at random in such a way that the proportion of ensemble

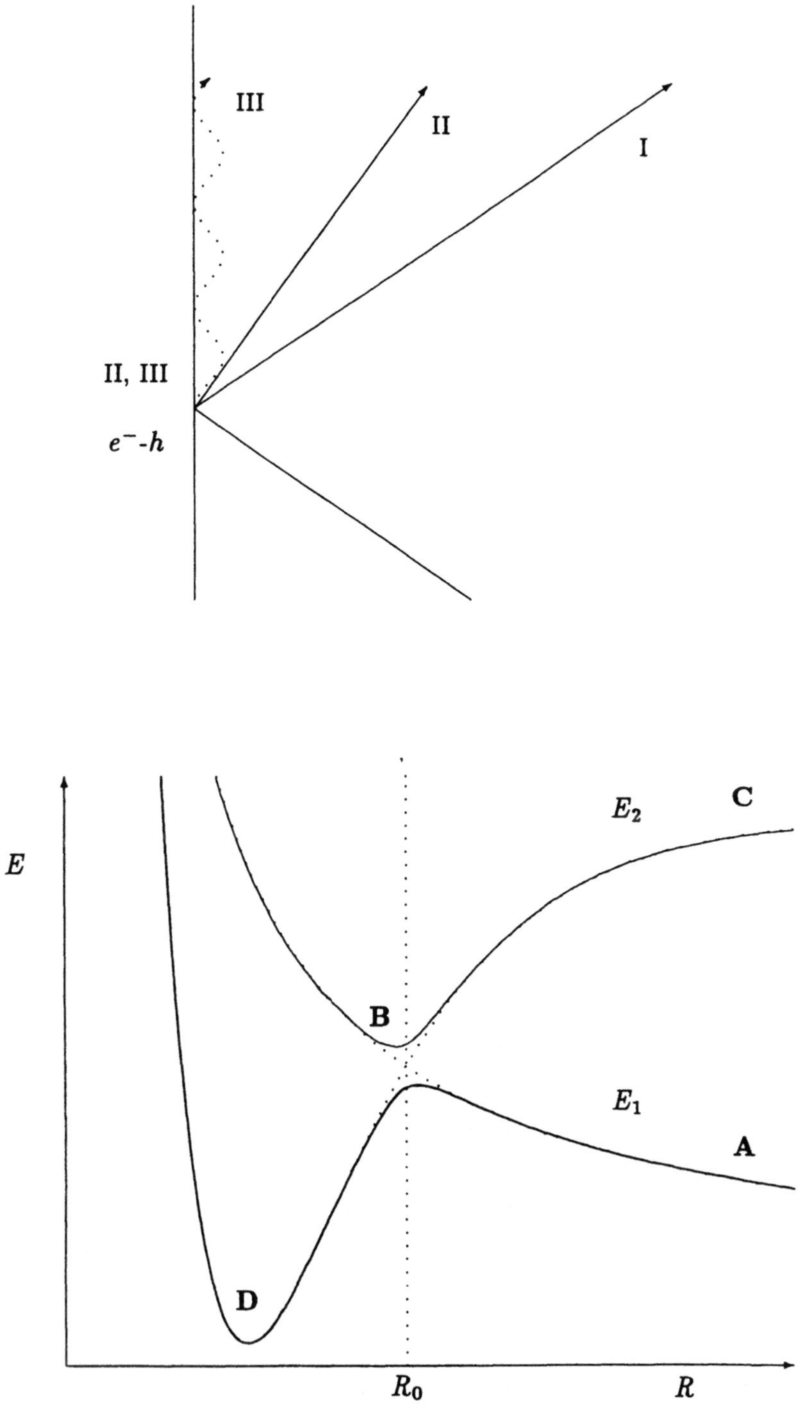

Fig. 5. A multichannel surface scattering problem. Example trajectories and potential energy curves.

members which branch into the different alternative channels is consistent with the statistical distribution of outcomes determined by the mixing coefficients of the dynamical electronic wavefunction.

There are four key ingredients of a practical, useful surface-hopping trajectory method.

(i) A swarm of independent trajectories, which can branch onto the different electronic energy surfaces, is propagated in such a way that the ensemble has the correct statistical distribution of occupation probabilities for the different adiabatic states consistent with the dynamical mixed-state electronic wave function. The dynamical properties can be obtained as averages over this ensemble of branching classical trajectories.

(ii) We must solve the time-dependent electronic Schrödinger equation for the dynamical electronic wavefunction so that these state occupation probabilities which will determine when, where, and how many classical trajectories in our ensemble will branch onto the different surfaces.

(iii) Once a trajectory leaves the coupling region and the dynamical wavefunction associated with this trajectory becomes a stationary mixture of the adiabatic states, we must turn off the hopping between the different surfaces and propagate the nuclear dynamics adiabatically until the next coupling region is encountered, where transitions between states must again be switched on.

(iv) Unless there are degeneracies resulting from special symmetry considerations, hops between surfaces will usually occur across a finite energy gap between the adiabatic curves. This change of energy associated with the electronic transition must be absorbed by the nuclear motions, and an approach for redistributing this energy amongst the nuclear degrees of freedom, which satisfies energy and angular momentum conservation laws, must be provided.

To make these methods tractable we must be able to perform all the necessary calculations "on the fly" i.e. using instantaneous local information about the dependence on nuclear coordinates of various quantities. In other words we don't want to have to know such global information as the complete variation of the electronic wavefunctions and energies at all possible configurations of the nuclear coordinates in advance. It would be desirable if we could answer the various questions outlined above at each instant using just the information about the current nuclear configuration. In the rest of this section we describe how these different features can be addressed in a viable numerical algorithm.

3.2. ELECTRONIC PHASE COHERENCE

The methods we use to solve the time-dependent electronic Schrödinger equation are based on the approach outlined in section 2.3, in which we choose an appropriate electronic basis set and integrate the expansion coefficient equations of motion in eqn (17) numerically. The solutions of this system of equations (or equivalently the adiabatic coefficient equations of motion in eqn (21)) will in general be oscillatory complex functions of time. Care must be taken to integrate these phase factors correctly, especially when hops between potential energy surfaces are allowed. The accurate treatment of electronic phase coherence is particularly important when the nuclear trajectory passes through successive regions of strong coupling soon after one another in time. The first passage into a strong-coupling region will start these expansion coefficients oscillating. If these initial oscillations do not have time to decay before the trajectory enters the next strong-coupling region, the oscillations in the expansion coefficients arising from each passage into a strong-coupling region may interfere with one another. The time delay between successive passages through the strong-coupling regions will influence the beat frequency associated with the interference terms in the expansion coefficient dynamics, and this delay time will be determined by the velocity of the trajectory in between the successive regions of strong coupling. These interference effects resulting from electronic coherence will make the probability of branching between surfaces an oscillatory function of the delay time and thus of the classical trajectory velocity.

These electronic phase coherence effects can be readily explored for the linear diabatic surface model discussed in section 2.2. This was first done by Stükelberg [39]. Instead of assuming that the energy gap between the diabatic curves varies linearly in time as we did in section 2.2, Stükelberg chose a quadratic model for the time dependence of the diabatic energy gap so $(\epsilon_2 - \epsilon_1)/\hbar = \alpha t^2 + \beta$ which, given the linear variation of the energy curves with R, is consistent with a parabolic time dependence for the nuclear trajectory $R(t) = vt + Ft^2/2m$, where $F = \sqrt{F_1 F_2}$. An example of such a trajectory is presented in Fig. 6.

Here we see that the trajectory encounters the crossing point R_0 on the incoming passage through the crossing region. It encounters R_0 a second time on the outgoing passage after passing through the turning point R_{T}. The time lag between these successive passages through R_0 will be determined by the incoming velocity and the model force on the trajectory F. Again making the simplifying assumptions that the diabatic functions and the coupling are independent of time, i.e. $\dot{\epsilon}_{12} = \dot{\phi}_1 = \dot{\phi}_2 = 0$, eqn (7) can be solved analytically. The solutions can be expressed in terms of Airy functions, and the asymptotic behaviour of these solutions gives the transition probability from the incoming ground state to outgoing excited state in the limit that $mv^2/2\epsilon^* \gg 1$ as

$$P = \frac{8\pi\epsilon_{12}^2}{|F_2 - F_1|\hbar v} \sin^2 \left[\frac{2}{3} \left(\frac{mv^2}{2\epsilon^*} \right)^{3/2} + \frac{\pi}{4} \right] \tag{22}$$

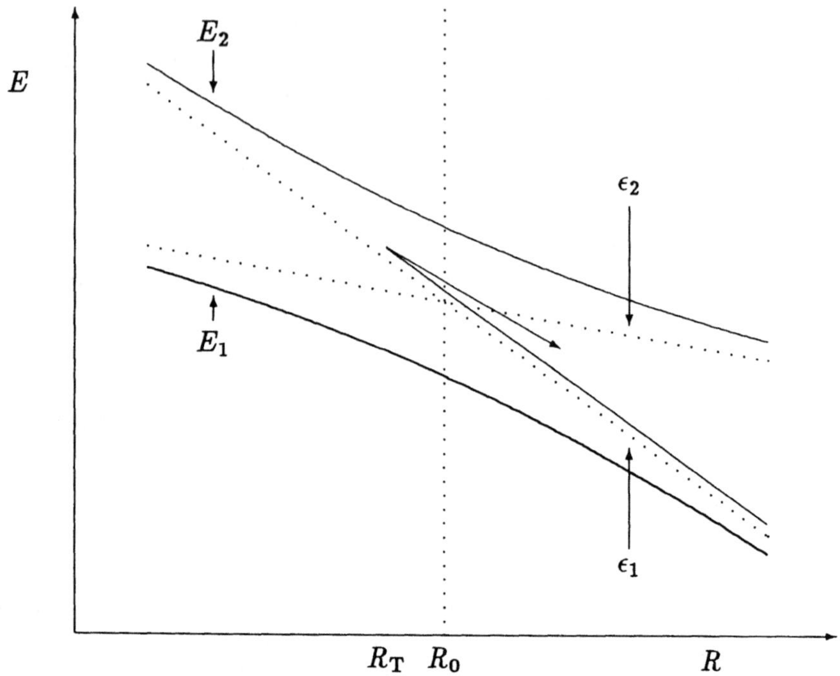

Fig. 6. Linear diabatic curve crossing and electronic phase coherence. The trajectory passes through strong-coupling region twice and interference effects give rise to Stükelberg oscillations in transition probability. R_T is the turning point of classical motion and R_0 is the position of the crossing.

where $\epsilon^* = (\hbar^2 F^4/2m|F_2 - F_1|^2)^{1/3}$. We see that this transition probability is a damped oscillatory function of the kinetic energy of the collision $\frac{1}{2}mv^2$. These oscillations in the transition probability are known as Stükelberg oscillations, and they result from the interference effects due to the electronic coherence [39–42]. To accurately represent these interference effects we must integrate the equations of motion for the expansion coefficients coherently through these successive coupling regions.

3.3. STOCHASTIC MODEL OF SURFACE-HOPPING DYNAMICS

In its earliest forms [10, 11] the surface-hopping trajectory methods could only be applied to very simple systems because of the need to know the complete dependence on nuclear coordinates of the various coupled potential energy surfaces. In these early versions one had to provide, in advance, a map of all regions of nuclear configuration space where the various interacting surfaces were strongly coupled. There is no unique way of locating these so-called "seam" regions in which hopping between surfaces can take place. For example, one could use the positions where the adiabatic energy surfaces are the closest to define the location of the hopping seams [10, 21–23], or alternatively the position of the regions of maximal electronic coupling as determined by $\langle\phi_k|\nabla_R|\phi_{k'}\rangle$ (see eqn (18)) could be used to define the seam location [11, 12, 15, 20, 24]. In addition to specifying the seam location function $S(\mathbf{R})$ one also needs to know the probability of hopping from surface to surface $P(\mathbf{R}, \dot{\mathbf{R}})$ as a trajectory crosses the seam joining these surfaces. As we saw in section 2.2, this hopping probability not only depends on where the trajectory crosses the seam but also on how quickly our trajectory is moving through the seam region. Specification of the functions $S(\mathbf{R})$ and $P(\mathbf{R}, \dot{\mathbf{R}})$ requires knowledge of how the wavefunctions and energies vary with nuclear geometry over all space, and an alternative approach which does not require this global information must be devised if a tractable approach to nonadiabatic dynamics in many-body systems is to be developed.

The simplest way one might imagine to obtain the correct statistical distribution of trajectories amongst the different electronic states is to compare the square magnitudes of the expansion coefficients $|a_k|^2$ at each nuclear timestep with a uniform random number between 0 and 1, and allow the occupied state identity to hop at random between the various competing electronic states. This approach clearly gives the correct statistical distribution but it suffers from the same problems as the "best trajectory" methods, described in section 2.4, in that if a trajectory emerges from a region of coupling with, say, a strong mixture of two adiabatic states, the above approach will give a trajectory which hops between the two component states even outside the strong-coupling region and the force on the nuclei will effectively be a weighted average of the two surfaces as discussed above. This is the reason that surface-hopping methods need a way of specifying the location of the hopping seams i.e. an approach for switching on, and switching off the hopping between surfaces. One way around this problem is to reset the expansion coefficients (occupation probabilities) to the appropriate unit vector after each hop [6, 23, 24] but this approach destroys all phase coherence information as we shall see in section 4.

Recently Tully [12] has devised a very effective scheme for switching the surface hopping. The approach is very attractive as it does not require specifying the location of the hopping seams in advance, and the expansion coefficient equations of motion are integrated coherently throughout the entire dynamics. The approach makes use of the fact that the only places where hopping can take place are where the state occupation probabilities $|a_k|^2$ are changing rapidly with time. Outside the hopping seams the expansion coefficients become independent of time as discussed in section 2.3 and the nuclear motion should be adiabatic in these regions. Tully's idea is thus to devise a new hopping probability which effectively includes a hopping switch, by making the hopping probability proportional to the velocity of change of the occupation probabilities, i.e. $\mathrm{d}|a_k|^2/\mathrm{d}t$. The new hopping probability must preserve the correct statistical distribution of occupations consistent with the solution of the electronic time-dependent Schrödinger equation.

It is most convenient to derive Tully's algorithm in terms of the electronic density matrix elements which are related to the basis set expansion coefficients by the expression $\rho_{ij}(t) = a_i^*(t)a_j(t)$. The diagonal elements of this matrix give the state occupation probabilities and the off-diagonal elements have information about the relative phases of the expansion coefficients.

To proceed we will consider the rate of change of the occupation probabilities $\dot{\rho}_{ii}$. First let us rewrite the expansion-coefficient equations of motion in eqn (17) in the form

$$\frac{\mathrm{d}a_i}{\mathrm{d}t} = \sum_k d_{ik} a_k \tag{23}$$

where $d_{ik} = -iE_{ik}/\hbar - \langle \phi_i | \partial/\partial t | \phi_k \rangle$. We can use this result together with the fact that $\rho_{ik} = \rho_{ki}^*$ and $d_{ik} = -d_{ki}^*$ to show that

$$\begin{aligned} \dot{\rho}_{ii} &= \frac{\mathrm{d}(a_i^* a_i)}{\mathrm{d}t} = 2\mathrm{Re}\left(a_i^* \frac{\mathrm{d}a_i}{\mathrm{d}t}\right) = 2\mathrm{Re}\left(\sum_k \rho_{ik} d_{ik}\right) \\ &= -2\mathrm{Re}\left(\sum_k \rho_{ki} d_{ki}\right) . \end{aligned} \tag{24}$$

Suppose our ensemble consists of N systems; then at any instant, the number of ensemble members in state i is $N_i = \rho_{ii} N$. If, during a timestep Δt, the occupation probability of state i changes by $\Delta\rho_{ii}$, then the number of ensemble members in state i will change by $\Delta N_i = \Delta\rho_{ii} N$. This change in the number of ensemble members in state i results both from trajectories already in state i hopping out of this state, and from trajectories in other states hopping into state i during the time interval Δt. We will assume that the occupation probabilities change according to simple first-order kinetics so that the fractional increase in the number of trajectories in state i during Δt can be written as

$$\Delta\rho_{ii} = \frac{\Delta N_i}{N} = -\sum_k g_{i\to k}\rho_{ii} + \sum_k g_{k\to i}\rho_{kk} . \tag{25}$$

We need to construct a model for the transition probabilities $g_{i\to k}$ in which this master equation representation is consistent with the equations of motion for the occupation probabilities in eqn (24). By multiplying and dividing eqn (24) appropriately by the occupation probabilities and using the first-order approximation $\Delta\rho_{ii} = \dot{\rho}_{ii}\Delta t$, we can obtain two alternative expressions for the fractional increase in the number of trajectories in state i during Δt as

$$\Delta\rho_{ii} = -\sum_k \frac{2\Delta t \mathrm{Re}\left(\rho_{ki} d_{ki}\right)}{\rho_{ii}} \rho_{ii} = -\sum_k \tilde{g}_{i\to k}\rho_{ii} \tag{26}$$

$$= \sum_k \frac{2\Delta t \mathrm{Re}\left(\rho_{ik} d_{ik}\right)}{\rho_{kk}} \rho_{kk} = \sum_k \tilde{g}_{k\to i}\rho_{kk} \,. \tag{27}$$

Equation (26) can be viewed as supposing that the change $\Delta\rho_{ii}$ results only from transitions out of state i and into all other states k with the model transition probabilities being $\tilde{g}_{i\to k}$. Similarly eqn (27) can be interpreted as saying that $\Delta\rho_{ii}$ results purely from transitions from other states k into state i, these transition probabilities being $\tilde{g}_{k\to i}$. The master equation (25) suggests that both these processes affect the change $\Delta\rho_{ii}$. We want to interpret the $\tilde{g}_{i\to k}$'s in our model ensemble as positive-definite transition probabilities. We see that the sign of the $\tilde{g}_{i\to k}$ is determined by the quantity $\mathrm{Re}\left(\rho_{ki} d_{ki}\right)$. From eqn (24), we can think of this quantity as a component of the velocity of density out of state i. If this component is positive, we allow transitions out of i and into k. If this component is negative, we do not allow transitions in this direction and set the transition probability to zero. In this situation, i.e. when $\mathrm{Re}\left(\rho_{ki} d_{ki}\right) < 0$, we have that $\mathrm{Re}\left(\rho_{ik} d_{ik}\right) > 0$ by virtue of the antisymmetry of the d matrix. From eqn (24), we see that $\mathrm{Re}\left(\rho_{ik} d_{ik}\right) > 0$ is a component of the velocity $\dot{\rho}_{ii}$ which represents the population in state i increasing due to transitions from k into i. We choose our positive-definite model transition probabilities to include a masking Heaviside function in the component of the velocity out of the occupied state being considered.

$$g_{i\to k} = \frac{2\Delta t \mathrm{Re}\left(\rho_{ki} d_{ki}\right)}{\rho_{ii}} \theta\left(-\mathrm{Re}(\rho_{ki} d_{ki})\right) \tag{28}$$

Substituting this form into eqn (25), we find that our model for the fractional increase in the number of trajectories in state i in Δt is

$$\Delta\rho_{ii} = -\sum_k \frac{2\Delta t \mathrm{Re}\left(\rho_{ki} d_{ki}\right)}{\rho_{ii}} \theta\left(-\mathrm{Re}(\rho_{ki} d_{ki})\right) \rho_{ii} + \sum_k \frac{2\Delta t \mathrm{Re}\left(\rho_{ik} d_{ik}\right)}{\rho_{kk}} \theta\left(-\mathrm{Re}(\rho_{ik} d_{ik})\right) \rho_{kk} \,. \tag{29}$$

As discussed above, when one of these two terms switches on, the other switches off and the result is identical to summing the two results we obtained from the equation of motion $\dot{\rho}_{ii}$, i.e. eqns (26) and (27) but including the Heaviside mask so that only transitions out of the occupied state are considered. Tully called this the "fewest-switches" approach since it supposes that the change in the occupied state density results only from transitions out of this state. This ensures that the number of hops in the ensemble dynamics is minimal, and as derived above it is guaranteed to give our model ensemble the correct statistical distribution of occupied states consistent with the mixed solution of the time-dependent Schrödinger equation.

We note that the hopping probabilities defined in eqn (28) have the correct scaling with time step Δt. As Δt is made smaller, the probability of making a hop goes down proportionally, but there will be more time steps in a given interval and hence more chances to try to make these less probable hops.

3.4. *AD HOC* VELOCITY RESCALING AND ENERGY REDISTRIBUTION IN SURFACE HOPPING

The remaining question we need to address, to establish a tractable surface-hopping algorithm, is how to redistribute the energy amongst the nuclear degrees of freedom when a hop occurs across a finite energy gap between two adiabatic curves. This electronic energy must be redistributed into the classical nuclear motions in such a way as to conserve the total energy of the coupled electronic-nuclear system. The angular momentum of the total system must also be conserved when these surface hops take place.

The simplest approach to redistributing the electronic gap energy between the quantum electronic subsystem and the classical nuclear subsystem involves adjusting the nuclear velocities so that the electronic energy gap is made up by nuclear kinetic energy. The main difficulty with this approach is that we need some way of deciding which components of the multidimensional velocity vector of the nuclear subsystem are coupled more strongly to the electronic subsystem, and so can exchange more energy to make up the electronic energy gap than other nuclear velocity components. As this adjustment procedure is *ad hoc* and non-unique a variety of different alternatives have been suggested as discussed earlier. Herman [15, 16] has presented a less *ad hoc* justification for the velocity adjustment procedure we shall now outline.

An effective way of deciding along which direction to adjust the nuclear velocities is motivated by considering the expansion-coefficient equations of motion eqn (17) and eqn (18). We see from the definition of the semiclassical approximation for the nonadiabatic coupling matrix elements $C^{\mathrm{SC}}_{kk'}$ in eqn (18) that if the nuclear velocity vector $\mathrm{d}\mathbf{R}_I/\mathrm{d}t$ is perpendicular to the coupling vector $\mathbf{D}_{kk'} = \langle\phi_k|\nabla_{\mathbf{R}}|\phi_{k'}\rangle$ then $C^{\mathrm{SC}}_{kk'}$ will be zero. Thus nuclear motions in this perpendicular direction will not couple amplitude transfer between the coefficients through eqn (17). On the other hand, if the velocity vector, $\mathrm{d}\mathbf{R}_I/\mathrm{d}t$, is parallel to the coupling vector $\mathbf{D}_{kk'}$, amplitude transfer between the coefficients a_k and $a_{k'}$ will be maximal. Nonadiabatic hops between states k and k' occur when there is large amplitude transfer between coefficients a_k and $a_{k'}$; thus we assume that these transitions are driven predominately by the velocity components along the direction of the nonadiabatic coupling vector $\mathbf{D}_{kk'}$. Tully and Preston and others [10, 15] thus suggested that, when a nonadiabatic hop between nondegenerate electronic states occurs, the electronic gap energy be redistributed amongst the components of the nuclear velocity vector in proportion to their projection in the direction of the nonadiabatic coupling vector $\mathbf{D}_{kk'}$. We thus resolve the nuclear velocity vector $\dot{\mathbf{R}}$ into components parallel and perpendicular to $\mathbf{D}_{kk'}$

$$\begin{aligned}\dot{\mathbf{R}}^{||} &= (\dot{\mathbf{R}}\cdot\hat{\mathbf{D}}_{kk'})\hat{\mathbf{D}}_{kk'}\\ \dot{\mathbf{R}}^{\perp} &= \dot{\mathbf{R}}-(\dot{\mathbf{R}}\cdot\hat{\mathbf{D}}_{kk'})\hat{\mathbf{D}}_{kk'}\end{aligned}$$

where $\hat{\mathbf{D}}_{kk'} = \mathbf{D}_{kk'}/|\mathbf{D}_{kk'}|$, as indicated in Fig. 7, so that $\dot{\mathbf{R}} = \dot{\mathbf{R}}^{||} + \dot{\mathbf{R}}^{\perp}$ and $\dot{\mathbf{R}}^{||}\cdot\dot{\mathbf{R}}^{\perp} = 0$. Thus, we choose the new nuclear velocity vector $\dot{\mathbf{R}}^{\mathrm{new}}$ so that the component perpendicular to $\dot{\mathbf{R}}$ is unchanged during the transition whilst the component in the $\dot{\mathbf{R}}^{||}$ direction is adjusted to take up the electronic energy gap. Thus we write $\dot{\mathbf{R}}^{\mathrm{new}} = \dot{\mathbf{R}}^{\perp} + \Delta_{||}\dot{\mathbf{R}}^{||}/|\dot{\mathbf{R}}^{||}|$ where the parallel velocity scaling factor $\Delta_{||}$ is determined from

$$\frac{1}{2}M\dot{\mathbf{R}}^{\mathrm{new}}\cdot\dot{\mathbf{R}}^{\mathrm{new}} = \frac{1}{2}M\dot{\mathbf{R}}\cdot\dot{\mathbf{R}} - \Delta E$$

as $\Delta_{||} = \sqrt{\dot{\mathbf{R}}^{||}\cdot\dot{\mathbf{R}}^{||} - 2\Delta E/M}$ where $\Delta E = E^{\mathrm{new}} - E^{\mathrm{old}}$.

The remaining quantity we need to calculate in order to determine how best to adjust the nuclear velocities so as to conserve energy is the nonadiabatic coupling vector $\mathbf{D}_{kk'} = \langle\phi_k|\nabla_{\mathbf{R}}|\phi_{k'}\rangle$. An expression for the coupling vector is readily obtained by making use of the result

$$\nabla_{\mathbf{R}}\langle\phi_{k'}|\hat{\mathcal{H}}|\phi_k\rangle = \nabla_{\mathbf{R}}E_k\delta_{k'k} \tag{30}$$

where we work with the adiabatic basis states $\{\phi_k(\mathbf{r};\mathbf{R})\}$. Writing out the derivative on the left hand side we readily find that the diagonal elements of the force matrix are the Hellman-Feynman forces

$$\mathbf{F}_{kk} = -\langle\phi_k|\nabla_{\mathbf{R}}\hat{\mathcal{H}}|\phi_k\rangle = -\nabla_{\mathbf{R}}E_k \tag{31}$$

and the off-diagonal elements are related to the coupling vector so

$$\mathbf{F}_{k'k} = -\langle\phi_{k'}|\nabla_{\mathbf{R}}\hat{\mathcal{H}}|\phi_k\rangle = (E_{k'}-E_k)\langle\phi_{k'}|\nabla_{\mathbf{R}}|\phi_k\rangle \quad k \neq k' \tag{32}$$

or

$$\mathbf{D}_{k'k} = \mathbf{F}_{k'k}/(E_{k'}-E_k)\,, \tag{33}$$

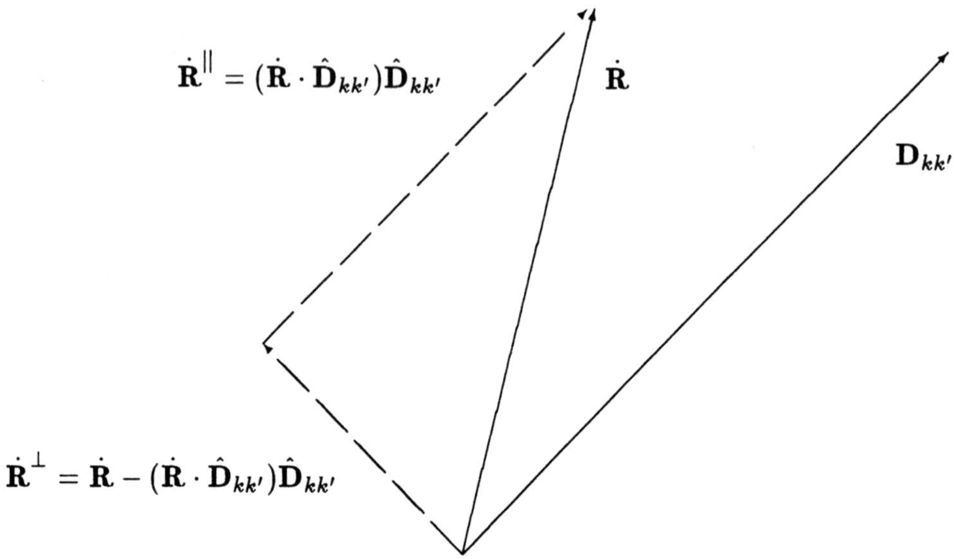

Fig. 7. The resolution of nuclear velocity vector $\dot{\mathbf{R}}$ into components parallel and perpendicular to the nonadiabatic coupling vector $\mathbf{D}_{kk'}$.

that is the coupling vector points in the direction of the off-diagonal force matrix elements.

The main objection to the approach outlined here for redistributing the quantum subsystem energy gap amongst the classical degrees of freedom during a nonadiabatic transition is that it is *ad hoc* and based on the assumption that the mechanism for energy exchange during a nonadiabatic transition involves coupling to only the classical particle velocities, i.e. the excess quantum system energy is dumped into the classical subsystem kinetic energy only, and there is no direct influence on the configuration of the classical system arising from the nonadiabatic transition. Thus there is no redistribution of quantum subsystem energy gap into classical system potential energy. In principle we might suppose that the classical subsystem moves to both a different position and different velocity as a result of the quantum subsystem changing state. Another issue that we have not addressed here is angular momentum conservation, and there are mechanisms within this *ad hoc* velocity adjustment approach for obtaining angular momentum conservation [10]. The self-consistent trajectory optimization techniques outlined in section 4 provide a rigorous approach for obtaining trajectories of the coupled quantum-classical system which conserve both energy and angular momentum. A variety of other approaches exist for adjusting the momentum so as to give conservative dynamics during surface-hopping trajectory, and the reader is referred to various articles for descriptions of these methods [13, 20, 21]

3.5. SUMMARY

Having outlined all the key ingredients of a surface-hopping algorithm which should provide a tractable approach for treating condensed-phase coupled quantum-classical systems, we are in a position to summarize the complete algorithm. To be explicit, we will present the algorithm in the adiabatic basis, though an arbitrary electronic representation can be employed.

(i) The ensemble is prepared by specifying initial conditions for the positions and velocities $\mathbf{R}^{(i)}$ and $\dot{\mathbf{R}}^{(i)}$ of the classical degrees of freedom for each ensemble member (i). The initial state of the quantum subsystem is specified by choosing initial values of the expansion coefficients $\{a_k^{(i)}\}$ for each ensemble member. The quantum subsystem can thus be prepared in a pure state k by choosing $\rho_{kk}^{(i)} = a_k^{(i)*} a_k^{(i)} = 1$ for all ensemble members. Alternatively we can prepare the quantum subsystem in a coherent mixed state in which we choose the initial coefficient values of various states $\{a_k^{(i)}, a_l^{(i)}, a_m^{(i)}, \ldots\}$ non-zero. In this coherent case however, each ensemble member has the same set of initial coefficients. The ensemble can also be prepared in an incoherent mixed state in which the mixing coefficients of each ensemble member are different.

(ii) Next the adiabatic eigenstates and energies are obtained for the current nuclear configuration by diagonalizing the electronic hamiltonian $\hat{\mathcal{H}}_{\rm el}{}^{(i)} \psi_k^{(i)} = E_{\rm el}^{(i)} \psi_k^{(i)}$ where $\hat{\mathcal{H}}_{\rm el}{}^{(i)} = \hat{\mathcal{K}}(\mathbf{r}) + \mathcal{V}_{\rm eN}(\mathbf{r}, \mathbf{R}^{(i)})$. Methods for obtaining these eigenstates are discussed in section 5.2.

(iii) The equations of motion for the classical variables are evolved through a timestep Δt over the currently occupied potential surfaces for the different ensemble members. The equation of motion for the $l^{\rm th}$ atom in the $i^{\rm th}$ ensemble member has the form

$$M_l \ddot{\mathbf{R}}_l^{(i)} = -\nabla_{\mathbf{R}_l^{(i)}} \mathcal{V}_{\rm NN}(\mathbf{R}^{(i)}) - \langle \psi_k^{(i)} | \nabla_{\mathbf{R}_l^{(i)}} \mathcal{V}_{\rm eN}(\mathbf{r}, \mathbf{R}^{(i)}) | \psi_k^{(i)} \rangle \,. \tag{34}$$

Here we suppose that the $k^{\rm th}$ state is occupied by trajectory i.

(iv) Next we evolve the equations of motion for the quantum expansion coefficient of each ensemble member through a timestep Δt. As discussed in section 2.3, in the adiabatic representation of eqn (20), these equations have the form

$$\frac{{\rm d}a_k^{(i)}}{{\rm d}t} = -\sum_{k'} \langle \psi_k^{(i)} | \frac{\partial}{\partial t} | \psi_{k'}^{(i)} \rangle \exp\left[-\frac{i}{\hbar} \int_0^t (E_{k'}^{(i)}(t') - E_k^{(i)}(t'))\, {\rm d}t' \right] a_{k'}^{(i)} \,.$$

(35)

There are a variety of numerical questions which must be addressed when computing the various quantities in the above equations and accurately integrating these equations of motion. The difficulties with integration arise because of considerable differences in timescale for the variations in the nuclear and expansion-coefficient variables. With the adiabatic representation, the expansion-coefficient variables change slowly outside the transition regions. In these regions of strong coupling, however, some of the expansion coefficients can change very rapidly, so care must be taken in these regions to ensure accurate integration. Various numerical considerations are discussed in sections 4.2 and 5.2.

(v) After the expansion coefficients have been evolved through a timestep, the hopping probabilities $g_{k\to k'}$ at the new nuclear configuration can be computed. If state k is currently occupied by trajectory i, this trajectory is allowed to hop to state k' with probability

$$g_{k\to k'}^{(i)} = \frac{2\Delta t \mathrm{Re}(\rho_{k'k}^{(i)} d_{k'k}^{(i)})}{\rho_{kk}^{(i)}} \theta\left(-\mathrm{Re}(\rho_{k'k}^{(i)} d_{k'k}^{(i)})\right). \tag{36}$$

This is done by selecting a uniformly-distributed random number ξ such that $0 < \xi < 1$. Then a hop to state 1 is made if ξ lies in the range $0 < \xi < g_{k1}$, a hop to state 2 is made if ξ satisfies $g_{k1} < \xi < g_{k1} + g_{k2}$, hopping to state 3 occurs if $g_{k1}+g_{k2} < \xi < g_{k1}+g_{k2}+g_{k3}$ etc. The newly-selected state $\psi_{k'}^{(i)}$ will be used in the next timestep as the occupied state of the quantum subsystem. As such, this new quantum distribution will determine the forces on the classical degrees of freedom for the next timestep.

(vi) If no hop occurs we return to step (ii) and continue evolution of the coupled system with the quantum distribution $\psi_k^{(i)}$ providing the forces. On the other hand, if it is decided that trajectory i must hop to the new state $\psi_{k'}^{(i)}$ then this trajectory must begin evolving over the new energy surface $E_{k'}(\mathbf{R})$. In general, at the current position there will be a finite energy gap $E_{k'}(\mathbf{R}) - E_k(\mathbf{R}) = \Delta E$. This excess quantum subsystem energy is redistributed into the classical subsystem kinetic energy by adjusting the velocities according to the prescription given in section 3.4 so that the total energy is conserved. If the new state is higher in energy than the old state, $\Delta E > 0$, then energy must be taken from the classical kinetic energy by velocity reduction. If this energy is greater than that available in the relevant velocity components then no hop is made.

4. Self-consistent energy-conserving nonadiabatic dynamics

The algorithms outlined above involve exact solutions of the separate time-dependent quantum and classical subsystems, together with a sensible approximate model for the coupling between these subsystems. As noted in section 2.4, the formally correct way to treat such a coupled classical-quantum system should involve the self-consistent determination of the classical trajectory, due to the fact that the classical trajectory depends on the time evolution of the quantum wavefunction which in turn depends on how the classical variables change with time. The need for self consistency has been removed in the algorithm outlined in section 3.5 by the choice of the model of the coupling between the classical and quantum subsystems. Removing the self consistency leads to a nonconservative algorithm and the need for *ad hoc* techniques for redistributing the energy and angular momentum between the classical and quantum degrees of freedom to enforce the conservation laws when hops between surfaces take place.

The need for self consistency between the classical trajectory and time-dependent wavefunction in a rigorous semiclassical nonadiabatic description was first formalized by Pechukas [35, 36]. This self-consistent theory has recently been developed into a tractable energy conserving algorithm for nonadiabatic dynamics in condensed systems by Webster *et al.* [6]. In this section we outline Pechukas' self-consistent theory and describe the implementation of these ideas by Webster and co-workers. We discuss the phase coherence problems with Webster's implementation and present a new energy-conserving self-consistent algorithm which correctly handles the electronic phase coherence.

4.1. PECHUKAS THEORY

Pechukas' theory of nonadiabatic collisions is based on Feynman's path-integral formulation of quantum mechanics [65, 66], in which the central quantity of interest is the quantum propagator

$$K(\mathbf{x}'', \mathbf{R}'', t''|\mathbf{x}', \mathbf{R}', t') = \langle \mathbf{x}'', \mathbf{R}''| \exp[-\frac{i}{\hbar}\hat{\mathcal{H}}(t'' - t')]|\mathbf{x}', \mathbf{R}'\rangle \tag{37}$$

which gives the probability amplitude for the system described by coordinates $\mathbf{x}$ and $\mathbf{R}$ to move from the point $\mathbf{x}', \mathbf{R}'$ at time t' to point $\mathbf{x}'', \mathbf{R}''$ at time t''. The specific example that Pechukas considers is molecular scattering, in which $\mathbf{R}$ describes the relative separation between the two molecules, which we will ultimately treat classically, and $\mathbf{x}$ describes the internal states of the molecules, for example their electronic, vibrational or rotational degrees of freedom, which are to be treated quantum mechanically. The hamiltonian is split up into the classical coordinate kinetic energy piece, and all the rest, as follows

$$\hat{\mathcal{H}}(\mathbf{x}, \mathbf{R}) = \frac{P_{\mathbf{R}}^2}{2M} + \hat{h}(\mathbf{x}, \mathbf{R}) = \hat{\mathcal{H}}_0 + \hat{h}\ . \tag{38}$$

In general, we are not interested in the complete detailed description of the quantal degrees of freedom afforded by the full propagator; rather we really only want to address the question, "What is the probability amplitude for the quantum subsystem to change from state α to state β as the classical subsystem evolves from $\mathbf{R}', t'$ to $\mathbf{R}'', t''$?". Pechukas thus considers the reduced propagator which answers this question, by simply projecting the full propagator onto an appropriate basis of internal quantum states $\{\phi_i(\mathbf{x}, \mathbf{R})\}$ which are parameterized by the classical coordinates, i.e.

$$\begin{aligned} K_{\beta\alpha}(\mathbf{R}'', t''|\mathbf{R}', t') &= \int \mathrm{d}\mathbf{x}''\mathrm{d}\mathbf{x}'\phi^*_\beta(\mathbf{x}'', \mathbf{R}'')K(\mathbf{x}'', \mathbf{R}'', t''|\mathbf{x}', \mathbf{R}', t')\phi_\alpha(\mathbf{x}', \mathbf{R}') \\ &= \langle \mathbf{R}''\beta|\exp[-\frac{i}{\hbar}\hat{\mathcal{H}}(t''-t')]|\,\mathbf{R}'\alpha\rangle \end{aligned} \tag{39}$$

$$\begin{aligned} &= \int \mathrm{d}\mathbf{R}_1 \ldots \mathrm{d}\mathbf{R}_{P-1}\langle \mathbf{R}''\beta|\exp[-\frac{i}{\hbar}\hat{\mathcal{H}}\epsilon]|\,\mathbf{R}_{P-1}\rangle \\ &\quad \langle \mathbf{R}_{P-1}|\exp[-\frac{i}{\hbar}\hat{\mathcal{H}}\epsilon]|\,\mathbf{R}_{P-2}\rangle \\ &\quad \ldots \langle \mathbf{R}_1|\exp[-\frac{i}{\hbar}\hat{\mathcal{H}}\epsilon]|\,\mathbf{R}'\alpha\rangle\,. \end{aligned} \tag{40}$$

Here $\langle \mathbf{R}''\beta| = \int \mathrm{d}\mathbf{x}''\phi^*_\beta(\mathbf{x}'', \mathbf{R}'')\langle \mathbf{R}''|$, and we have broken the time interval up into P slices $\epsilon = (t''-t')/P$ and inserted $P-1$ resolutions of the identity in terms of the classical coordinate position states i.e. $\int \mathrm{d}\mathbf{R}_i|\mathbf{R}_i\rangle\langle\mathbf{R}_i| = 1$. If ϵ and $|\mathbf{R}_j - \mathbf{R}_{j-1}|$ are small, we can substitute the following short-time approximation

$$\langle \mathbf{R}_j|\exp[-\frac{i}{\hbar}\hat{\mathcal{H}}\epsilon]|\mathbf{R}_{j-1}\rangle \simeq \langle \mathbf{R}_j|\exp[-\frac{i}{\hbar}\hat{\mathcal{H}}_0\epsilon]|\mathbf{R}_{j-1}\rangle \exp[-\frac{i}{\hbar}\hat{h}(\mathbf{x}, \mathbf{R}_j)\epsilon] \tag{41}$$

into eqn (40) and obtain the following path-integral expression for the reduced propagator in the limit as $\epsilon \to 0$, thus

$$K_{\beta\alpha}(\mathbf{R}'', t''|\mathbf{R}', t') = \int_{\mathbf{R}',t'}^{\mathbf{R}'',t''} \mathcal{D}[\mathbf{R}(t)]\exp[\frac{i}{\hbar}\mathcal{S}_0[\mathbf{R}(t)]]\,T_{\beta\alpha}[\mathbf{R}(t)] \tag{42}$$

where

$$\begin{aligned} T_{\beta\alpha}[\mathbf{R}(t)] &= \lim_{\epsilon\to 0}\langle\beta|\exp[-\frac{i}{\hbar}\hat{h}(t'')\epsilon]\exp[-\frac{i}{\hbar}\hat{h}(t''-\epsilon)\epsilon] \\ &\quad \ldots\exp[-\frac{i}{\hbar}\hat{h}(t')\epsilon]|\alpha\rangle \end{aligned} \tag{43}$$

and

$$\mathcal{S}_0[\mathbf{R}(t)] = \int_{t'}^{t''} \mathrm{d}t\ \tfrac{1}{2}M\dot{\mathbf{R}}(t)^2\,. \tag{44}$$

Here $\hat{h}(t) = \hat{h}(\mathbf{x}, \mathbf{R}(t))$ is the hamiltonian excluding the kinetic energy of the classical coordinates evaluated with these classical coordinates in the tth configuration. Eqn (42) says that the reduced probability amplitude for the quantum subsystem to make a transition from state α to state β as the classical subsystem moves from $\mathbf{R}'$ to $\mathbf{R}''$ in the time interval $t' \to t''$ is the sum over all classical coordinate paths $\mathbf{R}(t)$ which start at $\mathbf{R}'(t')$ and finish at $\mathbf{R}''(t'')$, each path's contribution to the sum being weighted by the functional weight

$$W[\mathbf{R}(t)] = \exp\left[\frac{i}{\hbar}S_0[\mathbf{R}(t)]\right] T_{\beta\alpha}[\mathbf{R}(t)] \, .$$

The quantity $T_{\beta\alpha}[\mathbf{R}(t)]$ is best interpreted as the overlap of the quantum subsystem wavefunction $\alpha(t,t')$ which is started out at t' in eigenstate $\phi_\alpha(\mathbf{x}, \mathbf{R}'(t'))$, and has been propagated under the influence of the time-dependent hamiltonian according to

$$i\hbar\frac{\partial\alpha(t,t')}{\partial t} = \hat{h}(\mathbf{x}, \mathbf{R}(t))\alpha(t,t') \tag{45}$$

until time t'', with eigenstate $\phi_\beta(\mathbf{x}, \mathbf{R}''(t''))$ at this final time, i.e. $T_{\beta\alpha}[\mathbf{R}(t)] = \langle\phi_\beta(t'')|\alpha(t'',t')\rangle$. Alternatively, due to time reversibility, the same result is obtained if we start in state $\phi_\beta(t'')$, propagate the dynamical wavefunction $\beta(t,t'')$ backwards in time until $t = t'$ and overlap with $\phi_\alpha(\mathbf{x}, \mathbf{R}'(t'))$. In general, $T_{\beta\alpha}[\mathbf{R}(t)] = \langle\beta(t,t'')|\alpha(t,t')\rangle = \langle\beta(t',t'')|\alpha(t',t')\rangle$ independent of t, the time where we choose to evaluate the overlap of the dynamical wavefunction.

Equation (42) is an exact full quantum expression for the reduced propagator. Pechukas proceeds by recognizing that the most important contributions to the reduced propagator come from semiclassical paths for which the phase $\Phi[\mathbf{R}(t)]$ of the weight functional $W[\mathbf{R}(t)] = A[\mathbf{R}(t)]\exp[i\Phi[\mathbf{R}(t)]]$ is stationary with respect to variations in the path, i.e.

$$\frac{\partial\Phi[R(t)]}{\partial\mathbf{R}(t)} = \frac{\partial}{\partial\mathbf{R}(t)}\left\{\frac{S_0[\mathbf{R}(t)]}{\hbar} + \mathrm{Im}\ln T_{\beta\alpha}[\mathbf{R}(t)]\right\} = 0\, . \tag{46}$$

Applying the rules of functional calculus it is readily shown that the semiclassical paths $\bar{\mathbf{R}}(t)$ satisfy the following equation

$$M\ddot{\bar{\mathbf{R}}}(t) = -\mathrm{Re}\left\{\frac{\langle\bar{\beta}(t,t'')|\partial\hat{h}/\partial\bar{\mathbf{R}}(t)|\bar{\alpha}(t,t')\rangle}{\langle\bar{\beta}(t,t'')|\bar{\alpha}(t,t')\rangle}\right\} \tag{47}$$

where the bars above the wavefunctions indicate that they are the solution of the time-dependent Schrödinger equation with hamiltonian $\hat{h}(\mathbf{x}, \bar{\mathbf{R}}(t))$. If

$\hat{h}$ depends on $\bar{\mathbf{R}}$ through the potential $\mathcal{V}(\mathbf{x}, \bar{\mathbf{R}}(t))$, then eqn (47) looks like a Newtonian equation of motion with a quantum force given by

$$\mathbf{F}_{\mathrm{Q}}^{\mathrm{P}}(t) = -\mathrm{Re}\left\{ \frac{\langle \bar{\beta}(t,t'')|\nabla_{\bar{\mathbf{R}}(t)}\mathcal{V}(\mathbf{x}, \bar{\mathbf{R}}(t))|\bar{\alpha}(t,t')\rangle}{\langle \bar{\beta}(t,t'')|\bar{\alpha}(t,t')\rangle} \right\} . \tag{48}$$

In order to determine this force at any instant t, we need to know the entire trajectory $\bar{\mathbf{R}}(t)$, since we must have the time record of the hamiltonian $\hat{h}(\mathbf{x}, \bar{\mathbf{R}}(t))$ in order to compute the forward- and backward-propagated wavefunctions and their overlap at the relevant time. Pechukas shows that the energy and angular momentum are conserved along $\bar{\mathbf{R}}(t)$ with the above choice for the quantum force.

The quantum force in eqn (48) is nonlocal in time, i.e. we need to know the entire time history of the system over the interval $t' \to t''$ in order to determine the quantum force at some intermediate time t in the range $t' < t < t''$. This nonlocality in time leads to the necessity for a self-consistent solution, i.e. guess a trajectory, solve the resulting time-dependent Schrödinger equation, compute $\mathbf{F}_{\mathrm{Q}}^{\mathrm{P}}(t)$ from eqn (48), then solve eqn (47) to obtain a new trajectory with which we start the next iteration cycle.

Pechukas' theory outlined above and Tully's surface-hopping trajectory approach have two central common features. First, both have a way of deciding if a transition in the quantum subsystem is supposed to take place during some time interval. Pechukas' mechanism for answering this question is wrapped up in his transition amplitude $T_{\beta\alpha}[\mathbf{R}(t)]$, while Tully decides on the basis of his model of the ensemble dynamics which is incorporated in the specification of the g_{ik}. The other common feature to these theories is a prescription for the force exerted by the quantal distributions on the classical coordinates, whilst the quantum subsystem is undergoing a transition. Pechukas' quantum force $\mathbf{F}_{\mathrm{Q}}^{\mathrm{P}}(t)$ in eqn (48) is nonlocal in time and conservative as discussed above. Tully's quantum force, on the other hand, is local in time and determined by the instantaneous distribution of the occupied quantum state k as

$$\mathbf{F}_{\mathrm{Q}}^{\mathrm{T}}(t) = -\langle \phi_k|\nabla_{\mathbf{R}(t)}\mathcal{V}|\phi_k\rangle . \tag{49}$$

Tully's quantum force is not conservative during a transition and his dynamics are artificially forced to conserve energy and angular momentum by an intuitive, yet *ad hoc*, velocity adjustment procedure. By its construction, Tully's approach is a time-stepping algorithm. Pechukas' theory, on the other hand, is really a trajectory optimization procedure for refining an entire finite-time-interval trajectory of the classical coordinates.

4.2. IMPLEMENTATION OF PECHUKAS THEORY

In this section we first outline an implementation of Pechukas theory developed by Webster *et al.* which uses a self-consistent iterative approach to advance the nuclear coordinate through a timestep Δt. The problem with this approach is that the mixed-state electronic dynamics is only evolved coherently over a nuclear timestep Δt; beyond this time all electronic coherence is lost with Webster's algorithm. In the second part of this section we present a new implementation of Pechukas theory which evolves the electronic motion coherently for all time scales.

4.2.1. Incoherent self-consistent nonadiabatic dynamics. Webster and coworkers [6] have recently presented a time-stepping algorithm which implements the ideas behind Pechukas' theory. Their approach involves treating each classical timestep as though it was a Pechukas scattering event. Thus instead of optimizing an entire trajectory consisting of many nuclear configurations at subsequent points in the future, Webster *et al.* just optimize the configuration of the nearest-neighbour future time point given the current configuration of the classical variables. In this way, their algorithm preserves a nearest-neighbour nonlocality and, as such, the new configuration must be determined iteratively, but provided the classical timestep Δt is small enough, the algorithm is still rigorously conservative and no *ad hoc* velocity readjustment is necessary.

Webster's algorithm which incorporates Pechukas' ideas as well as aspects of surface hopping can be outlined as follows. We want to evolve the system from t to $t + \Delta t$. The initial nuclear configuration is $\mathbf{R}(t)$ and we work with the adiabatic basis $\{\phi_k(\mathbf{r}, \mathbf{R}(t))\}$ and the adiabatic state energies are E_k. We assume that the potential varies linearly with time over Δt and, for the initial iteration, we suppose that the potential variation over the current time step $t \to t + \Delta t$ is the same as that over the preceding time step i.e. $\Delta \mathcal{V}^{(0)} = \mathcal{V}(\mathbf{r}, \mathbf{R}(t)) - \mathcal{V}(\mathbf{r}, \mathbf{R}(t - \Delta t))$. As well as using numerical integration techniques to solve the expansion-coefficient equations of motion, MD techniques are also employed to integrate the nuclear equations of motion. Due to the fact that the expansion coefficients change very rapidly in the strong-coupling regions, we generally must use a smaller integration time step to solve the expansion-coefficient equations of motion while employing a larger time step to integrate the nuclear trajectories. Thus we break the classical time step up into N_s small time steps $\delta t = \Delta t / N_s$ in order to evolve the quantal degrees of freedom.

With the assumed linear variations of the potential discussed above, we can set up our initial guess for the time-dependent quantal hamiltonian matrix at the lth small timestep as

$$\mathcal{H}_{ij}^{(0)}(t + l\delta t) = E_i\delta_{ij} + l\delta t\langle\phi_i(t)|\Delta\mathcal{V}^{(0)}|\phi_j(t)\rangle/\Delta t\,. \tag{50}$$

Once this time dependence of the hamiltonian is specified, the Schrödinger equation can be integrated forward or backward over the interval Δt from whatever starting distribution is desired. If the quantum system is in eigenstate $\phi_\alpha(\mathbf{r}, \mathbf{R}(t))$ at time t then we can use the linear time-dependent model hamiltonian $\mathcal{H}_{ij}^{(0)}$ to evolve the quantum system to a mixed state $\psi_\alpha^{(0)}(\mathbf{r}, t+\Delta t) = \alpha^{(0)}(t + \Delta t, t)$ at time $t + \Delta t$. Next we diagonalize the linear approximate hamiltonian at $t+\Delta t$ and obtain the eigenstates $\{\phi_\beta^{(0)}(\mathbf{r}, t+\Delta t)\}$ at this time. The transition amplitudes between the initial occupied state α and the various states β at time $t + \Delta t$ are now computed from $T_{\beta\alpha}^{(0)} = \langle\phi_\beta(t+\Delta t)|\psi_\alpha(t+\Delta t)\rangle$. We decide which state will be the occupied state at $t + \Delta t$ by comparing the $|T_{\beta\alpha}^{(0)}|^2$ with a uniform random number $0 < \xi < 1$ as discussed in section 3.5. Webster makes this decision on the basis of only these initial iteration transition amplitudes. Once a final state is chosen, we can now determine the Pechukas force operative at time t from eqn (48) as

$$\mathbf{F}_{\mathsf{Q}}^{\mathsf{P}}(t)^{(0)} = -\mathsf{Re}\left\{\frac{\langle\beta^{(0)}(t, t+\Delta t)|\nabla_{\mathbf{R}}\mathcal{V}(\mathbf{x}, \mathbf{R}(t))|\alpha^{(0)}(t,t)\rangle}{\langle\beta^{(0)}(t, t+\Delta t)|\alpha^{(0)}(t,t)\rangle}\right\} \tag{51}$$

where $\beta^{(0)}(t, t+\Delta t) = \psi_\beta^{(0)}(t)$ is the mixed state which is obtained using the $\mathcal{H}_{ij}^{(0)}$ to reverse propagate eigenstate $\phi_\beta^{(0)}(t+\Delta t)$ back to time t. The Newton-like equations (47) with this force are now integrated for a step Δt with standard MD methods to give the new classical coordinate configuration $\mathbf{R}^{(1)}(t + \Delta t)$. We can now refine our choice of the linear time dependence of the hamiltonian and use the variation of the potential over Δt to define $\Delta\mathcal{V}^{(1)} = \mathcal{V}(\mathbf{R}^{(1)}(t + \Delta t)) - \mathcal{V}(\mathbf{R}(t))$ with the new linear approximation to the time-dependent hamiltonian matrix being

$$\mathcal{H}_{ij}^{(1)}(t + l\delta t) = E_i(t)\delta_{ij} + l\delta t\langle\phi_i(t)|\Delta\mathcal{V}^{(1)}|\phi_j(t)\rangle/\Delta t\,. \tag{52}$$

As before we can propagate our basis states at $\mathbf{R}(t)$ or $\mathbf{R}^{(1)}(t + \Delta t)$ forward or backward in time to give approximate mixed states which can be used in eqn (51) to determine the next iteration of the force $\mathbf{F}_{\mathsf{Q}}^{\mathsf{P}}(t)^{(1)}$, and using this in eqn (47) gives the new iteration of the configuration $\mathbf{R}^{(2)}(t + \Delta t)$. A number of criteria could be used to determine convergence, such as continuing until successive iterations of the configuration are within some tolerance, but Webster suggests an energy conservation criterion in which we continue iterating until the work done by the quantum force $\mathbf{F}_{\mathsf{Q}}^{\mathsf{P}} \cdot \dot{\mathbf{R}}\Delta t$ matches the energy gap $E_\beta(t+\Delta t) - E_\alpha(t)$, so we iterate until $Z_F = |\mathbf{F}_{\mathsf{Q}}^{\mathsf{P}} \cdot \dot{\mathbf{R}}\Delta t + E_\alpha(t) - E_\beta(t + \Delta t)|$ becomes less than some tolerance.

With Webster's algorithm, as outlined above, we see that there are no special tricks to turn off the hopping outside the regions of strong coupling as was necessary with Tully's algorithm and why we chose the hopping probabilities proportional to the components of the velocity of the occupation probabilities. The reason for this difference is that at the end of each classical timestep Webster puts the quantum system into the selected eigenstate and effectively sets this state's occupation probability to 1 and all other state occupation probabilities to zero. Many workers [20–24] have proposed similar solutions to turning off the hopping after the trajectories emerge from regions of strong mixing, but these techniques do not keep track of the dynamical adjustments to the mixing coefficients that are taking place on the classical time scale. During each classical time step they are doing mixed-state propagation and should be accurately representing the changes in the mixing coefficients on this quantum time scale. At the end of each classical dynamical time step, however, they reset the coefficient vector to the appropriate unit vector consistent with the changes in the dynamical wavefunction that have occurred as a result of the propagation from the previous time step's pure state to the end of the current time step's mixed state. Outside the regions of strong coupling, the probabilities of transition to other states will not grow much during Δt so the system will correctly stay in the selected eigenstate rather than undergo many spurious hops to other probable states. With Webster's algorithm the hops are switched off by throwing away information about the relative importance of these other states once we get out of the current time step during which transition to these other states may occur.

The first problem we note with Webster's time-stepping implementation of Pechukas' theory is that, by forcing the system back into an eigenstate at the end of each Δt and only propagating as a mixed state during each Δt, the natural changes in the mixture of states over a long time (many Δt's) are not allowed to take place. With Tully's algorithm, on the other had, we coherently integrate the dynamical mixture of states for all times and as such must achieve the correct statistical distribution of occupied states for all times.

Another related problem with Webster's implementation can be understood by considering the evolution of a two-state system whose wavefunction we represent as a vector of expansion coefficients for the two-state basis

$$\underline{\psi}(t) = \begin{pmatrix} a_1(t) \\ a_2(t) \end{pmatrix} = \begin{pmatrix} 1 \\ 0 \end{pmatrix} \tag{53}$$

where we have chosen the phase appropriately and put the system in state 1 at time t i.e. $|a_1(t)|^2 = 1$. The quantum propagator which evolves this state vector to time $t + \Delta t$ in this two-state basis is the matrix of transition amplitudes, so the evolved wavefunction is

$$\underline{\psi}(t+\Delta t) = \underline{\underline{T}}^{\Delta t}(t)\underline{\psi}(t) = \begin{pmatrix} T_{11}(t) & T_{12}(t) \\ T_{21}(t) & T_{22}(t) \end{pmatrix} \begin{pmatrix} 1 \\ 0 \end{pmatrix} = \begin{pmatrix} T_{11}(t) \\ T_{21}(t) \end{pmatrix} . \tag{54}$$

If we now evolve the quantum system through a second timestep we will have

$$\begin{aligned} \underline{\psi}(t+2\Delta t) &= \underline{\underline{T}}^{\Delta t}(t+\Delta t)\underline{\psi}(t+\Delta t) \\ &= \begin{pmatrix} T_{11}(t+\Delta t) & T_{12}(t+\Delta t) \\ T_{21}(t+\Delta t) & T_{22}(t+\Delta t) \end{pmatrix} \begin{pmatrix} T_{11}(t) \\ T_{21}(t) \end{pmatrix} \\ &= \begin{pmatrix} T_{11}(t+\Delta t)T_{11}(t) + T_{12}(t+\Delta t)T_{21}(t) \\ T_{21}(t+\Delta t)T_{11}(t) + T_{22}(t+\Delta t)T_{21}(t) \end{pmatrix} . \end{aligned} \tag{55}$$

This process can be repeated and the quantum system evolved to long time, and this is entirely equivalent to evolving the expansion-coefficient equations of motion. From the above we see that the probability that the system is in state 1, for example, at time $t + 2\Delta t$ is

$$\begin{aligned} P_1(t+2\Delta t) &= |T_{11}(t+\Delta t)T_{11}(t) + T_{12}(t+\Delta t)T_{21}(t)|^2 \\ &= |T_{11}(t+\Delta t)|^2|T_{11}(t)|^2 + |T_{12}(t+\Delta t)|^2|T_{21}(t)|^2 \\ &\quad +2\mathrm{Re}(T_{11}(t+\Delta t)T_{11}(t)T_{12}^*(t+\Delta t)T_{21}^*(t)) . \end{aligned} \tag{56}$$

The Monte Carlo selection of the occupied state at the end of each timestep with Webster's algorithm can also be modelled in terms of a transition matrix. With this MC procedure, we start out with a probability vector

$$\underline{\tilde{P}} = \begin{pmatrix} 1 \\ 0 \end{pmatrix} . \tag{57}$$

Webster chooses the MC transition matrix elements to be $\tau_{\beta\alpha} = |T_{\beta\alpha}|^2$ so the probability vector after one step is

$$\begin{aligned} \underline{\tilde{P}}(t+\Delta t) &= \underline{\underline{\tau}}^{\Delta t}(t)\underline{\tilde{P}}(t) \\ &= \begin{pmatrix} |T_{11}(t)|^2 & |T_{12}(t)|^2 \\ |T_{12}(t)|^2 & |T_{22}(t)|^2 \end{pmatrix} \begin{pmatrix} 1 \\ 0 \end{pmatrix} \\ &= \begin{pmatrix} |T_{11}(t)|^2 \\ |T_{21}(t)|^2 \end{pmatrix} . \end{aligned} \tag{58}$$

Comparing this with the probability amplitude vector in eqn (54) we see that, for a single step, the occupation probabilities with Webster's MC procedure reproduce the quantum occupation probabilities. After two steps of the MC procedure the probability vector becomes

$$\underline{\tilde{P}}(t+2\Delta t) = \underline{\underline{\tau}}^{\Delta t}(t+\Delta t)\underline{\tilde{P}}(t+\Delta t)$$

$$= \begin{pmatrix} |T_{11}(t+\Delta t)|^2 & |T_{12}(t+\Delta t)|^2 \\ |T_{12}(t+\Delta t)|^2 & |T_{22}(t+\Delta t)|^2 \end{pmatrix} \begin{pmatrix} |T_{11}(t)|^2 \\ |T_{21}(t)|^2 \end{pmatrix}$$

$$= \begin{pmatrix} |T_{11}(t+\Delta t)|^2|T_{11}(t)|^2 + |T_{12}(t+\Delta t)|^2|T_{21}(t)|^2 \\ |T_{21}(t+\Delta t)|^2|T_{11}(t)|^2 + |T_{22}(t+\Delta t)|^2|T_{21}(t)|^2 \end{pmatrix} . \quad (59)$$

Comparing $\underline{P}(t + 2\Delta t)$ with the correct quantum result for the quantum occupation probability at $t + 2\Delta t$ in eqn (56), for example, we see that Webster's MC procedure drops the cross-interference terms like

$$2\mathrm{Re}(T_{11}(t+\Delta t)T_{11}(t)T^*_{12}(t+\Delta t)T^*_{21}(t)) .$$

With Tully's surface-hopping approach we propagate the expansion-coefficient complex amplitudes correctly and compute instantaneous transition probabilities from them as they are needed, so these important interference terms are always correctly represented.

Webster's MC state selection procedure and related techniques can not give the correct occupation probabilities as functions of time because it fails to evolve the state amplitudes coherently. Webster's approach does give coherent dynamics for the quantum subsystem during the classical time step Δt, but beyond this time scale the dynamics is completely incoherent. If the properties of the system are such that the effects of the coherent dynamics of the quantum subsystem are fast and over during a time step, then Webster's algorithm will be useful. In general, however, there is no *a priori* way of knowing the time scale for the effects of coherent quantum dynamics. Further there is certainly no guarantee that the classical coordinate equations of motion can be solved accurately with a time step which is large enough to include all coherent quantum dynamical time scales. Also, Webster's assumed linear time dependence of the Hamiltonian during Δt will only provide an accurate model Hamiltonian with which to integrate the coherent quantum dynamics if Δt is small enough. There is thus a limited window in the size of Δt: it has to be small enough so that the assumption of linear time dependence of the quantal Hamiltonian is accurate, and large enough so that coherent quantum effects occur over a short time compared to Δt.

4.2.2. Phase-coherent, energy-conserving nonadiabatic dynamics. The obvious way around the problems outlined above is to build a hybrid algorithm which combines the long-time coherent integration of the quantal degrees of freedom of Tully's model surface-hopping algorithm together with the "nearest-neighbour-in-time" nonlocal self-consistent adjustment of the nuclear trajectory, to give rigorous energy conservation from Pechukas' theory. Here we sketch how such a hybrid algorithm might be implemented and refer the reader to Ref. [67] for details and test applications.

We use Webster's linear approximation for the assumed time dependence of the quantal hamiltonian during Δt i.e. for our zeroth iteration we use $\mathcal{H}_{ij}^{(0)}$ in eqn (50). We diagonalize this approximate Hamiltonian at each small time step $t + l\delta t$ to obtain approximate eigenstates $\phi_k^{(0)}(t + l\delta t)$ and energies $E_k^{(0)}(t + l\delta t)$ in terms of the basis set of adiabatic basis at time t. These approximate adiabatic energies and states are used to propagate the expansion-coefficient equations of motion in eqn (35) through Δt. At each small time step during this propagation, i.e. at times $t + l\delta t$, we evaluate the hopping probabilities $g_{k \to k'}$ in eqn (36) for this zeroth iteration and we make stochastic state occupation decisions at each small step. We continue this propagation with the linear approximate Hamiltonian and stochastic state selection procedure until time $t + \Delta t$ where some final state has ultimately been chosen. We now propagate the newly selected eigenstate at $t + \Delta t$ back to time t. This is easily accomplished within our instantaneous adiabatic expansion-coefficient representation by simply establishing a unit coefficient vector at time $t + \Delta t$ representing the new pure state. We reverse the direction of time and propagate eqn (35) backwards to time t. The back-propagated coefficient vector $\{b_j^{(0)}\}$ can now be used to compute the zeroth order iteration for the Pechukas force from eqn (51).

If the occupied state at time t is $\phi_k(t)$ and the mixed state at time t obtained by back propagation from the final adiabatic state $\phi_{k'}^{(0)}(t + \Delta t)$ is

$$\psi_{k'}^{(0)}(t) = \sum_j b_j^{(0)}(t)\phi_j(t), \tag{60}$$

then eqn (51) gives the Pechukas force as

$$\mathbf{F}_{\mathsf{Q}}^{\mathsf{P}}(t)^{(0)} = -\mathsf{Re}\left\{ \frac{1}{b_k^{(0)}(t)} \sum_j b_j^{(0)}(t) \langle \phi_j(t) | \nabla_{\mathbf{R}} \mathcal{V}(\mathbf{x}, \mathbf{R}(t)) | \phi_k(t) \rangle \right\} . \tag{61}$$

We evolve the nuclei under the influence of this force, according to eqn (47), to configuration $\mathbf{R}^{(1)}(t+\Delta t)$ and get a new linear approximation for the time dependence of the Hamiltonian $\mathcal{H}_{ij}^{(1)}$ as in eqn (52), and iterate the above steps to convergence. Note that we don't recompute the hopping probabilities and perform a new state selection at each iteration. This is only done at the zeroth iteration as suggested by Webster.

Because we continue the forward propagation of the expansion coefficients coherently from one Δt to the next and we use the $g_{k \to k'}$'s as our hopping probabilities, the coherence problems of Webster's implementation are avoided and the self-consistent iteration procedure generates conservative dynamics.

An alternative to the above procedure, which accounts for the possibility of multiple hops during each Δt, involves prematurely halting the classical propagation at the point where the first small-step hop takes place, i.e. $t+l\delta t$ say. We then use the linear approximate hamiltonian to back propagate from this point and iterate the Pechukas force to convergence as discussed above. This procedure gives an algorithm with a self-adjusting nuclear timestep [67].

5. Test problems and applications

5.1. TESTS OF SURFACE-HOPPING DYNAMICS METHODS FOR EXAMPLE ONE-DIMENSIONAL TWO-STATE PROBLEMS

In this section we follow Tully, and check the performance of the surface-hopping algorithms developed in sections 3 and 4 by comparing the results of calculations using these methods with results from exact full quantum calculations, in which both the electronic and nuclear parts of the problem are treated quantum mechanically for some simple one-dimensional, two-state examples.

To begin, we return to the central result of section 2.3, i.e. eqn (13), in which we formally decoupled the electronic and nuclear degrees of freedom. We can rewrite this equation in a matrix formulation as follows

$$\left[\underline{\underline{\hat{\mathcal{K}}}}_R + \underline{\underline{\hat{\mathcal{V}}}}\right] \underline{\chi} = \underline{\underline{\hat{\mathcal{H}}}}\,\underline{\chi} = \mathcal{E}\underline{\chi}\,. \tag{62}$$

The nuclear kinetic operator matrix is diagonal and in one dimension

$$\underline{\underline{\hat{\mathcal{K}}}}_{kk'} = \frac{\hbar^2}{2m}\frac{\partial^2}{\partial R^2}\delta_{kk'}$$

whilst the potential operator matrix elements are $\underline{\underline{\hat{\mathcal{V}}}}_{kk'} = E_{kk'} + \hat{C}_{kk'}$. $\underline{\chi}$ is a vector of coefficients for the different electronic basis functions and these coefficients are themselves functions of nuclear coordinates giving the full electron-nuclear wavefunction as in eqn (11).

The classical trajectories which move over the coupled electronic surfaces in our surface-hopping trajectory calculations correspond to time evolving mixtures of the nuclear eigenstates which are the solutions of eqn (62). These dynamical nuclear mixed states are the wavepackets which satisfy the time-dependent Schrödinger matrix equation,

$$i\hbar\frac{\partial\underline{\chi}}{\partial t} = \underline{\underline{\hat{\mathcal{H}}}}\,\underline{\chi} \tag{63}$$

which for a one-dimensional, two-state problem has the form

$$i\hbar\frac{\partial}{\partial t}\begin{pmatrix}\chi_1(R,t)\\ \chi_2(R,t)\end{pmatrix} = \left[\begin{pmatrix}\hat{\mathcal{K}}_R & 0\\ 0 & \hat{\mathcal{K}}_R\end{pmatrix} + \begin{pmatrix}\mathcal{V}_{11}(R) & \mathcal{V}_{12}(R)\\ \mathcal{V}_{21}(R) & \mathcal{V}_{22}(R)\end{pmatrix}\right]\begin{pmatrix}\chi_1(R,t)\\ \chi_2(R,t)\end{pmatrix} \tag{64}$$

where

$$\hat{\mathcal{K}}_R = -\frac{\hbar^2}{2m}\frac{\partial^2}{\partial R^2}.$$

Equation (64) is exact, and all we need do to test the surface-hopping methods is set up a model potential matrix $\underline{\underline{\mathcal{V}}}$ and some initial wavepacket vector $\underline{\chi}$ which looks like a classical particle, and propagate it according to eqn (64) over the coupled electronic surfaces represented by $\underline{\underline{\mathcal{V}}}$. It is most convenient to specify the model potential matrix in the diabatic representation by just choosing model functions $\mathcal{V}_{ij}(R)$. For each value of R, diagonalization of $\underline{\underline{\mathcal{V}}}$ gives eigenvalues $E_k(R)$ which are the adiabatic potential energy surfaces which we need to compute the classical forces $\partial E_k/\partial R$, and a matrix of eigenvectors

$$\underline{\underline{P}} = \begin{pmatrix} P_{11} & P_{12} \\ P_{21} & P_{22} \end{pmatrix} = (\underline{\psi}_1, \underline{\psi}_2) \tag{65}$$

(as in eqn (3)) which we use to compute the coupling matrix elements

$$D_{12} = \langle\psi_1|\frac{\partial}{\partial R}|\psi_2\rangle = \underline{\psi}_1^* \cdot \frac{\partial}{\partial R}\underline{\psi}_2 = P_{11}^*\frac{\mathrm{d}P_{12}}{\mathrm{d}R} + P_{21}^*\frac{\mathrm{d}P_{22}}{\mathrm{d}R}. \tag{66}$$

Note that the electronic integrals become a simple dot product or sum in our two-state basis. All the quantities necessary to set up the expansion coefficients and nuclear equations of motion for the surface-hopping trajectory calculations are thus available and these equations can be readily integrated numerically given some initial condition.

The full quantum wavepacket propagation calculations are also performed numerically by fast Fourier transform (FFT) techniques [68–70]. With this approach we establish the initial shapes of the ground- and excited-state nuclear wavepackets on a discrete grid of R values. We diagonalize the potential matrix at each point on the grid thus $\underline{\underline{\mathcal{V}}}(R) = \underline{\underline{P}}^{\mathrm{T}}(R)\underline{\underline{E}}(R)\underline{\underline{P}}(R)$ where $\underline{\underline{E}}(R)$ is the diagonal matrix of adiabatic state energies at R. To evolve the wavepacket vector

$$\underline{\chi}(t) = \begin{pmatrix} \chi_1(R,t) \\ \chi_2(R,t) \end{pmatrix} \tag{67}$$

from $t \to t + \Delta t$ we apply a short-time approximation to the two-state quantum propagator as follows:

$$\begin{pmatrix} \chi_1(R,t+\Delta t) \\ \chi_2(R,t+\Delta t) \end{pmatrix} = \underline{\underline{M}}(R)(\mathrm{FFT})^{-1}\exp[-\frac{i}{\hbar}\underline{\underline{\mathcal{K}}}\Delta t](\mathrm{FFT})\underline{\underline{M}}(R)\begin{pmatrix} \chi_1(R,t) \\ \chi_2(R,t) \end{pmatrix} \tag{68}$$

In this expression,

$$\underline{\underline{M}}(R) = \underline{\underline{P}}^{\mathrm{T}}(R) \exp[-\frac{i}{2\hbar}\underline{\underline{E}}(R)\Delta t]\underline{\underline{P}}(R)$$

is the potential part of the propagator which we apply by transforming to the adiabatic representation. The elements of the vector $\underline{\chi}$ are the nuclear wavefunctions on the diabatic surfaces which are transformed to the wavefunctions on the adiabatic surfaces $\underline{\psi}$ by

$$\underline{\psi}(R,t) = \underline{\underline{P}}(R)\underline{\chi}(R,t)\,. \tag{69}$$

We apply the nuclear kinetic operator by Fourier transforming the nuclear wavefunctions on the diabatic surfaces to k-space where the kinetic operator becomes diagonal and

$$\mathcal{K}_{ij} = \frac{\hbar^2 k^2}{2m}\delta_{ij}\,.$$

The split-operator expression for the short-time propagator in eqn (68) is accurate to order $(\Delta t)^3$.

In all the examples we present below, the mass $m = 2000$ au and the initial condition for the surface-hopping trajectory ensemble is a δ-distribution in both position and momentum space in a pure electronic state, i.e. all ensemble members are started at the same point with the same velocity on the jth electronic surface so we choose $a_j^{(i)} = 1$ for all i. In our wavepacket calculations we model this initial condition for the classical trajectories as a narrow Gaussian distribution centred at the same point in phase space as the initial classical ensemble. Thus we make the jth element of the vector $\underline{\chi}$

$$\chi_j(R,0) = \exp[ik(R-R_a)]\exp\left[-\frac{(R-R_a)^2}{\sigma^2}\right] \tag{70}$$

and all other elements zero for a classical ensemble starting at R_a with momentum $P = \hbar k$. In all our examples the potentials are such that $\sigma = 1$ au gives a wavepacket which is initially localized on the length scale of the potential variation and we choose this value for the width parameter in all our calculations.

We have used the methods discussed above to propagate wavepackets and model surface-hopping trajectory ensembles using the algorithm presented in section 3.5 for simple 1D two-state example problems depicted in Figs. 8 and 9. Tully [11] used these examples to test how his fewest-switches surface-hopping trajectory algorithm performed in various situations.

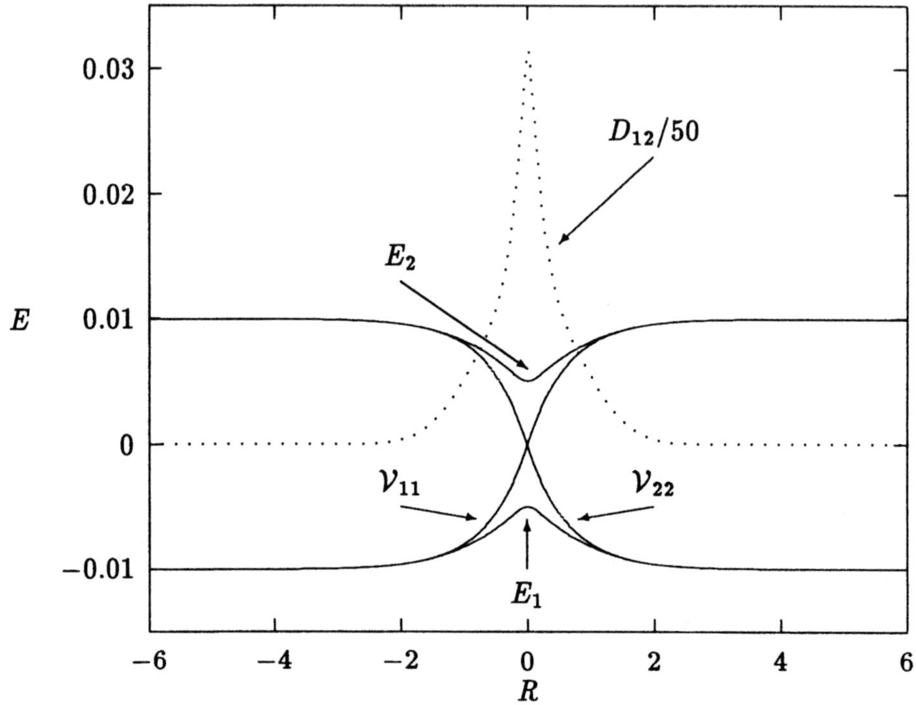

$$
\begin{aligned}
\mathcal{V}_{11}(R) &= A[1 - \mathrm{e}^{-BR}] \quad R > 0 \\
\mathcal{V}_{11}(R) &= -A[1 - \mathrm{e}^{-BR}] \quad R < 0 \\
\mathcal{V}_{22}(R) &= -\mathcal{V}_{11}(R) \\
\mathcal{V}_{12}(R) &= \mathcal{V}_{21}(R) = C\mathrm{e}^{-DR^2}
\end{aligned}
$$

Fig. 8. Model surfaces of simple avoided-crossing example. E_i are the adiabatic energy curves and $\mathcal{V}_{ii}$ are the diabatic potential curves. D_{12} is the nonadiabatic coupling vector for this problem.

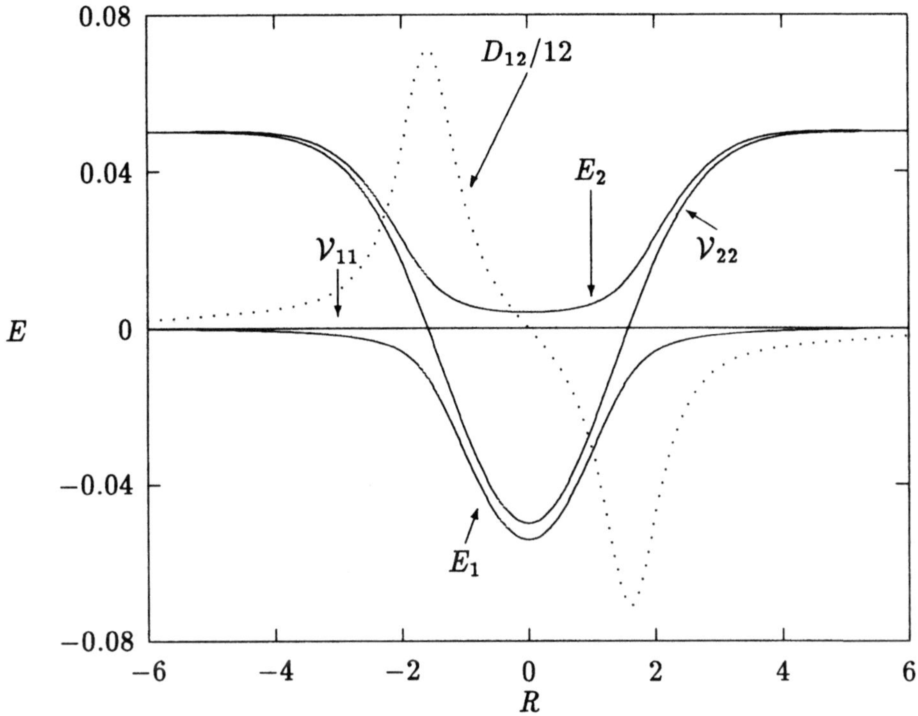

$$
\begin{aligned}
\mathcal{V}_{11}(R) &= 0 \\
\mathcal{V}_{22}(R) &= -A\mathrm{e}^{-BR^2} + E_0 \\
\mathcal{V}_{12}(R) &= \mathcal{V}_{21}(R) = C\mathrm{e}^{-DR^2}
\end{aligned}
$$

Fig. 9. Model surfaces of successive avoided-crossing example. E_i are the adiabatic energy curves and $\mathcal{V}_{ii}$ are the diabatic potential curves. D_{12} is the nonadiabatic coupling vector for this problem.

5.1.1. Simple avoided crossing. The variation of the potential matrix with nuclear coordinate for the first of these examples is presented in Fig. 8. In this example, the diabatic surfaces (diagonal elements of $\underline{\underline{V}}$) cross at the origin and a Gaussian off-diagonal potential form, centred at this crossing point, is assumed. The resulting adiabatic curves repel one another in the strong-coupling region ($R_0 = 0$) giving rise to a barrier in the ground state and a well in the excited state. The coupling matrix element $D_{12}(R)$ is strongly peaked at R_0.

In Fig. 10 we present results for the dynamical probability density computed with both wavepacket propagation (square amplitude of nuclear wavefunctions on ground and excited adiabatic electronic surfaces $|\psi_i(R)|^2$) and fewest-switches surface-hopping trajectory calculations (histograms of dynamical ensemble density on each surface). We started both the wavepacket and trajectory swarm far to the left of R_0 in the asymptotic region and heading towards the coupling region with high momentum ($k = 15$ au).

As the packet passes through the coupling region (follow the curves from left to right through $R_0 = 0$) we see excited-state density grow in as the ground-state density is depleted. We see that the distribution of the trajectory swarm remains very narrow as the ensemble passes through the interaction region and the mean position of the swarm tracks the position of the wavepacket centre very closely on both surfaces. The excited-state swarm propagates at a slower velocity than the swarm on the ground state, due to the fact that nuclear kinetic energy is removed when the ensemble members make the transition up to the excited electronic state. Just after the swarm leaves the region around $R = 0$, there are slightly too many trajectories on the excited state. Some of these excited-state trajectories subsequently fall down to the ground state again but they remain slightly delayed from the rest of the ground-state swarm due to the finite time they have spent travelling more slowly over the excited-state surface.

Despite the dramatic differences in the widths of the wavepackets as compared to the narrow spatial extent of the swarms, the time dependence of the transmission and reflection probabilities computed for the full quantum and semiclassical calculations presented in Fig. 11 are in good qualitative agreement. These quantities are computed by integrating the wavepacket density over the regions to the left or right of the origin or simply counting trajectories in these regions for the surface-hopping results. The quantum results change more gradually than the abrupt variation of the semiclassical results due to the differences in widths of the distributions. The overall time scale and asymptotic values, however, are in excellent agreement for this high-momentum collision.

Next we consider propagation, in this simple avoided-crossing example, at a decreased momentum ($k = 8.2$ au) chosen so that the total energy is just below the dissociation threshold of the well on the upper-state surface.

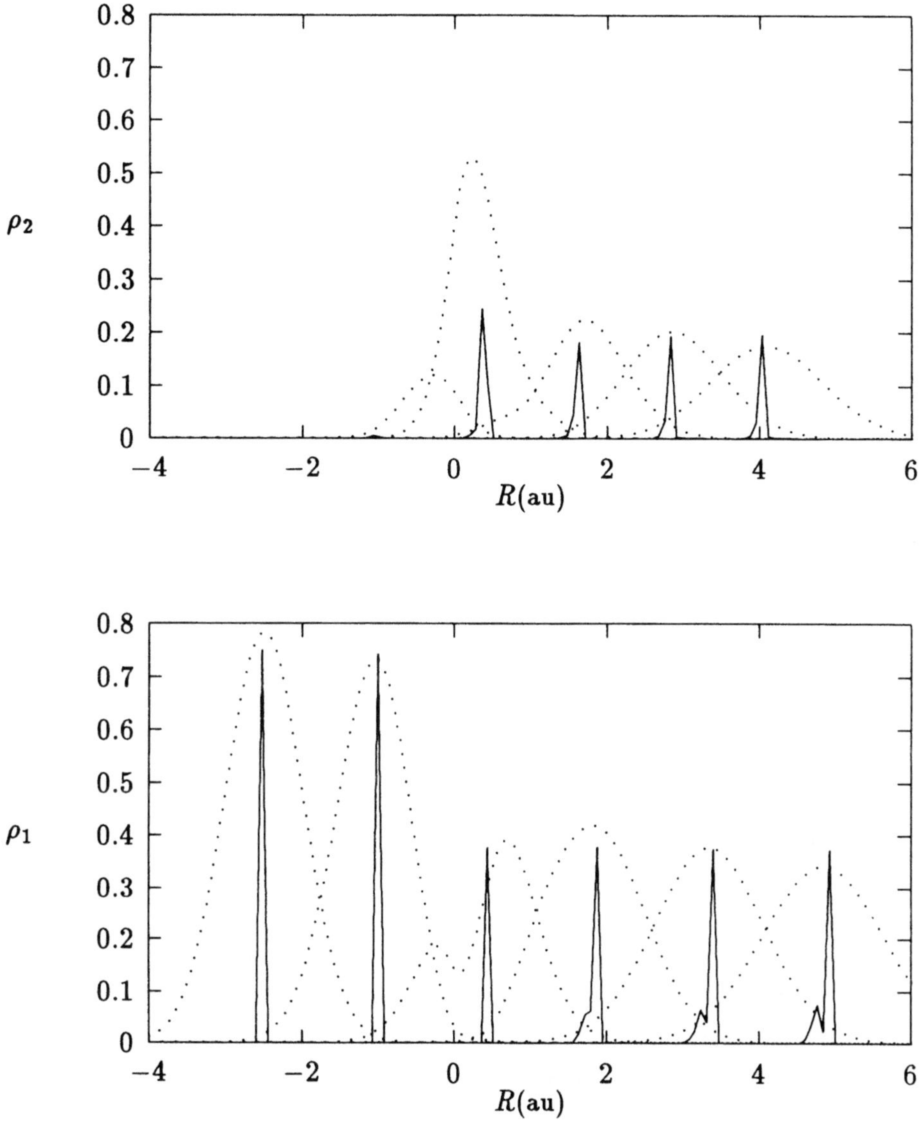

Fig. 10. Dynamical quantum and semiclassical distributions for simple avoided-crossing example ($k = 15\,\text{au}$, $\mathcal{K} = .0563\,\text{au}$). Dotted lines are full quantum wavepacket propagation results, solid curves are surface-hopping trajectory swarm density. Distributions on ground- and excited-state surfaces, ρ_1, ρ_2, are presented. $R(0) = -4\,\text{au}$, $a_1(0) = 1$, $a_2(0) = 0$. Packets and trajectories move from left to right as time evolves.

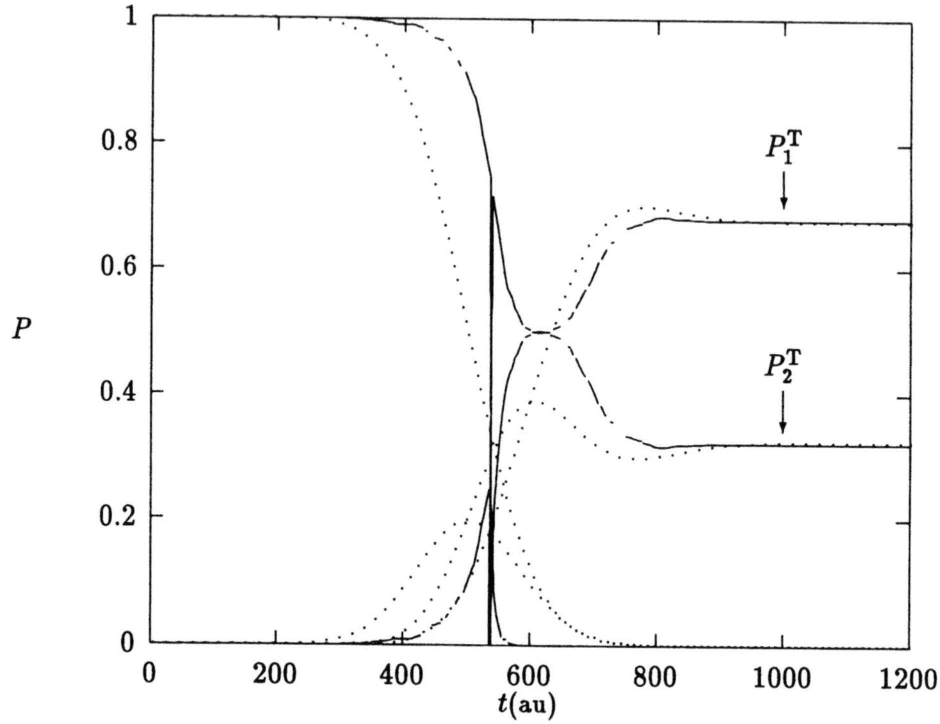

Fig. 11. Transmission and reflection probabilities as functions of time for the two-state simple avoided-crossing problem with high momentum. Wavepacket propagation (dotted lines) and classical trajectory calculations (solid lines).

Trajectories promoted to the excited state will thus oscillate back and forth in the excited-state well and stay in the strong-coupling region until all excited-state density eventually relaxes back to the ground state.

The early time evolution involves the swarm/packet impinging on the coupling region and some density having been promoted to the excited state. The swarm rolls up the right hand side of the excited-state well and the hopping turns off as the trajectories move away from the origin. In Fig. 12 we see the later time behaviour of the classical and quantum probability densities as functions of time. The upper-state swarm rolls back down towards the bottom of the excited-state well. When the excited-state swarm reaches the strong-coupling region again it is travelling backwards and the trajectories which drop back down onto the ground state start heading off with negative velocity as a reflected pulse on the ground-state surface. The position of the excited-state swarm is always localized between the classi-

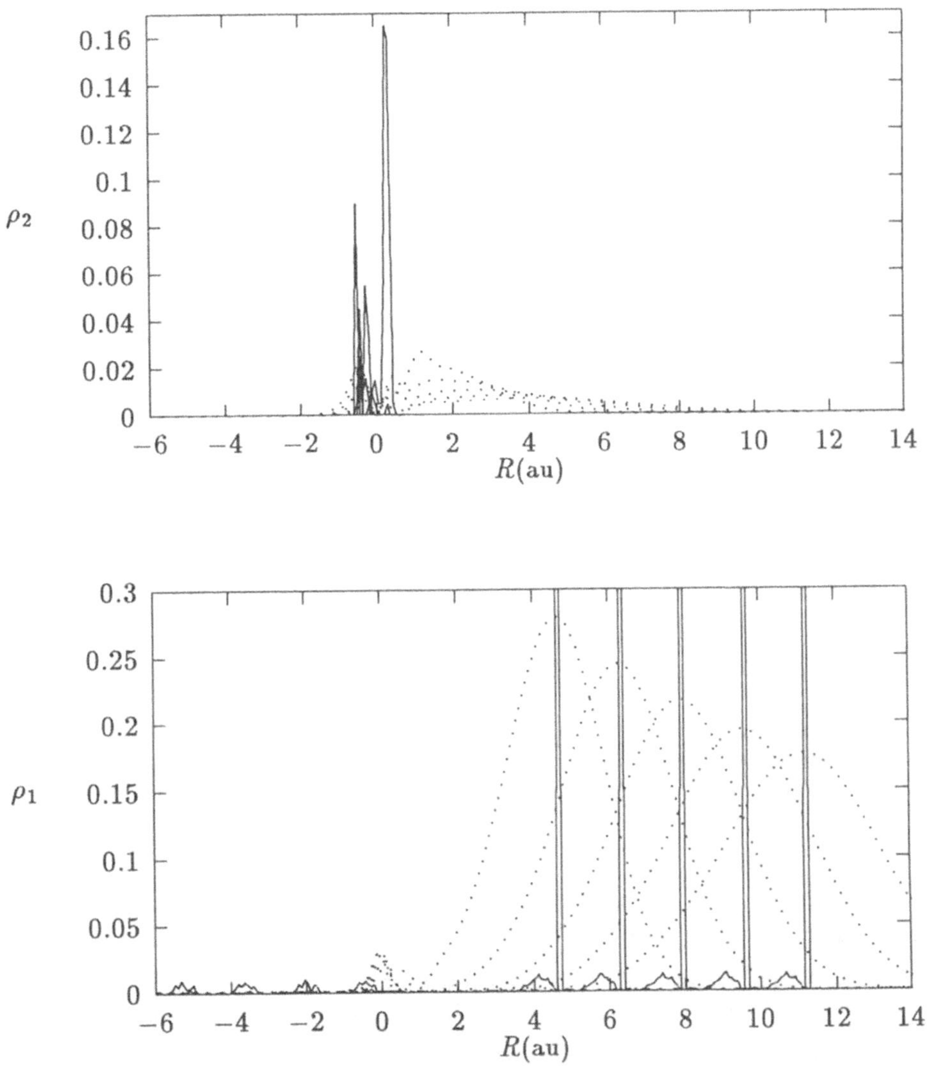

Fig. 12. Dynamical quantum and semiclassical distributions for simple avoided-crossing example (k = 8.2 au, $\mathcal{K}$ = .0168 au). Results for $t = 2000 \rightarrow 4000$ au.

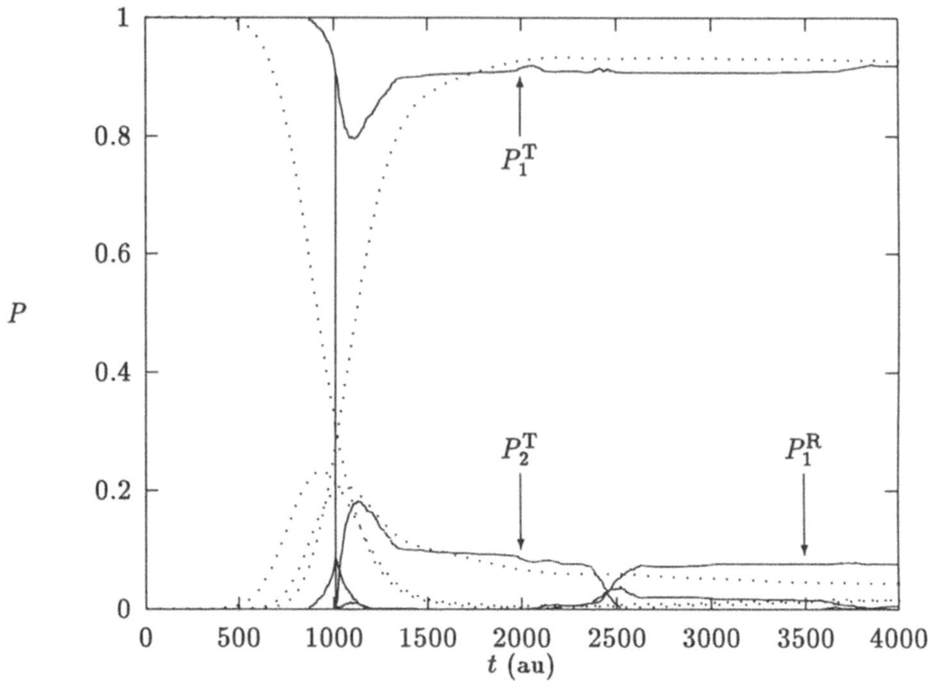

Fig. 13. Transmission and reflection probabilities as functions of time for the two-state simple avoided-crossing problem with low momentum. Wavepacket propagation (dotted lines) and classical trajectory calculations (solid lines).

cal turning points of the motion in the excited-state well. The excited-state wavepacket, however, tunnels deep into the side of the excited-state well.

Figure 13 compares the transmission and reflection probabilities as functions of time. The asymptotic values of the semiclassical and quantum ground-state transmission probability agree very well. At around $t = 2500$ au we see the effect of the return passage of the excited-state swarm through the coupling region. P_2^T falls to zero while P_1^R rises to its asymptotic value associated with back reflected pulse dropping down from the excited state. The quantum result for P_2^T (dotted line just below the solid classical curve) decays very slowly to zero because of the broad excited-state tunnelling packet which has very little density in the strong-coupling region. It will eventually decay to zero and the quantum wavepacket result for the ground-state reflection probability will grow until it is close to the classical P_1^R result. The time scales for these changes are longer due to the widths of the wavepacket compared with the trajectory swarm distribution and due to the strong quantum tunnelling of the excited-state wavepacket.

5.1.2. Successive avoided crossings. Model potential curves exhibiting a dual avoided crossing are displayed in Fig. 9. Dynamical probability density results for an initial momentum of $k = 30\,\mathrm{au}$ are presented in Fig. 14 for this potential. When the swarm encounters the first coupling region at $R \sim -2\,\mathrm{au}$ we see a small amount of density couple onto the excited state. When the ground-state swarm encounters the second coupling region at $R = 2\,\mathrm{au}$, a larger amount of density is coupled onto the excited-state surface and it hops up in front of the first coupled swarm component which was slowed down because of the kinetic energy taken from the nuclei during the first upward nonadiabatic electronic transition.

In Fig. 15 we plot the transmission and reflection probabilities as functions of time. If we follow the semiclassical curves from reflection to transmission probabilities through the break where the swarm crosses the origin, the resulting continuous curves show how the diagonal elements of the electronic density matrix, averaged over the swarm, vary with time. During the first strong-coupling encounter, occurring at a time of a little over $t = 400\,\mathrm{au}$, the ground-state occupation probability drops by 20%, consistent with the small number of trajectories hopping to the excited state initially. The second coupling region is encountered at a little over 600 au where the ground-state occupation probability falls by a further $\sim 50\%$. The numbers of trajectories, on either surface, subsequently show damped oscillation to asymptotic values of the transmission coefficients which agree remarkably well with the wavepacket propagation results.

The average frequency of the population oscillations corresponds roughly to the mean energy gap in the strong-coupling region. When the time spent between strong-coupling encounters exactly matches an integer multiple of mean oscillations, the expansion coefficients and their velocities at the second encounter will be very close to the same values they had at the first encounter. If the coupling vector is opposite in sign, the same number of ensemble members which made upward transitions at the first encounter will fall back to the ground state at the second encounter. On the other hand, if the time spent between encounters exactly matches a half integer multiple of mean coefficients oscillations, the opposite signed coefficient velocities combined with the opposite sign of $\mathbf{D}_{12}$ will result in repeated upward transitions for both encounters leading to the net depletion in ground-state population observed in Figs. 14 and 15. Thus we see constructive interference between strong-coupling encounters at the momentum in this example. If we increase the momentum to $k = 52\,\mathrm{au}$ (see results in Fig. 16) we see the destructive interference situation where all the downward transitions at the second encounter almost completely cancel all the upward transitions at the first, and

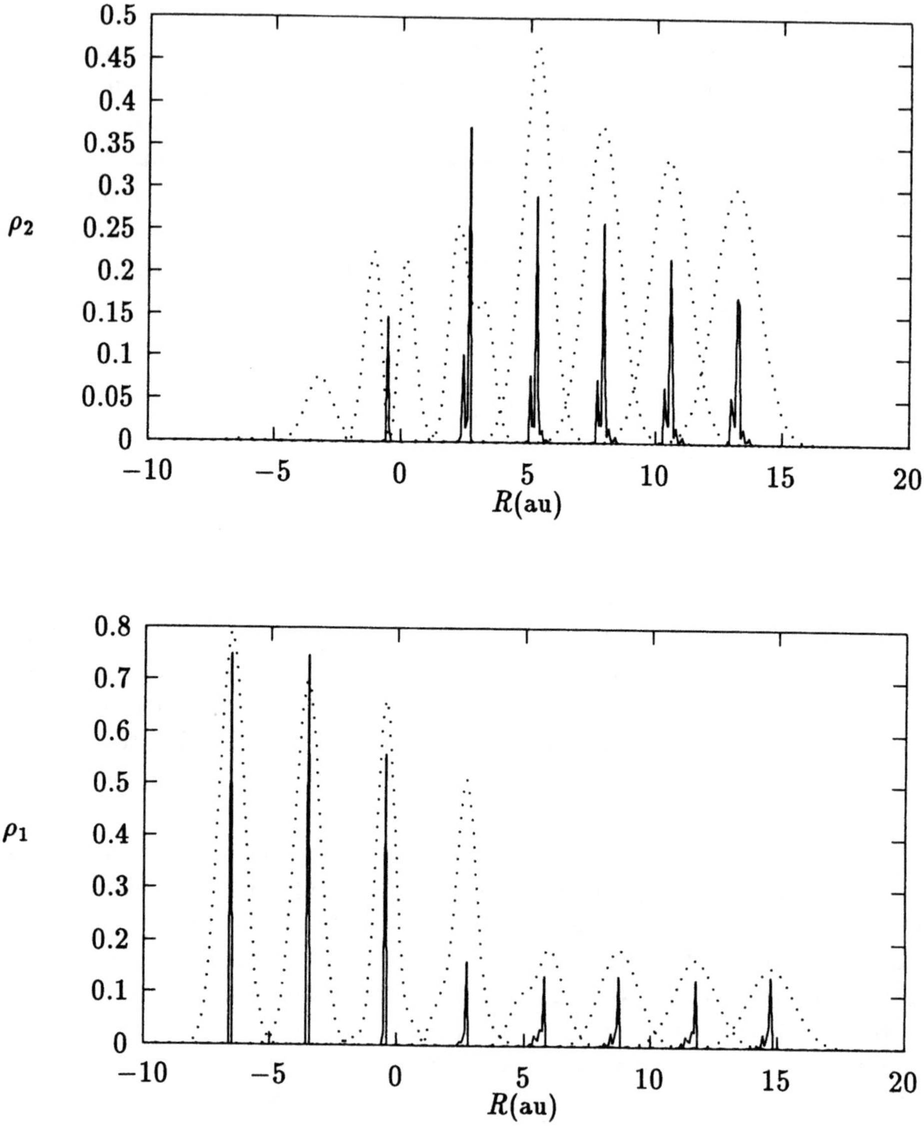

Fig. 14. Dynamical quantum and semiclassical distributions for successive avoided-crossing example ($k = 30\,\text{au}$, $\mathcal{K} = .225\,\text{au}$). Dotted lines are full quantum wavepacket propagation results, solid curves are surface-hopping trajectory swarm density. Distributions on ground- and excited-state surfaces, ρ_1, ρ_2, are presented. $R(0) = -8\,\text{au}$, $a_1(0) = 1$, $a_2(0) = 0$.

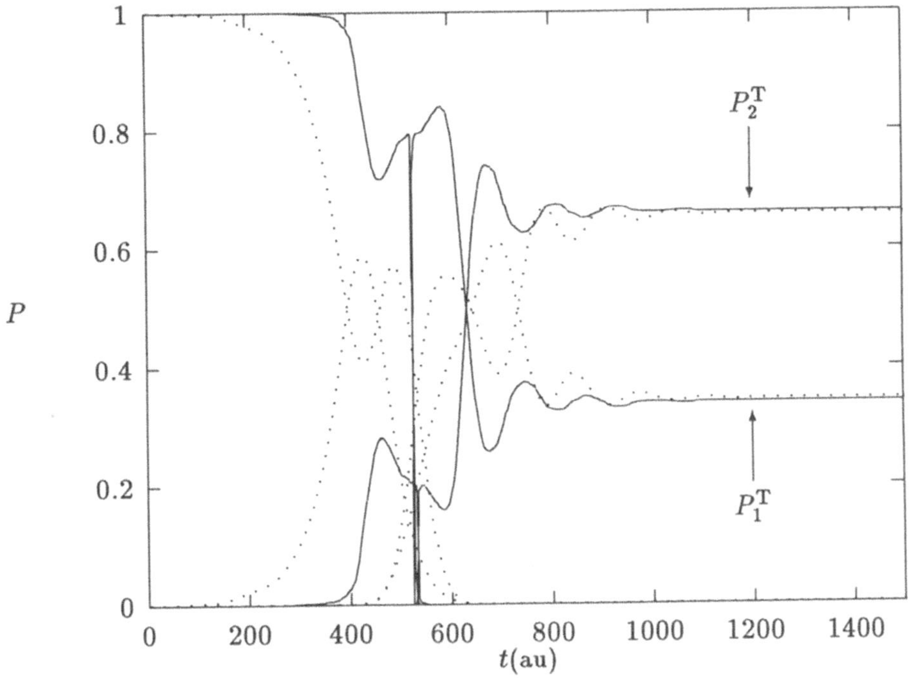

Fig. 15. Transmission and reflection probabilities as functions of time for the two-state successive avoided-crossing problem with high momentum $k = 30$ au. Wavepacket propagation (dotted lines) and classical trajectory calculations (solid lines).

there is no net effect of passing through the coupling region. At this high momentum, the coefficients can oscillate exactly once between encounters.

The phenomenon discussed above is the mechanism behind the Stükelberg oscillations in the branching ratio between the states as a function of collision momentum.

There are some circumstances where the fewest-switches surface-hopping trajectory method fails to reproduce the qualitative behaviour of the full quantum wavepacket results. These problems generally arise at low collision energies where quantum effects are expected to play a more significant role for the nuclear motion, and where the approximations due to the *ad hoc* velocity readjustment energy redistribution procedure are expected to be more severe.

With the self-consistent trajectory methods, electronic transition energy is redistributed into both the nuclear kinetic energy (velocity readjustment)

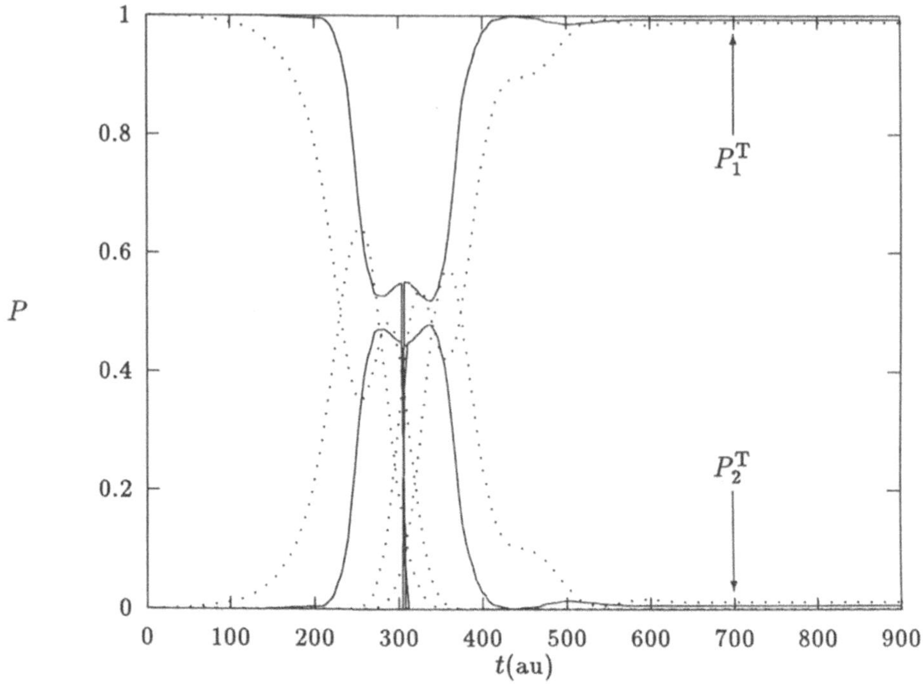

Fig. 16. Transmission and reflection probabilities as functions of time for the two-state successive avoided-crossing problem with very high momentum (k = 52 au, $\mathcal{K}$ = .676 au). Wavepacket propagation (dotted lines) and classical trajectory calculations (solid lines).

and the potential energy, so that both the positions and velocities at each time are altered to redistribute the electronic transition energy. With the *ad hoc* velocity readjustment procedure, on the other hand, all the energy is redistributed to the classical particle velocities which, as a result, will either slow down or speed up by excessive amounts, as this approach does not allow the electronic transition energy to be dumped into other energy components.

If these velocity errors are comparable in magnitude to the collision velocity, then the *ad hoc* velocity readjustment algorithm may give rise to qualitative disagreement between wavepacket and trajectory swarm results. An illustration of these problems for a low-momentum collision ($k = 13$ au) on the dual crossing surfaces is presented in Fig. 17. The energy of this trajectory is just below the dissociation threshold of the excited-state well. Thus in the long-time limit any transient excited-state density will roll back into the excited-state well and eventually be coupled back down to the ground state, appearing as either transmitted or reflected ground-state density. The momentum of the trajectory is large enough so that amplitude of the oscillatory motion in the excited-state well takes the trajectory swarm back and forth between the two strong-coupling regions several times. The excited state thus behaves as a "leaky bucket" of trajectories with two holes, one at either end. After the ground-state swarm passes through the interaction region at the right momentum so as to establish some excited-state density, the excited-state swarm behaves like a wave which sloshes back and forth over the two holes. The quantum phase coherence can be thought of in terms of the holes opening and closing in concert with one another. The dynamical evolution of the swarms in this case is qualitatively similar to that displayed in Figs. 12, 13 where there was only a single leak in the excited-state potential.

As discussed above, the interplay between the classical nuclear velocity and the phase of the expansion coefficient oscillations has a crucial influence on how much density branches onto the other state during each encounter with the strong-coupling regions. Due to its nature as a phase interference effect, this branching is extremely sensitive to the nuclear velocity and, if the errors in the velocity due to the *ad hoc* velocity readjustment become sizeable, significant disagreements between the full quantum wavepacket propagation and surface-hopping trajectory calculations are expected. The asymptotic values of the low-momentum trajectory transmission and reflection probabilities for the semiclassical calculations presented in Fig. 17 indicate that about 10% of the density is reflected on the ground state. The wavepacket calculations, however, give that all the density for this particular situation is in fact transmitted on the ground state with negligible reflected component. From the transmission and reflection probabilities in Fig. 17, we see the recycling of trajectories back and forth in the excited-state well and their decay back to the ground state. The rise of P_1^{R} at just before $t = 3000\,\mathrm{au}$ is associated with returning trajectories on the excited state spilling through the left-hand hole as a small backward-travelling pulse on the ground state. This occurs because of the special phase relationship between the expansion-coefficient oscillations and the velocity of the trajectory. When trajectories make the upward transition they are excessively slowed down by the velocity readjustment algorithm, because we insist on taking all the transition energy from the nuclear kinetic energy. This alters the phase relationship between the coefficient oscillations and the speed of the trajectory, resulting in the spurious pulse of reverse-travelling trajectories on the ground state. With the self-consistent trajectory methods, however, the trajectory velocity and position will be adjusted correctly during the transition so as to give the right phase relation i.e. the left hole will be closed when the swarm returns with the correct velocity, preventing the back-reflected

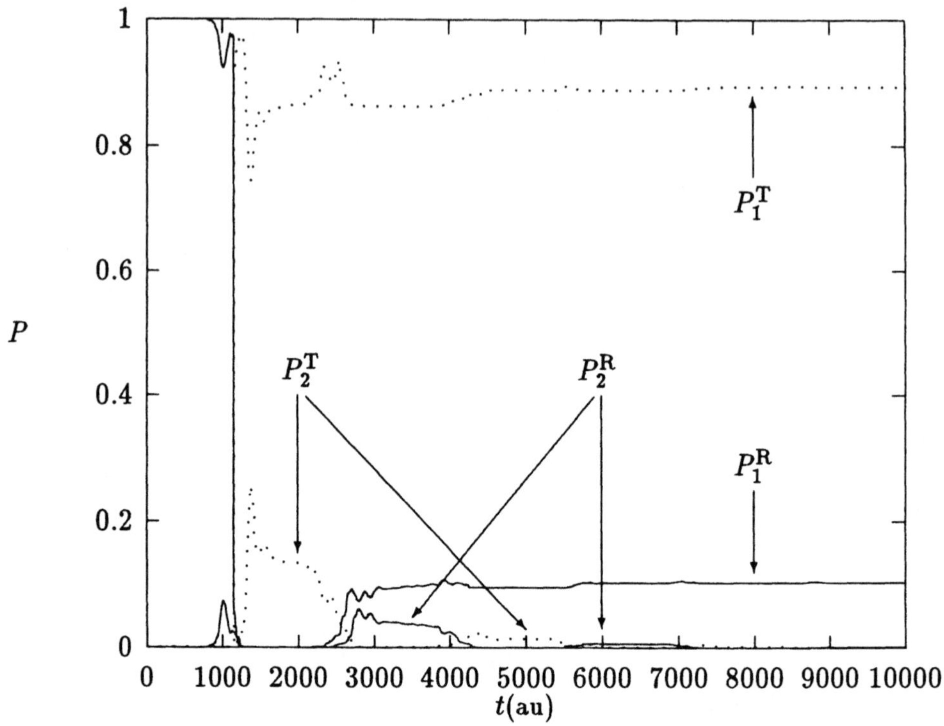

Fig. 17. Dynamical quantum and semiclassical distributions for successive avoided-crossing example at low momentum (k = 13 au, $\mathcal{K}$ = .0423 au). Only the classical trajectory transmission probabilities (dotted lines) and reflection probabilities (solid lines) are shown.

leakage [67].

5.2. APPLICATIONS: RELAXATION OF EXCITED EXCESS ELECTRONS IN LIQUIDS

5.2.1. Dynamical assumptions and techniques. The nonadiabatic dynamics algorithms presented in sections 3 and 4 have recently been used in a number of condensed-phase applications probing the relaxation of excess electrons in fluids. In these applications, the excess electron is treated quantum mechanically and the solvent degrees of freedom are described classically. The assumption here is that the low-energy excess electron can be singled out from all the molecular electrons for special treatment because its energy is so much smaller than the energies of the valence electrons, and it thus moves so much slower than these particles. We assume, however, that the excess electron still moves much faster than the nuclear coordinates, so that we can make a Born-Oppenheimer separation for the nuclear and excess electronic time scales.

The excess electron thus interacts with the average charge distribution of the valence electrons through pseudopotentials [71] which are constructed under the assumption that the valence electron distribution equilibrates instantly to the current position of the excess electron. These pseudopotentials include terms which model polarization of the molecular charge distribution, orthogonality of the excess electronic wavefunction with those of the valence electrons, Coulomb interactions, and the effects of exchange between the excess electron and the valence electrons of the target molecules. In this way we assume that the potential surface over which the nuclei move can be constructed from the usual intermolecular potential arising from the interactions of the ground-state molecular charge distributions, together with the excess electron-molecule interaction energy obtained by averaging over the charge distribution of the currently occupied excess electronic eigenstate.

With the dynamical assumptions outlined above, all that is needed to implement the various nonadiabatic dynamics methods described in this Chapter is a way to compute the instantaneous adiabatic basis set of eigenstates of the effective one-electron hamiltonian provided by each configuration of the solvent molecules along the trajectory. Webster *et al.* [6] have developed an efficient split-operator block Lanczos [72] technique for computing these instantaneous adiabatic eigenstates, which we use as a basis set to expand the time-dependent wavefunction of the excess electron. With these techniques, the eigenstates are computed on a regular grid in three dimensional space, although other-shaped grids have been recently explored [73].

Once the adiabatic basis at time t is computed, the integrals we need are obtained as sums over the grid points. The central quantities we must compute are the matrix elements of the potential at different configurations from which we compute $\langle\phi_i(t)|\Delta\mathcal{V}|\phi_j(t)\rangle$ (see eqns (50) and (52)) and matrix elements of the force, $-\langle\phi_i(t)|\nabla_{\mathbf{R}}\mathcal{V}|\phi_j(t)\rangle$. These force matrix elements can be used to compute the electron-molecule forces for eqn (49) or eqn (61) as well as the nonadiabatic coupling vector from eqn (33). Diagonalization of the approximate linear time-dependent Hamiltonian in eqns (50) or (52) gives eigenvalues $E_i(k)$ and eigenvectors $\underline{\alpha}_i(k)$ at electronic time step k. These eigenvectors are the expansion coefficients which give the electronic eigenstates at time $t + k\delta t$ in terms of the basis states at time t. These eigenvectors and eigenvalues can be used together with a finite difference approximation to the time derivative to obtain

$$\langle\phi_i|\frac{\partial}{\partial t}|\phi_j\rangle(k) = \frac{1}{\delta t}\left[\delta_{ij} - \sum_r \alpha_{in}(k)\alpha_{jn}(k-1)\right] \tag{71}$$

which we use to compute the terms in the expansion-coefficient equations of motion, eqn (21) or eqn (35).

5.2.2. Results and discussion. Depending on the nature of the electron-molecule interactions, the equilibrium excess electronic eigenstates can be strongly localized as in the case of hard-sphere solvents like helium, or polar solvents like water and ammonia, or at the other extreme, the eigenstates can be highly delocalized as found in fluids of highly polarizable spherical molecules such as in the heavier rare gases or highly branched alkanes [74–80]. Varying degrees of localization between these extremes are exhibited by other solvents, and we can change the spatial extent of the excess electronic states in a given solvent by modifying the conditions of density, temperature and pressure. The nature of the density fluctuations in some fluids is such that the electronic state can fluctuate dynamically between localized and delocalized states. Transport properties such as the electron mobility will be strongly influenced by the extent of localization of the electronic states [81–83].

The nonadiabatic dynamical relaxation of excited excess electrons in dense fluid helium [8, 9] and in water [6, 7] have been studied using various techniques outlined in this Chapter. Webster *et al.* [7] used the incoherent self-consistent nonadiabatic dynamics algorithm described in section 4.2.1 to study the relaxation of excess electrons injected into unperturbed water configurations in an attempt to explain the two-phase time evolution of the experimental transient absorption spectrum of the hydrated electron [84–87].

In these experiments, the excess electron is produced by photoionizing one of the water molecules. The calculations assume that the photoelectron is ejected with sufficient kinetic energy to escape from the other photoionization fragments so that the observed transient absorption spectrum involves the rearrangement of representative unperturbed fluid configurations to accommodate the relaxing excess electron, and it is not influenced by the presence of any other ionization fragments. This assumption is supported by the experimental observation that geminate recombination processes seem to take place on a 20 – 40 ps time scale [88–90] whereas the time scale for the transient evolution of the hydrated electron absorption spectrum is on the order of 1 ps.

Due to the size and complexity of these nonadiabatic dynamics calculations, only relatively small numbers of stochastic surface-hopping trajectories can be followed in these model studies; ensemble sizes of ~10 – 20 trajectories are typical with current computational resources. Fortunately the features observed in these small numbers of trajectories are quite ubiquitous for the systems studied. In Webster's hydrated electron nonadiabatic relaxation studies, for example, about 50% of the trajectories show rapid relaxation from the initially injected pure state, reaching a state whose spectrum resembles the equilibrium hydrated absorption spectrum in ~100 – 200 fs. In the rest of the trajectories the system gets hung up in an excited electronic state for 500 – 600 fs before reaching the ground state. Recent studies by Rossky and co-workers, in which the intramolecular vibrations of the water molecules are included, indicate similar dynamical features except that the relaxation time scales are shorter by about a factor of two.

When the transient absorption spectrum is computed by averaging histograms of the electronic absorption/emission frequencies out of the currently occupied state and weighting by the oscillator strength in accord with the Fermi golden rule, the two-phase transient absorption spectrum dynamics observed experimentally is reproduced qualitatively in the calculations: an initial broad band at low energies gives way to a narrower band at higher energies which is characteristic of the equilibrium hydrated electron absorption spectrum. The time scales for these changes are reasonably well reproduced by the calculations including the intramolecular vibrational motion. However, the energies of the calculated absorption bands differ quite considerably from experiment with the position of the calculated equilibrium absorption spectrum being about 1 eV higher than observed experimentally. This points to inadequacies in the electron-molecule pseudopotential used in these calculations. It is not known at this point whether the agreement with experiment, as far as relaxation time scale is concerned, is realistic, or results from fortuitous cancellations of effects associated with the problems with the pseudopotential and other dynamical approximations, such as ignoring electronic phase coherence on the time scale of the nuclear motion, or assuming an initial pure state rather than a more realistic mixed initial state.

This two-phase spectral evolution is observed in the transient absorption spectrum of many fluids which exhibit localized equilibrium solvated-electron states; for example it was first observed in methanol and ethanol [91, 92] where the relaxation time scales are much longer (~20 – 50 ps) indicating that molecular reorientation may be coupled to the nonadiabatic excess electronic relaxation process in these more complex molecular fluids.

The nonadiabatic dynamics calculations on excess electronic relaxation in dense fluid helium by Space and Coker [9] also show this two-phase transient spectral evolution under certain conditions of temperature and density. These workers use Tully's fewest-switches algorithm, which accurately describes electronic phase coherence but their approach for redistributing the nonadiabatic transition energy involves assuming that the energy is dumped uniformly into the kinetic energy of the solvent atoms in a sphere centered on the electron, and a few solvent diameters across.

In dense fluid helium the electron becomes localized by excluded volume interactions. At high solvent densities ($\rho^* \sim 0.9$), Space and Coker found that most of their nonadiabatic dynamics trajectories exhibited a simple rapid decay associated with injecting the electron into an initial state which was highly localized in a small cavity or bubble in the solvent. At these high solvent densities the fluid looks like a "honeycomb" of small disconnected bubbles and the available states around the injection energy are all spherical s-like states localized in these small cavities. The rapid relaxation under these circumstances is associated with the electron simply expanding the bubble in which the initial state is localized until the electron-solvent forces inside the bubble balance the external solvent-solvent forces. The electron remains in the same roughly spherical diabatic state throughout this rapid relaxation. The calculated spectral evolution in this case involves a band at high energies shifting continuously to lower energies and broadening to the equilibrium solvated-electron spectrum in about 200 fs.

At lower solvent densities in helium ($\rho^* = 0.5$), Space and Coker find a very similar situation to the behavior observed in water, in that about 50% of their trajectories exhibit fairly rapid relaxation (~ 0.2 ps) whilst the rest of the trajectories hang up in an excited state for quite a long time ($\sim 0.5 - 1$ ps) before relaxing into the ground state. Under these conditions, the calculated absorption spectrum shows the characteristic two-phase transient absorption spectrum with the same relaxation time scale as observed in water. At these lower solvent densities the fluid structure is much more open than at higher densities; thus some of the available states at the injection energy are highly confined and trapped in small bubbles which are not connected to the rest of the fluid. Injection into these states will result in a simple rapid relaxation as discussed above. Because of the inhomogeneity of the solvent structure at lower densities, there are also regions of the fluid which appear as connected solvent cavities, and there will also be excited states near the injection energy with nodes which reside in these larger connected void regions. If the state we choose to place the electron in has this form, then as the system relaxes in the presence of the excited-state forces there will be a strong nonadiabatic interaction, as the excess electronic state becomes a strong mixture of this excited state and the nodeless ground state. Due to the more diffuse nature of the states, and the prolonged nonadiabatic interaction under these circumstances, the relaxation time scale is considerably lengthened [8, 9].

It is interesting to note that the excess electronic relaxation time scales and processes in fluids as vastly different in character as helium and water are very similar. This probably results because the molecular motions involved in electronic relaxation in these fluids of small molecules are translational i.e. either helium atoms or water molecules are simply pushed up against their neighbors as the void is opened to accommodate the electron. In the more anisotropic alcohols, as mentioned above, molecules have to be rotated out of the way in order to open up a void large enough to solvate the electron, and this takes considerably longer than the small molecular translations needed to open up a solvent cavity large enough in water or helium to accommodate the electron. There will be longer time scale reorientations of the water molecules to reach the final equilibrated state of the electron and solvent but we believe these will have little effect on the hydrated electron absorption spectrum, as such reorientational motion probably influences both the ground- and excited-state energy in much the same way.

6. Conclusions

In this Chapter we have presented the formal theoretical principles underlying the semiclassical description of a dynamical system in terms of a pair of coupled subsystems: one which we describe using the laws of classical mechanics, and the other which we choose to treat quantum mechanically. Traditional molecular dynamics simulation methods have been restricted to modelling systems which undergo adiabatic ground-state dynamics. In these calculations the quantum degrees of freedom are incorporated in the ground-state Born-Oppenheimer potential surface over which the nuclei move. However, many important chemical and physical processes, particularly those involving electronic excitation, proceed via nonadiabatic dynamical pathways in which motion of the classical subsystem results in transitions and mixing of the states of the quantum subsystem.

We have explored the various issues which are crucial for an accurate description of these nonadiabatic phenomena, and presented several numerical algorithms which handle these different issues with varying degrees of success. The common thread through all these techniques is their applicability to large condensed-phase systems in which most of the degrees of freedom are described classically and relatively few require a detailed quantum description. Indeed the principal limitation of the techniques described here is that only a small number of quantum degrees of freedom can be treated. The solvated-electron system we considered in the last section provides an example of how the methods described in this Chapter can be implemented for a quantum solute in a large fluctuating classical solvent bath. Obvious extensions of these ideas include quantum mechanical intramolecular vibrational motions [3, 4] and rotations [93, 94] coupled to classical solvent dynamics.

As excited-state *ab initio* quantum calculations become more reliable, the nonadiabatic dynamical framework detailed in this Chapter will be directly applicable to exploring the excited-state relaxation dynamics of larger molecular systems in the presence of classical environmental motions. This development will pave the way for studies of many intriguing phenomena in excited-state chemistry and physics.

Acknowledgements

The author gratefully acknowledges financial support for this work from the National Science Foundation (Grant No. CHE-8913780 and No. CHE-9058348), the Petroleum Research Fund administered by the American Chemical Society (Grant No. 22238-AC6), and a generous allocation of supercomputer time from the National Center for Supercomputing Applications. The author also wishes to thank Doris Mei for her assistance in the preparation of the manuscript.

References

[1] R. E. CLINE and P. G. WOLYNES, *J. Chem. Phys.* **86**, 3836 (1987).

[2] D. P. ALI and W. H. MILLER, *J. Chem. Phys.* **78**, 6640 (1983).

[3] M. F. HERMAN and E. KLUK, Nonadiabatic scattering problems in liquid-state vibrational relaxation, in *Dynamical processes in condensed matter*, edited by M. W. EVANS, p. 577, Wiley, New York, 1985.

[4] M. F. HERMAN, *J. Chem. Phys.* **87**, 4779 (1987).

[5] J. N. ONUCHIC and P. G. WOLYNES, *J. Phys. Chem.* **92**, 6495 (1988).

[6] F. WEBSTER, P. J. ROSSKY, and R. A. FRIESNER, *Comput. Phys. Commun.* **63**, 494 (1991).

[7] F. J. WEBSTER, J. SCHNITKER, M. S. FRIEDRICHS, R. A. FRIESNER, and P. J. ROSSKY, *Phys. Rev. Lett.* **66**, 3172 (1991).

[8] B. SPACE and D. F. COKER, *J. Chem. Phys.* **94**, 1976 (1990).

[9] B. SPACE and D. F. COKER, *J. Chem. Phys.* **96**, 652 (1992).

[10] J. C. TULLY and R. K. PRESTON, *J. Chem. Phys.* **55**, 562 (1971).

[11] J. C. TULLY, Nonadiabatic processes in molecular collisions, in *Dynamics of Molecular Collisions, Part B*, edited by W. H. MILLER, p. 217, Plenum, New York, 1976.

[12] J. C. TULLY, *J. Chem. Phys.* **93**, 1061 (1990).

[13] W. H. MILLER and F. F. GEORGE, *J. Chem. Phys.* **56**, 5637 (1972).

[14] W. H. MILLER, *Adv. Chem. Phys.* **25**, 69 (1974).

[15] M. F. HERMAN, *J. Chem. Phys.* **81**, 754 (1984).

[16] M. F. HERMAN, *J. Chem. Phys.* **81**, 764 (1984).

[17] M. F. HERMAN, *J. Chem. Phys.* **76**, 2949 (1982).

[18] M. F. HERMAN and E. KLUK, *Chem. Phys.* **91**, 27 (1984).

[19] M. F. HERMAN, *J. Chem. Phys.* **79**, 2771 (1983).

[20] J. R. STINE and J. T. MUCKERMAN, *J. Phys. Chem.* **91**, 459 (1987).

[21] J. R. STINE and J. T. MUCKERMAN, *J. Chem. Phys.* **65**, 3975 (1976).

[22] J. R. STINE and J. T. MUCKERMAN, *J. Chem. Phys.* **68**, 185 (1978).

[23] N. C. BLAIS and D. G. TRUHLAR, *J. Chem. Phys.* **79**, 1334 (1983).

[24] N. C. BLAIS, D. G. TRUHLAR, and C. A. MEAD, *J. Chem. Phys.* **89**, 6204 (1988).

[25] L. J. DUNNE, J. N. MURRELL, and J. G. STAMPER, *Chem. Phys. Lett.* **112**, 497 (1984).

[26] L. J. DUNNE, *Int. J. Quantum Chem.* **32**, 31 (1987).

[27] C. W. EAKER, *J. Chem. Phys.* **87**, 4532 (1987).

[28] S. SAWADA and H. METIU, *J. Chem. Phys.* **84**, 227 (1986).

[29] B. JACKSON and H. METIU, *J. Chem. Phys.* **85**, 4129 (1986).

[30] D. DEHARENG, *Chem. Phys.* **110**, 375 (1986).

[31] P. G. WOLYNES, *J. Chem. Phys.* **86**, 1957 (1987).

[32] P. G. WOLYNES, *J. Chem. Phys.* **87**, 6559 (1987).

[33] H. NAKAMURA, *J. Chem. Phys.* **87**, 4031 (1987).

[34] M. BAER, The theory of electronic nonadiabatic transitions in chemical reactions, in *Theory of Chemical Reaction Dynamics*, edited by M. BAER, volume II, p. 219, CRC, Boca Raton, 1985.

[35] P. PECHUKAS, *Phys. Rev.* **181**, 166 (1969).

[36] P. PECHUKAS, *Phys. Rev.* **181**, 174 (1969).

[37] L. D. LANDAU, *Phys. Z. Sow.* **2**, 46 (1932).

[38] C. ZENER, *Proc. Roy. Soc. Lond. A* **137**, 696 (1932).

[39] E. C. G. STUECKELBERG, *Helv. Phys. Acta* **5**, 369 (1932).

[40] M. S. CHILD, Electronic excitation: Nonadiabatic transitions, in *Atom-Molecule Collision Theory*, edited by R. B. BERNSTEIN, p. 427, Plenum, New York, 1979.

[41] M. S. CHILD, *Molec. Phys.* **28**, 495 (1974).

[42] E. E. NIKITIN, *Theory of elementary atomic and molecular processes in gases*, Clarendon, Oxford, 1974.

[43] E. E. NIKITIN, Theory of nonadiabatic transitions. Recent development of the Landau - Zener (linear) model, in *Chemische Elementarprozesse*, edited by H. HARTMANN, Springer, Berlin, 1968.

[44] E. E. NIKITIN and S. Y. UMANSKII, *Theory of slow atomic collisions*, Springer Verlag, New York, 1984.

[45] D. A. MICHA, *J. Chem. Phys.* **78**, 7139 (1983).

[46] P. K. SWAMINATHAN, B. C. GARRETT, and C. S. MURTHY, *J. Chem. Phys.* **88**, 2822 (1988).

[47] W. H. MILLER and C. W. MCCURDY, *J. Chem. Phys.* **69**, 5163 (1978).

[48] C. W. MCCURDY, H. D. MEYER, and W. H. MILLER, *J. Chem. Phys.* **70**, 3177 (1979).

[49] H. D. MEYER and W. H. MILLER, *J. Chem. Phys.* **70**, 3214 (1979).

[50] H. D. MEYER and W. H. MILLER, *J. Chem. Phys.* **72**, 2272 (1980).

[51] M. H. MITTLEMAN, *Phys. Rev.* **122**, 449 (1961).

[52] J. B. DELOS, W. B. THORSON, and S. K. KNUDSEN, *Phys. Rev. A* **6**, 709 (1972).

[53] G. D. BILLING, *Chem. Phys. Lett.* **30**, 391 (1975).

[54] D. KUMOMOTO and R. SILBY, *J. Chem. Phys.* **75**, 5164 (1981).

[55] D. J. DIESTLER, *J. Chem. Phys.* **78**, 2240 (1983).

[56] Z. KIRSON, R. B. GERBER, A. NITZAN, and M. A. RATNER, *Surf. Sci.* **137**, 527 (1984).

[57] Z. KIRSON, R. B. GERBER, A. NITZAN, and M. A. RATNER, *Surf. Sci.* **151**, 531 (1985).

[58] S. SAWADA, A. NITZAN, and H. METIU, *Phys. Rev. B* **32**, 851 (1985).

[59] K. J. SCHAFER, J. D. GARCIA, and N. H. KWONG, *Phys. Rev. B* **36**, 1872 (1987).

[60] Z. KOTLER, R. KOSLOFF, and A. NITZAN, *Chem. Phys. Lett.* **153**, 483 (1988).

[61] M. AMAROUCHE, F. X. GADEA, and J. DURUP, *Chem. Phys.* **130**, 145 (1989).

[62] A. T. AMOS, K. W. SULSTON, and S. G. DAVIDSON, *Adv. Chem. Phys.* **76**, 335 (1989).

[63] J. C. TULLY, *J. Chem. Phys.* **60**, 3042 (1974).

[64] J. C. TULLY, Collisions involving electronic transitions, in *State to State Chemistry*, edited by P. R. BROOKS and E. F. HAYES, volume 56 of *ACS Symposium*, p. 206, ACS, ACS, Washington, DC, 1977.

[65] R. P. FEYNMAN, *Statistical Mechanics*, Benjamin, Reading, 1972.

[66] R. P. FEYNMAN and A. R. HIBBS, *Quantum Mechanics and Path Integrals*, McGraw-Hill, New York, 1965.

[67] D. F. COKER and L. XIAO, A new algorithm for phase coherent, conservative nonadiabatic dynamics, unpublished.

[68] D. KOSLOFF and R. KOSLOFF, *J. Comput. Phys.* **52**, 35 (1983).

[69] R. KOSLOFF, *J. Phys. Chem.* **92**, 2087 (1988).

[70] M. D. FEIT, J. A. FLECK, JR., and A. STEIGER, *J. Comput. Phys.* **47**, 412 (1982).

[71] L. SZASZ, *Pseudopotential theory of atoms and molecules.*, Wiley, New York, 1985.

[72] J. K. CULLUM and R. A. WILLOUGHBY, *Lanczos algorithms for large symmetric eigenvalue computations*, Birkhauser, Boston, 1985.

[73] G. MARTYNA and B. J. BERNE, *J. Chem. Phys.* **89**, 4516 (1988).

[74] D. F. COKER, D. THIRUMALAI, and B. J. BERNE, *J. Chem. Phys.* **86**, 5689 (1987).

[75] D. F. COKER and B. J. BERNE, Quantum Calculations on Excess Electrons in Disordered Media, in *Excess Electrons in Dielectric Media*, edited by C. FERRADINI and J. P. JAY-GERIN, pp. 211–257, CRC Pres, Boca Raton, 1991.

[76] B. SPACE, D. F. COKER, H. LIU, B. J. BERNE, and G. MARTYNA, *J. Chem. Phys.* **97**, 2002 (1992).

[77] D. CHANDLER, Y. SINGH, and D. M. RICHARDSON, *J. Chem. Phys.* **81**, 1975 (1984).

[78] D. LARIA and D. CHANDLER, *J. Chem. Phys.* **87**, 4088 (1987).

[79] A. L. NICHOLS III, D. CHANDLER, Y. SINGH, and D. M. RICHARDSON, *J. Chem. Phys.* **81**, 5109 (1984).

[80] A. L. NICHOLS III and D. CHANDLER, *J. Chem. Phys.* **84**, 398 (1986).

[81] A. L. NICHOLS III and D. CHANDLER, *J. Chem. Phys.* **87**, 6671 (1987).

[82] D. HSU and D. CHANDLER, *J. Chem. Phys.* **93**, 5075 (1990).

[83] S. BASAK and M. H. COHEN, *Phys. Rev. B* **20**, 3404 (1979).

[84] A. MIGUS, Y. GAUDUEL, J. L. MARTIN, and A. ANTONETTI, *Phys. Rev. Lett.* **58**, 1559 (1987).

[85] F. H. LONG, H. LU, and K. B. EISENTHAL, *Phys. Rev. Lett.* **64**, 1490 (1990).

[86] H. LU, F. H. LONG, and K. B. EISENTHAL, *J. Opt. Soc. Am.* **7**, 1511 (1990).

[87] F. H. LONG, H. LU, and K. B. EISENTHAL, *Chem. Phys. Lett.* **160**, 464 (1989).

[88] Y. GAUDUEL, S. POMMERET, A. MINGUS, and A. ANTONETTI, *J. Phys. Chem.* **93**, 3880 (1989).

[89] Y. GAUDUEL, S. POMMERET, A. MINGUS, and A. ANTONETTI, *Chem. Phys.* , to appear.

[90] Y. GAUDUEL, S. POMMERET, A. MINGUS, N. YAMADA, and A. ANTONETTI, *J. Am. Chem. Soc.* **112**, 2925 (1990).

[91] G. A. Kenney-Wallace and C. D. Jonah, *Chem. Phys. Lett.* **39**, 596 (1976).

[92] Y. Wang, M. K. Crawford, M. J. McAuliffe, and K. B. Eisenthal, *Chem. Phys. Lett.* **74**, 160 (1980).

[93] D. G. Taylor and H. L. Strauss, *J. Chem. Phys.* **90**, 768 (1989).

[94] J. E. Hunter III, D. G. Taylor III, and H. L. Strauss, *J. Chem. Phys.* **97**, 50 (1992).

Long Length-scale Aspects of Self Organization Phenomena

K. A. Dawson
Department of Chemistry
University College Dublin
Ireland.

ABSTRACT. Using very simple models it has now become possible to understand the properties of complex fluids, such as microemulsions and liquid crystals. An understanding of the basic science, combined with parallel processing methods, permits one to thoroughly describe phase diagrams, interfacial properties, and neutron scattering experiments.

1. Introduction: an apologia

Most of the contributions to this volume on computer simulation have quite a different perspective to that which I shall present. Thus, most of the Chapters emphasize continuum simulations in which some attempt is made to accurately represent the intermolecular potentials of the system. Since I shall make no such attempt I feel that some sort of 'apology' is required, as well as an explanation of how this material fits into the general scheme of the book.

Broadly speaking the ideas and techniques I shall describe are appealing because of their relative simplicity. The aim is to strip away the details, laying bare the essentials, and thereby clarifying the dominant physical phenomena. The models are very simple, even primitive. Consequently, it is usually possible by good algorithm design, dedication to efficient coding and exploitation of the computer hardware to be fairly confident that one has a carefully controlled approximation by simulation. This is often not the case in more sophisticated models where molecular detail is retained, and one finds that the amount of computer time required to develop and model equilibrium is simply too great.

So, in the end one must evaluate each new scientific project to determine which approach is more appropriate. If molecular detail is important, and there are no problems with equilibration and sampling, then continuum potential Monte-Carlo or molecular dynamics is appropriate. If there are generic features of the system to be elucidated that originate in a length scale

M.P. Allen and D.J. Tildesley (eds.), Computer Simulation in Chemical Physics, 379–396.

somewhat longer than the molecular size, or if equilibration and sampling promise to be difficult, you should probably consider using the techniques described in this Chapter.

Finally, I suppose one should admit that there is a certain sense in which some of us find working with lattice models simply good fun! I hope I will manage to convey that feeling to you as we proceed with the Chapter.

2. Universality and near universality

In this Chapter, I shall outline the current thinking on how best to model long length-scale phenomena in complex fluids. In some situations a novel strategy may be possible because, although some properties of a fluid may reflect the details of the intermolecular potential, those aspects of the liquid that originate in long-ranged correlations may be understood using very simple molecular models. Prominent examples of these extremes are the critical temperature (T_c) and exponents of a fluid at the critical point. Thus, the critical temperature, roughly the point at which entropic and energetic factors balance, is strongly affected by the strength of attractions between the molecules, a fairly short-ranged effect. On the other hand the critical exponent for the vanishing of the difference between liquid and gas densities is a consequence only of the fractal nature of the huge liquid and gas clusters at the critical point. Admittedly this is an extreme case because the long length scale, the correlation length ξ, actually diverges at the critical point and the critical exponents for all liquid-gas transitions are the same, irrespective of the molecules involved. This is a good example of a phenomenon named 'universality'. We will be more interested in less extreme cases, but the arguments for universality at the critical point are very simple, so that is where we will start [1]. Now, a change in density at some point in the fluid that we shall arbitrarily label as the origin is denoted $\delta\rho(0) = \rho(0) - \rho$ (ρ is the average bulk density), and will be correlated with a change ('fluctuation') at another position r, denoted $\delta\rho(r)$. The function that measures the correlations is $G(r) = \langle\delta\rho(0)\delta\rho(r)\rangle$ and at typical densities beyond the triple point, and distances beyond a few molecular diameters we typically find the form,

$$\begin{aligned} G(r) &= \langle\delta\rho(0)\delta\rho(r)\rangle \approx \frac{Ae^{-r/\xi}}{r} \qquad (r \gg a) \\ \xi &\approx a|T - T_c|^{-\nu} \end{aligned} \tag{1}$$

where a is essentially the molecular diameter, T_c is the critical temperature of the fluid and ν, the critical exponent, is a universal number for all simple fluids. Because of the properties of the exponential function (it is negligible for $r/\xi \gg 1$) we conclude that changes in densities at points nearer than about ξ are coupled in the sense that one affects the other. From the foregoing physical description of $G(r)$ it is reasonable to suppose that the mean square diameter of a correlated domain may be estimated by calculating the expectation value of r^2 with $G(r)$. Using eqn (1) this leads to

$$\frac{\int r^2 G(r)\,\mathrm{d}\mathbf{r}}{\int G(r)\,\mathrm{d}\mathbf{r}} = \xi^2. \tag{2}$$

The important thing to note here is that, as the temperature approaches T_c, ξ becomes very large and eqns (1) are increasingly well obeyed, for all simple fluids. Using eqn (2) we may interpret this by saying that the important contributions from the partition function arise from those configurations that have droplets or clusters of average r.m.s. diameter ξ. It is this length, rather than the molecular diameter, that dominates many observable properties. We should always remember that ξ is a new length scale that has emerged from statistical mechanics and it is a reflection of the balances between the intermolecular potential and the entropy of the system.

From the foregoing discussion we may deduce that, near the critical point, the behaviour of a fluid is essentially universal and various physical properties such as $\xi(T)$ acquire a generic form as functions of the underlying parameters such as temperature, pressure or concentration if the fluid has more than one component. Indeed, we may expect that, even away from the immediate vicinity of a critical point, there is much useful general comment that may be made about the liquid without specific knowledge about molecular detail. In such cases the essential feature of the system is that the length scale over which correlations dominate is much larger than the molecular or microscopic length scale. Later on we will see that such factors will pertain for some types of self-assembled fluid. However, for the moment we shall continue our program of deducing general information that will be important when we come to simulate complex, strongly interacting systems.

The first obvious point is that all practical simulations are carried out in a finite configuration space, with or without periodic boundary conditions. In either case we can sample only those configurations that fit within the finite space ('simulation box') and therefore see only correlated structures that are comparable to or smaller than this box size. Evidently, since the dominant structures in a statistical ensemble tend to be of extent ξ, the simulation box must have a somewhat larger volume than ξ^3. We next turn to the issue of how long it will take to calculate a good average by computer simulation. The argument we use is very simplistic, but it is essentially correct, and can be made fully rigorous if one desires. When a density disturbance of wavelength λ is imposed on the uniform fluid the energy increases by an amount

$$\mathcal{H} \propto \sum_{\mathbf{k}} \left(k^2 + \xi^{-2}\right) |\delta\rho_k|^2 \tag{3}$$

where $\delta\rho_k$ is the Fourier component of the density disturbance of wavenumber $k = 2\pi/\lambda$. The system relaxes back to equilibrium with a restoring force that is proportional to the derivative of the energy, though thermal fluctuations will tend to keep it excited. We ignore the thermal effects and, since the equation of motion of a density disturbance is just

$$\begin{aligned} \delta\dot{\rho}(t) &\approx \tau(k)^{-1}\delta\rho(k) \\ \tau(k) &\propto (k^2 + \xi^{-2})^{-1}, \end{aligned} \tag{4}$$

we expect the characteristic decay of the correlation function,

$$\langle \delta\rho_k(0)\delta\rho_k(t)\rangle \approx \delta\rho_k(0)\delta\rho_k(t) \propto \mathrm{e}^{-t/\tau(k)} \tag{5}$$

where $\tau(k)$ is that relaxation time associated with the decay of the density disturbance of wavenumber k. Now we begin to see a potential problem if we wish to simulate systems where the correlation length is long. Thus, if we write $x = k\xi$,

$$\begin{aligned} \tau(k) &= \xi^2 f(x) \\ f(x) &= (1 + x^2)^{-1} \end{aligned} \tag{6}$$

and consider disturbances of comparable or larger extent than the correlation length, then $f(x)$ is of order unity and may be ignored. The relaxation time of such a disturbance is of order ξ^2, and this may be exceedingly large if ξ itself is large. I might also mention that if one performs a conserved-particle-number simulation, such as one has in the canonical ensemble, the effective energy in eqn (3) has an additional multiplicative term of k^2 and, following the same arguments as above, we find that the relaxation time scales as the fourth power of ξ. This is intuitively reasonable because we know that it will be much easier to equilibrate a system where we are free to create and destroy particles, as in the grand ensemble.

These phenomena are called 'critical slowing' and are the origin of many anomalous mechanical properties as one approaches the critical point of a fluid. However, even somewhat away from the critical point we find that systems with large correlation length relax very slowly. Now Monte-Carlo simulation is simply a method for generating independent samples in a Markov chain, and the dynamics may therefore be represented by a Langevin equation similar to that above. We see that one must wait a time of order ξ^2 before an essentially independent configuration is generated by a conventional computer simulation. So, when we come to calculate observables from the simulation we will be able to use only a fraction $(1/\xi^2)$ of the computed configurations as independent steps in a Markov chain. In fact there have been a number of quite successful attempts to improve this situation, most of them based on some form of collective move or update in the simulation. Frenkel, in Chapter 4, has discussed one, and the technique of Fourier acceleration is an alternative.

So far we have observed one fundamental origin of long time scales in a strongly correlated system, so-called critical slowing down. The reader should note carefully that this phenomenon has nothing to do with the 'simulation being stuck in a metastable (local) minimum', another common problem with the Monte-Carlo technique of sampling. Indeed, that the phenomena are unrelated is immediately obvious from the incremental energy (eqn (3)) that is being minimized to cause the density disturbance to disappear. Since we have a purely quadratic minimum there is no possibility of a metastability phenomenon. If there are many activation barriers on the potential surface then sampling and equilibration is difficult. Some ideas for approaching this sort of problem do have some merit, for example simulated annealing, but I shall not discuss these matters in the present Chapter.

The principal conclusions that I wish to draw from these considerations are that long time scales can arise from long length scales with or without conserved mass or metastability and that, where such problems arise, we should make every effort to simplify the model in order to be able to effectively simulate it. Usually this will mean the replacement of the full intermolecular potential by some very simple set of effective interactions. These may be calculated only for discrete positions and orientations of the underlying 'molecules'. No matter what one does to improve the situation, there is usually a problem of insufficient computer time being available to calculate reasonable averages. For this reason I shall also outline some ways of applying parallel processing to speed up the simulation.

We now turn to a simple concrete example, the lattice-gas transcription of the nearest neighbour Ising model.

3. The nearest-neighbour Ising model as a computational model of simple fluids

Before entering into details let me emphasize that such models are only quantitatively useful when the correlation length is somewhat larger than the molecular length scale. If this situation pertains then the very same observations about time and length scales that led us to abandon a molecularly detailed treatment permit us to argue that such a treatment is not really necessary. Thus, we have observed that the correlation function of all simple fluids takes on the same basic form (eqn (1)) irrespective of molecular detail, and with a characteristic dependence on temperature that also seems to bear little relation to the detailed molecular picture. In essence, if the dominant correlations occur on a length scale considerably larger than the molecular diameter, we are free to considerably simplify the molecular treatment. For example, we might readily replace a van der Waals potential with a spherical-square-well potential providing we make the well depths comparable, thus ensuring that the same attractive energy scale is being used. An even more primitive approximation involves using only discrete values of the intermolecular potential.

To achieve this let us divide configuration space into cubes of side a, the diameter of the atom or near-spherical molecule we wish to describe. At the centre of each such cube we assign a lattice index $n = (x, y, z)$ and assign a two-valued variable $\rho_n = 0, 1$ which indicates whether or not a particle occupies that cube. Note that we will not be able to include an infinity of neighbouring configurations where the particles do not fit into the cubes. However, as we have already argued, if ξ is large enough this should not affect the calculation of bulk properties too much.

The energy function or hamiltonian may be written,

$$\mathcal{H} = \frac{1}{2} V \sum_{n,n'}^{\mathrm{NN}} \rho_n \rho_{n'} \tag{7}$$

where $|V|$ is just the well depth of the equivalent spherical square well potential, and NN indicates that the double summation is over all nearest-neighbour pairs. Since we wish to work in the grand ensemble we may simply add the term in the chemical potential, and then sum over all possible values of the variables, ρ. The effective hamiltonian is then just

$$\mathcal{H} = \frac{1}{2} V \sum_{n,n'}^{\mathrm{NN}} \rho_n \rho_{n'} - \mu \sum_{n} \rho_n. \tag{8}$$

Largely for historical reasons, one often replaces the density-type variables ($\rho_n = 0, 1$) with spin variables ($\sigma_n = 2\rho_n - 1 = -1, +1$) and the hamiltonian is viewed as that of an equivalent model of classical electron spins, named the Ising model. The simple case described above is trivial and may be written out directly. As we shall see, merely extending these ideas to account for interactions between rigid amphiphilic rods leads to the Widom-Wheeler [2, 3] model which is considerably more complex. From the point of view of computer simulation, rather than the application of analytical methods, we need not pursue this discussion much further. We have the system energy written as a function of the binary variables $\rho_n = 0, 1$ and can proceed to write a computer program.

4. Wheeler-Widom-type models as lattice fluids

Having discussed the simplest possible example of a lattice fluid we can now proceed to more complex cases. One natural extension of the simple lattice-gas or Ising model is possible if we have a number of components in the mixture. Here we may transcribe the fluid problem into an m-component so-called Potts model. For example, the case of a three-component mixture of species A, B and C may be described by a three-state spin variable, located at the vertices of a lattice. The spin variable takes on values $\tau_n = -1, 0, 1$ corresponding, respectively, to species A,B and C. We will not pursue such models in this Chapter, merely pointing out that they have been applied to simple and self-assembled fluids.

Instead we now turn to another type of model, the Widom-Wheeler model, that has been quite influential in recent thinking about self-assembled fluids specifically. The reason is that, despite the fact that it describes three-component mixtures of oil, water and amphiphile, by means of clever approximations one again ends up with an Ising model, but with more extended interactions.

I will not go into too much detail about setting up the model because that is explained in detail elsewhere. Rather I will attempt to focus on aspects of the simulations that are more generally useful to those of you that wish to use such methods to study other complex lattice models. I will simply point out that, as before, one imagines breaking up configuration space into little cubes, in this case each the length of an amphiphile molecule. Each of these cubes may be occupied by either hydrophobic or hydrophilic material. The hydrophobic material may come from amphiphile tails or oil whilst the hydrophilic material may originate in head groups of amphiphiles or water. With this restriction the amphiphile molecules must lie at the boundary of regions of oil and water and an individual molecule may therefore adopt only those angles and positions consistent with the underlying cubic symmetry. The fact that there are only two types of cubes means that one may label the centres of the cubes with with Ising variables, a positive value representing water and a negative value representing oil.

The fact that the molecules adopt only a few orientations with respect to each other means that it is a straightforward matter to relate the potential energy values to coupling constants of the equivalent Ising model. Although I have not explicitly shown how all this is carried out in practice I think the example of the much simpler nearest-neighbour Ising models is sufficient to illustrate the sort of thing I have in mind. Therefore I shall now just write out the lattice-model hamiltonian in the grand ensemble representation,

$$\mathcal{H} = \frac{J}{2} \sum_{n,n'}^{\mathrm{NN}} \sigma_n \sigma_{n'} + \frac{M_1}{2} \sum_{n,n'}^{\mathrm{LNNN}} \sigma_n \sigma_{n'} + \frac{M_2}{2} \sum_{n,n'}^{\mathrm{DNN}} \sigma_n \sigma_{n'} \tag{9}$$

where LNNN and DNN indicate sums over linear next-nearest neighbours and over diagonal nearest neighbours respectively. I should note that the couplings are elsewhere related to the potential values [2, 3] and that I have assumed equal volume fractions of oil and water to obtain a hamiltonian that is even with respect to inversion. Now we are ready to begin the Monte-Carlo studies of the model.

5. Some tricks of the trade; Ising-like models

My aim in this section is to give some guidance to the reader who is familiar with Monte-Carlo simulation, but not perhaps with the details of optimizing a lattice-model simulation [4]. I shall assume that we are going to study a two-state (Ising) model but most of the ideas are generalizable to a multistate Potts model, with an attendant loss of efficiency.

This section may be thought of as consisting of two parts. The first deals mainly with simple methods of speeding up and compacting the simulation program in the computer. The second part deals with recent thinking on how best to exploit vector and parallel machines in lattice model simulations. Lattice models involve only highly regular data structures and are ideal for exploiting, or learning about, such machines.

5.1. TABLE LOOK-UP

We begin by constructing a table of all possible interactions. In the case of lattice models this is particularly simple since the range of the potential is typically only a few lattice spacings and the molecules are permitted only a small number of orientations, generally defined by the lattice axes. A spin configuration is updated by changing a $+1$ (or -1) to a -1 (or $+1$) on any given site. The associated change in energy can therefore be only one of a small number of values. This permits us to list all possible exchange factors in a table and avoids the need to calculate any exponentials during the simulation. For the case of the nearest-neighbour model the possibilities are $\exp\{-2nJ\}$ where $n = 6, 4, 2, 0, -2, -4, -6$.

5.2. BIT PACKING

If one simply writes a standard Fortran code then the spin variables are declared as integers and occupy a full computer word. This is wasteful since the variables may be specified by one bit of information. One solution is called 'bit-packing' meaning simply that we assign a number of dynamical variables to a single computer word. Having done this we will have saved a considerable amount of real memory, possibly avoiding unwanted swapping problems. However to make Monte-Carlo moves we may have to be able to address the spin variables by operating on individual bits in a given word. Most manuals contain some guidance on bit operations. However, the reader should ensure that their use will not ultimately slow the program possibly by a failure to vectorize the bit operations.

5.3. UNVECTORIZABLE CODE

The failure to properly vectorize certain Fortran statements is a serious problem with some compilers. For example, the I.B.M. 3090 was unable to effectively vectorize if the code contained a conditional statement or subroutine call. One may use an alternative piece of code to replace conditionals and write 'in-line' code instead of subroutine calls.

Other common practices to avoid in writing a vectorized program include the use of recursive relations such as

```
      DO 1 J = 3, N
         A(J) = B(J) + A(J-2)
1     CONTINUE
```

The values of elements of the A array on the left and right side cannot be calculated simultaneously, so either the compiler will flag the piece of code or, if you force vectorization, produce nonsense. You should also note that the code,

```
      DO 1 J = 3, N
         A(J) = B(J-2)
         B(J) = C(J)
1     CONTINUE
```

has a 'dependency', meaning that we try to use the values of B(J) to calculate A(J) before we have calculated them. The meaning is clear to a scalar machine, but confusing to a vector processor. Finally, I/O statements can be a problem in vectorizing a code. The best way to deal with this is simply to move them to the very beginning or end of the program and read or write out everything before beginning or after ending the calculational or time-consuming loop.

5.4. COMPUTATIONAL INDEPENDENCE

The most important step in speeding up the program is to ensure perfect vectorization of the computationally intensive 'inner' loop. For example, to calculate the energy of flipping a given spin in the nearest-neighbour Ising model we need to consider the spin variables of the surrounding spins. In a succeeding Monte-Carlo step we will then move to one of the adjoining sites and calculate the energy to flip it, again by considering the nearest-neighbour sites. This procedure is satisfactory for a scalar machine, but there is clearly an implied dependency which will prevent vectorization. Thus, we will be attempting to simultaneously calculate the flip energies of neighbouring sites, a meaningless task since each calculation requires a properly defined state for the surrounding spins. To proceed we need to use an algorithm where the dependencies have been removed. I shall describe a method that is used for the nearest-neighbour model because it is basic and is readily generalizable. The method consists of defining two sublattices, 'black' and 'white', the black sublattice constructed by choosing every other spin on a given plane and the white sublattice consisting of the remaining spins. A given plane then has a colouring pattern that looks like a draughts or chequers board. This is the origin of the name, chequerboard decomposition.

The basic idea underlying the lattice decomposition is quite general. It is clear that for more extended interactions one must break up the lattice in more complex manners to ensure that every spin on a given sublattice has no direct interaction with any other on the same sublattice. Also the number of sublattices becomes dependent on the size of the lattice we wish to simulate. The method for dividing up the lattice for the case of the Widom model has been outlined elsewhere. The next point that we must address is how to usefully implement the procedures on commonly accessible computers. I shall not discuss the special-purpose or SIMD (single instruction multiple data) machines that are at present being used for these and other lattice calculations in particle physics because they require quite different treatment and are not commonly available.

5.5. GENERAL-PURPOSE PARALLEL AND VECTOR PROCESSORS

We have seen that the independent sublattices may be dealt with separately, and separate inner loops may therefore be assigned to each. This also means that a single sweep through each of these sublattices may be carried out on a separate processor. At the end of each such a sweep we will have updated sublattices and the process may be repeated.

All of these procedures are easily implemented on general-purpose machines because there are no dangers associated with different processors trying to address the same parts of memory at the same time. We say that data integrity is ensured automatically. In fact all of the passing of data and other aspects of parallelism are not evident to the programmer. An automatic compiler such as the I.B.M. P.F.P compiler will produce Fortran code that will run in parallel providing the programmer has written a code that is explicitly parallelizable in the sense discussed above. Most of the simulations discussed in this paper were carried out using parallel methods on the I.B.M. 3090 600E machine.

5.6. PARALLELISM IN THE MODERN WORKSTATION ENVIRONMENT

Until fairly recently parallel computing was available only to that section of the scientific community [5] with access to a general-purpose multiprocessor main-frame, or the new massively parallel systems. More recently it has become possible to develop a parallel computing capacity using groups of R.I.S.C. workstations connected together with optical or conventional cable.

Two types of computer organization are presently being studied, termed loosely- or tightly-coupled clustering. Loosely-coupled clusters are composed of workstations in different locations, connected together with ethernet. Tightly-coupled clusters are generally in the same physical location, and are therefore designed to minimize the problems associated with time delays in communicating information between machines.

There are now numerous software options that will create a parallel environment from a cluster of workstations without detailed technical knowledge of the user. Most software of this type is based on the primary Unix functions of the workstations themselves, but they act at a variety of hardware and software levels.

The software options currently being studied include Cooperative Processes Software (C.P.S.), Linda, I.S.I.S., the ParaScope editor for generation of Fortran D code, ParaSoft Express and Parallel Virtual Machine (P.V.M). Little quantitative evaluation appears to be available, although various experiences with ParaSoft Express and P.V.M. are now becoming available. P.V.M. (2.4) is in the public domain, whilst ParaSoft Express is supported by ParaSoft Corporation. The following quantitative comments about lattice-model calculations are sometimes the result of simulations of clusters rather than the actual calculations.

The basic strength of P.V.M. is that it is quite flexible, allowing the programmer to fully control the parallel calculations. The disadvantages stem from that same flexibility in that data integrity in a distributed memory and message switching across the cluster are direct responsibilities of the programmer, There are no high-level automatic functions. P.V.M. is composed of daemons (Pvmd) on the workstations and the P.V.M. library. These daemons control the sending, termination and synchronization of various tasks. The PVM library is the interface between the program and the daemons, and the programmer can thereby request services from these daemons. For technical reasons the daemons communicate using the UDP communication protocol (rather than TCP/IP) and this permits any daemon to talk to all others with specified addresses. Such a choice also permits packets to be transmitted without many of the communications overheads associated with the other protocols. A master daemon distributes the information about the various workstations, their addresses and the allowed packet sizes for network transmission. In most current studies a separate copy of the executable file is kept on each workstation, thereby limiting the size of calculation that may be carried out on the parallel system. This will probably not be necessary as the software develops.

At this stage P.V.M. seems most appropriate for rather coarse-grained parallel applications in the sense that each piece of the computation must be of sufficient size so that communications and dpvm overheads are relatively small. In this sense it is no different to parallel applications on general-purpose machines. The results of our studies indicate that lattice sizes of around thirty cubed and larger are necessary before parallelism becomes useful. A cluster of about six Indigo R4000 workstations can consistently deliver a performance exceeding that of a Cray XMP, and is considerably faster in favourable circumstances. The more specialized P.V.M.E software and hardware studied on the RS6000 machines seems to be even more promising.

The other comparable software option, ParaSoft Express, is a hybrid which allows low-level message communication such as that in P.V.M. as well as that at higher levels. There is also an automatic parallelizing compiler for Fortran called ASPAR that seems to be highly successful with determinist 'DO' loops.

The pace of development is rapid and more specialized software and networking hardware has begun to emerge that will ultimately be more suited to large lattice Monte-Carlo simulations. However, the coarse-grained parallelization of the Widom lattice-model simulation has been shown to be successful, often resulting in a near theoretical speedup when lattices of size forty cubed and larger are used. As we have earlier commented, this success is based on the programmer being able to split the problem up into a group of regular non-interacting sublattices. Typically the number of such sublattices should be comparable to or less than the number of workstations in the cluster.

One must accept that, on purely technical grounds, the coarse-grained parallelism paradigm seems less suited to large lattice simulations than one of the massively-parallel SIMD machines. However, if the sublattices are large enough and overhead communications are minimized one can practically attain supercomputer performance with a cluster of less than ten mid-range workstations. Finally, one should note that the most rapid improvements in processor performances are becoming available quickly and cheaply on workstations, simply because of market forces. This situation seems likely to persist for the foreseeable future. This is, perhaps, the single most compelling reason for the majority of scientists to pursue parallel workstation environments.

6. Analysis of Monte-Carlo data: finite-size scaling in Monte-Carlo simulations

One of the advantages of lattice models is that they are so simple that one can carry out exhaustive simulations to determine the phase diagrams and properties of the phases as a function of the parameters. Is is possible to carry out rather complete analyses of the data to determine, for example, if a transition is first or second order or to calculate macroscopic dependencies of the parameters from finite-sized samples. We shall now discuss some of these techniques.

The techniques of finite-size scaling, phenomenological rescaling and Monte-Carlo renormalization are useful for determining the nature of a transition, and then studying it in detail if it is second order. I shall discuss only finite-size scaling in this Chapter.

6.1. SECOND-ORDER TRANSITIONS

The reader will recall that the principal characteristic of a second-order transition is that the correlation length diverges at the critical temperature with a power law behaviour. This should be distinguished from a first-order (discontinuous) transition where the correlation length remains finite. Now, given that the simulation is to be carried out in a finite-size cubical box it is clear that we can never see a true critical phenomenon because ξ can at most be the box-length, L. We can fairly accurately locate the critical temperature of the infinite system by extrapolation of the finite size 'transition' temperatures according to the following prescription. The correlation length of the finite system is as large at it can be, given the finite size of the box. We may assume, therefore, that the 'critical' temperature of the finite system ($T_c(L)$) is related to that of the infinite system (T_c) by eqn (1)

$$\left(\frac{L}{a}\right)^{-1/\nu} = T_c(L) - T_c. \tag{10}$$

This equation may be used to extrapolate the data from simulations of different system sizes, L, to obtain T_c. Similar arguments apply to all other divergent quantities, such as the heat capacity. In that case we have,

$$C_v \approx C\,|T - T_c|^{-\alpha} \tag{11}$$

where α is another critical exponent, and therefore,

$$C_v(L) \approx C\left(\frac{L}{a}\right)^{-\alpha/\nu}. \tag{12}$$

One should be aware that all of these arguments apply only to the 'critical region' which may involve very large system sizes. In such cases one would prefer to apply workstation parallelism in such a manner that only a single independent sublattice is stored and updated on each workstation. Unfortunately this means that the network communication to access the state of neighbouring sites may be impracticably long.

6.2. FIRST-ORDER PHASE TRANSITIONS

It is known from elementary discussions of Einstein fluctuations that, within one phase, one expects that values of macroscopic observables should be Gaussian distributed around the mean. This is nothing more than a restatement of the central limit theorem for 'large' variables. As one approaches a first-order transition (say where the phases are unrelated by symmetry) it is possible for two phases to be simultaneously metastable. One might then expect a double Gaussian peak. In fact one can show that, to a good approximation, the coexisting phases produce Gaussian peaks in the probability of finding a given energy that have precisely the same area. The resulting analysis has been outlined by Allen in Chapter 3 and will not be reproduced here. Suffice to say that the shift in the transition temperature (of order L^{-3} in three dimensions), and heat capacity (of order L^3) are now understood and full finite-size analyses are now readily carried out.

7. Equilibrium simulation of Ising lattice model of microemulsion and the systematic construction of the phase diagram of microemulsion

As I have attempted to emphasize throughout, the main advantage of lattice models is that they are sufficiently simple that we can hope to reliably reproduce the global phase diagram as a prelude to studying the properties of the phases. The advantage of having a global phase diagram is that one need not be overly concerned about the primitive nature of the interactions, since it is possible to locate the phase or phases of interest. The interaction parameters are therefore less significant in a quantitative role. Indeed, the interesting physical questions usually relate to variations of correlation lengths, surface tensions or other observables as a function of parameters. Even if one had an accurate molecular potential for oil, water and amphiphile, and could effectively carry out the simulations for the mixtures, it is not clear that the effort would lead to any real advance in understanding. For example, the basic phenomena of liquid and gaseous phases in equilibrium along a first-order transition, the latter ending in a critical point, are all faithfully reproduced by the nearest-neighbour Ising model. It turns out that if one chooses a reasonable value of coupling and then evaluates the liquid-gas interfacial tensions along the coexistence curve in a fluid with a fairly symmetrical phase diagram, there is really quite good agreement with experiment. I believe this is the sort of information that one should expect from most lattice-model studies. Thus, one may expect the correct topology of the phase diagram, and reasonable forms for the evolution of most observables with parameters, but certainly not precise values of non-universal quantities. At this stage I intend to give a very down-to-earth view of the construction of the phase-diagram of the model described in eqn (9). My reason for doing this is that it is one of the most complex models to have been elucidated. My comments are directed towards an aspiring practitioner, and certainly makes clear that Monte-Carlo simulation alone cannot solve the problem.

I begin by reminding you of the basic form of the hamiltonian, eqn (9). It turns out that this energy can be decomposed into a sum of independent energies for octahedral clusters, a technique that has been thoroughly discussed in the literature [4]. As a consequence, it is possible to determine which local arrangement of spins are stable for the various values of the couplings. If those local arrangements can be patched together to fill the whole lattice then that is the ground or zero-temperature state of the hamiltonian. Sometimes many states may be made by patching clusters in different ways, and this leads to a great degeneracy of ground states at a single set of parameters. Now, at any finite temperature most of these states acquire some local stability with respect to Monte-Carlo simulation. Hence we have a huge number of stable spin-states that are somewhat comparable in free-energy.

Our approach is to first determine the ground states and then use Monte-Carlo simulation to determine the loci of instability of the ground states by choosing various cuts on the finite-temperature phase diagram and locating the maximum in the heat capacity. For some part of the instability curve we find that the ordered phase undergoes a transition to the disordered phase. The presence of hysteresis is a warning that the transition may be first order, and should be investigated more closely. Where the loci of stability cross there is clearly more than one local minimum of the free-energy and one either calculates the free-energy for each phase by integration of the internal energy or simply relies on a test of relative stability. The latter consists of evaluating the time for break-up of the competing phases. There is no deep a priori reason for this to work, but it turns out to be a fairly reliable guide. In fact, the low-temperature phase diagram of the present model may be thoroughly studied using low-temperature expansion and a cross check is therefore possible with the conclusions of simulation. With this assurance we can then simply begin with the known phases present at low temperature and carry out much shorter runs to construct the phase diagram at higher temperature.

The phase diagram close to the order-disorder transition is more subtle because, although mean-field theory indicated that the transitions should be second order, there are no stable fixed points accessible to a renormalization group flow, and other considerations indicate that fluctuations may drive the transitions to be first order. The figures below illustrate some of these points. Firstly, Fig. 1 is the phase diagram. Figure 2 shows the energy and heat capacity calculated as one crosses the ferromagnetic boundary close to the apparent Lifshitz point. It seems that this is clearly a second-order

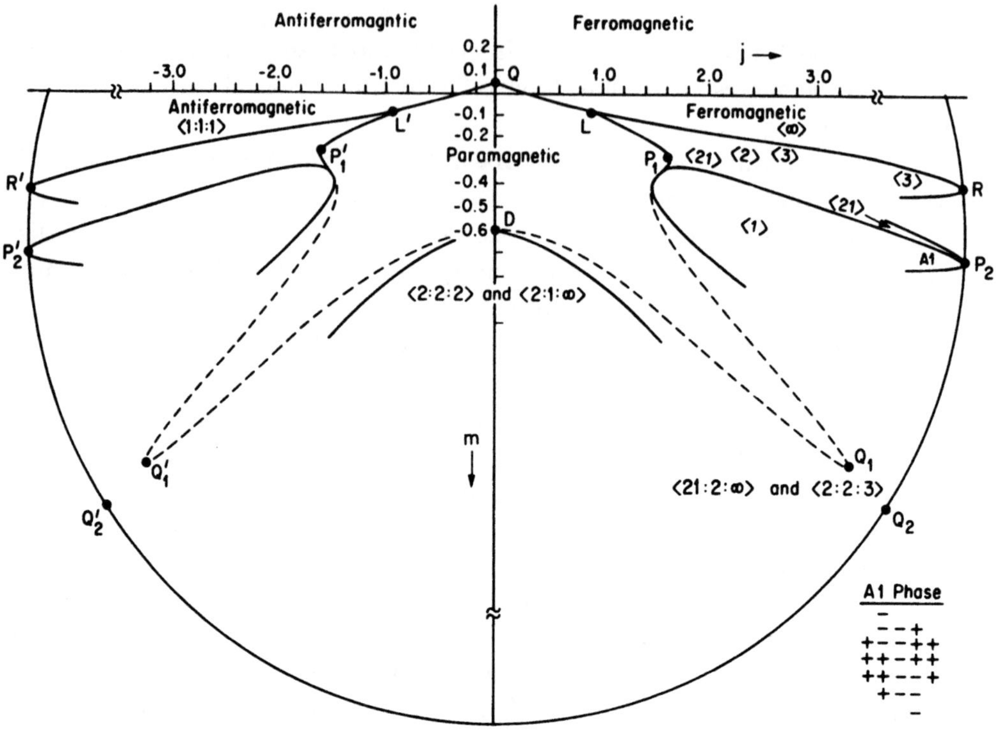

Fig. 1. Phase diagram of the Wheeler-Widom model. $j = J/k_{\rm B}T$, $m = M/k_{\rm B}T$ with $M_1 = M$, $M_2 = 2M$ in eqn (9).

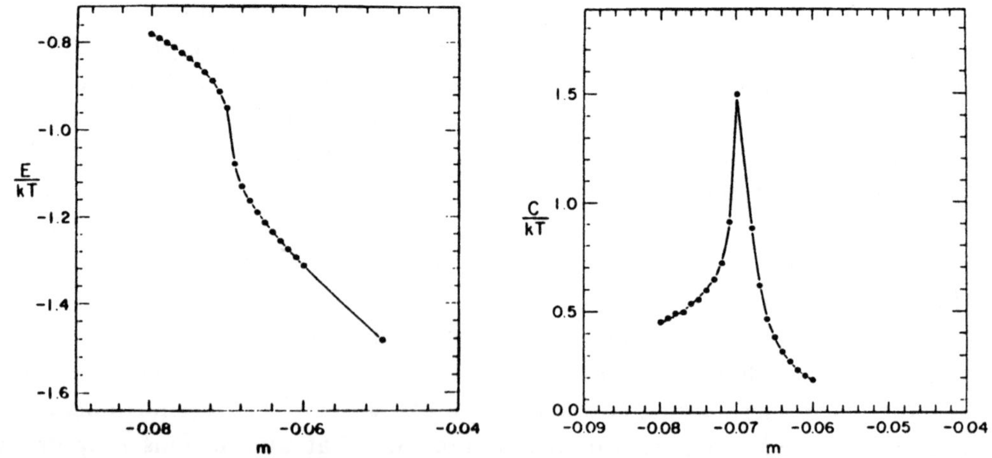

Fig. 2. Energy and heat capacity across the ferromagnetic-paramagnetic boundary for the Wheeler-Widom model.

transition, a point established by finite size scaling.

Figure 3 shows the energy and heat capacity for the lamellar to paramagnetic (microemulsion) phase transition, again close to the point L on the phase diagram. I should emphasize that for smaller systems this appears to be a continuous transition, and only when one studies such large sizes does it become apparent that the transition is clearly first order. Again finite-size scaling confirms this conclusion.

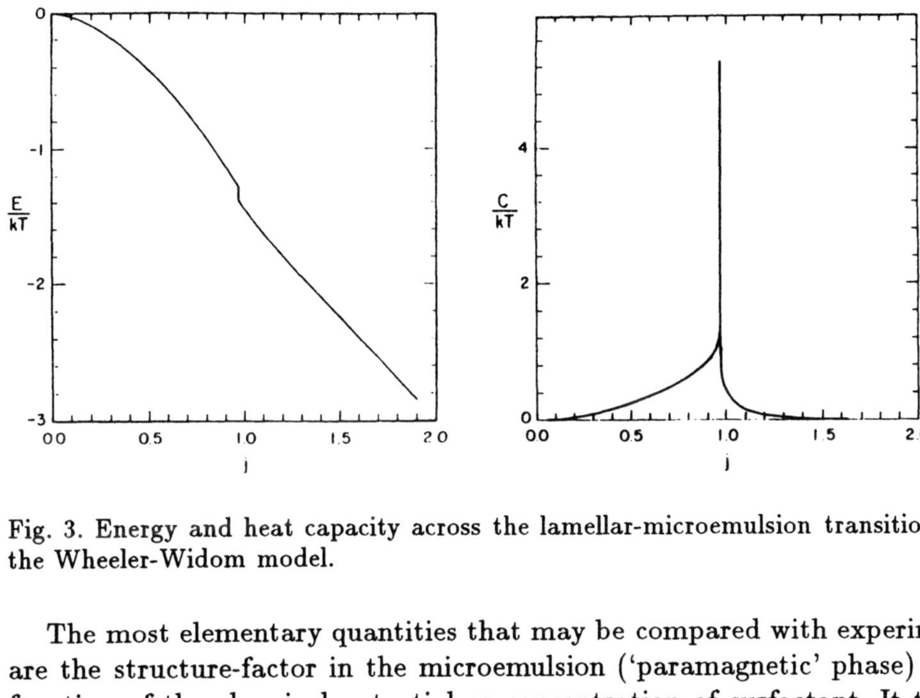

Fig. 3. Energy and heat capacity across the lamellar-microemulsion transition for the Wheeler-Widom model.

The most elementary quantities that may be compared with experiment are the structure-factor in the microemulsion ('paramagnetic' phase) as a function of the chemical potential or concentration of surfactant. It turns out that in the lattice model the simple form for the correlation function given in eqn (1) is no longer appropriate and, typically, a new length scale, d, emerges. Such a comment had been made by Teubner and Strey, based on experiment and a Landau theory. Thus,

$$G(r) \approx A \frac{e^{-r/\xi}}{r} \sin(2\pi r/d) \tag{13}$$

where d/ξ is roughly three for a good microemulsion. In these matters theory and experiment agree quite well. In fact there are a number of other phenomena such as interfacial tensions and wetting transitions where the model seems to agree quite well with experiment. However the discussion of these matters would require much more space than this Chapter provides.

8. Conclusions

The outline that I have presented above should illustrate the strengths and weaknesses of lattice models. Apart from the scientific advantages of being able to study very simple models that encompass the essential physical phenomena, one has a problem that is readily parallelizable. This should be contrasted with the case of continuum potentials where the data structures are not regular and difficult to parallelize in a highly efficient manner.

In any case the new developments in the creation of parallel workstation environments should have a qualitative effect in the sort of lattice problems that will be addressed by the broad scientific community in the coming few years.

References

[1] S. MA, *Critical Phenomena*, Plenum Press.

[2] B. WIDOM, *J. Chem. Phys.* **84**, 6943 (1986).

[3] K. A. DAWSON, M. D. LIPKIN, and B. WIDOM, *J. Chem. Phys.* **88**, 5149 (1988).

[4] K. A. DAWSON, B. L. WALKER, and A. BERERA, *Physica A* **165**, 329 (1990).

[5] G. VISSER, A. HOGGAT, and K. A. DAWSON, Computer Sciences Distributed and Cooperative Processing, in *Collected Papers of I.B.M. Prizes in Computation (1990)*, 1990.

Computer Simulation of Polymers

K. Kremer
IFF
Forschungzentrum Jülich
5170 Jülich
Germany

ABSTRACT. This Chapter gives a short overview of current Monte Carlo and molecular dynamics simulations of coarse-grained polymer models. For general statistical properties chemical details are not needed. The described methods and models apply to single isolated polymers as well as to melts and solutions. The problem of the dynamics of polymer melts is treated in some detail.

1. Introduction

Polymer simulations have been a very important tool in polymer science for many years. There are several reviews, which cover much of the field [1–6] including some of the technical aspects. It is the purpose of the present Chapter to combine both technical information about algorithms and methods together with typical examples and applications. To this extent the Chapter will be more general than some of the other contributions in this volume (see Chapters 4 and 6). General discussions of the Monte Carlo (MC) and molecular dynamics methods can be found in the contributions of Tildesley (see Chapters 1 and 2) as well as in several books and reviews [7–15].

Polymers, however, require special attention and consideration. They are, for the purposes of the present Chapter, chain molecules which are built of repeat units. Typical examples of monomers which are connected to chains, are

PE	$(CH_2)_N$	polyethylene
PDMS	$(Si(CH_3)_2O)_N$	polydimethylsiloxane
PS	$(CH_2(CH(C_6H_5))_N$	polystyrene
BPAPC	$(C_6H_4C(CH_3)_2C_6H_4CO_3)_N$	bisphenyl A polycarbonate

These are just four examples ranging from the most simple case of PE to the technically very important but more complicated BPAPC, which is used

M.P. Allen and D.J. Tildesley (eds.), Computer Simulation in Chemical Physics, 397–459.

among other things for compact discs. At a very first glance it is tempting and would be very interesting to perform a full scale MD (molecular dynamics) simulation, which includes all the chemical details. A valid question, however, is to what extent can this be done or is it necessary at all? The Table indicates that this is not possible in general. The short list of monomers illustrates that there are just far too many different polymers that can be chemically synthesized. Thus one would be restricted to the study of a very few model systems. However, even then it is not always clear how to determine the force fields properly. In many cases they are not very well known. In particular the complicated interchain or polymer-solvent interactions are only poorly understood at present.

Thus it is not possible to perform such a simulation properly. Even if one were able to overcome such difficulties, the situation would have improved only marginally, if at all. The reason lies in the huge variation of the relevant time and length scales, as Fig. 1 illustrates. The typical times range from

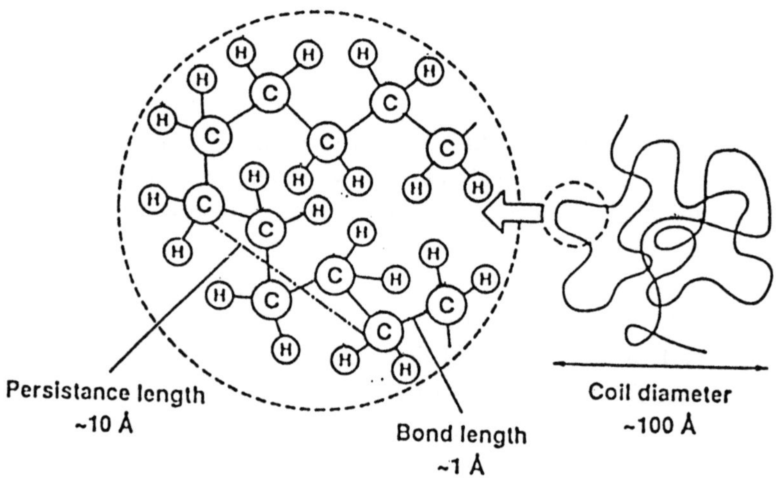

Fig. 1. A sketch of the different length scales in polymer problems for the simple example of polyethylene [3, 16]

10^{-13} s for local oscillations, requiring MD timesteps of about 10^{-14} s, up to seconds or hours for the diffusion time of a chain, and to much longer times for collective phenomena or relaxation near the glass transition. A time of the order of the chain diffusion time is needed to relax the chain conformations. In a similar way we have to deal with length scale variations that range from 1 Å to several hundreds of Å's. For a reasonably sized system, no computer will cover this range of $\mathcal{O}(10^{14})$ timesteps. Even semi-detailed models would require too much time. The reptation study which will be discussed later in the Chapter took around 1000h of cpu time [17–19]. A similar study with Roe's PE model [20, 21], a unified atom model, with a C-C spring constant reduced by a factor of 7, would have taken roughly 500000h of cpu time on a Cray YMP [6, 16]. It is easy to imagine the requirements for more complicated systems. Despite these tremendous complications the situation is better than the analysis suggests. To what extent do we really need the chemical details? Fig. 1 gives a caricature of the various relevant length scales. It also illustrates universality. For properties which are governed by length scales larger than a few Kuhn lengths (the stiffness length), the properties are independent of the chemistry. For example, in the limit of long chains of N monomers the mean squared end-to-end distance $\langle R^2(N)\rangle = AN^{2\nu}$. ν is a universal exponent, which is 0.588 in a good solvent for $d = 3$ and 0.5 in the melt. The chemical details are hidden in the prefactor A. This suggests a relatively satisfactory situation, namely for most questions of interest one can confine the simulation to the simplest models and for computational purposes the fastest models. In many cases we just need the monomer-monomer excluded volume and the chain connectivity. These models are often called coarse-grained models. Later on, we will face the problem of "mapping properties back to a given chemical system". This problem will be mentioned in a number of places throughout this Chapter.

Firstly, the Chapter will consider some general points on the various simulation approaches. Then methods for calculating static properties, mainly for single chains, will be discussed. After that MC and MD approaches to study solution and melt dynamics, glasses and networks are described. The conclusions give a short perspective on future directions as well as a short overview on other related work, which is not discussed in detail.

2. Polymer simulations: general considerations

Typically, there are three classes of models used for computer simulations of polymers, see Fig. 2. Sometimes combinations of these models are used. The simplest model, and the most widely used one, is the simple self-avoiding walk (SAW) on a lattice [2]. Each lattice point can be occupied only once. It is trivial to introduce nearest-neighbour energies and other generalizations. The second model system would be the direct generalization for continuous space, the pearl necklace model. The chain consists of hard spheres of diameter σ_0 and a fixed bond length ℓ_0. The third variant is mainly used in MD simulations. The monomers consist of point particles which interact with, in most cases, a purely repulsive Lennard-Jones interaction. For the bonded nearest-neighbours along the chain an additional spring potential is added which together with the Lennard-Jones repulsion determines the bond length ℓ. The purpose of a simulation is to generate statistically independent conformations or to follow the time evolution of a given global

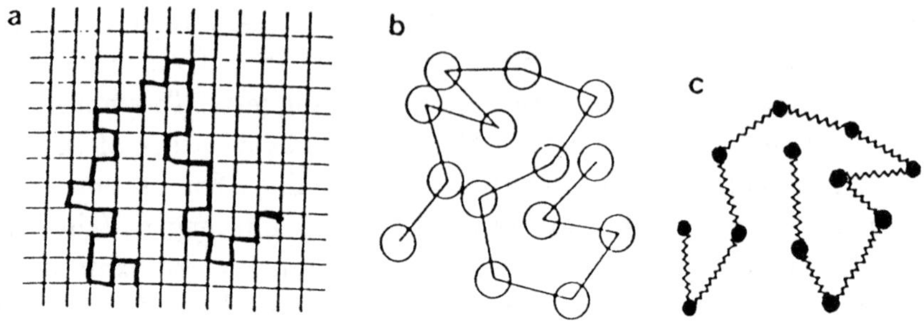

Fig. 2. Typical models for polymers used in simulations

configuration. The most natural ansatz would be to simply perform a MD simulation, where Newton's equations of motion

$$d^2\mathbf{r}_i/dt^2 = -\nabla\mathcal{V}_i(\{\mathbf{r}_i\}) \tag{1}$$

are solved numerically. $\mathbf{r}_i$ is the position of monomer i, and $\mathcal{V}_i$ is the potential from the interaction with all the other monomers. However, this direct approach cannot be used for isolated chains. The reason lies in the structural properties of the chains. A linear polymer without excluded volume interactions has the structure of a random walk [22]. Its dynamics can be described by eigenmodes, Rouse modes, which decouple [23]. Thus, solving eqn (1) exactly without excluded volume would never equilibrate the chain. In the case of a SAW, the Rouse modes are no longer eigenmodes of the chain. However, the deviations come mainly from the long-range contacts (monomers which are far apart along the chain) which still obey the excluded volume constraint if they approach each other in space. However, the chains are fractal objects with a fractal dimension $d_f = 1/\nu$. For long chains the density decays with a power law. Consequently, the necessary long-range interactions, which cause the swelling of the chains, are very infrequent. Thus the natural MD approach can only be used for a chain interacting directly with solvent, other chains or for chains with long-range interaction potentials. To avoid this problem one has to couple the chains to a stochastic process. This can typically be done using the Monte Carlo method. (For details of the Monte Carlo methods see Chapter 1. Another technique, for use in conjunction with the MD method, would be to introduce local stochastic jumps. This would result in a so-called dynamic MC simulation. Such local changes follow the characteristic Langevin dynamics of the chains. This should, however, be

TABLE I
An illustration of the timescales in polymer simulations.

	Liquid	Polymer melt
particles	N_{TOT}	N_{TOT}
density	ρ_0	ρ_0
cpu time per timestep	$\propto N_{\mathrm{TOT}}$	$\propto N_{\mathrm{TOT}}$
motion distance to equilibrate	$\propto \rho_0^{-1/d} \propto N_{\mathrm{TOT}}^0$	$\propto \langle R^2(N)\rangle^{1/2} \propto N^{1/2}$
physical relaxation time	$\tau \propto N_{\mathrm{TOT}}^0$	$\tau_N \propto N^{2\ldots3.4}$ (varies with chain length)
cpu time	$T \propto N_{\mathrm{TOT}}$	$T \propto N_{\mathrm{TOT}} \cdot N^{2\ldots3.4}$

avoided whenever possible. To illustrate this, let us compare a simple liquid or a lattice gas and a polymer melt of the same density and total number of particles. A timestep for MD is explicit in the algorithm, while for MC it can be taken as one attempted move per particle of the system. Of course, the comparison does not hold for critical phenomena within the liquid; there critical slowing down also significantly increases the cpu time. The exponent N^x, $x = 2\ldots3.4$ can actually be interpreted as a dynamical exponent for the critical slowing down [23]. (In the language of critical phenomena the inverse chain length corresponds to $(T - T_c)/T_c$. Thus the relaxation time follows $((T - T_c)/T_c)^z$ with $2 \leq z \leq 3.4$.) This causes significant problems since typical N values of interest in most cases start at around 50 to 100. Thus, there is one important rule that should be taken into account before starting a polymer simulation: if possible, avoid algorithms with slow physical dynamics. There are a few exceptions, which will be discussed later. However, if one wants to study chain dynamics there is no other possibility than to follow the slow realistic dynamics.

Following these general remarks we now discuss a few methods in terms of their applications. Naturally, this covers only a small number of the problems which have been investigated by many authors in the past.

3. Algorithms for static properties

3.1. STATIC METHODS FOR ISOLATED CHAINS

Having demonstrated that the methods which follow the realistic dynamics are not very useful for single isolated chains, we now try to generate the conformations by a purely stochastic process. In order to do so, we require of course that the random numbers which are available are uncorrelated [2, 10]

First, let us consider simple Self-Avoiding-Walk (SAW's) on a lattice. In order to fulfil the excluded volume requirement each lattice site can only be occupied once. Since this is the only requirement, each available conformation of an N-step walk has the same probability. If we fix the first step, then each new step is taken with probability $1/(q-1)$, where q is the coordination number of the lattice ($q = 4$ square lattice, $q = 5$ triangular lattice, $q = 6$ simple cubic lattice etc.). The most naive way of generating conformations with equal probability is called simple sampling.

By this procedure the walks are grown purely stochastically. Each time an attempted new bond hits a site which is already occupied, one has to start at the very beginning. If one did not do this, different conformations would have different probabilities. As we will see later, this would introduce an effective attraction among the monomers. With this method, each conformation is taken randomly out of the $q(q-1)^{N-1}$ possible random walk paths, which do not include direct back-folding. However, the total number of SAW's on a lattice is given by [22]

$$Z(N) = c_0\, q_{\text{eff}}^N\, N^{\gamma-1} \qquad N \gg 1 \tag{2}$$

with $q_{\text{eff}} < q - 1$ and c_0 a number of order unity. γ is a purely dimension- but not lattice-dependent critical exponent ($\gamma \approx 7/6(d = 3), 4/3(d = 2)$) Typical numbers for q_{eff} are 2.6385 (square), 2.879 (diamond, $q = 4$) and 4.6835 (simple cubic) [2]. Thus the success rate $A(N)$

$$A(N) = A_o(q_{\text{eff}}/\,(q-1))^N\, N^{\gamma-1} \tag{3}$$

decays exponentially. This attrition has already been discussed and recognized in the very first paper by Rosenbluth *et al.* [24]. A typical number for A is e.g. $A(100) = 0.03$ on the diamond lattice. This illustrates that this trivial first attempt is only useful to get a first and rough estimate for both q_{eff} and γ. The first improvement was already suggested in 1955 by Rosenbluth and Rosenbluth [24]. The idea of the inversely restricted or biased sampling is to look ahead for at least one step in order to overcome the attrition. More modern approaches look several steps ahead or implement this idea in a dynamical scheme (see Chapters 4 and 6). For simplicity and clarity we confine ourselves here to the basic idea and methods.

Before attempting a new move, one first counts all empty neighbouring cells, which we call jump sites. Then the new step is chosen randomly out of the available steps. This procedure is illustrated in Fig. 3. Especially for $d = 3$ there is almost no attrition left for $N \lesssim 1000$ (diamond lattice) to $N \lesssim 10000$ (fcc lattice) [25]. Now, however, the different conformations have different probabilities. Configurations with locally higher density appear too often since their probability is higher. There is another problem to keep in mind; for lattices with $q - 1 > 2$ ($q - 1 = 2$ only holds for the

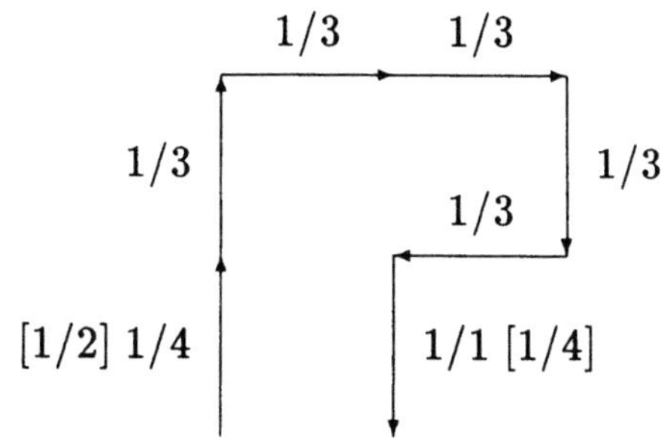

Fig. 3. An illustration of the probabilities and the irreversibility of walks generated by the inversely restricted sampling [2].

honeycomb lattice) the probability of a given conformation might depend on the direction of the growth of the chain, Fig. 3.

In order to sample the simple SAW's where each conformation has the same probability, one has to correct for the bias introduced. If q_i is the number of available jump sites for step i, the corrected weight $G(\{q_i\})$ of the N-step walk is given by

$$G(\{q_i\}) = \prod_{i=2}^{N} \left(\frac{q_i}{q-1} \right) \tag{4}$$

Averages are then given by (e.g. for $\langle R^2(N) \rangle$):

$$\langle R^2(N) \rangle = \sum_{\text{walks}} R^2(\{q_i\})\, G(\{q_i\}) \Big/ \sum_{\text{walks}} G(\{q_i\}) \,. \tag{5}$$

This approach is conceptually very simple and it is very tempting to use it in general. There are, however, severe limitations, which we want to discuss in more detail now. In order to estimate the quality of a sample generated by the Rosenbluths' method, we go back to the number of jump sites q_i. Since $q - 1 - q_i$ sites are occupied by other monomers, which are not nearest neighbours along the chain, we can interpret them as nearest-neighbour contacts or nearest-neighbour energies [26]. With

$$E(\{q_i\}) = \sum_{i=2}^{N} (q - 1 - q_i) \tag{6}$$

we obtain for the corrected walk

$$\left\langle E^{\text{SAW}} \right\rangle = \sum_{\text{walks}} E(\{q_i\})\, G(\{q_i\}) \Big/ \sum_{\text{walks}} G(\{q_i\}) \tag{7}$$

The originally generated sample, however, from which we want to calculate $\langle E^{\mathrm{SAW}} \rangle$ can also be analyzed using

$$\left\langle E^{\mathrm{BS}} \right\rangle = \sum_{\mathrm{walks}} E(\{q_i\}) \Bigg/ \sum_{\mathrm{walks}} 1 \,. \tag{8}$$

From this we arrive at a simple argument for the general problem of inversely restricted sampling. Energies are extensive quantities, giving $E^{\mathrm{SAW}} \propto a_{\mathrm{SAW}} N$ and $E^{\mathrm{BS}} \propto a_{\mathrm{BS}} N$ with $a_{\mathrm{BS}} > a_{\mathrm{SAW}}$. The typical energy distribution is given by a Gaussian with mean value of E and a width, which is given by the square root of the heat capacity C, with

$$C \propto \left\langle (E - \langle E \rangle)^2 \right\rangle = \begin{cases} c_{\mathrm{SAW}} & N \\ c_{\mathrm{BS}} & N \,. \end{cases} \tag{9}$$

Since the heat capacity itself is an extensive quantity, the width of the energy distributions grow with $\sqrt{N}$, while their mean values are proportional to N. c_{SAW} and c_{BS} are the corresponding prefactors. Using this argument it can

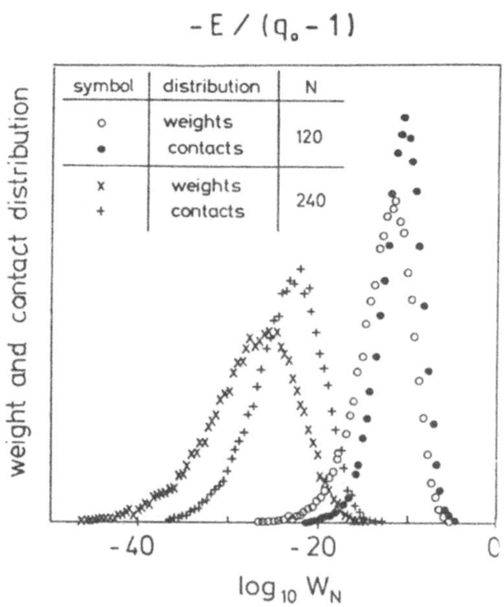

Fig. 4. Typical distribution functions of weights G and nearest-neighbor contacts E^{SAW}, eqn (4) and eqns (7,8) from Ref. [27].

actually be shown that the errors, at a fixed number of generated walks, grow exponentially with the length N of the walk; for details see Ref. [26]. Fig. 4 gives a sketch of the distribution of weights and weighted energies for $N = 120, 240$. As we will see later there are more modern and much better ways of generating simple SAW's. There are still applications, where this approach either independently or in combination with other methods, is very useful. The bias introduced by the sampling can be interpreted as an attractive interaction between the monomers. Thus a first attempt is to use the Rosenbluth scheme for the collapse transition of an individual chain.

To see this we go back to eqn (5). If we introduce a Boltzmann weight for the interaction, we can write for any quantity $\langle \mathcal{A} \rangle$

$$\langle \mathcal{A} \rangle = \frac{\sum_{\text{walks}} \mathcal{A}\, G(\{q_i\}) \exp(-E(\{q_i\})/k_{\text{B}}T)}{\sum_{\text{walks}} G(\{q_i\}) \exp(-E(\{q_i\})/k_{\text{B}}T)} \,. \tag{10}$$

For the Θ-temperature, roughly speaking the temperature where the volume exclusion and the attraction between the monomers just cancel each other, we find (for $N \lesssim 1000$)

$$\left[G(\{q_i\}) \mathrm{e}^{-E(\{q_i\})/k_{\text{B}}\Theta} \right]_{\text{average}} \approx 1\,. \tag{11}$$

Thus introducing the standard bias almost exactly samples conformations at the Θ point (only $d = 3$!). To be more precise, the effective interaction, introduced by the algorithm, corresponds to a temperature that is weakly above the Θ-point. For details see Refs. [25, 28].

To illustrate this application let us briefly discuss the collapse transition of linear and star polymers. Before we proceed there is one important note of caution to be made. For all methods where chains are grown, one typically uses all chain lengths which occur during the generation for averaging. For example, while generating chains of length $N = 200$, all chains of length $N = 100$ which are generated, independently of whether they finally reach $N = 200$ or not, are taken for averaging. There is nothing wrong with this; however, the errors which occur in the averaging are correlated since both samples contain to some extent the same short chains. If poor random numbers are used this might lead to wrong results. In any case the error is not given by the scatter of the data! To estimate the error one has to use the fact that the chains are not self averaging. The distribution $\langle R^2 \rangle$ has a width $\Delta \approx \langle R^2 \rangle$ thus for m generated chains the error is roughly $\Delta/\sqrt{m}$, which might be larger than the scatter of the data suggests.

Typically polymer chains can be solvated in organic solvents. The quality of the solvent, however, varies with temperature or other parameters such as the chemical details or the pressure. There are three different regimes for the asymptotic properties of isolated chains [22]. With temperature T as the variable parameter one finds ($d = 3$)

$$\langle R^2(N) \rangle \propto N^{2\nu} \begin{cases} \nu = 0.588 & T > \Theta \\ \nu = 1/2 & T = \Theta \\ \nu = 1/3 & T < \Theta \end{cases} \,. \tag{12}$$

For good solvents the chains are expanded or swollen. At Θ, where the repulsion and the attraction just cancel, the average conformation is that of a random walk ($\nu = 1/2$), while in the case of a poor solvent the chain becomes a collapsed compact object. This interesting behaviour is rather difficult to study experimentally, since one has always to work at finite concentration, which will eventually lead to a phase separation between solvent and polymers. Following de Gennes we can write down the Flory theory for the free energy density f of polymer solution as

$$f/T = \phi \ln \phi/N + \frac{1}{2}(1 - 2\chi)\phi^2 + \frac{1}{6}\phi^3 + \cdots . \tag{13}$$

ϕ is the monomer density and N the length of the chain. The first term is given by the translational entropy of the chains. The ϕ^2-term describes the two-body interaction, while χ, the Flory Huggins parameter, is usually thought to be proportional to $1/T$ ($\chi = 1/2$ corresponds to $T = \Theta$). The ϕ^3 term describes the repulsive 3-body interaction. For dilute solutions the ϕ^2 term will dominate the free energy. However, for $\chi = 1/2$, the ϕ^3 term survives. In the classical picture, where only binary interactions play a role, $\chi = 1/2$ corresponds to the Θ-point. Before going into the details, let us briefly discuss the phase diagram (see Fig. 5). If we disregard the ϕ^3 term the second virial coefficient A_2 between chains and monomers would vanish. Within a scaling picture, as we can see from Fig. 5, the Θ point is a tricritical point. One can write $R^2(N)$ in a crossover scaling form

$$\langle R^2(N,T)\rangle = N f(N^{\varphi}\tau) \tag{14}$$

where φ is the tricritical crossover exponent, $\varphi = 1/2$, and τ the reduced temperature, $\tau = (T - \Theta)/\Theta$. With $x = N^{\varphi}\tau$ we then have

$$f(x) \propto \begin{cases} x^{(2\nu-1)/\varphi} & x \to +\infty \\ x^0 & x \to 0 \\ x^{(2/3-1)/\varphi} & x \to -\infty \end{cases} \tag{15}$$

resulting in the asymptotic behaviour of eqn (12). It should be noted that such a scaling description is strictly valid only in the limit $N \to \infty$, with $N^{\varphi}\tau = x$ finite. It is actually one of the classic research areas in statistical physics to analyse finite size corrections to scaling. This scaling description leads to a very intriguing, simple and powerful physical picture, the blob picture, as illustrated in Fig. 6. At a given temperature τ near the transition temperature $\tau = 0$ the Θ-correlation length, ξ_B, is given by the average diameter of a chain of $N_B = |\tau|^{-1/\varphi}$ monomers. For $N \gg N_B$ the scaling variable $x \to \pm\infty$, denoting a good or poor solvent. For $N \ll N_B$ we get $x \to 0$, thus our chain is in the Θ-regime. Near Θ very long chains on length scales smaller than ξ_B are like random walks, while on scales larger than ξ_B

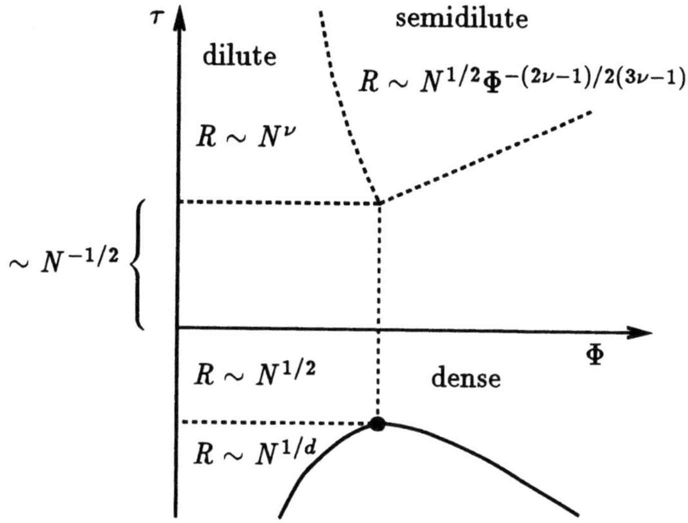

Fig. 5. A sketch of the phase diagram of a polymer solution near $T = \Theta$; $\tau = (T - \Theta)/\Theta$. Φ is the volume fraction. See e.g. Ref. [22].

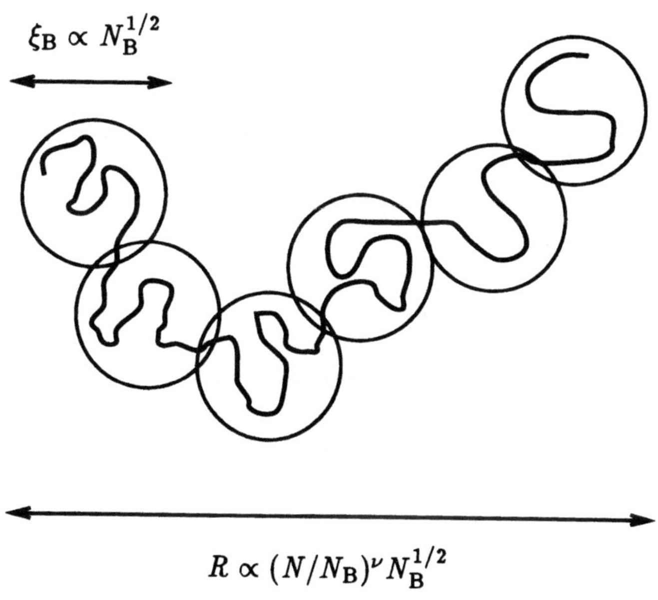

Fig. 6. An illustration of de Gennes' blob picture for a chain near Θ.

they are either good or poor solvent chains constructed of blobs of diameter ξ_B. This length scaling can directly be verified in a computer simulation. So far we have disregarded any influence of the higher-order interactions. Simulations, however, are already showing us that A_2 for monomers is slightly negative at the "experimental" Θ temperature, which is defined through the virial between chains or using the behaviour of $\langle R^2(N)\rangle$ [29]. A more precise treatment of the Θ-point within a renormalization group calculation shows that for single linear chains the repulsion three body leads to logarithmic corrections, such as [29, 30]

$$\langle R^2(N,\Theta)\rangle \propto N(1 - 37/(363 \ln N)) . \tag{16}$$

Although the corrections are probably not very important in the daily life of a polymer chemist, they are very interesting from a theoretical and conceptual point of view. On the other hand they are very difficult to detect. Thus we need to look for a system where the higher-order interactions are more important. In what follows, I will describe a recent investigation of the collapse transition of star polymers, as an example of how one might proceed [27, 28]. The aim of the investigation is to confirm the standard Θ-scaling and to search for effects of the higher-order interactions. To do this we will use the previously described scaling concepts extensively to analyse the data.

To generate chains near the Θ point we use the inversely restricted sampling for chains and star polymers on a face centred cubic lattice (fcc). The fcc lattice has coordination number 12; thus it easily allows us to consider star polymers, which grow out of one centre with up to $f = 12$ arms. First let us study the collapse transition point of a linear chain, i.e. $f = 1$. Chains of up to $N = 900$ monomers were generated using the biased sampling method. The physical quantities of interest were calculated using eqn (10). Firstly, we have to determine the asymptotic Θ-point. As Fig. 5 illustrates this transition point is only defined for $N \to \infty$. We use the scaling eqn (14) with

$$\frac{\langle R^2(N,T)\rangle}{N} = f(N^{\varphi}\tau) = f(x) . \tag{17}$$

The scaled end-to-end distance is a function of a single variable $x = N^{\varphi}\tau$. Thus $\langle R^2\rangle/N$ is a constant for fixed x. Plotting $\langle R^2(N,T)\rangle/N$ against T should yield a set of curves, which cross at $T = \Theta$. This is, however, not exactly the case, because of finite size effects. Fig. 7 shows these curves for $450 \le N \le 900$.

Since a clean extrapolation of Θ from Fig. 7 is impossible, we use the idea that $f(x) = \text{const}$ corresponds to a specific temperature for any N. Plotting $N^{-1/2}$ against T for several values of $f(x)$ should yield a set of straight lines, which cross at $N^{-1/2} = 0$. Fig. 8 shows such a plot. This

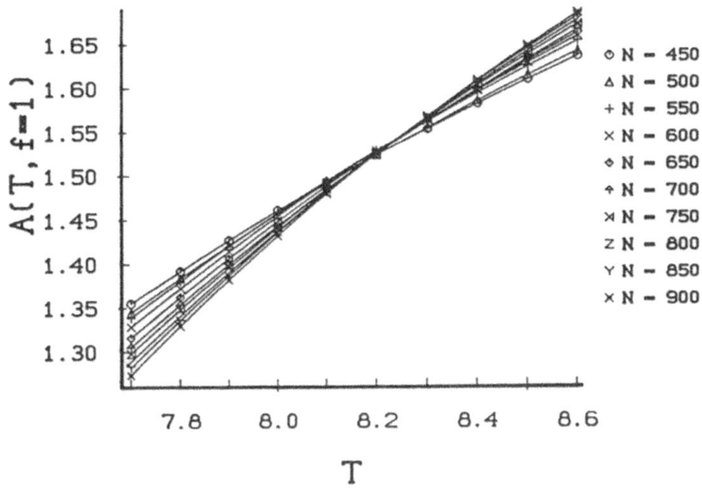

Fig. 7. A plot of $\langle R^2(N,T)\rangle/N$ as a function of temperature T for chains between $N = 450$ and $N = 900$ on the fcc lattice [28].

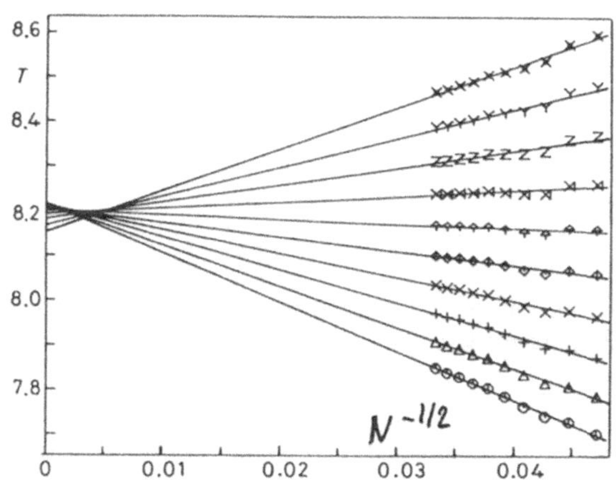

Fig. 8. An extrapolation of the crossing point of $f(x)$ = const for various chain lengths [28].

gives a rather precise estimate of $\Theta = 8.20 \pm 0.02$ for the fcc lattice. The lines cross at finite N, which is probably a finite size problem. This cannot be explained, however, by taking logarithmic corrections into account. Using this estimate of Θ one gets a very nice scaling plot for $\langle R^2 \rangle$ and other second moments (see Fig. 11). The chains are certainly not perfect random walks. If we consider higher moments we detect deviations. For a Gaussian random walk the second cumulant vanishes:

$$U = 1 - \frac{\langle R^4 \rangle}{\frac{5}{3}\langle R^2 \rangle^2} = 0\,. \tag{18}$$

This is not the case for the present chain lengths. Thus we find a tricritical scaling for second moment quantities. However, we are not able to detect the influence of the higher order interactions. The way to do this is to pack chains together, for example in stars. Fig. 9 shows typical snapshots of stars with $f = 10$ arms and arm-lengths of $N = 50$, 100, 200 [31]. The figure

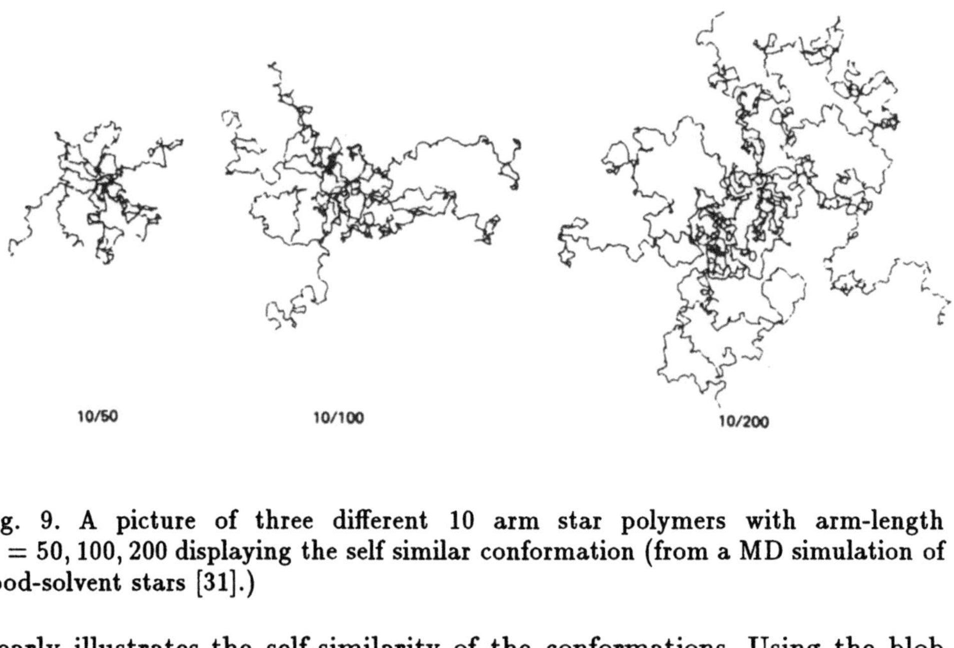

Fig. 9. A picture of three different 10 arm star polymers with arm-length $N = 50, 100, 200$ displaying the self similar conformation (from a MD simulation of good-solvent stars [31].)

clearly illustrates the self-similarity of the conformations. Using the blob picture we can directly derive the density as a function of the distance from the centre of the star, as well as the end-to-end distance. A star can be described as a semi-dilute polymer solution, where the screening length ξ is a function of the distance from the centre. For f arms $\xi^2(r)$ is given by the available surface area per arm, r^2/f thus

$$\xi(r) \propto r f^{-1/2} \tag{19}$$

consequently one directly derives for the centre-end distance

$$\langle R^2(N,f)\rangle \propto N^{2\nu} f^{1-\nu}\,, \tag{20}$$

for the density decay $\rho(r)$

$$\rho \propto r^{(1-3\nu)/\nu} f^{(3\nu-1)/2} \tag{21}$$

and for the g-factor

$$g(f) = \lim_{N\to\infty} \frac{\langle R^2(N,f)\rangle}{\langle R^2(Nf,1)\rangle} \propto f^{1-3\nu}\,. \tag{22}$$

Although these laws are strictly valid only for large N and many arms, in a good solvent it turns out that the above scaling holds for $f \geq 3$ [31–33]. We will return to this point in the context of the dimerization algorithm and of the MD simulations. Now let us return to the collapse transition. From the phase diagram one would expect Θ to be the same for stars and linear chains in the limit $N \to \infty$, f fixed. Experiments which measure the second virial coefficient, however, find a Θ-shift, namely $T(A_2 = 0) < \Theta$. To analyse this

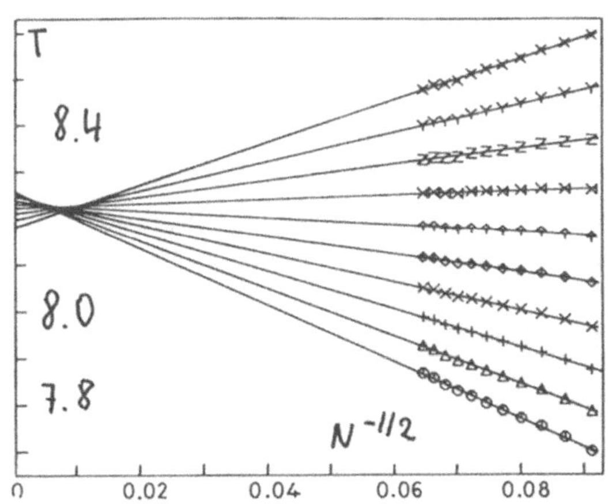

Fig. 10. As for Fig. 8 but for stars with $f = 8$ arms and arm-lengths of up to $N = 240$ monomers [28].

we look at single stars on an fcc lattice using the Rosenbluth scheme. Fig. 10 shows the extrapolation for a star polymer of $f = 8$ arms with up to $N = 240$ monomers per arm. Both the linear polymer and the star give the same Θ-temperature. Even more, using $\varphi = 1/2$ for both systems we observe the expected tricritical scaling of $\langle R^2(N,f)\rangle/N$ against τN^{φ}, as Fig. 11 shows.

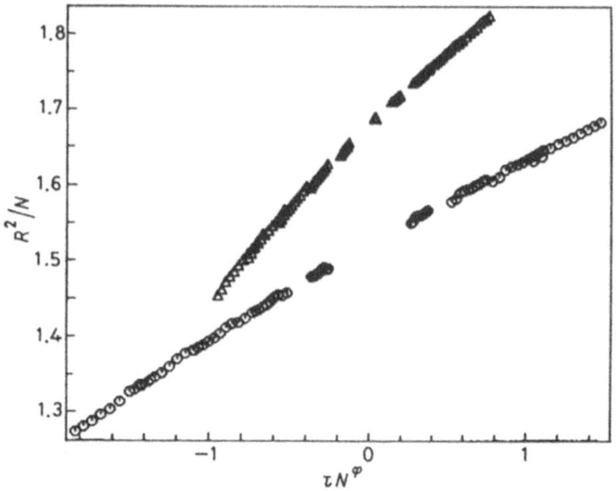

Fig. 11. The tricritical scaling of the $\langle R^2(N, f, T)\rangle$ for $f = 1$ (circles) and $f = 8$ (triangles) [28].

Needless to say, star polymers with intermediate number of arms give the same result. How can this be understood in the light of the experiments? To see this we look at the g-factor, eqn (22). If we replace the good-solvent ν by $\nu = 1/2$, $g \propto f^{1-3\nu} \propto f^{-1/2}$. This argument should hold for many arms. On the other hand there is an old argument by Candau *et al.* [34] derived from the random walk properties of a Θ-chain. They argue that at Θ each pair of arms can be taken as a random walk, giving $g = (3f - 2)/f^2$. This assumes that the chains are completely invisible to each other at Θ, leading to diverging densities for large f. If only two-body terms were important in the free energy this would be fine; however, the higher-order interactions cause the star to expand. For good solvents all expansions of the properties of stars to first order start with a term of order $(f - 1)$; the simplest star is the two-arm star. At Θ at least two branches are needed which repel each other in order to build such a "star". Since two chains together see another pair we can conjecture that the "minimal star" needs $f = 4$ branches leading to the ansatz

$$g(f) \propto (f - 3)^{-1/2} \tag{23}$$

which is a direct consequence of the 3-body interactions. Thus, measuring g at the linear polymer Θ temperature should yield a crossover from the Candau scaling to the scaling of eqn (23). Fig. 12 gives a detailed comparison that the general picture of the tricritical scaling is confirmed by both linear and star polymers using the inversely restricted sampling.

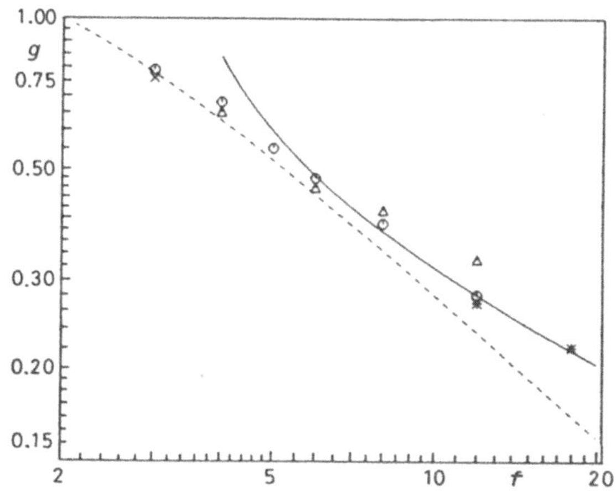

Fig. 12. The g-factor for simulation and experiment at $T = \Theta$ as determined from the linear polymer simulation/experiments. The circles are from simulation, while the other symbols are for experiments of different polymers. The full line gives the scaling of eqn 23, while the dotted one corresponds to the Candau rule. For details see Ref. [28].

It is important to remember that data of such quality could only be generated because the bias and the Boltzmann factor almost cancelled each other, eqn (11). There are many other applications of this idea. Another example is the adsorption of a polymer near a surface or an interface [35–39]. This is an interesting transition of a d to a $d-1$ dimensional object. Especially for the Θ-chains this yields interesting results [39], which are both experimentally and theoretically of high relevance. For details we refer to the recent book of E. Eisenriegler [40] and the original simulation papers. Again the clue to success was that the bias directed the system to an effective temperature or interaction constant, which is near to the critical value. There are several extensions of this approach throughout the literature. Meirovitch uses a method where he scans for possible future directions by looking several steps ahead [37, 38, 41]. Recently, the idea was incorporated in a very interesting, general dynamical scheme, called the conformational biased Monte Carlo, CBMC, procedure. This approach is discussed in some detail in Chapters 4 and 6.

We still have to devise a good and efficient approach to generate SAW's from scratch. A very interesting unbiased method is the *dimerization* approach due to Alexandrowicz [42] and Suzuki [43]. The idea is illustrated in Fig. 13. Originally, the idea is to begin by generating many short SAW's of length N_0. These walks can then be combined randomly. Thus for every

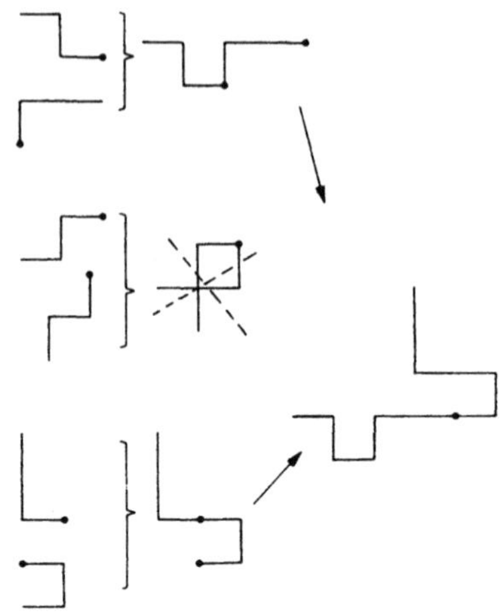

Fig. 13. Illustration of the dimerization method [2].

step the success rate P_{bin} for binary assembly of walks of length N_1 and N_2 is

$$P_{\text{bin}}(N_1 + N_2 = N) = Z(N)/(Z(N_1)Z(N_2)) . \tag{24}$$

Using $Z \propto q^N N^{\gamma-1}$ we get

$$P_{\text{bin}}(N_1, N_2) = N^{\gamma-1}/(N_1 N_2)^{\gamma-1} . \tag{25}$$

With $N_1 + N_2 = N$, it is obvious that P_{bin} is maximal for $N_1 = N_2 = N/2$. The original procedure, however, is not very efficient. Alexandrowicz generated a few thousand short walks, which had to be stored. A more direct procedure is to generate walks using simple sampling and map the dimerization procedure onto this process [27, 33]. To do this we define the basic unit as N_1. After a SAW of N_1 monomers is generated we continue. If the next N_1 steps are violating the EV condition within the second piece, we start at N_1 again. Only an exclusive overlap with the first N_1 monomers makes it necessary to start at the very beginning. In this way a hierarchical procedure is defined which is very efficient and gives very good results for both the exponents γ and ν. This can now be readily generalized to branched polymers such as stars. Batoulis and Kremer used this approach for the study of the good-solvent properties of star polymers. In this case, in particular, the dependence of γ on the number f of arms of the star is interesting and very difficult to estimate analytically. Using this method we were able to give precise results for $\gamma(f)$ but also $\langle R^2\rangle$, $\langle R_G^2\rangle$ and the hydrodynamic radius

$$\left\langle \frac{1}{R_H} \right\rangle = \frac{1}{N^2} \sum_i \sum_{i \neq j} \left\langle \left| \frac{1}{r_{ij}} \right| \right\rangle \qquad \text{where } r_{ij} = |\mathbf{r}_i - \mathbf{r}_j| \,. \tag{26}$$

In this case, very strong corrections to scaling were observed, which, to some extent, explain the deviations of experiments such as light scattering from the asymptotic power law behaviour. Fig. 14 shows the extrapolation of the hydrodynamic radii as a function of $N^{-1/2}$, which was thought to be the correction to scaling for this quantity [33]. The correct form is now known to be $N^{-(1-\nu)} = N^{-0.41}$ [44].

So far we have only discussed methods, which generate walks from scratch. It is clear that this is only be useful in the very dilute regime. Besides the methods discussed, there are many other approaches, which start with a given conformation and manipulate it dynamically. There are many methods which can be found in the literature. For details we refer to Ref. [2]. Here we confine ourselves to two examples of methods, which do not change the length of the chains, the pivot method and the generalized reptation methods. Neither of these methods can be used for dynamics, although there are ocasional examples of this in the literature.

3.2. DYNAMIC METHODS FOR STATIC PROPERTIES

A conceptually very simple approach is given by the *pivot* algorithm [2, 45, 46]. For this algorithm, as illustrated in Fig. 15, a point on the chain is chosen randomly. Then one part of the chain is rotated at random. As for dimerization this can easily be done for both lattice and off-lattice systems. If we have a chain of length N and a pivot point at position N_1 then the acceptance rate p is given by the probability that the new conformation has no overlaps. On a lattice this needs a simple $\mathcal{O}(N)$ check, while in continuous space this is a little more difficult. Madras and Sokal [46] showed that this approach is ergodic in the sense that this method potentially reaches all possible SAW conformations, as the previously discussed methods do as well. Madras and Sokal claim an $N^{0.2}$ power law (where N is number of moves) for the relaxation time of $\langle R^2(N)\rangle$. They used the method to obtain very precise estimates of the exponent ν, $\nu = 0.9496 \pm 0.0007$ $(d = 2)$ and $\nu = 0.592 \pm 0.003$ $(d = 3)$. There are, however, a few points that need to be clarified. If an energy is introduced, to study monomer-monomer interactions, the acceptance rate of the moves is dramatically reduced. This also holds for solutions, especially for $d = 2$. Another important point is that this method relaxes large length scales very fast, while short-range bond orientations need a longer time. For example, nearest-neighbour bond orientation correlations need a time of up to $\mathcal{O}(N^2)$ moves to relax completely. In spite of these

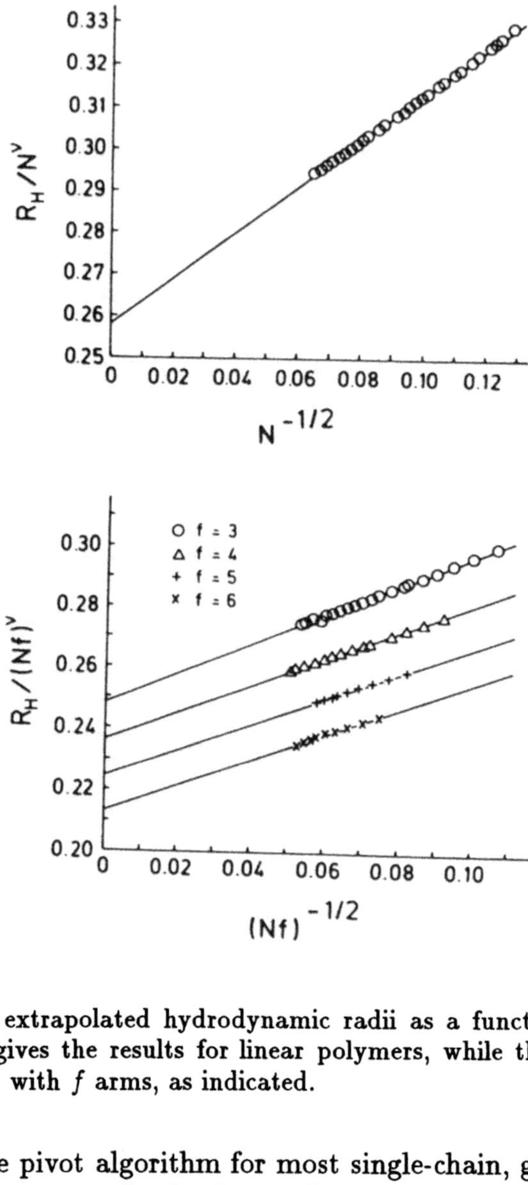

Fig. 14. The extrapolated hydrodynamic radii as a function of $N^{-1/2}$ [33]. The upper curve gives the results for linear polymers, while the lower corresponds to star polymers with f arms, as indicated.

problems the pivot algorithm for most single-chain, good-solvent problems is the fastest approach. So far we have studied methods which are only successful for isolated chains or dilute solutions, where the chains do not overlap. Most experimentally interesting cases, however, deal with many chain problems in the semidilute to dense regime. Since the concentration C^* where the chains start to overlap is

$$C^* \propto N/\langle R^2\rangle^{d/2} \propto N^{1-d\nu}\,, \tag{27}$$

which drastically decays with N, dilute solutions are very difficult to analyse. Thus we are looking for a method which allows an effective simulation of

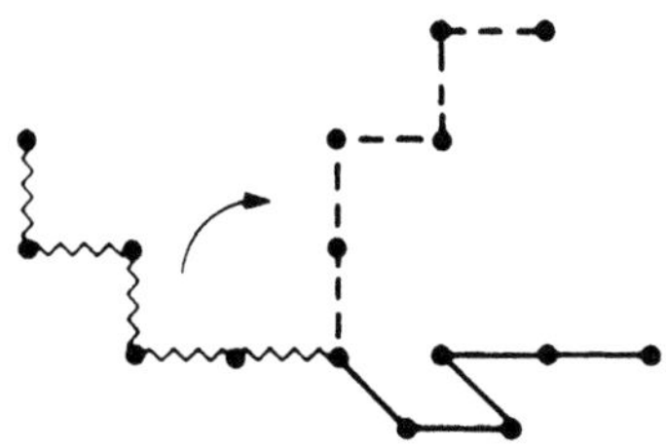

Fig. 15. An illustration of the pivot algorithm [2].

semi-dilute to dense solutions.

3.2.1. Many-chain methods: generalized reptation algorithms. The generalized reptation algorithms are probably among the most effective ways to simulate dense solutions. However, also for single chains, especially in the collapsed regime for $T < \Theta$, they are very useful. The method is explained for a single chain. The generalization to many chain systems as well as to continuous space is straightforward. This is probably the best method to date for chains above C^*, the overlap concentration. The original idea of the slithering snake goes back to Kron in 1965 and Wall and Mandell 1975 [47, 48]; their original approach is illustrated in Fig. 16.

One randomly takes an end monomer and tries to add it at the other chain-end in a random direction. If the new chain fulfils the SAW condition, the move is accepted, otherwise rejected. In this way the chain moves like a snake along its own contour. (The algorithm should not be confused with the reptation problem in polymer melts!). For this method, detailed balance is fulfilled trivially. The extension to many chains as well as interacting chains is trivial. One can also directly introduce a grand canonical version by allowing the chain length to fluctuate. The use of pointers allows for very fast and efficient codes. How fast is the relaxation of a chain conformation in cpu time? To estimate this, let us look at the position of the middle monomer of the chain at the beginning. This point is fixed in space. Now we start the algorithm. A measure of the relaxation time T_N is given by the time until this space point is not on the chain for the first time . For random walks this first passage time follows the same power law as the one-dimensional diffusion along the chains. Since each move only requires a constant N-independent number of operations we get $T_N \propto N^2$ for random walks. This can easily be confirmed by a simulation as a good check on the

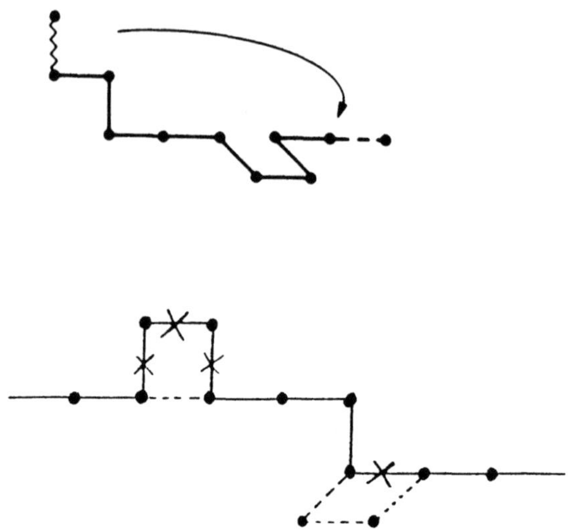

Fig. 16. An illustration of the original slithering snake algorithm (above) and the generalized reptation algorithm (below).

program. Thus this method is $\mathcal{O}(N)$ faster than algorithms with realistic dynamics. Somewhat more complicated is the situation for SAW's. With a definition for the time unit as one attempted move (independent of N) one expects for the diffusion constant $D \propto N^{-1}$. With $R^2 \propto N^{2\nu}$ and $DT_N \approx R^2$ we expect $T_N \propto N^{1+2\nu}$. This is, however, not true. Numerically one finds for the autocorrelation function of $\langle R^2 \rangle$

$$T_N^{\text{cpu}} \propto \begin{cases} N^{2.15} & d = 2 \\ N^{2.02} & d = 3 \end{cases} \tag{28}$$

which is clearly faster [49, 50]. This result, although good for simulations, is not yet understood. The explanation is probably in the difference of diffusion times and first passage times. In addition, as should be kept in mind, because of the SAW condition, SAW's are correlated objects!

The original method has, however, certain disadvantages, which become especially important for dense solutions. First, it is not ergodic in the sense that there is a subclass of conformations, which cannot be reached by the algorithm of Fig. 16. If both ends of the chain are rolled-in like in a snake, then there is no way out of this situation just by end exchange. Also for dense systems on a lattice in the limit of high density there is a tendency towards alignment of the chain ends, since the available vacancies are then located near the ends. To overcome most of these problems one can generalize the algorithm as shown in Fig. 16. A bond is taken randomly at one place in the chain and replaced by a set of bonds from another place. Such methods are typically combined with kink jump algorithms. What is important, however, is that the number of blocking conformations is dramatically reduced; while the relaxation time follows the same characteristics as for the original slithering snake algorithm. A generalisation to continuous space, although possible, is slightly more difficult than the original method. This algorithm was recently used to study the asphericity of SAW rings in $d = 3$ [50]. Fig. 17 shows a typical plot of the asphericity parameter

$$A_3 = \frac{d}{d-1}\langle \mathrm{Tr}\,(\mathbf{Q}^2)/(\mathrm{Tr}\,\mathbf{Q})^2 \rangle.$$

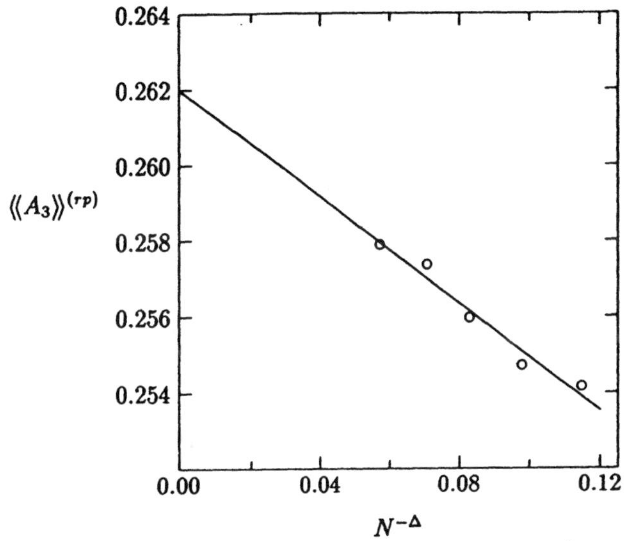

Fig. 17. The asphericity of ring polymers in $d = 3$ [50].

Here $\mathbf{Q}$ is the inertia tensor of the polymer. This definition, where the full expression is averaged, has the advantage that it takes the influence of fluctuations properly into account compared to the traditional form $A_3 = \langle \ldots \rangle / \langle \ldots \rangle$ of averaging numerator and denominator separately.

In the case of dense systems there are different extensions and further generalizations of this approach. For high density ($\rho = 1$), the approach by Olaj and Lautschbauer and Mansfield [51, 52], the so-called chain breaking method, is suitable [2]. This method gives a polydisperse set of chains. The polydispersity can, however, be manipulated and reduced by the introduction of a chemical potential. A more recent approach is given by Geyler *et al.* [50]. In this case, one creates loops of moves, which conserve the chain lengths. It should be noted, however, that the earlier papers on this method contain errors in the method as well as the results. The dynamical interpretation of the data is certainly in error.

So far, we have discussed a variety of different Monte Carlo methods which can be used to study static properties of polymer chains. Now, we present examples of methods for the simulations of polymer dynamics and their applications.

4. Simulations of polymer dynamics

So far, we only discussed methods which can be used to study static properties. If we want information about dynamics, which covers the time range of up to the order of the chain relaxation time, we have to follow the slow physical path. There are many tricks to vectorize or simplify the algorithms, but these only manipulate the prefactors of the power laws shown in Table I. It is certainly tempting to avoid this problem, but all attempts to do so necessarily include a model of dynamics which is far beyond the molecular basis. So far, we are still discussing very basic questions. Thus any simulation which does not start from these basic principles is, in the context of dynamics, extremely questionable, unless it can show the compatibility of the model with experiment [2,3,5,6]. Even so, this does not prove the model to be valid.

Thus we are faced with the problem of a naturally very slow process and we are forced, consequently, to extremely simple models. Later we will discuss a reptation study by MD [17–19], which took between 1000–1500 h cpu time on a CRAY YMP for a highly optimized code. At the present time, the same study including equilibration runs can be accomplished using the bond fluctuation Monte Carlo method in a few hundred hours of cpu time, due to the optimal vectorization of the MC program, where no distances have to be calculated. However, a similar study of the same effective chain length for a more microscopic model, such as the unified atom model of Rigby and Roe would take of the order of 500000 h cpu time [3, 6, 16]. A direct and conceptually simple method of studying dynamics is the molecular dynamics approach. This method is very versatile and has great potential. In most cases, however, it is more time consuming than the stochastic Monte Carlo methods. Direct Langevin dynamics (Brownian Dynamics) is usually even more time consuming and will not be discussed here.

Firstly, let us consider dynamical Monte Carlo methods. Since MC is a stochastic method, it does not have an intrinsic natural time-scale. In addition, as an aside, there are deep problems of principle if one wants to study hydrodynamics with dynamical MC methods. For the properties of melts and for model dynamics in more dilute systems, the so-called free draining limit, this is not included and not needed. To test the applicability of a method we have to find a basic model for dynamics. This model is the so-called Rouse model [23]. This is still the only model on a "molecular level", which can be solved analytically. The Rouse model treats the dynamics of a harmonic random walk in the overdamped limit. All the complicated interchain and intrachain interactions are summarized in the viscous background and the heat bath. Thus, we totally disregard excluded volume and topology constraints beyond the plain chain connectivity. This model can easily be solved analytically. With $\mathbf{r}_i$ being the position of the ith monomer, the Rouse equation of motion is written as

$$\zeta \mathrm{d}\mathbf{r}_i/\mathrm{d}t = -k(2\mathbf{r}_i - \mathbf{r}_{i+1} - \mathbf{r}_{i-1}) + \mathbf{f}_i(t)\,. \tag{29}$$

ζ is the monomeric friction constant, k the bond spring constant and $\mathbf{f}_i(t)$ the heat bath with $\langle \mathbf{f}_i \rangle = \mathbf{0}$. ζ and $\mathbf{f}_i(t)$ are coupled via the fluctuation dissipation theorem $\langle \mathbf{f}_i(t)\mathbf{f}_i(0)\rangle = 2d\zeta k_{\mathrm{B}}T\delta(t)\delta_{ij}$. Thus the resulting deviation from free diffusion is only due to the connectivity of the chain. In the discrete version with open ends the Rouse modes $\mathbf{x}_p$ are given by

$$\mathbf{x}_p(t) = \frac{1}{N}\left\{\sum_i \mathbf{r}_i(t)\cos\left(\frac{p\pi(i-1)}{N-1}\right)\right\} - \frac{1}{2N}(\mathbf{r}_i(t) + \mathbf{r}_N(t)) \tag{30}$$

with relaxation times

$$\tau_p = \tau_1/p^2, \qquad \tau_1 = \frac{\zeta N \langle R^2\rangle}{3\pi^2 k_{\mathrm{B}}T} \propto N^{1+2\nu}\,.$$

Within this model the eigenmodes decay exponentially, $\langle \mathbf{x}_p(t)\mathbf{x}_p(0)\rangle/\langle \mathbf{x}_p^2\rangle = \exp(-t/\tau_p)$, resulting in an overall chain diffusion constant

$$D = k_{\mathrm{B}}T/\zeta N \tag{31}$$

for the whole chain. More interesting is the mean square displacement of the individual monomers $\Delta r^2(t) = \langle(\mathbf{r}_i(t) - \mathbf{r}_i(0))^2\rangle$. There is a very simple way of deriving this. The motion of the single monomer is given by the single monomer diffusion constant $k_{\mathrm{B}}T/\zeta$. For $\langle \Delta r^2(t)\rangle$ each monomer has to carry along as many other monomers n, as there are within the sphere of radius Δr. Thus $\langle \Delta r^2(t)\rangle \propto (k_{\mathrm{B}}T/\zeta)\langle \Delta r^2\rangle$, since the chain is a random walk ($n = \langle \Delta r^2\rangle$). This leads directly to

$$\Delta r^2(t) \propto \begin{cases} t^{1/2} & t < \tau_N \quad (= \tau_{\mathrm{Rouse}}) \\ t^1 & t \geq \tau_N\,. \end{cases} \tag{32}$$

This general picture is illustrated in Fig. 18. In this figure the kinks are of course artificially sharp and in reality we will observe a smooth crossover. This simple model describes the dynamics of polymer melts of short chains

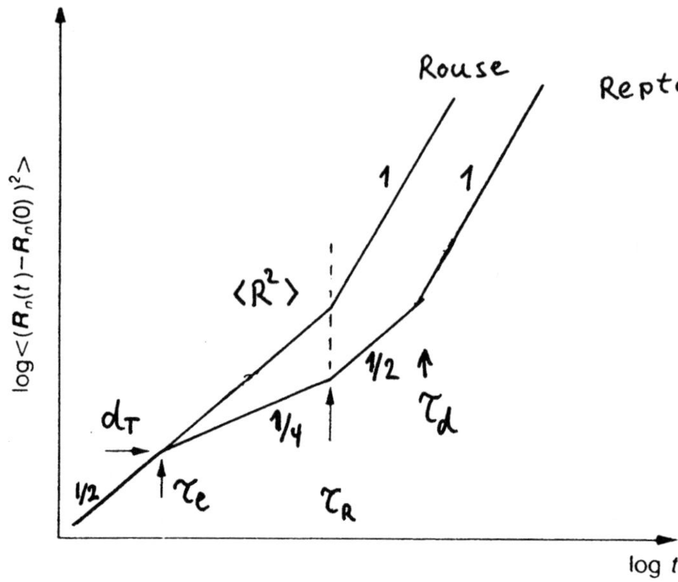

Fig. 18. A sketch of the monomer mean square displacements for the standard Rouse model and for the reptation model.

rather well (for an explanation of "short" see section 5 on reptation). All the complicated chain-chain interactions seem to be represented reasonably by a heat bath and a frictional background.

This model now gives a basis for the dynamic interpretation of stochastic algorithms. Besides the standard requirements for MC procedures (detailed balance etc.) we need the following property [2]: any algorithm with stochastic or artificial dynamics must have local moves and must yield Rouse dynamics for random walks in order to be valid to study the dynamics of polymer chains. We will concentrate on lattice models; however, the generalization to continuous space is obvious. The standard procedure is illustrated by the typical kink jump method, Fig. 19. The algorithm follows the procedure: (i) select monomer at random; (ii) select type of local motion at random; (iii) check the acceptance from topology; (iv) eventually implement a Metropolis check; (v) decide on acceptance; (vi) start over again. Each attempted move affects $\mathcal{O}(N^{-1})$ monomers. Thus the moves are local. If we use the heat bath picture, then, at any time, each monomer is affected equally on average by the random force and the heat bath. Therefore it is natural to define a MC timestep as one attempted move per monomer of the whole system. This defines up to a prefactor the physical time unit τ_0. In most cases the prefactor is set to one and only in direct comparison to experiment adjusted. In a simulation we are, however, mainly interested in

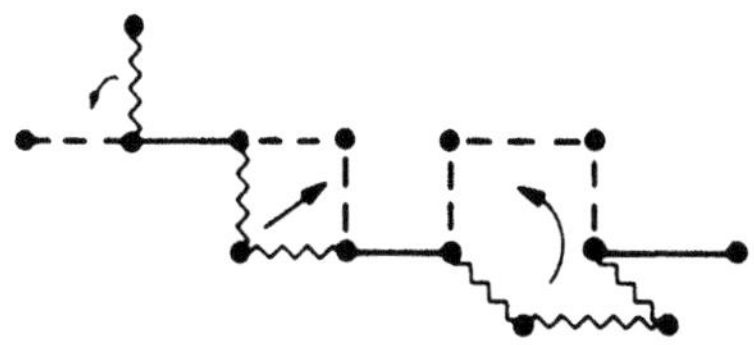

Fig. 19. Typical kink jump moves for a dynamical Monte Carlo simulation.

SAW's. To check the validity of a method for dynamics we have, however, to go back to RW's. This leads in general to a different set of local move rules. To avoid this, the test is performed on NRRW's, nonreversing random walks. NRRW's are walks which do not allow direct back-folding but do not have any long-range excluded volume (actually one could take any finite range EV along the chain). Such walks behave on large length scales like random walks, while on short scales along the chain they have the very same local conformations as ordinary SAW's and consequently give rise to the same rules for a move. Thus for any local move algorithm, we can find a NRRW, which conserves the local move rules but allows us to check for the Rouse dynamics of the algorithm. After this more general discussion we give some simple rules to construct algorithms, which fulfil the above criteria.

There is a very simple rule for move selection, which almost guarantees Rouse dynamics for NRRW's. The set of allowed moves must contain moves which create new bond vectors inside the chain [2, 6]. If we take the two moves illustrated in Fig. 19, there are 2-bond moves and 3-bond moves. With $\{\mathbf{b}_i\}$ as the set of bond vectors, the 2-bond moves only exchange bonds but do not change the set of bonds present. For rings, an algorithm only containing such moves would never equilibrate, while for open chains new bonds can only diffuse from the freely rotating ends into the system. This would increase the relaxation time at least by a factor of N. In order to achieve Rouse dynamics local rotations, such as the 3-bond move, are necessary. Then the set of bonds $\{\mathbf{b}_i\}$ is not conserved. The minimal number of bonds involved in such crankshaft moves varies from 2 in continuum or on the fcc lattice to, for example, 4 on the diamond lattice. An example for the use of such an algorithm is the motion of a chain in a rigid straight tube (Figs. 20 and 21) [53]. There we have to distinguish between the parallel mean square displacement of the monomers $g_{||} = \langle (r_i^{||}(t) - r_i^{||}(0))^2 \rangle$ and

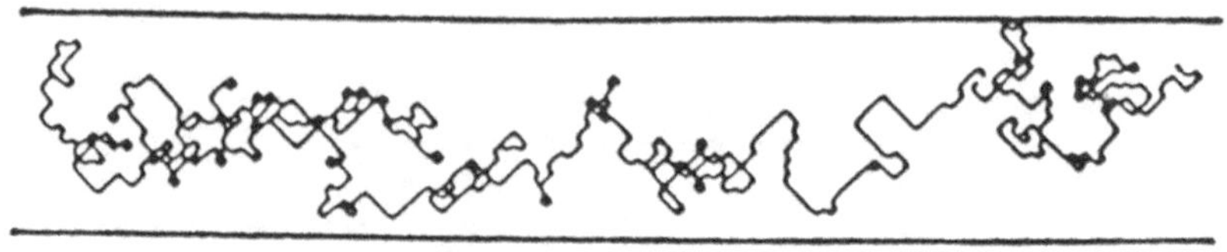

Fig. 20. The conformation of a chain of $N = 800$ monomers in a straight tube of diameter $d_{\mathrm{T}} = 32$ on the diamond lattice [53].

$g_\perp$. Fig. 21 gives the two g's for different chain lengths and tube diameters from a MC simulation on the diamond lattice. $g_\perp$ is expected to follow the Rouse $t^{1/2}$ regime (actually $t^{1/(1+2\nu)} \approx t^{0.54}$ since we are dealing with good-solvent chains) until the monomers hit the walls of the tube. Then $g_\perp$ is expected to be constant. So that, $g_\perp$ should be a function of the tube diameter but not of the chain length N. For $g_\parallel$ the picture is different. Again we expect an N-independent $g_\parallel \propto t^{1/2}$ for short times, but for $t > \tau_{\mathrm{Rouse}}$, $g_\parallel \propto t$ with $D \propto 1/N$. This crossover should only depend on N and not on the tube diameter d_{T}. Fig. 21 shows that both expected phenomena are present. Before we go onto dense systems let us look at $d = 2$. There we run into severe trouble if we only use the standard kink jump method. For the 180° rotation of the 3-bond move we again only exchange bonds. For $d = 2$ a method is needed which allows for a fluctuating bond length [2, 6, 54–56].

4.1. BOND FLUCTUATION ALGORITHM

In order to circumvent the above difficulty, one might try to construct a chain of small spheres and allow the bond angle and the bond length to vary. One could then increase the density and still keep the system mobile. An off-lattice simulation of such a model in which the bond angle but not the length was allowed to vary has been tried, e.g. [57]. Unfortunately no technical details of the simulations are given. However in order to obtain a reasonable acceptance rate, it was necessary to restrict the maximum size of the attempted bond angle rotation to a small value. For such a model, the excluded volume constraints usually time consuming to evaluate. Distances have to be calculated at every time step between several monomers. This results in the loss of the main advantage of the MC methods compared with the MD methods. However it is possible to obtain a very effective algorithm by combining the ideas from the standard lattice MC methods with the notation that the one should allow the bond angle and even the bond length

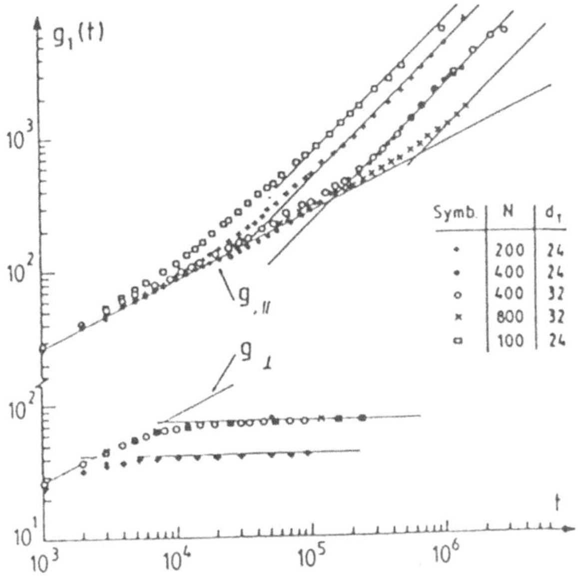

Fig. 21. The mean square displacements of monomers of different chain lengths parallel and perpendicular to the tube for two different tube diameters as indicated [47].

to vary. This new approach is the bond fluctuation method of Carmesin and Kremer [56, 58].

Fig. 22 illustrates the method for $d = 2$ and $d = 3$. Each monomer consists of 2^d lattice sites. In addition to the excluded volume interaction, the bond length l is restricted to a maximum extension to avoid bond crossing. On the square lattice, one has the constraint that $1 < l < \sqrt{16}$. For $d = 3$, the situation is slightly more complicated. In this case a set of 108 different bonds are allowed [54, 55, 59]. Since each monomer occupies 2^d sites, but every jump only requires 2^{d-1} empty sites, the method works effectively at high densities. It also suffers less from the nonergodicity problems due to blocked conformations than the standard methods do. For the 3-d study of the dynamics of polymer melts, which have already been mentioned, densities as large as 0.5 were used, although higher densities are also possible. Such a density corresponds to a much higher effective density than for the same number of occupied sites using the standard lattice models. This can easily be seen from the static properties. Kolinski *et al.* [60, 61] find for SAW's on a simple cubic lattice a static screening length for the excluded volume interaction of about 12 monomers, indicating that they are in a semi-dilute

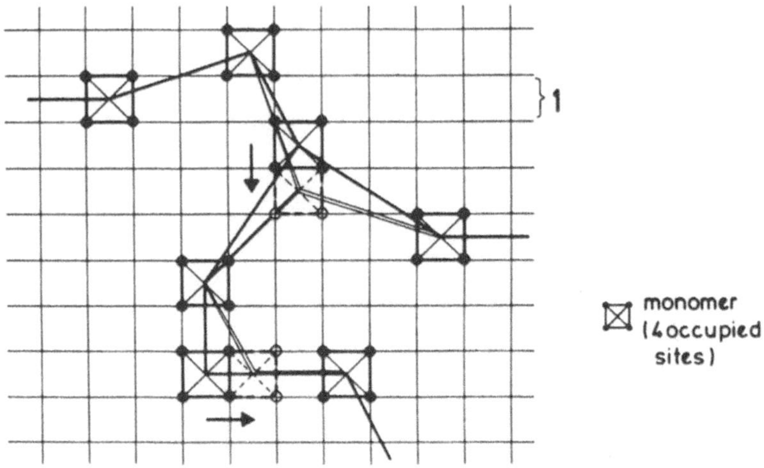

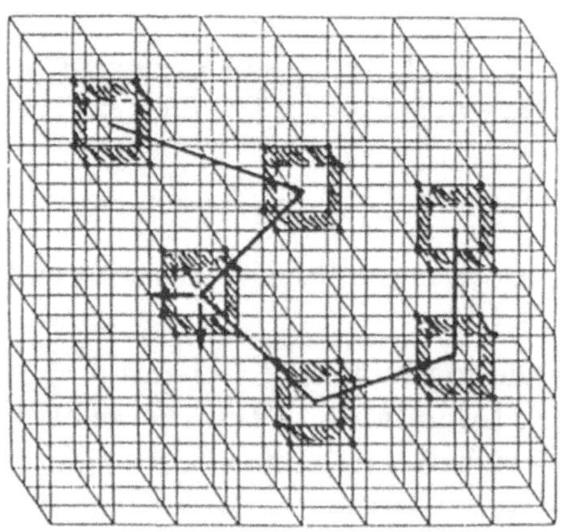

Fig. 22. An illustration of the bond fluctuation model for a 2-d branched polymer and [56] and for a 3-d linear polymer [59]. Typical elementary moves are indicated.

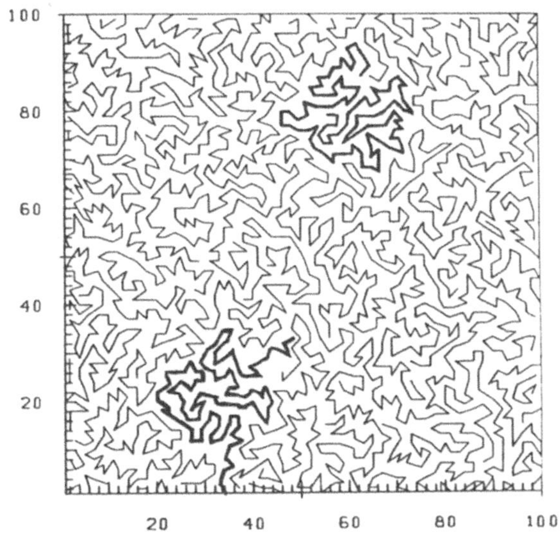

Fig. 23. A 2-d polymer melt with $N = 100$ and $\Phi = 0.8$ for an almost completely equilibrated sample [58].

regime, whereas Paul *et al.* [59, 62] find for $\phi = 0.5$ ($d = 3$) a value of about 2. This means that for the chains on a scale larger than a trimer the excluded volume interaction is already screened. An optimized vector program gave approximately 1.7×10^6 attempted moves per second for one processor on the Cray YMP [54, 55] for the same density. Since the characteristic motion distance is now smaller than a bond length, the method will be slower for isolated single chains. A first serious application is the study of a 2-d melt of chains [58]. There one would expect the chains to segregate for entropic reasons. In a Monte Carlo study using the bond fluctuation model chains of up to $N = 100$ monomers at a density of up to 80% occupied lattice sites where investigated. Fig. 23 shows the segregation procedure clearly. Much more interesting, however, is the dynamics of the melt. Since the chains cannot cut through each other, they have to move along each other. On the other hand, for $d = 2$, the chains are compact objects with $\langle R^2(N) \rangle \propto N \propto R^d$. The simple Rouse model does not take into account any motion constraints besides the connectivity. From an analysis of the neighbouring chains one finds a typical 2-d soft-sphere liquid with an average of 6 chains surrounding a given chain. A first test is to measure the Rouse modes and the relaxation spectrum. As expected, the Rouse modes are eigenmodes. More surprising is that the autocorrelation function shows an almost perfect single exponential decay. This allows one to define the relaxation time as given in Fig. 24 for

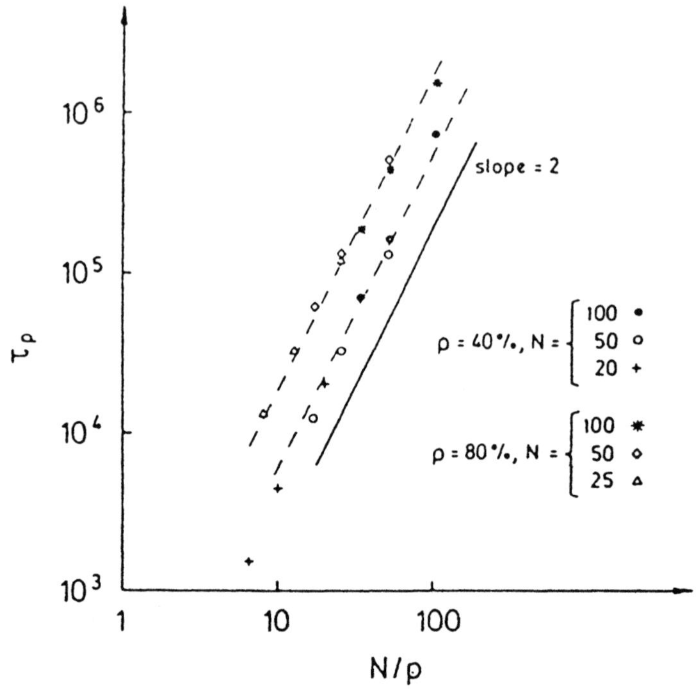

Fig. 24. The relaxation times of the Rouse modes of a 2-d polymer melt for different chain lengths and densities against N/p, p being the mode index [58].

different chain lengths and two different densities. On the other hand we can also measure the diffusion constant D_0 of the chains directly from the mean square displacements. Both the mode relaxation and D_0 give a monomeric friction constant ζ, which should be consistent, if the Rouse model describes the motion. For $d = 2$, however, one finds that $\zeta_{\text{MOD}} > \zeta_D$, which means that the apparent Rouse behaviour is not a real Rouse relaxation. For the present case ζ_{MOD} is governed by the conformational relaxation of the chain, while D_0 is governed by shape fluctuation without really reorganizing its internal conformational structure. Problems of 2-d melts/solutions are not only of academic relevance. For example, chains in a bilayer of phospholipids are thought to be 2-dimensional.

More relevant to most actual experiments is of course the study of 3-d melts, where one wants to look at the problem of chain reptation, namely the motion of a chain in a melt of identical chains. This can be studied using a variety of numerical methods. Besides standard kink jump Monte Carlo and the bond fluctuation method another alternative is to perform a molecular dynamics simulation. We begin by describing this technique.

4.2. MOLECULAR DYNAMICS

The simulations using the MD method employ the bead-spring model [63], where each monomer is weakly coupled to a heat bath. Each polymer chain consists of N monomers of mass m connected by an anharmonic spring. The monomers interact through a shifted Lennard-Jones potential given by

$$v^0(r) = \begin{cases} 4\epsilon\left[\left(\frac{\sigma}{r}\right)^{12} - \left(\frac{\sigma}{r}\right)^{6} - \left(\frac{\sigma}{r_c}\right)^{12} + \left(\frac{\sigma}{r_c}\right)^{6}\right] & r \leq r_c \\ 0 & r > r_c \end{cases} \tag{33}$$

where r_c is the interaction cutoff. For the polymer melts, we choose $r_c = 2^{1/6}\sigma$ so that the interaction is purely repulsive. For monomers which are connected we add an additional attractive interaction potential of the form [64]

$$v^{\mathrm{bond}}(r) = \begin{cases} -0.5kR_0^2 \ln\left[1 - (r/R_0)^2\right] & r \leq R_0 \\ \infty & r > R_0 \end{cases} . \tag{34}$$

For melts the parameters $k = 30\epsilon/\sigma^2$ and $R_0 = 1.5\sigma$ are chosen to be the same as in Ref. [63].

Denoting the total potential of monomer i by $\mathcal{V}_i$, the equation of motion for monomer i is given by

$$m\frac{\mathrm{d}^2\mathbf{r}_i}{\mathrm{d}t^2} = -\nabla\mathcal{V}_i - \Gamma\frac{\mathrm{d}\mathbf{r}_i}{\mathrm{d}t} + \mathbf{W}_i(t) . \tag{35}$$

Here Γ is the bead friction which acts to couple the monomers to the heat bath. $\mathbf{W}_i(t)$ describes the random force acting on each bead. It can be written as a Gaussian white noise with

$$\langle \mathbf{W}_i(t) \cdot \mathbf{W}_j(t') \rangle = \delta_{ij}\delta(t - t')6k_{\mathrm{B}}T\,\Gamma, \tag{36}$$

where T is the temperature and k_{B} is the Boltzmann constant. This approach is necessary in order to stabilize the runs. MD runs for polymers typically exceed the stability limits of a microcanonical simulation. Using the fluctuation dissipation theorem we can define a canonical ensemble. We have used $\Gamma = 0.5\tau^{-1}$ and $k_{\mathrm{B}}T = 1.0\epsilon$ in most cases and $m = 1$. This value of Γ was large enough to stabilize the run but small enough not to produce Rouse-like behaviour on length and time-scales of the order of a few bond lengths. In the following we describe the parameters of a typical large simulation [17–19]. For most of the runs the equations of motion are then solved using a third- or fifth-order predictor-corrector algorithm with a timestep $\Delta t = 0.006\tau$. Here $\tau = \sigma(m/\epsilon)^{1/2}$. The program can be vectorized for a Cray supercomputer following the procedure described by Grest *et al.* [65] except that an additional interaction between monomers which are connected was added. The cpu time per step increased approximately linearly with N. On the Cray YMP at Jülich our program runs at a speed of $140000/(MN)$ timesteps per second where M is the number of chains of N monomers in the melt. On the XMP, the program is approximately 30% slower. However we have recently increased the effective speed of the algorithm by nearly a factor of two. We replaced the predictor corrector scheme by the Verlet algorithm and increased the timestep to 0.013τ. Also for the noise term a simplification is possible. One can use equally distributed random numbers, which have the mean value and the second moment required by eqn (36) [66].

Most of the systems studied contained $M = 20$ chains of length $N = 10$–200. For $N = 400$, the sample contained only $M = 10$ chains. However this sample was used only to demonstrate the primitive chain for the reptation tube and was not used for dynamical analysis since the chains saw themselves through the periodic boundary conditions. To check for finite size effects we also ran a large sample of $M = 100$ chains of length $N = 200$. Within the error we saw no deviations from the smaller samples. With this choice of parameters, the average bond length between nearest-neighbours that are connected is found to be 0.97σ. The maximum extension of the bonds are monitored during the course of the simulation in order to assure that no bond crossing occurred.

4.3. MONTE CARLO VERSUS MOLECULAR DYNAMICS FOR MELTS

From the previous discussion the question arises which method is best to use for studying the dynamics of dense polymeric systems. We think that the choice should be between the bond fluctuation method or the molecular dynamics or variants of these two approaches. Both allow the study of dense systems. A highly vectorized version of the bond fluctuation algorithm [54, 55] is now available. For the purely athermal case, which means that only the excluded volume interactions are taken into account, we can compare the cpu time to reach the crossover time τ_e (see section 5) as a measure of the relative speed of the two algorithms. By doing this we estimate that the bond fluctuation method is approximately a factor of 10 faster than the molecular dynamics method on the Cray XMP or YMP. However the inclusion of additional interactions has a much stronger effect on slowing down the bond fluctuation method than it does the MD method. The main speed advantage of the MC comes from the simple acceptance test for the moves compared to the complicated force calculations of MD. The question about which method to use depends on the particular question under consideration and whether it is better to work in the continuum or not and whether it is acceptable to have stochastic dynamics on all time-scales. Using MD, the Rouse behaviour for short chains or early time for long chains is a consequence of the interactions, while it is built in explicitly in MC. If one is interested in studying the density dependence of the entanglement length in the free draining limit [59, 62] or polymer glasses [67–69], then the bond fluctuation method is more appropriate. However if one wants to shear a sample and, for example, calculate the stress relaxation one should use a continuum model. Similarly, if one is interested in the behaviour of gels or polymer networks under swelling or elongation or the forces between polymer brushes [70], then a continuum simulation using MD is probably the more appropriate way to proceed. Generally speaking, it seems that the MD method is more flexible. At the present time, however, for those problems where it is applicable, the bond fluctuation MC method is the fastest.

In the following we now discuss simultaneously the results of a recent MD study [17–19] and the results from recent Monte Carlo simulations [59, 62].

5. Reptation

5.1. MONOMER MOTION AND RELAXATION FUNCTIONS

The dynamics of polymer melts is observed experimentally to change from an apparent Rouse-like behaviour to a dramatically slower dynamics for chains exceeding the characteristic length N_e. There are several theoretical models which try to explain this behaviour. However, only the reptation concept of Edwards and de Gennes [23] and variants of this approach take the non-crossing of the chains explicitly into account. This approach is most widely used. It is the only one, which at least qualitatively, can account for a wide variety of different experimental results, such as neutron spin echo scattering, diffusion and viscosity. While it cannot explain all experimental data it does remarkably well, particularly considering its conceptual simplicity. For short chains the topological constraints do not play an important rôle. The dynamics is reasonably well described by the Rouse model. This experimental and numerical observation is surprising and still not understood, since the presence of all the other chains around a given chain is taken into account only by a stochastic heat bath and a background friction. For longer chains, exceeding a characteristic length N_e, one observes a dramatic slowing down to $D \propto N^{-2}$ and $\eta \propto N^{3.4}$. The idea of the reptation model is that the topological constraints imposed by the surroundings on each chain cause a motion along the polymer's own coarse-grained contour. The diameter of the tube, in which the chain is constrained, is the diameter of a subchain of length N_e, namely $d_T \propto N_e^{1/2}$. The chains follow the Rouse relaxation up to the time $\tau_e \propto N_e^2$. For longer times the constraints are supposed to become dominant and the chain is supposed to move along its own contour. In order to leave the tube, the chain has to diffuse along the tube a distance of the order of its own contour length, $d_T N/N_e$. In order to leave this original tube the chain needs a time $\tau_d \propto N^2(N/N_e)$, giving

$D \propto N^{-2}$ and $\eta \propto N^3$. The difference between the predicted and the measured exponent for η is still not completely understood. For the mean square displacements of the monomers $g_1(t)$ the model predicts the standard Rouse behaviour $g_1(t) \propto t^{1/2}$ for $t \leq \tau_e \propto N_e^2$; the Rouse relaxation along the tube with $g_1(t) \propto t^{1/4}$ for $t \leq \tau_N \propto N^2$; the diffusion along the tube with $g_1(t) \propto t^{1/2}$ for $t \leq \tau_d \propto N^3$; and finally the free diffusion of the chains. These four distinct regimes are a direct consequence of the reptation model. For a more general discussion of this and the competing but less successful concepts, and a rather complete list of references, see Refs. [6, 17–19]. In this Chapter we review a few results, which have been obtained recently by MD and by MC simulations.

One way to characterize the transition from Rouse to reptation is to investigate the diffusion constant $D(N)$,

$$D(N) = \frac{k_B T}{N\zeta}, \quad N < N_e \tag{37}$$

$$D(N) \propto N^{-2}, \quad N \gg N_e . \tag{38}$$

Here ζ is the monomeric friction coefficient. Several forms exist in the literature for the prefactor of the second expression [71, 72]. In order to be able to make a significant contribution by simulation, the data must be capable of covering this crossover. Following eqns (37) and (38), $D(N)N$ should define a plateau, giving the monomeric friction or mobility. To compare results from different simulations and also experiment, a plot of $D(N)/D_{\text{Rouse}}(N)$ versus N/N_e or M/M_e respectively should give one universal curve since N_e is thought to be the only characteristic length of the crossover. Here $D_{\text{Rouse}} = k_B T/\zeta N$ and M_e is the experimental entanglement mass. This mapping is important for our understanding, since experiment and simulation use different methods to estimate M_e or N_e. In simulations N_e is determined by the crossover towards the $t^{1/4}$ regime in $g_1(t)$, while normally, experimentally, the plateau modulus from the stress relaxation function is used. The scaling of the different data onto one curve shows, that for both cases the same length scale is used. Fig. 25 shows the results for simulation and experiment [6, 17–19, 59]. The MD [17–19] simulations were performed at a density of $\rho = 0.85\sigma^{-3}$ and the MC simulations at two different volume fractions $\phi = 0.4, 0.5$. The experimental data are NMR measurements of Pearson *et al.* [73]. For PE, the entanglement molecular weight M_e is taken from the plateau modulus using the Doi Edwards theory [23]. This gives $M_e = 1350$ or equivalently $N_e = 96$ monomers, as indicated in the Figure. Both simulations and the experiment show a clear crossover from the Rouse $D(N) \propto N^{-1}$ behaviour towards the $D(N) \propto N^{-2}$ regime. Similar results have been obtained by Kolinski *et al.* [60, 61] using standard lattice MC on the simple cubic lattice. For $N = 800$ and $\phi = 0.5$ their data give $N_e =$

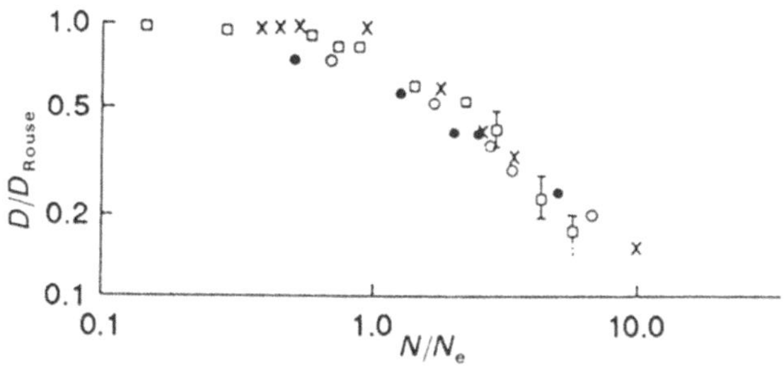

Fig. 25. The diffusion constant normalized to the Rouse diffusion constant for MD data and MC data at two different densities and for PE, as a function of N/N_e, from [59].

110–120. Thus the available simulations typically reach 6–7 entanglement lengths. From these numbers it is clear that the simulations are in a position to analyse the crossover towards the reptation regime in some detail. However they do not cover the fully entangled regime. As mentioned earlier, a key signature of the reptation in simulations is the mean square displacement of the monomers and the centre of mass of the whole chain. Since all computer simulations so far still can only deal with rather short chains, we have, for the analysis, to confine ourselves to the innermost monomers in order to identify the long chain behaviour. The motion of the ends shows that they have an enhanced mobility, which would dominate the data. Thus we define

$$g_1(t) = \frac{1}{5} \sum_{i=N/2-2}^{N/2+2} \langle (\mathbf{r}_i(t) - \mathbf{r}_i(0))^2 \rangle \tag{39}$$

and

$$g_3(t) = \langle (\mathbf{r}_{\rm cm}(t) - \mathbf{r}_{\rm cm}(0))^2 \rangle \tag{40}$$

where $\mathbf{r}_i$ is the position of the $i^{\rm th}$ monomer and $\mathbf{r}_{\rm cm}$ the centre of mass of the chain. From eqn (40) we get $6D(N) = \lim_{t\to\infty} g_3(t)/t$. Fig. 26 shows the results for $g_1(t)$ of the bead spring MD simulations of Kremer and Grest [17–19] for a variety of chain lengths. Fig. 27 shows more recent MC results for just one chain length ($N = 200$ at $\phi = 0.5$) of Paul *et al.* [59], but with increased precision in the longer time regime.

In Fig. 26, for short times, all the data collapse onto one single curve, indicating that initially the monomers do not feel the constraints of the full chain to which they belong. Only for longer times does one find the typical

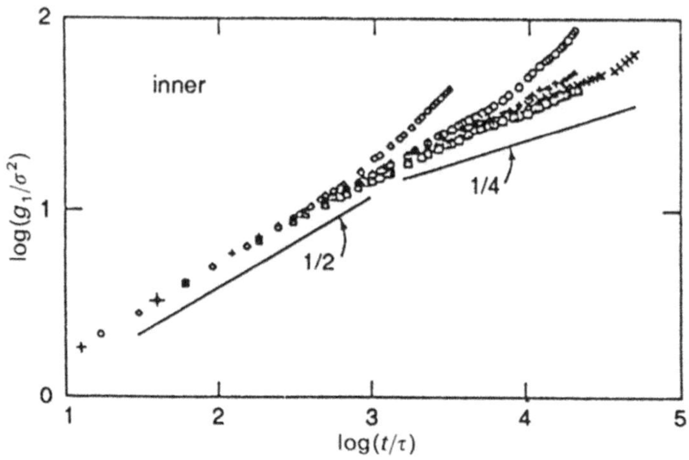

Fig. 26. The mean square displacement $g_1(t)$ of the inner monomers for a variety of chain lengths $N = 25, 50, 100, 150, 200$. MD simulations from Refs. [17–19].

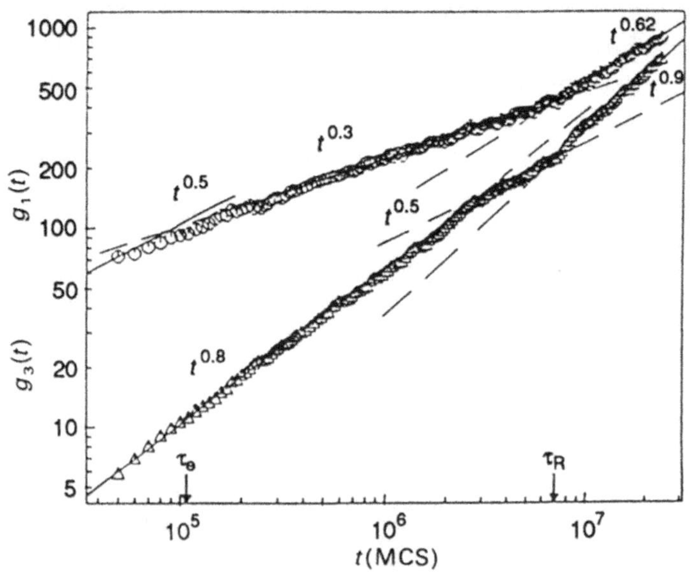

Fig. 27. The mean square displacements $g_1(t)$ of the inner monomers and $g_3(t)$ of the centre of mass of the chain of $N = 200$ at a density of $\Phi = 0.5$ from bond fluctuation MC. The region between τ_e and the Rouse time τ_N ($\equiv \tau_R$) is shown in detail [62].

crossover to a slower motion. It is important to notice that this crossover occurs within the error bars at the same times and monomer displacements, independent of chain length for the longer chains. It indicates a characteristic length N_e (the entanglement length) and time τ_e (the Rouse relaxation time of a subchain of length N_e). For longer times the slope in Fig. 26 is around 0.28, a little larger than the $t^{1/4}$, expected from the reptation model. Using for the crossover $g_1(\tau_e) = 2\langle R_g^2(N_e)\rangle$ the data of Fig. 26 give $N_e \approx 35$ and $\tau_e \approx 1800\tau$ for the MD data. For the MC simulation only one chain length is shown. Following Fig. 25, where the same analysis used for the MD data was performed for the simulations with the bond fluctuation model at $\phi = 0.5$, we get a value of $N_e = 30$. Within the reptation model one would expect the $t^{1/4}$ regime to hold up to the Rouse time τ_N of the chain. From Fig. 26 this looks reasonable. The increase in slope for the older $N = 150$ data is certainly consistent with this picture, but those data were not accurate enough in order to make a definitive statement. However the data in Fig. 27 are accurate enough to clearly exhibit a crossover towards a steeper slope at the Rouse time τ_N of the chains. Using $N_e = 30$ the time ratio between the crossover times towards the slowing down (τ_e) and the increase of the slope of $g_1(t)$ (Rouse time $\tau_N \equiv \tau_R$) should be given by $(N/N_e)^2 = (200/30)^2 \approx 44$. This is in excellent agreement with the data. Even more importantly the data are also accurate enough to display the slowing down in the motion of the centre of mass of the chains. After the initial effective slope of 0.8, it decays to 0.5. Finally, it again increases to a value slightly below 1 at about the Rouse time of the chains, again as expected by the reptation model.

An interesting question is, whether we can distinguish between the reptation model and other models discussed in the literature. A direct test would be to simulate 200 chains each of length $30N_e$ until the reptation relaxation time τ_d is reached. While this is conceptually straightforward, it simply is impossible using current computers. Thus in order to give a preference to one of the models, we must look more carefully at a variety of different quantities. In Schweizer's mode coupling approach the characteristic length is the radius of gyration. The data however show a clear signature of an N-independent characteristic length given by N_e. An even clearer picture arises if we consider the individual Rouse modes and also look for anisotropy in the motion of the monomers. As for the mean square displacement of the monomers, the autocorrelation function of the Rouse modes of the whole chain mixes the motion of inner and outer monomers. For chains in the melt, the Rouse modes are eigenmodes of the chains and their relaxation time should follow the Rouse scaling. For $N/p < N_e$ we expect this standard Rouse behaviour. This is observed for short chains, as shown in Fig. 28. The observed friction coefficient $\zeta = (17 \pm 2)\tau^{-1}$ for the MD simulations is in excellent agreement with the value of $\zeta = 16\tau^{-1}$ obtained from the diffusion constant [17–19]. There is however some deviation from the single exponential decay for the

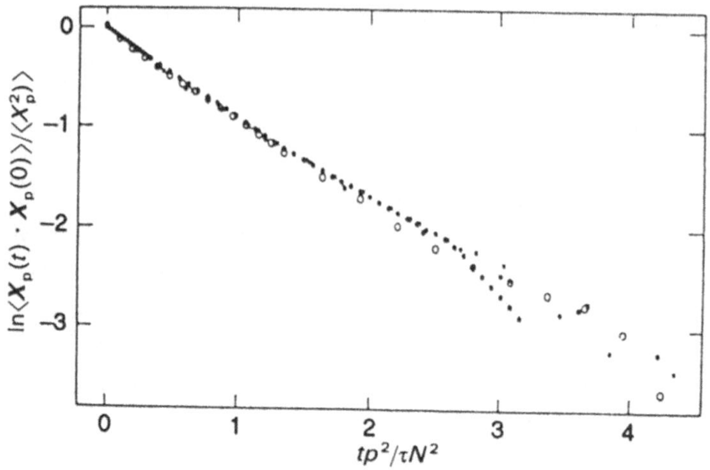

Fig. 28. An autocorrelation function of Rouse modes of a chain with $N/N_e < 1$ [17–19].

very short times. For longer chains deviations from the Rouse behaviour occur. Although the data tend to agree more with the reptation model than other models, it is difficult to distinguish on this basis alone. For the reptation model the relaxation of the modes is expected to follow the eqn. (30) for times up to τ_e for $N/p > N_e$. Then the relaxation rate should increase by a factor $N^2 N/N_e$. This again is very difficult to observe for the whole chain because of the strong influence of the more mobile ends of our relatively short chains. Recently we carried out an additional mode analysis, similar to the analysis of the mean square displacements for the MD data of 100 chains of $N = 200$. We studied the relaxation function of the Rouse modes of the innermost monomers. Following eqn (30) we calculated

$$\mathbf{x}_{\mathrm{in},p}(t) = \frac{1}{N/p} \sum_{i=N/2-N/2p}^{N/2+N/2p-1} \mathbf{r}_i(t) \cos\left(\frac{p\pi(i-1)}{N-1}\right) - \frac{1}{2N/p}[\mathbf{r}_{i_0}(t) + \mathbf{r}_{i_{\mathrm{end}}}(t)] \tag{41}$$

where i_0 and i_{end} are the first and the last monomer of the centrepiece of the original chain. The analysis was carried out for $N/p = 30$ to $N/p = 100$. Fig. 29 gives the Rouse scaling plot for the mode relaxation.

Although we do not find a single exponential decay in the Rouse regime, all the data fall onto one common curve reasonably well . At a time, which is approximately the same for all modes, the individual relaxation functions

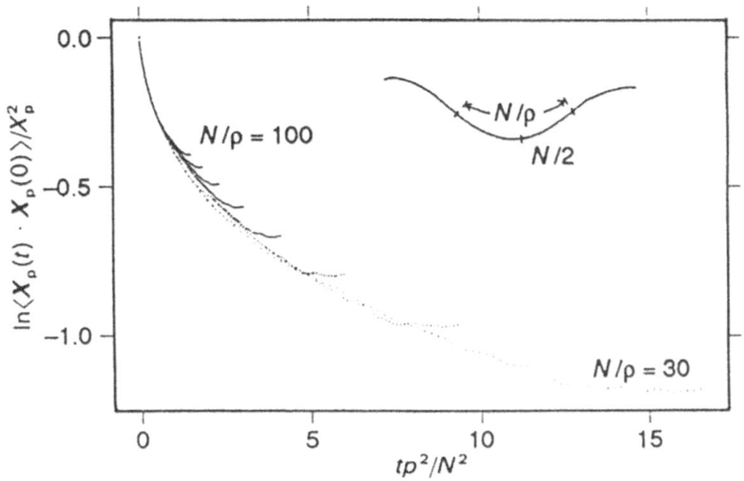

Fig. 29. An autocorrelation function of the Rouse modes of an inner subchain, as defined in eqn (41) for $N = 200$ from the data of Refs. [17–19]. The construction is indicated in the figure [6].

deviate from the Rouse curve towards a much slower decay. This time is about a factor of 4–5 larger than τ_e, as deduced from the mean square displacement $g_1(t)$ (Fig. 26), in agreement with the ideal Rouse model [62]. However the strong non-exponential character at shorter times is surprising and not completely understood and certainly deserves further investigations. However the fact that the crossover time for the modes is well defined and mode independent, rules out the coupling scheme recently put forward by Ngai *et al.* [74]. They propose a smooth p-dependent crossover.

Finally for our discussion of the dense systems, let us ask to what extent can one follow the motion of the monomers along the contour of the chains. For the chain lengths available we have to restrict ourselves to the primitive chain. For the original chain of $N = 200$ from our MD data the tube diameter d_T is only $(N/N_e)^{1/2} \approx 2.4$ times smaller than the mean end-to-end distance of the chain itself. To improve this we [17–19] constructed a primitive chain using

$$\mathbf{R}_1 = \frac{1}{N_e}\sum_{i=1}^{N_e}\mathbf{r}_i\,, \qquad \mathbf{R}_2 = \frac{1}{N_e}\sum_{i=3}^{N_e+2}\mathbf{r}_i\,, \ldots \qquad . \tag{42}$$

The primitive chain (PC) feels the constraints of the tube much more strongly. From $g_1(t)$, now evaluated for the PC, we find a tube diameter $d_{T,PC} \approx 5\sigma$. If we now simply plot conformations of the PC, separated by a constant time increment, we can directly observe the confinement into the tube. Fig. 30

gives plot of 20 conformations spaced 1200τ apart for a total elapsed time of 24000τ (about $13\tau_e$) for $N = 200$. For this time we expect an inner monomer of the PC to travel a distance of $(3g_1(\tau_{max}))^{1/2} \approx 13\sigma$ compared to the tube diameter $d_t \approx (3g_1(\tau_e))^{1/2} = 5\sigma$ if the chains did not reptate [17–19]. The figure clearly shows the confinement of the chain to a tube. Visualizing the reptation concept qualitatively gives a profound picture of the underlying physical mechanism. We also looked at a variety of other quantities, such as the motion along the direction perpendicular to the tube. These investigations support the impression, which one gets from Fig. 30. Note that the data do not agree with the mode coupling approach of Schweizer [75] since the motion is anisotropic.

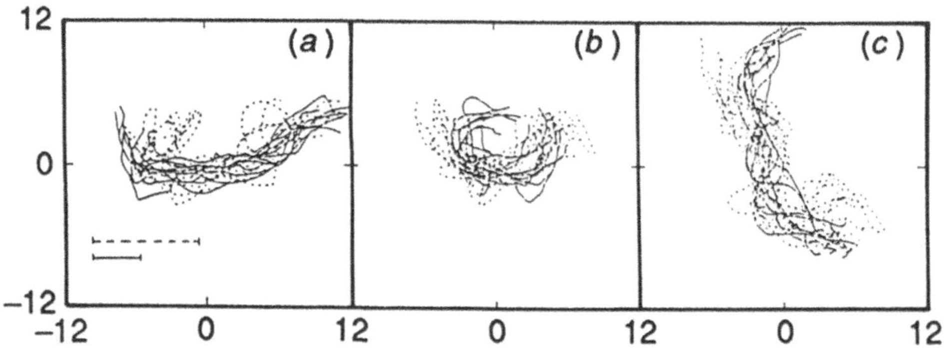

Fig. 30. Plot of the conformations of the PC of a chain of $N = 200$ from MD simulations. The two bars indicate the estimated tube diameter and the estimated motion distance from isotropic motion [17–19].

Kolinski *et al.* [60, 61] show similar plots and perform a similar analysis for their lattice MC simulation. However they claim that even for their longest chain of $N = 800$, $N/N_e \approx 7$, on the sc lattice there is no evidence for reptation. We strongly disagree with their interpretation, since they disregard the tube diameter and the enhanced mobility of the ends. Thus, when carefully analyzed, data for chains of about 6 to 7 entanglement lengths clearly show good evidence for the validity of the reptation concept.

5.2. COMPARISON TO EXPERIMENT

Although we have demonstrated that the reptation model describes the motion of a polymer chain in a melt rather well, there are other important open questions about this system. The first of these is a precise estimate of the entanglement length. Experimentally N_e is usually determined by the plateau modulus for very long chains ($N \gg N_e$), while in simulations $g_1(\tau_e)$ is used [23]. A second question, which is directly related to the first, is what can we learn about experimentally interesting linear polymer melts besides the now obvious statement that reptation seems to work. To be able to make quantitative predictions for experimental systems, we need to map the model chain onto a particular chemical species. Once this has been done, the first question will automatically be answered. A similar mapping is necessary to compare experimental results of one system with another.

When the static properties of polymers are discussed, the persistence lengths are usually mapped onto each other. However, this mapping is not unique and does not take into account any dynamics. The persistence length mapping only gives the limit of the smallest statistical unit with which we can compare. It is also known from recent experiments that simply mapping the persistence lengths is not sufficient [76]. To make this mapping unique, we must use some additional information. What is needed is to fix the mapping in a way that both static and dynamic properties can be interpreted. Since the quantity that governs the behaviour of long entangled polymers is obviously the entanglement length, N_e, or entanglement molecular weight, M_e, a reasonable procedure to follow is to compare chains in units of N_e. Two chains are equivalent in length if $N/N_e, M/M_e$ is the same. This is the procedure we have already adopted to compare the three simulation methods. Here we are simply extending it to all linear chain polymer melts. To include dynamics, one needs a way to identify the relationship between the time-scales in the simulation and experiment. For simulations it is convenient to compare the extrapolated short chain diffusion constants. This can clearly be achieved by plotting D/D_{Rouse} against N/N_e as is shown in Fig. 25 for PE using a number of simulations. One can then simply equate the Rouse diffusion constant with that determined from the simulation for an equivalent number of monomers. For systems where the diffusion constant is not available, one can compare the bead friction for long chains, which determines the viscosity in the Rouse limit. In comparing two simulations, the easiest procedure is to simply calculate the time at which $g_1(t)$ crosses over from $t^{1/2}$ to $t^{1/4}$, though the other two methods can also be used as well. Once this mapping has been completed, it is possible to use the simulation results to predict the crossover time τ_e, the crossover distance $g_1(\tau_e)$, or the tube diameter, for a variety of experimental systems. For polymers whose monomeric units do not contain large side groups we expect this mapping to be quite accurate. There are some recent indications that strong local interchain attractive potentials might result in an additional splitting of the short time scales. This however does not affect the presently discussed examples [K. Kremer, in preparation].

Here we would like to review a few of the interesting predictions from such a mapping. Details of the mapping and the results are presented in an extended table which can be found in [18, 19], comparing the MD simulations with the systems PS, PE, PDMS, PTHF, PEP and PI at a variety of temperatures. From this mapping, we find that the crossover time τ_e varies considerably from one polymer to another as one would expect. Some typical values we predict for τ_e are 5.5×10^{-5} s for PS at 485 K compared to 3.2×10^{-9} s for PTHF at a comparable temperature, 500 K. For PDMS at 273 K, $\tau_e = 4.1 \times 10^{-7}$ s while it decreases to 1.0×10^{-7} s at 373 K. These times shed some light on the long-standing discussion about whether neutron spin-echo scattering can be used to see the predicted plateaus in the time-dependent single chain scattering function $S(q,t)$ or not. Our estimates of τ_e for PDMS and PTHF at the experimental temperatures show that the neutron spin echo experiments on PTHF should have seen a a deviation from Rouse behaviour, which they did. However for PDMS this is not the case, since the times were beyond the resolution of the experiment, which was around 10^{-8} s at that time [17–19].

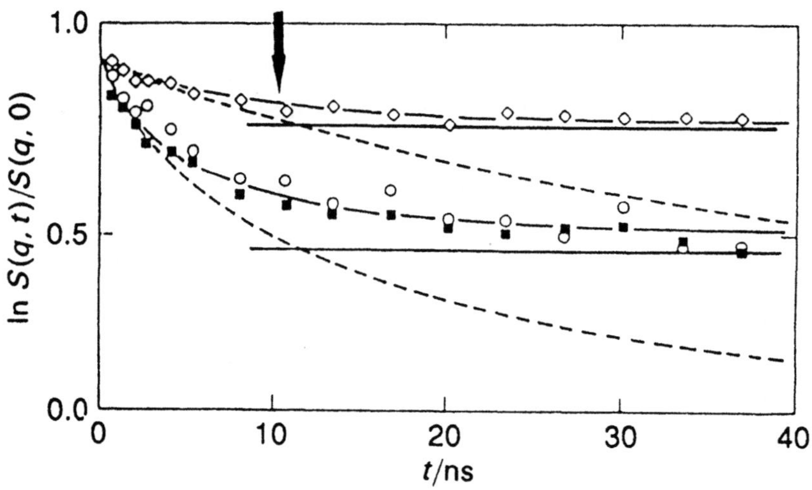

Fig. 31. The dynamical structure function $S(q,t)$ from neutron spin echo for PEP compared to the crossover time extracted from the simulation. The crossing of the Rouse curve and the plateau level gives the approximate experimental crossover time [17–19].

Recently Richter *et al.* [77] have studied the polyethylene-propylene (PEP) samples more thoroughly. They used an improved spin echo machine with a resolution of 4×10^{-8} s. For PEP we predict a crossover time of 1.0×10^{-8} s, which is within the resolution of the new instrument. For times longer than τ_e but shorter than τ_d (the actual longest time [53] depends on the wave vector q), the single chain scattering function, $S(q,t)$, should deviate from the simple Rouse decay and approach a constant of $S(q, t \gg \tau_e) = 1 - q^2 d_T^2/36$. This is because for these times the scattering function only measures a smeared density in the tube. If we extrapolate from this plateau value to short times, the crossing time with the Rouse relaxation curve should be τ_e. Fig. 31 shows this for two scattering vectors for PEP. The agreement with the data is excellent, displaying the overall consistency of the numerical and experimental results combined with the reptation concept.

5.3. THE CROSSOVER FROM SEMI-DILUTE AND DEVIATIONS FROM ROUSE

In the course of our investigation up to now, we have confined our attention to dense melts, where we found good agreement with the reptation model. For short times, however, not all the data fit the Rouse model perfectly. One way to examine this in more detail is to study the density crossover from solution to melt in the free draining limit. Experimentally this is may not be possible because of the effects of hydrodynamics, which strongly influence the dynamics. Since MD is not suitable for this problem (see section 1), the bond fluctuation algorithm was used [59, 62].

The first question concerns the entanglement length related to the static screening length ξ, derived from the screening of the excluded volume interaction [59]. In a semi-dilute solution of density Φ the excluded volume screening length ξ is proportional to $\Phi^{-\nu/(3\nu-1)}$, the typical distance between strands. ν is the critical exponent for the the linear extension of the chain in the dilute regime, $\langle R_G^2(N)\rangle \propto N^{2\nu}$.

On length scales up to ξ the chains show the single chain good-solvent behaviour with an exponent of $\nu = 0.588$. For scales larger than ξ one has a melt of blobs of diameter ξ and the chains behave like random walks with $\nu = 0.5$. At the density $\Phi^* \propto N^{-(3\nu-1)}$, where R_G and ξ are comparable, the coils just start to overlap.

In order to analyze the data within a crossover scaling scheme, one has to properly identify the scaling variables. On a lattice with fixed bond length l the lattice constant can be taken as the fundamental length scale. In this case, and for the bead spring models, the average bond length $\langle l^2\rangle^{1/2}$ is density dependent. To derive a proper scaling for the linear dimensions of the chains, this effect has to be included. Physically the variation of l shows that the number of monomers in each correlation volume ξ^3 not only depends on Φ but also on $l(\Phi)$. This causes an N-dependent shift of Φ^* in addition to the standard scaling ansatz. Identifying $\langle l^2(\Phi)\rangle^{1/2}$ as the basic unit of length, the crossover scaling function for the radius of gyration, and similarly for the end to end distance, can be written as

$$\frac{\langle R_G^2(N)\rangle}{(N-1)^{2\nu}\langle l^2\rangle_{\Phi=0}} = \langle l^2(\Phi)\rangle f_{RG}[(N-1)(\Phi l^3)^{1/(3\nu-1)}] \, . \tag{43}$$

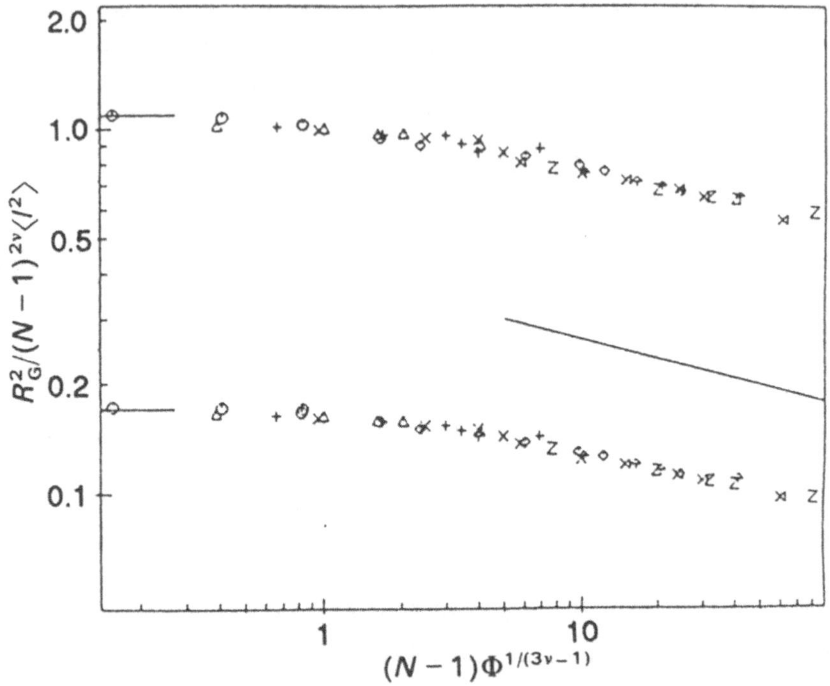

Fig. 32. The crossover scaling for $\langle R^2(N,\Phi)\rangle$ (upper part) and $\langle R_G^2(N,\Phi)\rangle$ for densities ranging from $\Phi = 0.025$ to 0.5 for chain lengths between $N = 20$ and 200. Each symbol corresponds to a different chain length [59].

Fig. 32 shows the scaling of both R_G and R [59]. We note that the naive approach, which does not take the bond length variation into account, does not give a satisfactory scaling. The data are precise enough to show this effect clearly, even though $\langle l^2\rangle$ only varies from 6.943 ($\Phi = 0.5$) to 7.469 ($\Phi = 0.025$). In Fig. 32 the entire crossover from the dilute single chain limit to the dense regime is shown. In a similar way, one can construct the crossover scaling for the static structure factor $S(q)$ of the chains. For the present MC data of Paul *et al.* [59] the screening length $\xi(\Phi)$ can directly be determined from $S(q)$ giving $\xi(\Phi = 0.1) = 21.2$ and $\xi(\Phi = 0.5) = 6.46$ for the highest density considered. For $\Phi = 0.5$, the screening only reaches the next nearest-neighbours along the chain, clearly showing that these data are in the melt density limit.

Based on these results, the data should also show the crossover from the dilute Rouse regime to the melt-reptation regime. One assumption in most theories for the dynamics of semi-dilute systems in the free draining limit is that $N_e(\Phi)$ is just a multiple of the number of monomers in a correlation volume ξ^3, namely proportional to $\xi^{-1/\nu}(\Phi)$. As in Fig. 32, one can make

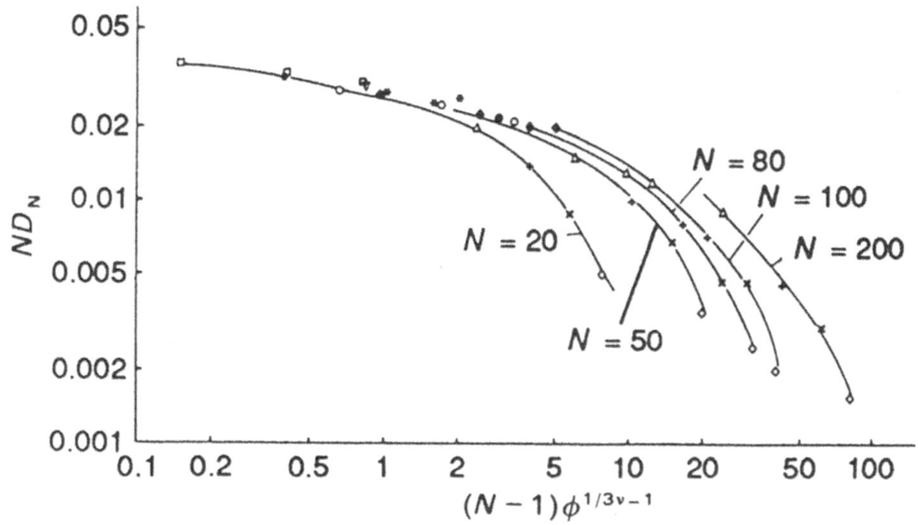

Fig. 33. A naive scaling of the diffusion constants without normalizing for the density-dependent bead friction. The symbols are the same as in Fig. 32, [59].

a scaling plot of the diffusion constant $D(N, \Phi)$. Fig. 33 shows the result, which does not display any reasonable scaling. To understand this, one has to consider the density dependence of the monomeric friction. Since the MC simulations are athermal, Paul *et al.* use the mobility W [59, 62] to set the time-scale. W can be determined from the mean square displacements of the monomers

$$g_1(t) = \langle l^2 \rangle W t^{0.54} \tag{44}$$

in the short time limit. The power law for $g_1(t)$ takes the difference between $\nu = 0.59$ for the dilute solution in good solvent and $\nu = 0.5$ for the melt into account [2]. By plotting $g_1(t)$ we can determine W. Increasing density certainly causes a reduction of accepted moves. Thus one might interpret W as a simple function of the acceptance rate A for attempted moves. This however is not the case, as Fig. 34 shows. The mobility W clearly decays more strongly than the acceptance rate A. In a recent simulation of glassy polymers, it was found that the mobility is zero, while the acceptance rate was still finite [67]. The acceptance of a move does not automatically give rise to a net diffusion of the chains. The surrounding chains cause many moves to be correlated, averaging to a vanishing contribution to the diffusion constant D. For times less than τ_e this causes deviations from the Rouse behaviour.

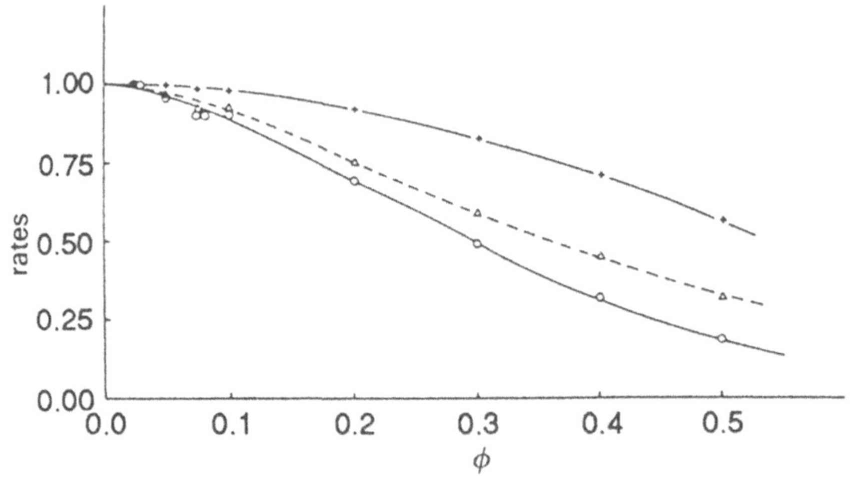

Fig. 34. Density dependence of the acceptance rate of the moves $A(\Phi)/A(\Phi = 0)$ and mobility $W(\Phi)/W(\Phi = 0)$ extrapolated to large N [62].

Now taking the mobility properly into account, one can again attempt to produce a crossover scaling plot for $D(N, \Phi)$. Fig. 35 gives the scaling plot, now with a diffusion constant normalized by the density-dependent mobility W and bond length l. For our earlier MD data [17–19], various crossover forms for the diffusion constants were compared with the data. It transpires that the formula of Hess [71, 78, 79] $D/D_{\text{Rouse}} = N_e/(N + N_e)$ described the data better than the standard reptation result $D/D_{\text{Rouse}} = 4N/(15N_e)$ of Graessly [72]. Again one finds that only the expression derived by Hess fits the data, leading to an N-independent function $N_e(\Phi) \approx 500(\Phi l^3)^{1/(3\nu-1)}$. At least in the crossover regime the enhanced constraint release, which Hess finds for reptating chains, describes the data.

Extending the above discussion one can now also a define a number of different relaxation times τ_x [62]. They all should follow the same general N-dependence, but have different prefactors. A typical time τ_1 might be defined by $g_1(\tau_1) = \langle R_G^2 \rangle$, the time when the motion of the monomers reaches the radius of gyration. Here we want to discuss another yet more interesting time. As seen in Fig. 27, $g_3(t)$ starts out from a power law $g_3(t) \propto N^{0.8}$, which clearly is in contradiction to the standard Rouse behaviour. Note that this is not an artifact of the MC method or of *low* density, since the same can be seen in the MD data [17–19]. It is important to note that the modes and the dynamic scattering function $S(q,t)$ for chains shorter than N_e only display very small deviations from Rouse. The effect on $g_3(t)$ prob-

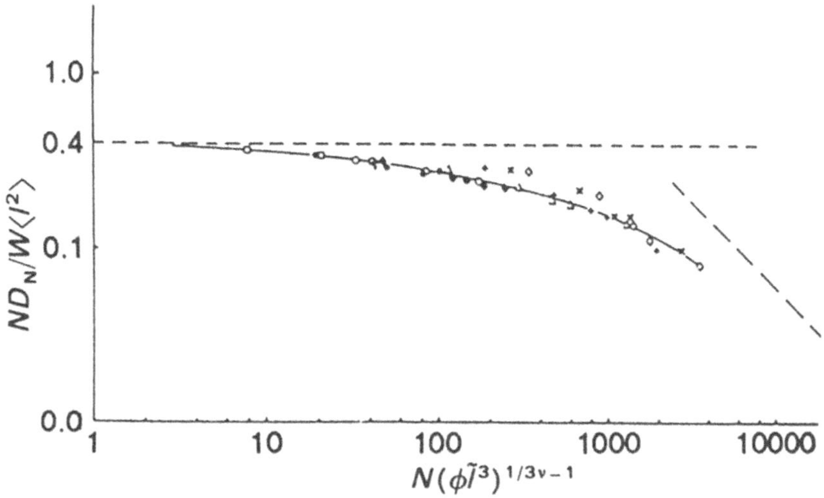

Fig. 35. Scaling of the density and N-dependent diffusion constant with the correct scaling of the lengths and the mobilities [62].

ably reflects the difference between the very local mobility, as dominated by the acceptance rate A, and the chain mobility as given by W. From the reptation theory one would expect this to hold up to the time τ_e. Then for $g_3(t)$ a crossover to a $t^{1/2}$ regime is predicted, which should hold up to τ_N, just as the $t^{1/4}$ regime for $g_1(t)$. Fig. 27 shows that this crossover can be observed for the chain lengths and densities studied up to now. For $t > \tau_N$ one expects $g_3(t) = 6D(N)t$. The data however show a smaller power of about 0.9. It is not clear *a priori*, whether this is still a result of a slow crossover or whether this smaller power persists and eventually might be a signature of an enhanced relaxation time compared to the usual reptation prediction. The physical picture, giving $g_3(t > \tau_N) = 6D(N)t$ comes from the diffusion of the chain along the tube. This motion is the same as performed in the standard slithering snake algorithm. If we follow Semenov [80], this diffusion is retarded due to the long life time of density fluctuations. These fluctuations, which allow the chain to reptate out of the tube, should only relax by reptation itself. The data from the present microscopic simulations are certainly not sufficient to clarify this point. What we however can do, is to look at the time τ_D, which is the time where $g_3(t)/t$ becomes constant. This is not achieved with very high accuracy, but good enough for the present discussion. The interesting question is, whether $\tau_D \approx \tau_N \propto N^2$ or $\tau_D \approx \tau_d \propto N^{3...3.4}$. To test this we plot the scaled relaxation time $\tau_D W(\Phi)/N^{1+2\nu}$ against the same scaling variable, x, as for the diffusion

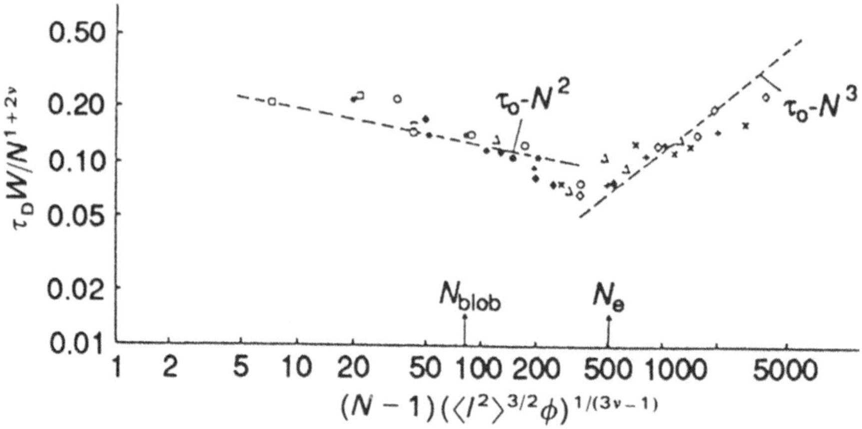

Fig. 36. Scaling of the diffusion relaxation time τ_D [62], symbols as for Fig. 35.

constant, $x = (N-1)(l^3\Phi)^{1/(3\nu-1)}$ in Fig 36. In the dilute, good-solvent limit $\tau_D \propto N^{1+2\nu}$, leading to a plateau for $x \to 0$. In the reptation picture one expects a slope of $N^2/N^{1+2\nu} = N^{-0.18}$ for both the semi-dilute Rouse regime and the reptation regime . If however τ_D shows the same behaviour as the reptation relaxation time, τ_{d}, one expects a turn upwards to the asymptotic $N^{3...3.4}/N^{1+2\nu}$ power for $x \approx 500$ [62]. (For the present MC model we have $N_e = 500x^{-1}$.) Fig. 36 indicates the latter! Although there is significant scatter in the data, the general trend is obvious. It is not clear, if this is related to the restricted diffusion along the tube or to the memory effects from the very short-time, enhanced diffusion. These certainly are very interesting and, for simulations, challenging problems which probably can only be settled with the next generation of supercomputers. A detailed high precision *short*-chain, *short*-time simulation might be helpful, certainly is needed but it is not sufficient to settle this issue satisfactorily.

6. Extensions: glasses and networks

So far we only discussed melts far above the glass transition. In many technical applications, however, polymeric glasses are used. A rather simple way of analysing the glass transition is given by a modification of the bond fluctuation method [67–69, 81]. Up to now all bonds, which are allowed, have the same energy $\epsilon = 0$. However, by changing this we can devise a model which will not crystallize with decreasing temperature and will also have a frustrated ground state. This is what is required in order to make sure that a decay into a glass occurs. Here we will only briefly discuss the 2-d case. By the use of the very fast bond fluctuation program it is actually possible to reach experimentally relevant time-scales [67–69]. For 2-d one can, for example, choose the energy

$$\epsilon(\ell) = \begin{cases} 0 & \ell = \sqrt{10} \\ \epsilon > 0 & \ell \neq \sqrt{10} \end{cases} . \tag{45}$$

This gives a ground state with one blocked site [67]. For $\rho > 80\%$, the system has no choice but to decay into a frustrated state as $T \rightarrow 0$. Fig. 37 shows a typical conformation of a frozen state. Even at this point, many

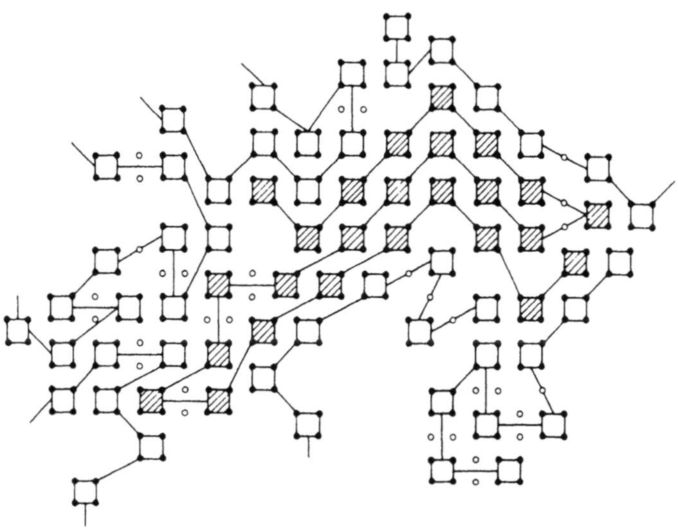

Fig. 37. A close-up view of a partially frozen 2-d polymer melt [67].

monomers are mobile enough to perform correlated forward and backward jumps, however, this does not contribute to the overall diffusion of the chains or the structural relaxation of the glass. Fig. 38 shows the typical behaviour of the energy $E(T)$ as a function of temperature for different densities. For $\rho = 0.8$ the energy does not decay to zero, but levels off at $E(T \rightarrow 0) > 0$. In a similar way it is also possible to introduce some elementary chemical details into models such as the bond fluctuation model on a rather coarse-grained level. First attempts on polycarbonate [BPA-PC] show that this is a very promising direction for future research [82, 83].

Another very interesting and also important subject is the study of polymer networks. There has been extensive research on networks for many years.

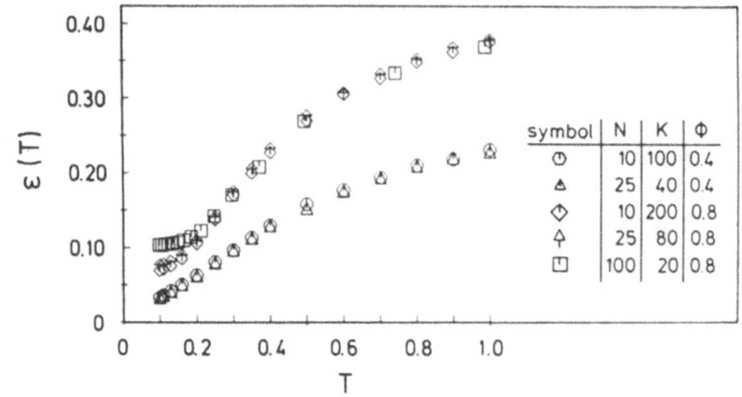

Fig. 38. The average energy E per bond as a function of temperature for different densities and chain lengths of the 2-d polymer glass model [67].

The main problem, however, that it is experimentally impossible to collect the necessary structural information in order to allow an unambiguous test of theoretical predictions. By using simulation, it is possible to give a detailed analysis of the structural properties [84–91] modifying the standard cluster analysis. Using a modification of the standard cluster analysis algorithms, as developed for percolation, it is possible to give the number of elastically active network strands and the elastically active cross-links exactly. This allows a direct test of current theoretical models. Of special interest, however, is the role of the melt entanglement length, as found in the reptation problem. If the entanglements are governed by knots, which "connect" different elementary circuits in the network, then they should show up in the modulus. In the affine network model, the modulus is given by [84, 90]

$$G^0 = \frac{\nu kT}{V} + T_e\,. \tag{46}$$

ν is the number of elastically active strands and V the volume. T_e is the socalled the trapping factor, which should only be non-zero, if the melt plateau modulus is nonzero. However, this requires $N_s > 2N_e$, where N_s is the strand length between two cross-linking points. In a recent simulation of end cross-linked chains [90], we calculated the modulus and compared it to the currently available models. The chains started as an equilibrated melt of length $N = 12$, 25 and 50. Then 1/4 of the ends were declared to be activated cross-linkers with a functionality $f = 4$. After the kinetics cross-linking was finished, which was a study in its own right [89], the relaxation of the network was studied. Fig. 39 shows the inverse modulus $1/G^0$ against the strand length N_s, which is monodisperse in the present case. Obviously, the melt entanglement length plays a very important role for the modulus of $N_s = 50$, while for $N_s = 12, 25$ the "measured" modulus is only weakly larger than the affine model predictions. This increase for short chains is due to the reduced conformational phase space available to the chains [84]. The above example is, to our knowledge, the first direct evidence for the fact that the melt entanglements can actually be taken as additional crosslinks.

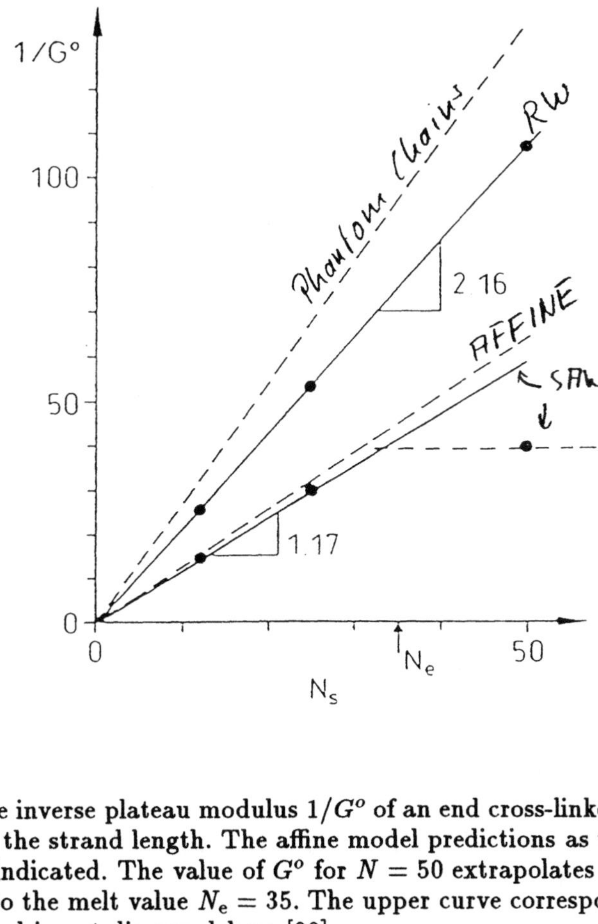

Fig. 39. The inverse plateau modulus $1/G^o$ of an end cross-linked polymer melt as a function of the strand length. The affine model predictions as well as the measured points are indicated. The value of G^o for $N = 50$ extrapolates to $N_s = 32$ in eq. 44 very close to the melt value $N_e = 35$. The upper curve corresponds to random walk networks and is not discussed here [90].

Those effects are even more important under strong elongational stress [91]. There knots dominate the nonlinear regime. However, such simulations, which take a completely mobile sample into account, are in their early stages. Future parallel computers should allow us to simulate much larger systems.

6.1. CHAIN WITH SOLVENT: THE ZIMM MODEL

Since the pioneering work of Kirkwood and coworkers [92, 93], it is well-known that the Brownian motion of long flexible macromolecules in dilute solution is dominated by hydrodynamic interactions, i.e. correlations in the displacements of monomers mediated by fast momentum transport through the solvent. This hydrodynamic interaction is taken into account through the off-diagonal elements of the diffusion tensor $\mathbf{D}_{ij}$, which describes the coupling of motion between monomer i and j. $\mathbf{D}_{ij}$ usually is taken as the Oseen tensor

$$\mathbf{D}_{ij} = \mathbf{D}_0\,\delta_{ij} + (1-\delta_{ij})\frac{k_\mathrm{B}T}{8\pi\eta_\mathrm{s}r_{ij}}\left\{\mathbf{1} + \frac{\mathbf{r}_{ij}\mathbf{r}_{ij}}{r_{ij}^2}\right\} . \tag{47}$$

$\mathbf{D}_0$ is the monomer diffusion constant, η_s the low Reynolds number shear viscosity of the solvent and $\mathbf{r}_{ij} = \mathbf{r}_i - \mathbf{r}_j$, the distance between monomers. Within the Rouse-Zimm model the equation of motion for a monomer, in its simplest form, is [23]:

$$\frac{\partial \mathbf{r}_i}{\partial t} = \sum_j \mathbf{D}_{ij}\left\{\frac{\partial \mathcal{V}_i}{\partial \mathbf{r}_i} + \mathbf{W}_i(t)\right\} . \tag{48}$$

This is a highly nonlinear equation of motion, which is extremely difficult to handle. We note that $\mathbf{W}_i(t)$ and $\mathbf{D}_{ij}$ are coupled via the fluctuation dissipation theorem in a nonlinear way! Thus at every step in a Langevin simulation one would have to calculate the square root of $\mathbf{D}$. This leads to severe numerical problems [94, 95]. Although eqns (47) and (48) are only rough approximations which can be improved in various ways, they are expected to take the leading, long-wavelength, low-frequency contributions properly into account. For the case of the chain without any excluded volume, Zimm solved this model in the so-called preaveraged version where $\mathbf{D}_{ij}$ is given by

$$\mathbf{D}_{ij} = \frac{1}{6\pi\eta_\mathrm{s}}\left\langle\frac{1}{|\mathbf{r}_i - \mathbf{r}_j|}\right\rangle \qquad i \neq j. \tag{49}$$

With the hydrodynamic radius $\langle R_\mathrm{H}^{-1}\rangle$ given by eqn (26), the diffusion constant of the chains is

$$D = \frac{D_0}{N} + \frac{k_\mathrm{B}T}{6\pi\eta_\mathrm{s}}\left\langle\frac{1}{R_\mathrm{H}}\right\rangle . \tag{50}$$

Since $\langle R_\mathrm{H}^{-1}\rangle^{-2}$ is proportional to the mean square end-to-end distance $\langle R^2\rangle = \langle(\mathbf{r}_1 - \mathbf{r}_N)^2\rangle$ or the mean square radius of gyration of the chain one expects

$$D \propto N^{-\nu} \tag{51}$$

where ν again is the universal exponent for the mean extension of the chain (the correlation length exponent). For random walks, for which the linearized Zimm model can be solved, $\nu = 1/2$. Consequently, the longest relaxation time changes from $\tau_N \propto N^2$ to $\tau_N \propto N^{3\nu}$. The dynamical scattering function of the individual chain $S(q,t)$ is now a function of the scaling variable q^3t instead of $tq^{(2\nu+1)/\nu} = q^4t\,(\nu = 1/2)$ for the Rouse case and the mean square displacement $g_1(t)$ (of the inner monomers of a chain, see eqn (39)) is supposed to follow a $t^{2/3}$ power law instead of the $t^{1/2}$ law for the simple Rouse chain. For the case of the good solvent, where the mutual excluded-volume interaction of the monomers is relevant, causing ν to change to $\nu = 0.59$ ($d = 3$), neither the simple Rouse model nor the Rouse Zimm model can be solved exactly. One has to rely on perturbation expansions or scaling arguments.

The experimental situation is not significantly different [96]. The assumption in such a simple model is to have an incompressible, homogeneous background which mediates the hydrodynamic interaction. This is certainly not true since typical organic solvents have roughly the same size as most monomers. Secondly, D and R_{H} are not easily accessible experimentally. In most cases, the short-time decay of the dynamical light scattering is used to determine R_{H} through eqn. (50), neglecting D_0 [96]. R_{H}, however, is known to display very strong corrections to scaling, because of the way short distances enter in eqns (49) and (50). Finally, the mean square displacements cannot be measured directly. The only direct microscopic experiments so far are neutron spin echo measurements [97].

Thus with a properly chosen simulation approach we should be able to test the validity of this rather crude model on a more microscopic level. To do this we performed an MD simulation of a single polymer chain in a large number of solvent particles, namely monomers, in a cubic box with periodic boundary conditions [94]. Similar attempts have been made previously, but were restricted to chains which were too short or for runs with poor statistics. The problems of such an approach have been already discussed in section 1. Of course we are now restricted to a microcanonical simulation [94]. To avoid the typical stability problems, about 700 independent starting states were generated integrating eqn (35) as equation of motion. Then for these states Γ was set to zero. The simulations ran at a number density $\rho = 0.864\sigma^{-3}$ and temperature $k_{\text{B}}T = 1.2\epsilon$. The parameters for the bond potential were $k = 7\epsilon\sigma^{-2}$, $R_0 = 2\sigma$. Three different systems of $M = 4096$, 4096 and 8000 monomers were simulated, containing a chain of $N = 30$, 40 and 60 monomers respectively. Newton's equations of motion were integrated with a timestep of $\Delta t = 0.004\tau_{\text{LJ}}$, $\tau_{\text{LJ}} = (m\sigma^2/\epsilon)^{1/2}$, using a 5th-order predictor corrector scheme. The microcanonical runs were confined to a maximum time of $t_{\max} = 100\tau_{\text{LJ}}$.

The chains are long enough to exhibit quite accurately the static asymptotic behaviour. For $S(q, t = 0)$ one expects the fractal scattering law $S(q, 0) \propto q^{-1/\nu}$, for q values measuring the internal structure of the chain. From this we obtain $\nu = 0.58 \pm 0.01$, which is in good agreement with the expected asymptotic value $\nu = 0.59$. In all cases the box sizes were sufficient to prevent direct self-interaction through the periodic boundary conditions. Fig. 40 shows a typical result for the mean square displacement of the inner monomers $g_1(t)$.

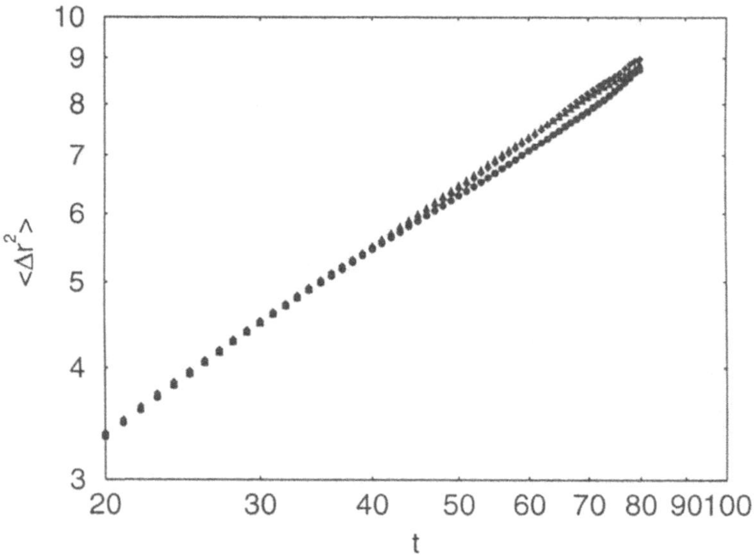

Fig. 40. A log-log plot of the mean square displacement $g_1(t)$ for the inner monomers of the chains of length $N = 60$ (+), 40 (△), 30 (○) (from [94]).

The expected scaling behaviour, namely a $t^{2/3}$ time-dependence can be followed down to distances of only $g_1 \approx 1.5\sigma^2$ (actually the best fit for the exponent is 0.70). This means that the hydrodynamic interaction already dominates the motion on microscopic scales of the order of a bond length ℓ ($\ell \approx 1\sigma$). It is also important to note that the amplitude is roughly independent of the chain length and the system size. A direct comparison with the Rouse model can be given by the chain scattering function $S(q, t)$. Fig. 41 compares the Rouse and the Zimm scaling for $N = 60$. The data clearly display Zimm scaling. At this point, we re-emphasize that the Zimm model gives only a very crude description of the local interaction.

Within this scheme one should be able to confirm the Kirkwood diffusion equation, (50). The diffusion constant, D, as well as η_s and $\langle R_H^{-1} \rangle$ can

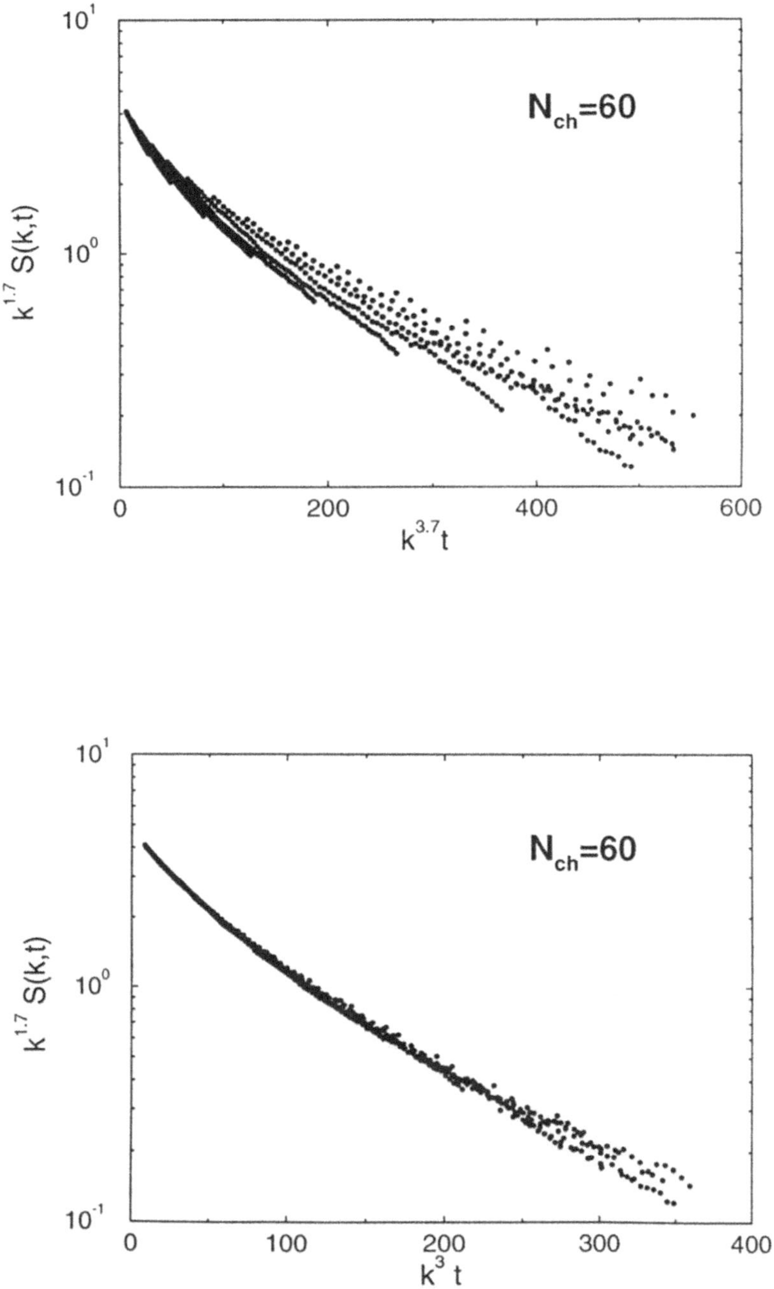

Fig. 41. The data collapsing plot of the decay of the dynamical structure function $S(k,t)$ for $N = 60$ in Rouse scaling (top figure) and Zimm scaling (bottom). For Rouse: $(\sigma k)^{1/\nu}S(k,t)$ is plotted against $(k\sigma)^{3.7}(t/\tau_{\rm LJ})$; for Zimm: against $(k\sigma)^3(t/\tau_{\rm LJ})$. We used $\nu = 0.59$ and restricted k to the scaling regime $0.7\sigma^{-1} \leq k \leq 3\sigma^{-1}$ (from Ref. [94]).

be measured directly. This does not, however, give a consistent picture; D_0 would have to be negative. In order to take the interactions properly into account, we have to incorporate, in the calculation of $\langle R_H^{-1} \rangle$, all the periodic images of the chains by an Ewald summation. The chain sees itself as another chain through the periodic boundary conditions. Only if this is included, which for the present box sizes reduces $\langle R_H^{-1} \rangle$ by approximately a factor of 2, do we arrive at a generally consistent picture. Since $\langle R^2 \rangle^{1/2}/L$ is roughly the same for all three systems, this influence on the diffusion does not show up in Fig. 40. Thus the conceptually very simple Zimm model provides an excellent description of the microscopic dynamics of chains in dilute solutions. Following this work we learned of a similar recent study of shorter chains [98]. Their results are similar, however they could not completely verify the dynamical scaling since the chains were too short. Also the Ewald summation was not used in this study. Subsequently, these authors also performed an Ewald analysis leading to similar results. For shorter distances, higher q vectors, they try to describe the dynamics with rescaled hydrodynamics [99]. It can been shown that the breakdown of the simple Zimm description occurs at a scale where the structure of the liquid becomes relevant. In this regime however a description using continuous hydrodynamics is highly questionable [44].

7. Further reading

So far I have given a rather detailed description of some technical aspects as well as some applications in which I have been involved in recent years. There are of course many other interesting applications, which cannot be discussed in detail here. Nevertheless, a few should be mentioned.

A different, equally important but even more complicated set of problems of polymers in solution concerns the properties of charged polymers, polyelectrolytes. While for neutral chains there are essentially two length scales besides the bond length ℓ, namely R_G and ℓ_p the persistence length, there are several additional length scales important for polyelectrolytes. These additional scales are related to the electrostatic interactions, namely the Bjerrum length, the Debye Hückel screening length and the electrostatic persistence length. It would be beyond the scope of the present Chapter to discuss these in detail. The long-range interaction causes significant complications for experiment, analytic theory and simulation. Until now we have been able to handle up to about 2000 ions (1000 chain monomers and 1000 counterions) [100, 101]. Again this is just the very beginning of what, in my opinion, is a fascinating area of coming polymer simulations. As for networks, there are significant experimental difficulties in examining the conformations of the chains. Simulations will be a key tool to bridge the gap between experiment and analytic theory.

One very active field of research covers the properties of chains in disordered media. There are two main cases. One is where the chain exists on a percolating cluster near p_c and the monomer size of the cluster is identical to the monomer size of a polymer, e.g. a self avoiding walk on a square lattice cluster. The conformation of the chains near p_c is still not understood, in spite of some very extensive recent research [102, 103]. A second aspect of this problem considers a chain in a disordered medium, where the elementary units of the disordered medium are much larger than the monomer of the chain. There the so-called localization of chains occurs [104, 105]. In larger holes one finds for entropic reasons an increased monomer concentration compared to the small hole.

Other interesting recent developments concern the phase separation of polymer mixtures. By the use of a quasi-grand canonical algorithm, where the type of the chains can be exchanged, it is possible to study the phase separation transition very effectively [2]. Actually, Monte Carlo simulations [106, 107] were the first to show clearly that a recent theory of Schweizer and Curro [108, 109], which proposed $T_c \propto N^{-1/2}$ instead of the commonly accepted $T_c \propto N^{-1}$ was wrong. In this simulation a short-range repulsive monomer interaction was introduced in the bond fluctuation model between different polymer species.

Another very interesting area concerns the study of polymer brushes and the comparison to experiment. This was first studied using MD simulations [110–112] and then also extensively using the Monte Carlo method [113–116]. Here mean field theory [117, 118], scaling theories [115, 116] as well as the question of the forces between two brushes, are all areas of especially high scientific and practical interest[119, 120]

In addition a huge body of recent work considers tethered surfaces or membranes. A discussion of these topics would, however, require and warrant a review of its own.

8. Conclusions

The present very short overview should give the reader an introduction to the possibilities in polymer simulations. The content is certainly strongly biased towards the author's own research interests. I nevertheless hope that the Chapter will be of some use to everybody interested in polymer simulations, whether just as an observer of the field or as a participating newcomer. Compared to the extremely large simulations of Ising models, used to obtain critical exponents to an accuracy of more than 3 digits, the computer simulations of polymers are just in their infancy. The reason is in the huge number of intrinsic degrees of freedom and the slow relaxation. The single isolated chain actually can be viewed as a critical phenomenon with all its complications and implications. Nevertheless, there has been significant progress over the last years, which is due to improved models and algorithms and to improved hardware. Currently the advent of massively parallel computers, with individual nodes which have the computational power of a modern workstation, will enable significant progress in many new areas. This can, however, only be achieved if the close contact with analytical theory is maintained: a view which I have try to convey throughout this Chapter.

Acknowledgements

The work described here was supported by the Deutsche Forschungsgemeinschaft (DFG) and a NATO travel grant. For the polymer melts (MD) and solutions cpu time was granted by the German Supercomputer Centre, HLRZ in Jülich. The work described here was performed in a pleasant and close collaboration with B. Binder, B. Dünweg, G. S. Grest, J. Batoulis, I. Carmesin, H. P. Wittmann, W. Paul, D. W. Heermann, E. R. Duering, M. Stevens and R. Everaers. Part of this Chapter was prepared during the author's visit to the Department of Chemical Engineering and Materials Science of the University of Minnesota, Minneapolis, as the G. T. Piercy Professor 1991.

References

[1] A. BAUMGÄRTNER, in *Application of the Monte Carlo method in statistical physics*, edited by K. BINDER, Springer Berlin and New York, 1984.

[2] K. KREMER and K. BINDER, *Comput. Phys. Rep.* **7**, 259 (1988).

[3] K. BINDER, J. Biorano, New York, Basel, Hong Kong, 1992.

[4] R. J. ROE, *Computer simulations of polymers*, Prentice Hall, Englewood Cliffs NJ, 1991.

[5] H. MARK and B. ERMAN, *Elastomeric Polymer Networks*, Prentice Hall, Englewood Cliffs NJ, 1992.

[6] K. KREMER and G. S. GREST, *J. Chem. Soc. Faraday Trans.* **88**, 1707 (1992).

[7] K. BINDER, *Monte Carlo methods in statistical physics*, Springer Berlin, New York, 1979.

[8] K. BINDER, *Applications of the Monte Carlo method in statistical physics*, Springer Berlin, New York, 1984.

[9] D. W. HEERMANN, *Introduction to computer simulation methods in theoretical physics*, Springer Berlin New York, 1986.

[10] K. BINDER and D. W. HEERMANN, *Monte Carlo Simulations in statistical physics*, Springer Belin New York, 1988.

[11] M. P. ALLEN and D. J. TILDESLEY, *Computer simulation of liquids*, Clarendon Press, Oxford, 1987.

[12] G. CICCOTTI and W. G. HOOVER, *Molecular dynamics simulations of statistical mechanical systems*, North Holland, Amsterdam, 1986.

[13] G. CICCOTTI, D. FRENKEL, and I. R. MCDONALD, *Simulations of liquids and solids*, North Holland, Amsterdam, 1987.

[14] W. G. HOOVER, Springer, Berlin New York, 1987.

[15] R. W. Hockney and J. W. Eastwood, *Computer simulations using particles*, Adam Hilger, Bristol, 1988.

[16] K. Kremer and G. S. Grest, *Computer simulations of polymers*, Prentice Hall, Englewood Cliffs NJ, 1991.

[17] K. Kremer and G. S. Grest, *J. Chem. Phys.* **92**, 5057 (1990).

[18] K. Kremer and G. S. Grest, *J. Chem. Phys.* **94**, 4103 (1991), erratum.

[19] K. Kremer and I. Carmesin, *Phys. Rev. Lett.* **61**, 566 (1990).

[20] D. Rigby and R. J. Roe, *J. Chem. Phys.* **87**, 7285 (1987).

[21] D. Rigby and R. J. Roe, *J. Chem. Phys.* **89**, 5280 (1988).

[22] P. G. de Gennes, *Scaling concepts in polymer physics*, Cornell University Press, Ithaca, NY, 1979.

[23] M. S. Doi and S. F. Edwards, *Theory of Polymer Dynamics*, Clarendon Press, Oxford, 1986.

[24] M. N. Rosenbluth and A. W. Rosenbluth, *J. Chem. Phys.* **23**, 356 (1955).

[25] J. W. Lyklema and K. Kremer, *J. Phys. A Math. Gen.* **19**, 279 (1985).

[26] J. Batoulis and K. Kremer, *J. Phys. A Math. Gen.* **21**, 127 (1988).

[27] J. Batoulis, PhD thesis, University of Mainz, 1987.

[28] J. Batoulis and K. Kremer, *Europhys. Lett.* **7**, 683 (1988).

[29] K. Kremer, A. Baumgärtner, and K. Binder, *J. Phys. A Math. Gen.* **15**, 2879 (1982).

[30] B. Duplantier, *J. Phys., Paris* **43**, 1982 (1991).

[31] G. S. Grest, K. Kremer, S. T. Milner, and T. A. Witten, *Macromolecules* **22**, 1904 (1989).

[32] G. S. Grest, K. Kremer, and T. A. Witten, *Macromolecules* **20**, 1376 (1987).

[33] J. Batoulis and K. Kremer, *Macromolecules* **22**, 4277.

[34] F. Candau, P. Rempp, and H. Benoit, *Macromolecules* **5**, 627 (1972).

[35] E. Eisenriegler, K. Kremer, and K. Binder, *J. Chem. Phys.* **77**, 6296 (1982).

[36] K. Kremer, *J. Chem. Phys.* **83**, 5882 (1985).

[37] H. Meirovitch and S. Livne, *J. Chem. Phys.* **88**, 4498 (1988).

[38] H. Meirovitch and S. Livne, *J. Chem. Phys.* **88**, 4507 (1988).

[39] F. van Dieren and K. Kremer, *Europhys. Lett.* **4**, 569 (1987).

[40] E. Eisenriegler, *Polymers at interfaces*, World Scientific, 1993.

[41] H. Meirovitch, *Macromolecules* **18**, 563 (1985), and references therein.

[42] Z. Alexandrovicz, *J. Chem. Phys.* **51**, 561 (1969).

[43] K. Suzuki, *Bull. Chem. Soc. Japan* **41**, 538 (1968).

[44] B. Dünweg and K. Kremer, *J. Chem. Phys.* (1992), submitted for publication.

[45] M. Lal, *Molec. Phys.* **17**, 57 (1969).

[46] N. Madras and A. D. Sokal, *J. Stat. Phys.* **50**, 109 (1988).

[47] A. K. Kron, *Polymer Science USSR* **7**, 1361 (1965).

[48] F. T. Wall and F. Mandel, *J. Chem. Phys.* **63**, 4592 (1975).

[49] K. Kremer, unpublished results.

[50] O. Jogadzinski, E. Eisenriegler, and K. Kremer, in press.

[51] O. F. Olaj and W. Lantschbauer, *Makrom. Chemie Rapid Comm.* **3**, 847 (1982).

[52] M. L. Mansfield, *J. Chem. Phys.* **77**, 1554 (1982).

[53] K. Kremer and K. Binder, *J. Chem. Phys.* **81**, 6381 (1984).

[54] H. P. Wittmann and K. Kremer, *Comput. Phys. Commun.* **61**, 309 (1990).

[55] H. P. Wittmann and K. Kremer, *Comput. Phys. Commun.* **71**, 343 (1992), erratum.

[56] I. Carmesin and K. Kremer, *Macromolecules* **21**, 2819 (1988).

[57] A. Baumgärtner, *Ann. Rev. Phys. Chem.* **35**, 419 (1984).

[58] I. Carmesin and K. Kremer, *J. Phys., Paris* **51**, 1567 (1990).

[59] W. Paul, K. Binder, D. W. Heermann, and K. Kremer, *J. Phys., Paris* **37**, II1 (1991).

[60] J. Sklonick, R. Yaris, and A. Kolinski, *J. Chem. Phys.* **88**, 1407 (1988).

[61] J. Sklonick and R. Yaris, *J. Chem. Phys.* **88**, 1418 (1988).

[62] W. Paul, K. Binder, D. W. Heermann, and K. Kremer, *J. Chem. Phys.* **95**, 7726 (1991).

[63] G. S. Grest and K. Kremer, *Phys. Rev. A* **33**, 3628 (1986).

[64] R. B. Bird, R. C. Armstrong, and O. Hassager, *Dynamics of Polymeric Liquids*, J. Wiley NY, 1971.

[65] G. S. Grest, B. Dünweg, and K. Kremer, *Comput. Phys. Commun.* **55**, 269 (1989).

[66] B. Dünweg and W. Paul, *Int. J. Mod. Phys.* **2**, 817 (1991).

[67] H. P. Wittmann, K. Kremer, and K. Binder, *J. Chem. Phys.* **96**, 6291 (1992).

[68] J. Baschnagel, K. Binder, and H. P. Wittmann, *J. Phys. Cond. Mat.* (1993), in press.

[69] J. Baschnagel and K. Binder, *Physica A* (1993), in press.

[70] M. Murat and G. S. Grest, *Phys. Rev. Lett.* **63**, 1074 (1989).

[71] W. Hess, *Macromolecules* **19**, 1395 (1986).

[72] W. W. Graessley, *Adv. Polym. Sci.* **47**, 67 (1982).

[73] D. S. Pearson, G. V. Strate, E. von Meerwall, and F. C. Schilling, *Macromolecules* **20**, 1133 (1987).

[74] K. G. Ngai, S. L. Peng, and J. Skolnick, preprint, 1991.

[75] K. G. Schweizer, *J. Chem. Phys.* **91**, 5802 (1989).

[76] D. Richter, personal communication.

[77] D. Richter, B. Farago, L. J. Fetters, J. S. Huang, B. Ewen, and C. Lartigue, *Phys. Rev. Lett.* **64**, 1389 (1990).

[78] W. Hess, *Macromolecules* **20**, 2589 (1987).

[79] W. Hess, *Macromolecules* **21**, 2670 (1988).

[80] A. N. Semenov, *Physica A* **166**, 263 (1990).

[81] H. P. Wittmann, K. Kremer, and K. Binder, *Makrom. Chemie Theor. Sim.* **1**, 275 (1992).

[82] W. Paul, K. Binder, K. Kremer, and D. W. Heermann, *Macromolecules* **24**, 6323 (1991).

[83] W. Paul, *AIP Conference Proceedings* **256**, 145 (1992).

[84] E. R. DUERING, K. KREMER, and G. S. GREST, *Phys. Rev. Lett.* **67**, 3531 (1991).

[85] E. R. DUERING, K. KREMER, and G. S. GREST, *Prog. Coll. Polym. Sci.* **90** (1992).

[86] M. SCHULZ and J. U. SOMMER, *J. Chem. Phys.* **96**, 7102 (1992).

[87] G. S. GREST and K. KREMER, *J. Phys., Paris* **51**, 2829 (1990).

[88] G. S. GREST and K. KREMER, *Macromolecules* **23**, 4994 (1990).

[89] G. S. GREST, K. KREMER, and E. R. DUERING, *Europhys. Lett.* **19**, 195 (1992).

[90] E. R. DUERING and G. S. G. K. KREMER AND, *Macromolecules* (1992), submitted for publication.

[91] R. EVERAERS and K. KREMER, in preparation, 1992.

[92] J. G. KIRKWOOD and J. RISEMAN, *J. Chem. Phys.* **16**, 565 (1948).

[93] J. J. ERPENBECK and J. G. KIRKWOOD, *J. Chem. Phys.* **29**, 909 (1958).

[94] B. DÜNWEG and K. KREMER, *Phys. Rev. Lett.* **66**, 2996.

[95] B. DÜNWEG, PhD thesis, University of Mainz, 1991, and preprint 1992.

[96] B. J. BERNE and R. PECORA, *Dynamic Light Scattering*, J. Wiley New York, 1976.

[97] D. RICHTER, K. BINDER, B. EWEN, and B. STÜHN, *J. Chem. Phys.* **88**, 6618 (1984).

[98] C. PIEREONI and J. P. RYCKAERT, *Phys. Rev. Lett.* **66**, 2992 (1991).

[99] C. PIERLONI and J. P. RYCKAERT, *J. Chem. Phys.* **96**, 8539 (1992).

[100] M. J. STEVENS and K. KREMER, preprint, 1992.

[101] K. KREMER, B. DÜNWEG, and M. J. STEVENS, *Physica A* (1993), to appear.

[102] S. B. LEE and H. NAKANISHI, *Phys. Rev. Lett.* **61**, 2022 (1988).

[103] Y. MEIR and A. B. HARRIS, *Phys. Rev. Lett.* **63**, 2819 (1989).

[104] A. BAUMGÄRTNER and M. MUTHUKUMAR, *J. Chem. Phys.* **87**, 3082 (1987).

[105] A. BAUMGÄRTNER, *Europhys. Lett.* **4**, 1221 (1987).

[106] H. P. DEUTSCH and K. BINDER, *Europhys. Lett.* **17**, 697 (1992).

[107] K. BINDER and H. P. DEUTSCH, *Europhys. Lett.* **18**, 667 (1992).

[108] K. S. SCHWEIZER and J. G. CURRO, *J. Chem. Phys.* **94**, 3986 (1991).

[109] K. S. SCHWEIZER and J. G. CURRO, *Phys. Rev. Lett.* **60**, 809 (1988).

[110] M. MURAT and G. S. GREST, *Macromolecules* **22**, 4054 (1989).

[111] M. MURAT and G. S. GREST, *Phys. Rev. Lett.* **63**, 1074 (1989).

[112] M. MURAT and G. S. GREST, *Macromolecules* **24**, 704 (1991).

[113] P. Y. LAI and K. BINDER, *J. Chem. Phys.* **95**, 9288 (1991).

[114] A. CHAKRABARTI and R. TORAL, *Macromolecules* **23**, 2016 (1990).

[115] S. ALEXANDER, *J. Phys., Paris* **38**, 983 (1977).

[116] P. G. DE GENNES, *Macromolecules* **13**, 1069 (1980).

[117] S. T. MILNER, T. A. WITTEN, and M. E. CATES, *Macromolecules* **21**, 2610 (1988).

[118] S. T. MILNER, T. A. WITTEN, and M. E. CATES, *Europhys. Lett.* **5**, 413 (1988).

[119] P. Y. LAI and K. BINDER, preprint, 1992.

[120] M. MURAT and G. S. GREST, preprint, 1992.

Computer Simulations of Surfactants

B. Smit
Shell Research B.V.
Koninklijke/Shell-Laboratorium Amsterdam
PO Box 3003
1003 AA Amsterdam
The Netherlands.

ABSTRACT. This Chapter is a review of the computer simulations of surfactant systems. The focus is on monolayers and on micelles as relatively simple examples of surfactant self assemblies. During the last decade simulations of monolayers have progressed to a level of sophistication where they can be used to interpret experimental data and even suggest new experiments. The computer simulation of micelles is much less advanced. We demonstrate that simulations of micelles are much more difficult then those of the corresponding monolayers, and we present some very simple models that are able to describe a micellar solution.

1. Introduction

Amphiphilic molecules such as surfactants, lipids, or fatty acids have an affinity for both water and oil. The polar head of the molecule is hydrophilic, while the hydrocarbon chain is responsible for the affinity for oil. At the oil/water interface these molecules can locate their hydrophobic chains in the oil phase and allow the polar head to be in the water phase. This situation is energetically more favourable than complete solution in either phase. The tendency for these surface-active agents to adsorb causes a reduction of the bare oil/water interfacial tension. Because of this property, surfactants are used in applications such as detergency and tertiary oil recovery [1].

The complexity of these systems makes it impossible to study surfactant systems with the same level of detail that would be used in a simulation of a simple liquid. It is therefore essential to make simplifications. As we will see, making a simplification is a subjective matter which reflects the interests of the individual researcher. For example, if we are interested in the chain conformations in a micelle, it is important to take into account the details of the intramolecular potentials of the molecules. Simulations with such realistic potentials are limited to times of the order of nanoseconds:

M.P. Allen and D.J. Tildesley (eds.), Computer Simulation in Chemical Physics, 461–472.

too short to observe surfactants leaving or entering a micelle. For realistic models, the observations are limited to the shortest time scale, e.g. the movements of surfactants within a micelle. If we choose to model a surfactant with a more idealized potential, the (effective) time span of the simulations is much larger, and it is possible to observe surfactants entering or leaving a micelle. However, since all the atomic details are lost, direct comparison with experimental data is impossible.

In this Chapter we focus on two examples of surfactant self assemblies: surfactant monolayers and micelles. This is a small selection out of the vast number of different assemblies that form in surfactant systems. For simulations that have been performed on bilayers (of importance in biological systems, since a bilayer is a simple model of a membrane) we refer the reader to the various review articles (for example, Refs. [2, 3]).

2. Surfactant monolayers

2.1. LANGMUIR MONOLAYERS

There is now a history of 100 years of experimental work [4] on Langmuir films. Langmuir films are monolayers of amphiphilic molecules on the water surface. These monolayers are studied by measuring their surface pressure-area isotherms. These experiments gave evidence for the existence of three phases: a gas phase, liquid phase, and solid phase of the monolayer. Some of the later experiments suggested that even more phases existed [5, 6], but these results were controversial because of the possible presence of impurities. It is only recently that these additional phases have been confirmed using novel experimental techniques [7–9]. Nowadays it is even possible to make direct observation of domains of the various phases characterized by different tilt directions of the chains (see for example Ref. [10]). At present, considerable effort is being spent in identifying the various phases [11]. One can expect that simulations will contribute to the characterization of these phases. As a result, monolayers are an attractive area of research, and several simulation studies of these systems have been reported.

One of the first simulations of a 'realistic' model of a monolayer was reported by Kox *et al.* [12]. The model surfactants were chains of seven sites. The water was modeled as an external field that constrained the head groups to move in a plane. The interactions between the tail segments consisted of a short-ranged repulsion, and a Lennard-Jones potential was used for the interactions between the head groups. Springs were used to fix the distance between neighbouring carbon sites (C) and the angle between the the C-C bonds. These simulations showed evidence for a liquid-gas phase transition. These early simulations were not sophisticated enough to observe other phase transitions.

The controversy about the existence of the LE (liquid-expanded) and LC (liquid-condensed) phases inspired several groups to study Langmuir monolayers using more detailed models [13, 14]. The models of these simulations are based on the pioneering work of van der Ploeg and Berendsen on bilayers [15]. The surfactants are represented with a pseudo-atom model, i.e. the methylene, the methyl, and head groups are treated as one atom. The water substrate is modelled with an external field so that the head groups cannot evaporate from the surface. These simulations yield information on the structure of the monolayers and the chain conformations. The differences between the models used by the various groups are small.

At high surface density and low temperatures, various solid-solid phase transitions can occur. For example, Lin *et al.* [16] used X-ray-diffraction to study a monolayer of heneicosanic acid ($CH_3(CH_2)_{19}COOH$) and observed first-order transitions between different lattice structures, and continuum transitions between phases characterized by different tilts of the molecules. Computer simulations at these high surface concentrations were performed by Karaborni and Toxvaerd [17, 18]. At these conditions, it is questionable whether a united-atom approach is sufficiently accurate [19, 20] (for more information on the importance of explicit hydrogen atoms see Ref. [21]). Karaborni and Toxvaerd used an anisotropic united-atom model to represent the interactions of the alkane backbone. This anisotropic potential is introduced to model the effects of the hydrogens. In Refs. [22, 23] it is shown that, at high densities, these anisotropic potentials give a significantly better equation of state than the united-atom model for alkanes. Furthermore, repulsive dipolar interactions were used to model the interactions between the head groups. The water was modelled using an external potential. This external potential was such that the chains could penetrate in the water phase. The simulations of this model of surfactants with 15 C segments showed two distinct phase transitions in the dense monolayer region. A transition between a liquid and solid was observed, and at higher densities a solid-solid transition. This solid-solid transition was characterized by a change from an all-trans conformation to conformations of the chains with kink defects . Interestingly, the solid-liquid transition was not observed when the simulations were repeated with a united-atom model.

Most of the previous results were for monolayers of hydrogenated fatty acids ($CH_3(CH_2)_nCOOH$). To study the relative importance of the various intermolecular interactions, experiments have been performed in which the H-atoms are replaced by the larger F-atoms [24]. These fluorocarbon chains are much stiffer than their hydrogenated counterparts. Furthermore, for the fluorocarbon chains the head group is smaller than the backbone of the chain, while for the hydrogenated chains the head group is larger. Whether a computer simulation can predict the subtle difference between these systems is an important test for the credibility of these methods. In fact, Rice and co-workers [25, 26] have shown that computer simulations can describe most of the experimental observations in these systems. The study in Ref. [26] predicts that the monolayer exhibits a phase in which the molecules tilt towards their nearest neighbours or towards their next-nearest neighbours depending on whether the chain length is odd or even. It is interesting to see whether these predictions will be verified by experiments in the near future.

2.2. LANGMUIR-BLODGETT FILMS

Monolayers that form at the air/water interface can be transferred to a solid substrate by dipping the substrate into the water. Depending on whether the substrate is hydrophobic or hydrophilic, the heads or the tails of the surfactants attach to the surface, respectively. When the substrate is dipped into the solution again, a second layer will attach. This procedure can be repeated, and films with over 1000 layers can be produced with this technique.

Most computer simulations have been performed on a single-layer Langmuir-Blodgett film [20, 27]. In a real system, there is a clear difference between water as a substrate (Langmuir monolayer) and a solid substrate (Langmuir-Blodgett film). In most simulations, however, both water and a solid substrate are modelled with an external potential that constrains the head groups to a plane. This makes a distinction between the two types of monolayers somewhat arbitrary. In fact, the distinction depends mainly on the experimental systems the authors chose to compare their results with. Many of the simulations discussed in the previous section could also be discussed in this section, and vice versa.

Cardini *et al.* [28] were the first to present computer simulations of a surfactant monolayer with the term "Langmuir-Blodgett film" in the title. The model used by Cardini *et al.* is identical to the one used by Bareman *et al.* [13] for a Langmuir monolayer (see the previous section), i.e. united atoms, no explicit head group, and the substrate modeled with an external field. The simulations yielded information on the structure of a relatively dense monolayer of arachidic acid ($CH_3(CH_2)_{18}COOH$). A more realistic model was introduced by Moller *et al.* [20]. The head group was explicitly taken into account (but the interactions between the head groups was modelled with a Lennard-Jones potential instead of a more realistic dipolar interaction). Furthermore, a detailed comparison was made between the united-atom model for the CH_2 backbone and a model in which the hydrogens were taken into account explicitly. The simulations of Moller *et al.* predict a transition of the tilt of the molecules, depending on the temperature. In the united-atom model such a transition could not be observed.

2.3. SELF-ASSEMBLED MONOLAYERS

Monolayers of surfactants on a gold or silica surface can form spontaneously by self assembly of the surfactant molecules from solution [29]. The structure of these monolayers is found to depend on the face of the gold crystal used, and the nature of the head groups of the surfactant. By altering the chain terminal group, the characteristics of the monolayer can be manipulated.

Detailed molecular dynamics simulations on these systems have been performed by Hautman and Klein [30–32]. The model used in these simulations is similar to the one used in the simulations of the Langmuir films (see section 2.1). The main difference is that the head group is now chemically bound to the substrate, and the underlying structure of the substrate will influence the structure of the monolayer. At present, the detailed nature of this bond is unknown, and different models are studied. The first study [30] used an artificially large, hard core for the S-S interactions to model the gold crystal lattice, Subsequent studies [32] used a corrugation potential. The lattice structure of the substrate does not necessarily correspond to the optimal structure for the head groups. This source of frustration leads to many interesting phenomena. For example, Hautman *et al.* [32] studied the structure of the exposed surface for various head groups; in particular the systems with X=$-CH_3$, –OH, and –C≡N were studied. As might be expected, the surface of the $-CH_3$ terminated layer does not differ much from that of an ordinary Langmuir monolayer. The –OH surface, however, shows a disordered (glass-like) one-dimensional network of hydrogen bonds. For the –C≡N surface, a ferroelectric structure is observed that is incommensurate with the gold lattice. The wetting properties have also been studied in Ref. [33].

Siepmann and McDonald [34] studied the behaviour of binary mixtures of alkyl thiols with different chain lengths. These simulations were made possible by the use of special Monte Carlo techniques (details of these techniques can be found in Chapter 4). Siepmann and McDonald observed segregation in mixtures of C_{10} and C_{20} chains.

The simulations on self-assembled monolayers are already at such a level of sophistication that they provide a molecular explanation of the experimentally observed phenomena.

2.4. THE OIL/WATER INTERFACE

The oil/water interface is important in many applications. In these systems, surfactants are often used to reduce the interfacial tension. From an industrial point of view, one is interested in the relationship between the structure and properties of these surfactants. These structure-property relationships can be obtained from experiments on surfactants in which the chemical structure is changed systematically [35]. Smit and co-workers [36, 37] developed a highly idealized model, based on that of Telo da Gama and Gubbins [38], to study the behaviour of surfactant molecules at the oil/water interface. In this work, Smit and co-workers tried to eliminate all molecular details. As a result oil, water, and surfactants were modeled with simple Lennard-Jones potentials using the same ϵ and σ for all interactions. Only the truncation of the potential depends on the type of interaction. Surfactants are constructed by connecting oil and water segments with harmonic springs. Using such an approach, it is impossible to obtain detailed molecular information, but the effects of tail length or branching on the efficiency of a surfactant in reducing the surface tension is predicted [37, 39].

3. Micelles

When surfactants are added to water they will form a monolayer at the water surface. The properties of such a layer have been discussed in section 2. A further increase of the surfactant concentration saturates the interface and the surfactants aggregate in the bulk. In these micelles, the surfactants point their hydrophilic head towards the water phase, and their hydrophobic tails form the interior of the micelle. The importance of these structures does not need any further discussion when we realize that it would be impossible to wash our hands with soap if micelles did not exist.

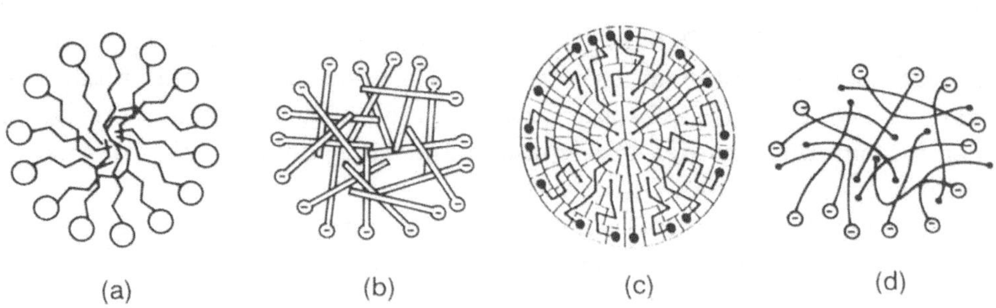

Fig. 1. Various views of a micelle based on Refs. (a) [40], (b) [41], (c) [42], and (d) [43].

Compared to our knowledge at a molecular level of the structure of a monolayer, our understanding of a micelle is still very poor (see Fig. 1 for a historic evolution of the structure of a micelle). Elementary questions are still unanswered. To what extent does water penetrate into the core of a micelle [44]? What is the nature of the hydrocarbon/water interface [45]? What type of interactions are essential for the formation of micelles [46]? In fact, there are still qualitative differences of opinion on how a micelle might look.

From a computer simulation point of view, micelles are extremely difficult to study. The problem becomes clear if we consider the time scales that are involved in a micellar solution. These have been experimentally estimated. It takes from between 10^{-8} s to 10^{-6} s for a surfactant to enter or leave a micelle, the fusion of two micelles occurs in a time span of between 10^{-5} s and 10^{-3} s, and the lifetime of a micelle is estimated at between 10^{-2} s and 1 s [47]. If we recall that a computer simulation with realistic potentials is limited to several nanoseconds, it is obvious that drastic assumptions have to be made to study micelles. If one wishes to use a realistic model, it is necessary to construct the micelle at the beginning of the simulation and study its evolution. With more idealized models, it turns out that one can actually observe the self assembly of micelles.

3.1. ISOLATED MICELLES

Computer simulations of micelles have been pioneered by Pratt and co-workers [48–50]. In these studies, a simple lattice model was proposed in which the surfactant chain can occupy the sites of a diamond lattice. The energy parameters are chosen so that the tail-tail interactions are attractive and the head-head interactions repulsive. In contrast with the conventional notion that a micelle is a spherical structure with the head groups localized on the outside of a sphere, these simulations showed that fluctuations in shape and size lead to a more irregular picture of the micelle. This irregularity is also present in the more recent studies of more realistic models.

The first molecular dynamics simulations of a micelle containing 'realistic' surfactants were performed by Haile and O'Connell [51]. In this study a dodecyl surfactant was modelled with a united-atom representation of the carbon chain (similar to the model of Ref. [15]) and the head group is modelled as an external potential that fixes the carbon chain to lie inside a sphere. The volume of this sphere was chosen so that the density of the core of the micelle corresponds to that of liquid butane. These simulations yield results on the conformations of chains in the micelle. However, these results are biased by the underlying assumption that the micelle can not fluctuate in size, and that the head groups are constrained to fixed points on the surface of a sphere. In view of the results of Pratt and co-workers, these assumptions may not be valid. In subsequent work, by Wood *et al.* [52], the surfactants were allowed to move on the surface of the sphere, but were still constrained by a harmonic spring to the outer layers of the sphere.

The results of these simulations were similar to those of the earlier work. This work was extended by Karaborni and O'Connell [53, 54]. Their model of water allowed the surfactants to move everywhere, but the tails have to cross an energy barrier to enter the water phase, and the head groups a barrier to enter the core of the micelle. Since the external fields are spherically symmetric, only small fluctuations in size and shape were observed in this model.

The effect of water is explicitly taken into account in Refs. [55–57]. In addition, these simulations start with a postulated micellar structure. Watanabe *et al.* [56, 57] simulated a sodium octanoate micelle in an aqueous solution for 0.2 ns. They show that the micelle was stable over the length of the simulation. Considering the time scales of the simulations compared to the relevant experimental time scales, it is impossible to be certain that such a micelle is thermodynamically stable.

If we compare the simulations of monolayers with those of micelles, it is evident that the latter are much more difficult. In a micelle, the interface between the hydrocarbons in the core of the micelle and the water is ill defined. This implies that the modelling of the water using an external field will be less successful. Furthermore, a micellar solution is characterized by its micelle size distribution [58], i.e. there is not one micelle that determines all properties, and we can expect that, on the experimental time scale, fluctuations in micelle size will be important. With these realistic models, a simulation of the time evolution of such a system over 0.2 ns requires a sizable amount of computer time. Since, 0.2 ns is too short to observe surfactants leaving or entering a micelle, these simulations are necessarily focused on the configurations of the surfactant chains in a micelle. Unfortunately, experimental information is scarce and is not specific enough to allow direct comparison with simulation, a problem already encountered with the monolayers.

Smit and co-workers [39, 59, 60] studied micelles with a coarse-grained model (see section 2.4). In this case, all molecular details are lost, but a much longer time scale can be studied. In Ref. [61] the simulations were started with a random distribution of surfactants in a water/oil system. It was observed that a monolayer was formed at the oil/water interface and that micelles were formed in the bulk. A snapshot of the system is shown in Fig. 2. Despite the simplicity of the model, the simulations required a large amount of computer time and were performed on a parallel computer [62]. Unfortunately, the number of micelles was too small to determine the micelle size distribution from these simulations. Such a distribution was obtained in subsequent work using a slightly different model [60]. In these simulations (equilibrium) dynamical processes such as the inclusion of single surfactants into micelles, the elimination of single surfactants from the micelles, the fusion of two micelles, and the slow breakdown of a micelle, were observed.

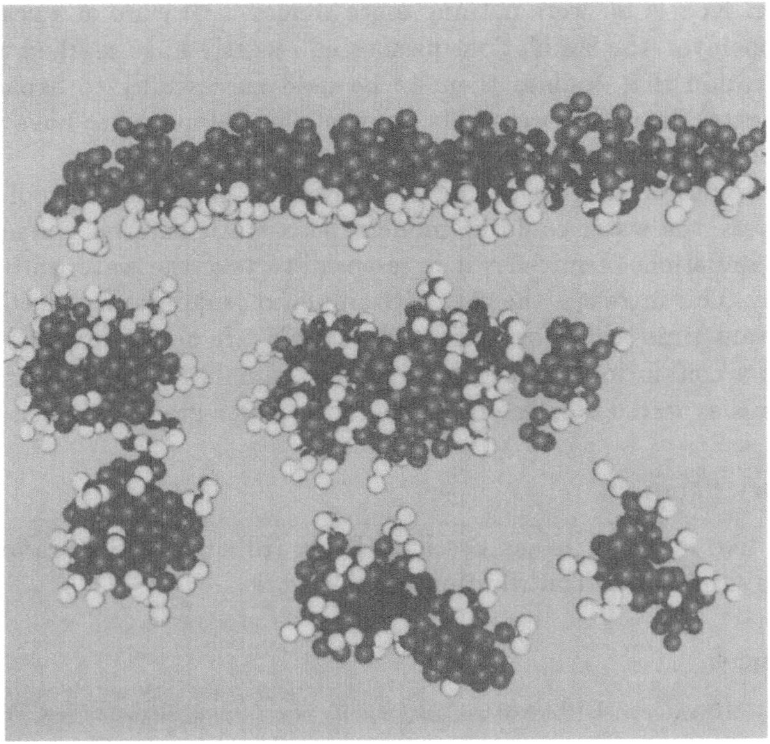

Fig. 2. A typical example of a configuration of surfactant molecules in an oil/water system for 1.5% surfactants. The snapshot shows only the surfactants at one of the monolayers and the surfactants in the water phase. For clarity, the surfactants in the oil phase and the oil and water particles are not shown. The hydrophilic segments are light grey and the hydrophobic segments dark. The surfactants have a head of two hydrophilic w particles and a tail of five hydrophobic o particles (h_2t_5).

This is an encouraging observation, since the experimental time scale for these processes spans more than 6 decades. These results suggest that it is possible to perform systematic studies of dynamical processes in micellar solutions. The first interesting results in this direction have been reported by Karaborni *et al.* [63]. Karaborni *et al.* used a similar oil/water/surfactant model to study the oil solubilization in surfactant solutions. On the basis of their molecular dynamics results they identified three different mechanisms of oil solubilization.

4. Concluding remarks

Looking back at a decade of computer simulations of monolayers, one can only conclude that the progress has been enormous. While the first simulations of Kox *et al.* were nothing more than a caricature of a real surfactant monolayer, the simulations performed recently have reached a level of sophistication that enables them to be used successfully to explain some experimental data. Encouragingly, some of these simulations have provided interesting suggestions for new experiments.

Simulations of micelles are far less advanced. In simulations of aqueous monolayers, the water could be modeled as a static external potential, but in the simulations of micelles it is essential to take the water into account explicitly. This increases the cost of a simulation significantly. Furthermore, the relevant time scales for realistic potentials, are not, at present, accessible with a conventional molecular dynamics simulation. As a consequence, progress is expected to come from the more coarse-grained models.

Acknowledgement

I would like to thank Klaas Esselink, Peter Hilbers, Sami Karaborni, and Nico van Os for their contributions to this work.

References

[1] K. L. Mittal and B. Lindman, *Surfactants in solution*, Plenum, New York, 1984.

[2] H. J. C. Berendsen, Biological molecules and membranes, in *Molecular Dynamics Simulations of Statistical Mechanical Systems*, edited by G. Ciccotti and W. G. Hoover, Proceedings of the 97th International Enrico Fermi School of Physics, 1986.

[3] O. G. Mouritsen, Computer simulation of cooperative phenomena in lipid membranes, in *Molecular Description of Biological Membrane Components by Computer Aided Conformation Analysis (vol. I)*, edited by R. Brasseur, CRC Press, Inc Boca Raton, FL, 1990.

[4] A. Pockels, *Nature* **43**, 437 (1891).

[5] S. Stälberg-Stenhagen and E. Stenhagen, *Nature* **156**, 239 (1945).

[6] M. Lunquist, *Chemica Scripta* **1**, 5 (1971).

[7] S. R. Middleton, M. Iwahashi, N. R. Pallas, and B. A. Pethica, *Proc. Roy. Soc. Lond. A* **396**, 143 (1984).

[8] B. M. Moore, C. M. Knobler, S. Akamatsu, and F. Rondelez, *J. Phys. Chem.* **94**, 4588 (1990).

[9] K. Kjaer, J. Als-Nielsen, C. A. Helm, P. Tippman-Krayer, and H. Möhwald, *J. Phys. Chem.* **93**, 3200 (1989).

[10] X. Qiu, J. R. Garcia, K. J. Stine, C. M. Knobler, and J. V. Selinger, *Phys. Rev. Lett.* **67**, 703 (1991).

[11] A. M. Bibo, C. M. Knobler, and I. R. Peterson, *J. Phys. Chem.* **95**, 5591 (1991).

[12] A. J. Kox, J. P. J. Michels, and F. W. Wiegel, *Nature* **287**, 317 (1980).

[13] J. P. Bareman, G. Cardini, and M. L. Klein, *Phys. Rev. Lett.* **60**, 2152 (1988).

[14] J. Harris and S. A. Rice, *J. Chem. Phys.* **89**, 5898 (1988).

[15] P. van der Ploeg and H. J. C. Berendsen, *J. Chem. Phys.* **76**, 3271 (1982).

[16] B. Lin, M. C. Shih, T. M. Bohanon, G. E. Ice, and P. Dutta, *Phys. Rev. Lett.* **65**, 191 (1990).

[17] S. Karaboni, S. Toxvaerd, and O. H. Olsen, *J. Phys. Chem.* **96**, 4965 (1992).

[18] S. Karaboni and S. Toxvaerd, *J. Chem. Phys.* **96**, 5505 (1992).

[19] J. P. Bareman and M. L. Klein, *J. Phys. Chem.* **94**, 5202 (1990).

[20] M. A. Moller, D. J. Tildesley, K. S. Kim, and N. Quirke, *J. Chem. Phys.* **94**, 8390 (1991).

[21] M. L. Klein, *J. Chem. Soc. Faraday Trans.* **88**, 1701 (1992).

[22] S. Toxvaerd, *J. Chem. Phys.* **93**, 4290 (1990).

[23] P. Padilla and S. Toxvaerd, *J. Chem. Phys.* **94**, 5650 (1991).

[24] S. Barton, A. Goudot, O. Boulossa, F. Rondelez, B. Lin, F. Novak, A. Acero, and S. A. Rice, *J. Chem. Phys.* **96**, 1343 (1992).

[25] S. Shing, N. Collazo, and S. A. Rice, *J. Chem. Phys.* **96**, 1352 (1992).

[26] N. Collazo, S. Shing, and S. A. Rice, *J. Chem. Phys.* **96**, 4735 (1992).

[27] G. Cardini, J. P. Bareman, and M. L. Klein, *Chem. Phys. Lett.* **145**, 493 (1988).

[28] I. L. Carpenter and W. J. Hehre, *J. Phys. Chem.* **94**, 531 (1990).

[29] R. G. Nuzzo and D. L. Allara, *J. Am. Chem. Soc.* **105**, 4481 (1983).

[30] J. Hautman and M. L. Klein, *J. Chem. Phys.* **91**, 4994 (1989).

[31] J. Hautman and M. L. Klein, *J. Chem. Phys.* **93**, 7483 (1990).

[32] J. Hautman, J. P. Bareman, W. Mar, and M. L. Klein, *J. Chem. Soc. Faraday Trans.* **87**, 2031 (1991).

[33] J. Hautman and M. L. Klein, *Phys. Rev. Lett.* **67**, 1763 (1991).

[34] J. I. Siepmann and I. R. McDonald, *Molec. Phys.* **75**, 255 (1992).

[35] N. M. van Os, G. J. Daane, and T. A. B. M. Bolsman, *J. Coll. Int. Sci.* **115**, 402 (1987).

[36] B. Smit, *Phys. Rev. A* **37**, 3431 (1988).

[37] B. Smit, A. G. Schlijper, L. A. M. Rupert, and N. M. van Os, *J. Phys. Chem.* **94**, 6934 (1990).

[38] M. M. Telo da Gama and K. E. Gubbins, *Molec. Phys.* **59**, 227 (1986).

[39] B. Smit, *Computer simulation of phase coexistence: from atoms to surfactants*, PhD thesis, Rijksuniversiteit Utrecht, The Netherlands, 1990.

[40] G. S. Hartley, *Trans. Faraday Soc.* **31**, 31 (1935).

[41] P. Fromherz, *Chem. Phys. Lett.* **77**, 460 (1981).

[42] K. A. Dill and P. J. Flory, *Proc. Nat. Acad. Sci.* **78**, 676 (1981).

[43] F. M. Menger and D. W. Doll, *J. Am. Chem. Soc.* **106**, 1109 (1984).

[44] F. M. Menger, *Acc. Chem. Res.* **12**, 111 (1979).

[45] S. H. Chen, *Ann. Rev. Phys. Chem.* **3**, 351 (1986).

[46] A. H. Beesley, D. F. Evans, and R. G. Laughlin, *J. Phys. Chem.* **92**, 791 (1988).

[47] J. Lang and R. Zana, Chemical relaxation methods, in *Surfactants in solution: new methods of investigation*, edited by R. Zana, 1987.

[48] S. W. Haan and L. R. Pratt, *Chem. Phys. Lett.* **79**, 436 (1981).

[49] B. Owenson and L. R. Pratt, *J. Phys. Chem.* **88**, 2905 (1984).

[50] B. Owenson and L. R. Pratt, *J. Phys. Chem.* **88**, 6048 (1984).

[51] J. M. HAILE and J. P. O'CONNELL, *J. Phys. Chem.* **88**, 6363 (1984).

[52] M. C. WOODS, J. M. HAILE, and J. P. O'CONNELL, *J. Phys. Chem.* **90**, 1875 (1986).

[53] S. KARABORNI and J. P. O'CONNELL, *Langmuir* **6**, 905 (1990).

[54] S. KARABORNI and J. P. O'CONNELL, *J. Phys. Chem.* **94**, 2624 (1990).

[55] B. JÖNSSON, O. EDHOLM, and O. J. TELEMAN, *J. Chem. Phys.* **85**, 2259 (1986).

[56] K. WATANABE, M. FERRARIO, and M. L. KLEIN, *J. Phys. Chem.* **92**, 819 (1988).

[57] K. WATANABE and M. L. KLEIN, *J. Phys. Chem.* **93**, 6897 (1989).

[58] Y. CHEVALIER and T. ZEMB, *Rep. Prog. Phys.* **53**, 279 (1990).

[59] B. SMIT, P. A. J. HILBERS, K. ESSELINK, L. A. M. RUPERT, N. M. VAN OS, and A. G. SCHLIJPER, *J. Phys. Chem.* **95**, 6361 (1991).

[60] B. SMIT, K. ESSELINK, P. A. J. HILBERS, N. M. VAN OS, L. A. M. RUPERT, and I. SZLEIFER, *Langmuir* (1993), in press.

[61] B. SMIT, P. A. J. HILBERS, K. ESSELINK, L. A. M. RUPERT, N. M. VAN OS, and A. G. SCHLIJPER, *Nature* **348**, 624 (1990).

[62] K. ESSELINK, B. SMIT, and P. A. J. HILBERS, *J. Comput. Phys.* (1992), in press.

[63] S. KARABONI, N. M. VAN OS, K. ESSELINK, and P. A. J. HILBERS, submitted for publication, 1992.

Parallel Computing and Molecular Dynamics Simulations

P. A. J. Hilbers and K. Esselink
Koninklijke/Shell-Laboratorium Amsterdam
(Shell Research B.V.)
PO Box 3003
1003 AA Amsterdam
The Netherlands.

ABSTRACT. Molecular dynamics simulations are an important means to study macroscopic phenomena at a microscopic level. Due to their huge demand for computing resources, implementations on parallel processor networks are an important topic of research. Some aspects of the design of parallel programs and of the implementations of such programs on a processor network are reviewed. Three general techniques for exploiting parallelism and their appropriateness in the context of molecular dynamics are discussed. We show that short-range and multi-particle potentials can efficiently be implemented when geometric parallelism is used. Our approach is demonstrated by showing some results of large scale molecular dynamics simulations on the nucleation properties of linear chain molecules. These results show that parallel processor networks open new perspectives for the study of large systems and problems which could not previously be dealt with.

1. Introduction

Computer simulations of physical systems play an important rôle in statistical mechanics. Usually properties of the system are described on the microscopic scale by the type of particle, the particle positions, and their momenta. Since the calculations look the same for all particles, parallel processing of these calculations is attractive. This fact has been exploited in several implementations of the molecular dynamics codes on parallel processor networks [1–8].

Apart from this inherent parallelism, parallel processor networks usually have the attractive feature that they can be expanded in size. Increasing the number of processors leads to more computing power and larger total memory size. As a consequence, such an extended system opens the way to study larger problems in a cost-effective way. Examples of new phenomena that could be studied on parallel processor networks include micelle formation, surfactant adsorption at interfaces, oil solubilization in water/oil/surfactant

M.P. Allen and D.J. Tildesley (eds.), Computer Simulation in Chemical Physics, 473–495.

systems, and nucleation processes. In this Chapter we review several aspects of the realization of a parallel computing application. Parallel molecular dynamics is used as an example, but the emphasis is on the generality of the methods applied. Following Ref. [9], we opt for a "how-to-do-it" approach and we hope that the text will be useful to people who want to use parallel processor networks to simulate the behaviour of many-particle systems.

In the next section, parallel processor networks are discussed from a historical perspective. We show how advances in hardware technology have resulted in the current class of parallel processor networks. These networks are compared to other types of modern computer architectures and it is shown that, at least from a hardware point of view, the parallel processor networks are to be preferred.

The design of algorithms which are well-suited to exploit the processing power of parallel processor networks is quite a different subject. It is an area of extensive research and in general it is considered a difficult task. We show that the restriction to some simple design disciplines, leads to a generally-applicable approach to solving large computational problems. In section 3, several of the concepts underlying this approach are outlined and their usefulness is demonstrated by some simple examples.

Parallel molecular dynamics is the subject of section 4 . We discuss three general techniques for exploiting parallelism and their appropriateness in the context of molecular dynamics. Some theoretical results concerning implementations using the geometric-parallelism method are given.

As an example of the possibility of studying new phenomena by using a parallel processor network, we present some recent results of large-scale molecular dynamics simulations on the nucleation properties of linear chain molecules in section 5. We end with some conclusions in the last section.

2. Computer architectures

From an application point of view there is always a demand for faster, larger, and, at the same time, cheaper computers. In the past the progress in computing power and size has been realized by improvements in the design of complex processors and memories. The cost of the equipment was dominated by the cost of the hardware components employed to build the processor. Due to the significant advances in chip technology this is no longer true. The cost and the performance of the current generation of computers are dictated by the communication wires and not by the logic [10–13].

- Most of the area of a chip is covered with wires on several layers, whereas the logic devices are only found in the lowest layers and rarely take up more than about five percent of the area. So, communication is expensive in its requirement for chip area.

- The capacitance of internal and inter-chip wires dominates device gate capacitance, and so most of the energy dissipated is used to drive the wires.
- The propagation time to drive a signal through a wire is long compared to the switching time of a transistor and this makes communication an important factor in determining speed.

These facts have been explored by all manufacturers of the current generation of supercomputers in building machines which use several small processors together in order to solve a single problem. Two main categories of organization of such collections of processors exist [4].

The first category concerns the Single Instruction Multiple Data (SIMD) machines. In a SIMD computer, a single master processor broadcasts program instructions to individual processing elements which carry out the instructions on their own local data. Well-known examples of this class of machines are the Connection Machine (CM) and the Distributed Array Processor (DAP).

Although these machines have been successfully applied in a number of fields, we concentrate on the second category of Multiple Instruction Multiple Data (MIMD) machines because of their generality. In a MIMD machine each processor executes its own program on its own data. There are several varieties of MIMD machines. The primary distinction is between machines with a small number of complex processors, and machines with a large number of simple processors. As long as the number of processors is small, it is technologically feasible to connect them to one large memory. These shared-memory machines, exemplified by the Cray Y-MP, are attractive because of the many similarities with traditional sequential computers. For this reason, many programming techniques developed for sequential machines can be directly applied to this architecture. They, however, have one great flaw: enlarging the number of processors inevitably leads to memory access problems. The solution for overcoming this problem is to distribute the memory over the available processors. This approach has led to the class of parallel processor networks consisting of a collection of small, usually identical processors together with a sparse, regular interconnection network. Each processor has its own small private memory and has a small number of communication links for exchanging messages with neighbouring processors. Well-known examples of such processors are Intel's iPSC/860 and INMOS's Transputer. There is a rich variety in communication networks because of the different number of possible communication links per processor and the huge number of possibilities for combining the links into a network. Usually combinations are obtained by aggregating large networks from smaller ones. In Fig. 1 some simple, small processor networks are shown. Apart from being processor networks themselves, their importance comes from being a basic building block for larger processor networks, such as a hypercube, a H-tree,

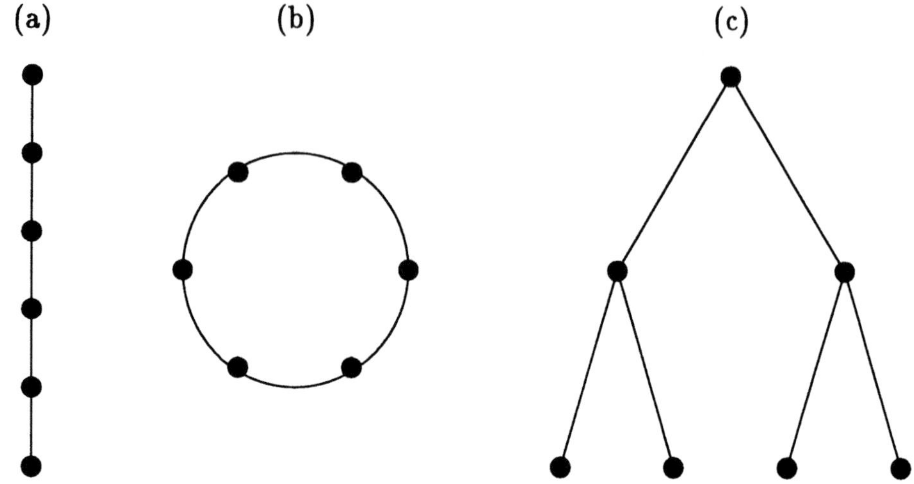

Fig. 1. Examples of processor networks. (a) A linear array of 6 processors. Apart from the two boundary processors each processor has two bidirectional links along which the processor can send and receive messages. It is rather simple to solve the boundary irregularity. By adding one link, a ring (b) is obtained. In (c) a tree is shown.

and a torus. An example of a torus is shown in Fig. 2.

Increasing the number of processors in Fig. 2 but keeping the structure invariant can be accomplished by adding a column of 6 processors. This is a general feature of processor networks with a sparse, regular interconnection pattern. They can be expanded to be as large as desired, with each part still operating at the same rate as when it was incorporated into a smaller system. This feature is absent in the shared-memory systems.

Important aspects of the communications network when analyzing its performance are its latency and its throughput. The latency is defined as the average time needed to deliver a single message in isolation in the network, and should be short. The throughput of a network is the total number of messages that can be handled per time unit in the network, and should be high. We do not discuss these aspects further in this Chapter, but interested readers are referred to Refs. [10, 14].

The design of programs for parallel processor networks is an area of extensive research; some ideas and principles are considered in the next section.

3. Parallel programming concepts

In the previous section, we noted that systems consisting of a collection of small processors, each with its own private store, and connected by a sparse, regular network, are optimal from the hardware point of view. The

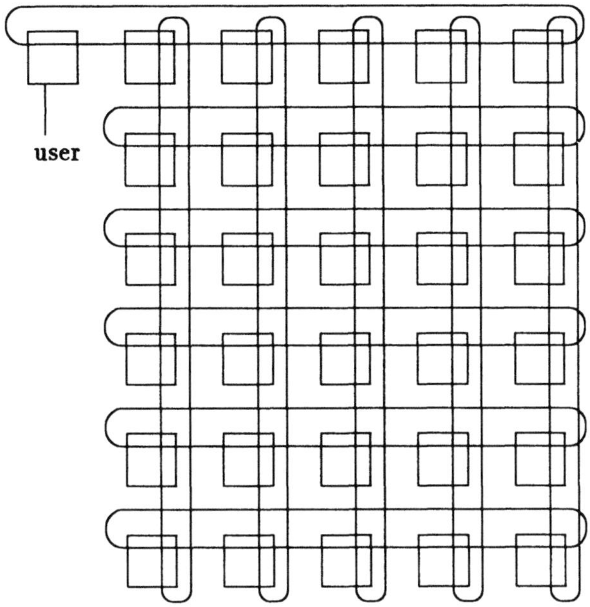

Fig. 2. A 6 × 5 torus network together with a connection to its environment. In both the horizontal and vertical direction the ring structure is apparent.

design of programs for such a parallel processor network that provides the maximum benefit from the available resources is thought to be a difficult task. In fact, in general, it is not known how to achieve this goal [15]. In this section, we review several aspects of the design of parallel programs and of the implementation of these programs on a processor network.

In recent years, much has been published on designing parallel programs [16–21]. These programs consist of many parts that can be executed simultaneously. Each such part is called a process.

In general, a process needs to communicate with some other processes. To that end, parallel programming languages contain language constructs for communication. Several kinds of communication constructs exist. For our examples of parallel programs we use a Pascal-based language, called Transputer Pascal [22], but all the ingredients that will be used can be found in almost every other parallel programming language.

Transputer Pascal is equipped with special constructs for processes and channels. A channel can be considered as a communication link between a sending and receiving process. A communication along a channel takes place when the sending process performs an output operation on the channel, while at the same time the receiving process performs an input operation on that channel. Execution of the operations is suspended until both processes are ready for their action. We denote the send operation of an expression e on channel out by

```
write(out, e),
```

whereas the operation on channel in to store the received value in the variable v is written as

```
read(in, v).
```

So, when processes s and r share a channel, called out in s and in in r, and they perform the communication operation at the same time, the result of

```
read(in, v)  ||  write(out, e)
```

where || denotes parallel composition, is equivalent to v := e plus synchronization.

When processes share a communication channel, then it is not necessary that the processors executing these processes also share a hardware communication link. If they do not share a link, extra communication processes are needed to propagate messages along some path of links between the processors. Such processes are called routers.

Because the task of a router is strictly specified, and because routing should be fast, in most new processor networks routers are realized in hardware, and, as a result the network might be considered as having a connection between every pair of processors. Generally, the time needed to transmit a message between two processors depends on the number of hardware links that the message has to traverse. Minimizing the network distance between two processes that need to communicate, is an important issue in the design of parallel programs. However, it is only a secondary step, because the first point of concern is the even distribution of work over the processors. Otherwise the answer to the minimization is easy: one would assign all the work to just one processor and no communications would have to take place at all. Moreover, having the work load evenly distributed and having a minimal distance, does not necessarily imply that an efficient implementation of a parallel program has been achieved. The missing aspect is the amount of time spent in communication versus the amount of time spent in computation. An implementation of a parallel program in which two processes are almost constantly communicating could, for instance, be improved by fusing these two processes.

It might seem that all these aspects make the design of parallel programs that can be efficiently implemented particularly difficult. There are, however, a number of underlying principles which can alleviate this task. In the previous section, we have emphasized the expansibility of the parallel processor networks. It is clear that we should design our programs so that they can profit from an increase in the number of processors, i.e. our programs should be scalable. It follows that when designing parallel programs we should not be too concerned by the size and structure of the processor network that the program will be executed on. One approach to this goal is to stick to some simple communication primitives that are applicable to any processor network. In the examples below we describe some of these primitives.

Although already indirectly mentioned, an important issue that needs more comment concerns the mechanisms to analyze the performance of parallel programs. Three measures are used to judge the effectiveness of a parallel implementation: *speedup*, *efficiency*, and *isoefficiency* [23, 24]. Let p be the number of processors being used, n the problem size, and $t(p,n)$ the time required to solve the problem on the processor network. Then, the speedup, $S(p,n)$, is

$$S(p,n) = \frac{t(1,n)}{t(p,n)}.$$

The efficiency, $E(p,n)$, is a measure for the utilization of every processor in the network:

$$E(p,n) = \frac{S(p,n)}{p}.$$

While speedup and efficiency are useful quantities, they are not sufficient to describe the scalability of a parallel algorithm when the number of processors is increased. For any fixed problem size, the amount of work per processor decreases with increasing number of processors. Hence, beyond some threshold, efficiency declines. To use an increasing number of processors efficiently, it is necessary for the work load, and hence the problem size, to increase correspondingly. The isoefficiency [23, 24] is a measure of this property. It is defined as the amount by which the problem size must increase as the number of processors increases in order to maintain a certain efficiency. Using the isoefficiency analysis applied to a small problem and a small processor network, one can deduce a reasonable estimate of a program's performance on a larger problem and a larger processor network.

In the remaining part of this section, we give some examples of these parallel programming techniques.

3.1. EXAMPLE 1: A RING ALGORITHM

Consider a ring network of 3 processors and assume that, at some time during the execution of a parallel program, a communication phase occurs. In this phase each processor has to perform two actions

- it has to send a value to its neighbour in the clockwise direction, and

- it has to receive a value from its neighbour in the counterclockwise direction.

A direct approach would lead perhaps to the following program being run on each processor

```
        write(next,val)
    ;   read(prev,v)
```

where `next` and `prev` are channels, `v` is some variable, and `val` the value that is sent. Unfortunately, execution of this small piece of code might lead to a situation in which all processors are blocked. Assume all processors are starting their communication phase; then we might have the following outcome. Processor 0 wants to send a message to processor 1, but at the same time processor 1 wants to send a message to processor 2, while processor 2 wants to send a message to processor 0. Since, in a communication, both processors have to participate, the processors are cyclically waiting for each other and the computation can not proceed. This is an example of a more general phenomenon called deadlock. It is essential that such a situation be avoided.

There are several solutions to this problem. One possibility is chain breaking: at least one processor is forced to listen first. Although correct, it is not the solution we prefer. The main concern with this approach is that the original problem does not specify any order in which the actions should be performed, but this solution has a strict ordering of actions. In general, restricting the order in which actions have to be performed may lead to less elegant and less efficient solutions. In fact the philosophy behind parallel computing is to take advantage of having no particular order to the actions by allowing them to occur in parallel. In particular this means that if we do not know anything about the relative processing speeds it is desirable to postpone choices for as long as possible.

In this case there is no need to restrict the order of actions, so we should leave any ordering out. One programming construct to specify that actions can be executed in parallel is the `cobegin-coend` pair. All statements within such a pair are executed in parallel with each other. Hence, the solution we prefer reads

```
cobegin write(next, val)
;       read(prev, v)
coend.
```

The correctness of this algorithm is obvious: a process that is suspended will be released when the other process involved in the communication is ready to participate. Notice that the value to be sent is not restricted to integers, reals, etc. It might also be a complete array. Sending an array as a whole has the advantage that only one communication has to be performed, while sending the array item-by-item, in general, leads to poor communications behaviour, due to the relatively long time needed to initiate a communication.

Although often encountered in a discussion of a ring algorithm we have just chosen the names `prev` and `next` for presentation purposes. A direct generalization is to consider these channel names, the variable, and the value involved in the communications, as formal parameters of a procedure called `turn`. As we show in discussing parallel molecular dynamics, actual calls to this procedure can then, for instance, be `turn(east,pvar,west,pval)` or `turn(north,pvar,south,pval)` for a torus network. This application of the extremely simple, small piece of parallel programming text is no coincidence. It can be found in some sort of functionality in each programming environment. For instance, the `cshift` of the Connection Machine's environment serves the same purpose.

Notice that in deriving this piece of text we did not use any details about the specific network the program should be executed on and we needed some kind of parallel thinking to arrive at a solution. In the next example we continue along these lines.

3.2. EXAMPLE 2: TREE ALGORITHM AND PROCESSOR FARM

In this example we consider a class of frequently-used programs. The programs are characterized by the fact that all communications to be performed are between one processor on the one hand and every other processor on the other hand. Examples of such programs are a broadcast, in which one processor transmits information to all other processors, and a global sum, in which information from all processors is combined into a single value. We only give some of the ideas behind this class of algorithms; a more general treatment of the subject can be found, for instance, in Ref. [21].

The global sum problem can be specified as follows. In a processor network each processor has locally calculated a value and all these values have to be summed over all processors in the network. A solution to this problem is to consider a spanning tree in the network rooted in processor `r`. For an example of a tree see Fig. 1. The algorithm we suggest is: all leaves (the outermost points) send their value to their parents in parallel, the parents sum the received values and their own value, and send the newly calculated values to their parents etc., until the value is at processor `r`. Reversing the communication directions leads to a broadcast program. The root starts transmitting a value to its children, these send the received value to their children, and so on, till the values are at the outermost points, the leaves of the tree.

Despite its simplicity, the tree algorithm is used in many applications. For instance, in parallel molecular dynamics simulations, the broadcast of the system parameters, the calculation of the total energy and the temperature etc., can be solved this way. Notice that the algorithm is generally applicable because in each processor network a spanning tree can easily be defined. Note also that the number of active processors decreases with the level of the tree. To overcome this decrease in performance, one usually combines a number of different global sums into one calculation. After sending their first local value, the processors start sending the next local value and so on. If the summed values have to be sent back to the processors this scheme guarantees a fair load balance.

When the number of global sums to be calculated is much larger than the number of processors, an alternative solution exists. Instead of letting the processors cooperate in one global sum, each processor is assigned a number of sums to be calculated on its own. The direct generalization of this parallel program, in which a group of processors work independently on their copy of a set of parameters assigned to them by a single control processor, is known by a number of names such as the replication algorithm and the processor farm [17, 19, 25]. The performance of this algorithm is good, as long as the amount of work for each processor is large. However, it has poor scalability properties. The single control processor readily becomes saturated when the number of processors is large, or the ratio between work and communication small. In the next section, we discuss two other parallel programming techniques that do not suffer from these limitations.

4. Parallel molecular dynamics

In this section, we describe two generally-accepted parallel programming techniques. In order to demonstrate the appropriateness of the techniques in computer simulations we use molecular dynamics as an example. In particular, we concentrate on several of the aspects of the performance of a parallel program obtained by applying these techniques.

In a molecular dynamics simulation at each timestep, the force on, and the new position of, a particle are calculated for each particle (see e.g. Chapter 2). Since the amount of work per particle does not differ much for any two particles, obtaining a good load balance in a molecular dynamics simulation is, in general, straightforward. We simply ensure that all processors have to do the calculations for (almost) the same number of particles. There are several ways of achieving this goal:

- a processor farm: one control processor assigns, possibly on demand, particles to the working processors;
- particle parallelism: each processor has been assigned a fixed set of particles for which it performs the calculations during the simulation;

- geometric parallelism: each processor has been assigned a fixed spatial part of the universe in which the particles move around.

In applying the first two techniques, one can be guaranteed that all processors have (almost) the same amount of work. In applying the last technique, an extra assumption is needed: during the whole simulation run, particles have to be homogeneously distributed over space. As discussed in the previous section, we do not consider the processor-farm technique as a preferred solution, although the simplicity of the method certainly has its charm [17].

Before discussing the two other methods, we begin by reviewing the kind of information that is needed to calculate the force exerted on a particle. A force on a particle can be written as a sum of terms depending on the coordinates of pairs of particles (such as the Coulombic term, the Lennard-Jones, and covalent binding terms), of triplets of particles (such as a bond-angle term), and of quadruplets of particles (such as a dihedral-angle term). By far the majority of these terms are the pair interactions [3, 26], while the amount of work per particle needed to calculate a term hardly depends on the type of term being considered. So the primary concern is to implement the pair interactions efficiently.

The first method we discuss is the particle-parallelism technique in which each processor has been assigned a fixed, equal number of particles. Since there is no spatial relationship between processors and particles, the new positions of the particles can only be calculated when all particle pair interactions have been checked. Therefore, each timestep, every two processors have to communicate with each other. Several strategies exist to realize this exchange efficiently, from which we consider two representative examples: the systolic loop [4, 18, 19, 27, 28] and the alternating circulation method [1]. Both strategies assume the processors to be connected in a ring, and in both cases, the ring algorithm described in the previous section plays an important rôle.

The simplest of the two methods is the systolic loop. In this method, each processor starts by sending the data of one of its own particles to the next processor in the ring, and at the same time receives from its other neighbour in the ring an 'alien' particle. Then the forces between this alien particle and the processor's own particles are calculated. Thereafter the processor sends the alien particle to its successor in the ring and receives a new alien particle from its predecessor. This whole scheme is repeated until the complete ring of processors has been visited, while the entire procedure has to be carried out for each of the particles of a processor. In algorithmic form the method reads

```
for i:=1 to N/P do
 begin turn(prev,anew,next,p(i))
 ;     calc_forces(anew)
```

```
;       aold:=anew
;       for j:=2 to P do
         begin turn(prev,anew,next,aold)
         ;       calc_forces(anew)
         ;       aold:=anew
         end
end,
```

where, apart from the alien variables, N is the total number of particles, p(i) is the data concerning particle i, P is the total number of processors, and calc_forces is the name of a procedure in which all N/P possible interactions between an alien particle and the particles residing on the processor are determined. (In each final turn a particle is sent back to its processor, so this last step might be avoided. Furthermore when a particle is back at its processor only N/P-1 possible interactions are to be considered. For simplicity we have assumed that the last turn is performed and that all N/P possible interactions are to be considered.)

We now consider the performance of this algorithm. Let α denote the time needed to perform a 'turn' on the data of one particle, and let β denote the average time to do one force calculation between two particles. Then we have

$$t(p,n) = \frac{n}{p} \cdot p \cdot \left(\alpha + \frac{n}{p} \cdot \beta\right) = n\left(\alpha + \frac{n\beta}{p}\right),$$

and hence the speedup and the efficiency of this program are

$$s(p,n) = \frac{n^2\beta}{n(\alpha + n\beta/p)} = \frac{p}{\alpha p/\beta n + 1} \qquad \text{and} \qquad E(p,n) = \frac{1}{\alpha p/\beta n + 1}.$$

So, a high speedup and efficiency are obtained for βn much larger than αp. This means that efficient implementations are achieved when either the number of particles is large or the ratio between one communication and one force evaluation is small. A disadvantage is that, because of the dependence on the ratio of the number of communications to the number of processors, this scheme has the tendency to become communication bound.

Instead of alternating communication and computation on a particle basis in the alternating-circulation method, communication and computation may be alternated on a timestep basis. At every timestep all the particle positions are distributed in one direction over half the ring. Then all the relevant pair interactions are evaluated, and, finally, partial forces circulate and accumulate over half the ring in the other direction. Note that in order to guarantee that all particle pair interactions can be handled, then only half the ring communication is necessary and sufficient. Furthermore, because the particle positions move only half the ring, the accumulation of forces is needed. If particle i visits the processor on which particle j is resident, then particle j does not visit the processor of particle i. So, the contribution of the force term $\mathbf{f}_{ij}$ to the force on i has be to taken care of by the processor of j. For an odd number of processors, an algorithm to implement this method is

```
        for q:=1 to P/2 do turn(prev,a(q),next,a(q-1))
    ;   calc_all_forces
    ;   for q:=P/2 downto 1 do
         begin turn(next,af,prev,f(q))
         ;     f(q-1):=f(q-1)+af
         end,
```

where a(q), $1 \leq$ q $\leq$ P/2, denote the particle arrays at processor p for its P/2 processors in a clockwise direction, a(0) denotes the particle array for its own particles, and f(q) is used for the forces in the same way. calc_all_forces is the name of a procedure in which, for each particle of a processor, the interaction between this particle and all other particles residing on or received by the processor is calculated. For an even number of processors a small adjustment to the algorithm is required.

If we compare the performance measures of this program to those of the systolic loop method, then there are some differences. Firstly, because complete particle arrays are being sent as a whole instead of item by item, there is a reduction in communication cost. Secondly, extra time is spent in the computational part of the procedure due to the accumulation of the forces. Thirdly, the amount of memory needed differs significantly in the two methods. After the introduction of the geometric-parallelism method we will give a more formal treatment of the performance of this algorithm.

Both methods are capable of handling the particle pair interactions, but for triplets and quadruplets of particles this is not the case. In the systolic-loop method, difficulties arise when such a set of particles has been assigned to more than two processors, since, in order to evaluate the potential, a processor needs all particle positions at the same moment in time. Without an extra allocation strategy, even the alternating circulation method might fail. We can see this by assuming the triplet $\{i, j, k\}$ is needed for a bond-angle calculation, while the particles reside on three different processors. During the pair-evaluation phase, the following situation might occur. Particle k visits the processor of i, particle j the processor of k, and particle i is received by the processor of j. Then, despite the storage of the data for half the total number of particles, none of the three processors possesses the triplets of positions at the same moment in time, as is required for the calculation. A solution, of course, is to follow the replicated-data scheme [28, 29] and to communicate and store all particle positions over the whole ring, but this would lead to both high memory requirements and a large communications overhead. The situation might get even worse when several iterations of the triplet evaluation are needed because of the application of constraints (see Chapter 2). The clue to the method proposed in Ref. [1] is to realize that the above scenario can only happen because the relevant pairs of the triplet are on distant processors. When assigning particles to processors it is important to minimize the communication distance between the processors involved in a triplet, a quadruplet, or in a constraint calculation. In general, this minimization problem is hard to solve, but, as shown in Ref. [1], in practical situations a small, but not necessarily minimum, value for the communications distance can be obtained. In effect, solutions are achieved by translating the minimization problem to the determination of the minimal bandwidth of the large sparse symmetric matrix of covalent bonds, by allowing a renumbering of the particles. For a number of practical situations the bandwidth turns out to be much smaller than half the number of processors. The alternating circulation method has been successfully implemented in a distributed memory machine especially designed for molecular simulations.

So far we have assumed that all particle pair interactions have to be calculated. In many instances a cutoff radius, r_c, is used so that only interactions between particles at a distance of at most r_c have to be considered. Because the movement of the particles excludes any spatial relationship between particles and processors in the two particle-parallelism methods, neither of them can obtain the optimum benefit from this simplification. The next parallel programming technique that we review is exactly directed to this relationship and is therefore known as the spatial-decomposition or geometric-parallelism method.

When this method is used, each processor has been assigned a fixed part of the total simulation space (the universe). During the whole simulation run the processor has to calculate all interactions for the particles inside its space. Since particles move during the run in and out of this space, processors need to redistribute the particles to guarantee that each one has the right subset.

The first issue we address concerns which part of the universe should be assigned to a processor. For the moment we again restrict ourselves to pair interactions and we assume that a cutoff radius r_c is used. Then, in order to calculate the forces on a particle in the processor's space, all particles in a sphere with radius r_c are needed for each of the processor's particles. From the analysis for the sequential case it follows that this can efficiently be achieved when applying the linked-list method [9]. The simulation universe is then divided into cells, and the search for interactions of particles in the subject cell is restricted to those cells within a certain small environment. As a result of the division into cells, the assignment problem can be reformulated so that cells should be assigned to a processor. Under the assumption of a homogeneous distribution of particles, a good load balance can be achieved by assigning an equal number of cells to all processors. Another important aspect of an efficient implementation is the minimization of the communication costs. These costs are influenced by several quite different factors: the number of cells, the size and shape of a cell, the number of processors, and the network connecting the processors. In general, the minimization problem is hard to solve. In the literature various approaches to its solution can be found. In Ref. [5] a mapping of the universe on a ring is considered, while 2-dimensional and 3-dimensional decompositions are the subject of e.g. Refs. [2, 3, 30]. In particular, under certain restrictions, optimal solutions to the minimization problem are given in Ref. [3]. For instance, if a cell is cubic with size r_c, then it has been shown that for square torus networks (see section 2) of up to 1024 processors the optimal way of assigning cells to processors is columnwise, i.e. a processor with coordinate (x, y) is responsible for all cells having the same (x, y)-coordinate. For larger processor networks more cubical shaped blocks than columns of cells become beneficial because of the ratio of surface area to volume [3, 26].

Next, we give some insights as to why columns of cells give good results for the communication costs for several classes of parallel processor networks. Suppose we have cubic cells of size r_c, and a processor has been assigned a column of cells. If we further assume that Newton's third law is applied, then each processor needs only the particle information of four other processors (see Fig. 3). So if the processor network is a torus, then the particle information can be received directly from the processors numbered 1 and 3. Processors numbered 2 and 4 seem to cause some extra problems, but notice that when each processor has sent its particle information to the north, the particle information of processor 4 is now also at processor 3, and that of processor 2 is also at the processor's west neighbour. All the communication actions needed to realize the scheme are shown in Fig. 3.

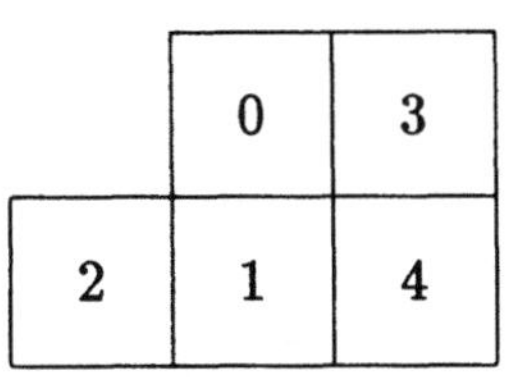


```
  turn(south,p(1),north,p(0))
; turn(west,p(2),east,p(1))
; turn(east,p(3),west,p(0))
; turn(east,p(4),west,p(1))
```

Fig. 3. Processor 0 with 4 neighbours and the communication scheme

After having applied this communication scheme, all pair interactions can be computed. During this force-evaluation phase, a processor stores the calculated forces for both its own particles and the communicated particles. Thereafter, the forces for the neighbouring particles are sent back using the same communication scheme but in the opposite direction. Notice that during this last phase the forces need to be accumulated for the same reason as in the alternating circulation method.

The main property of this scheme is that each communication is between two neighbouring processors. The straightforward generalization of this scheme for cells having a size that is not related to r_c has the same property. Consequently, the communication cost is proportional to the number of particles to be communicated and does not depend on the distance between the processors that have to exchange the particle information or on the total number of processors involved in the scheme. Hence this algorithm has precisely the scalability properties that we have discussed in the previous section. This is exemplified when we analyze the performance measures of this algorithm. In this analysis we assume that the cells are cubic, that the size of a cell is not related to the cutoff radius, and that multiple columns of cells in both the x- and y-directions are allowed on a processor. Furthermore, let S_α $(\alpha = x, y, z)$ denote the 'stretch' in the α-direction, i.e., the number of cells in that direction that have to be scanned for building the pair interaction lists.

In the determination of the execution time of the algorithm two terms need to be distinguished. Firstly, we have a term, $t_{\text{gp}}^{\text{com}}$, describing the contribution of the two communication phases, one phase for the transmission of the particle positions and one phase for the retransmission and accumulation of the forces. Secondly, we have a term, $t_{\text{gp}}^{\text{for}}$, describing the contribution from the calculation of the forces. Notice that $t_{\text{gp}}^{\text{com}}$ is just a global term, because all particles share the particle information that is communicated, while $t_{\text{gp}}^{\text{for}}$ has a contribution per particle. $t_{\text{gp}}^{\text{com}}$ depends on the number of processors involved in the communication environment and the time, γ_0, needed to communicate all the particle information between two processors:

$$t_{\text{gp}}^{\text{com}} = (2S_x S_y + S_x + S_y)\gamma_0 \,.$$

The term $t_{\text{gp}}^{\text{for}}$ is more complicated. For each of its particles in all its cells, a processor has to scan all particles in the communication environment of the cell to look for a possible interaction. To decide upon an interaction a distance calculation has to be performed, and only for those particles within the cutoff radius does a force have to be evaluated. Therefore, $t_{\text{gp}}^{\text{for}}$ has two sub-terms, one for the particles inside a sphere and one for the calculations of, say, the particles outside. The number of inside particles, $N_{\text{gp}}^{\text{IP}}$, is (on average) $\frac{2}{3}\pi r_c^3 \rho$, so the total number of outside particles, $N_{\text{gp}}^{\text{OP}}$, is

$$N_{\text{gp}}^{\text{OP}} = N^{\text{C}} \cdot r_{\text{ce}}^3 \rho - \frac{2}{3}\pi r_c^3 \rho \,,$$

where r_{ce} is the size of a cell and N^{C} is the number of cells in the communication environment of a cell

$$N^{\mathrm{C}} = 4S_xS_yS_z + 2(S_xS_y + S_xS_z + S_yS_z) + S_x + S_y + S_z + 0.5\,.$$

This leads to the following equation for the computational part

$$t_{\mathrm{gp}}^{\mathrm{for}} = (N_{\mathrm{gp}}^{\mathrm{OP}} \cdot \gamma_1 + N_{\mathrm{gp}}^{\mathrm{IP}} \cdot \gamma_2)\frac{n}{p}$$

in which γ_1 is the time needed to perform a distance calculation and γ_2 is the time needed for a force evaluation.

It is interesting to compare these terms with those for the alternating circulation method in which the same cutoff radius is used. Again the communication term is the simplest of the two

$$t_{\mathrm{acm}}^{\mathrm{com}} = \frac{p}{2}\gamma_0\,.$$

For small processor networks, and small simulation universes, the terms $p/2$ and $2S_xS_y + S_x + S_y$ will not differ much, but for the other cases the geometric-parallelism method is to be preferred when communication costs are considered.

Next we compare the computational part of both algorithms. Since in both methods the same inner particles are considered, the algorithms do not differ in the N^{IP}-part. But since in the alternating circulation method all communicated particles have to be considered, the difference in the N^{OP}-part is, in general, considerable. We have

$$N_{\mathrm{acm}}^{\mathrm{OP}} = \frac{p+1}{2} \cdot \frac{n}{p} - \frac{2}{3}\pi r_c^3 \rho\,,$$

so, for each of its particles, a processor has to calculate the distance with approximately half the total number of particles. In general, this makes it hard to achieve good scale-up properties if both the universe and the processor network size increase. On the contrary, if the cell size is chosen judiciously in the geometric-parallelism method, this number is very small and independent of the total number of particles.

In short, the geometric-parallelism method has very good scale-up properties, while the alternating-circulation method is useful for small processor network sizes and small simulation universes only. Since both the communication term and the N^{OP}-term can be considered as overheads we suggest the following MD performance measure

$$G_p = \frac{2\pi r_c^3 \rho n}{3p\tau}\,,$$

where τ denotes the (average) time needed to perform one timestep.

So far we have only considered pair interactions in combination with the geometric-parallelism method. We have seen that potentials involving triplets or quadruplets of particles require extra arrangements when applying the particle-parallelism method. The situation seems to be even more complicated when implementing such multi-particle potentials if the geometric-parallelism method is used. At some moment in time a quadruple of particles may all reside on one processor while some moments later they might be on four different processors. It is important to decide which multi-particle potentials a processor should evaluate.

Fortunately, the situation appears not to be too complicated. As in the alternating-circulation method, the clue is that particles involved in a multi-particle potential evaluation are geometrically close to each other. This property is explicitly formulated in the following theorem. Let M be a set of particles needed for the evaluation of a multi-particle potential. When the distance between each pair of particles in M is at most the cutoff radius of the pair potential, then there is exactly one particle in M that has all other particles in its pair evaluation list [31]. According to the theorem we can easily assign a particle and hence a processor for the evaluation of the multi-particle potential. Hence, when geometric parallelism is used, no extra communications are needed to evaluate the multi-particle potentials provided the distance constraint holds.

We end this section with some timing results for the scaling behaviour of the geometric-parallelism method. In the literature, e.g. Refs. [7, 8, 26], timing results are given for the Lennard-Jones interactions using a geometric-parallelism method. Here we assume that also bonding, bending, and torsion potentials are to be evaluated. Some of these results can also be found in Refs. [32, 33]. The processor network on which the simulations have been performed is a square torus network of 400 T800 transputers. In our implementation the size of a cell is not related to the cutoff radius. Furthermore, multiple columns of cells in both the x- and y-direction are allowed on a processor, and lists are built for the evaluation of the non-bonded, bonding, bending, and torsion interactions. By enlarging the cutoff radius by a small amount, the lists are used for several iterations before updating. In Fig. 4 the results of runs for a number of system sizes are shown. The first system consisted of linear chains of length 20, and the second one of length 50. The average density in these simulations was $2.24\,\sigma^{-3}$, which resulted in ~ 200 Lennard-Jones interactions per particle.

The timing results show clearly the linear dependence of the execution time on both the number of particles and the network size. Already for small numbers of particles a parallel machine is attractive, but its real power is shown at large system sizes. This feature of parallel machines opens new possibilities for the study of very large systems, and problems which could not previously be dealt with. We give an example of one such problem in

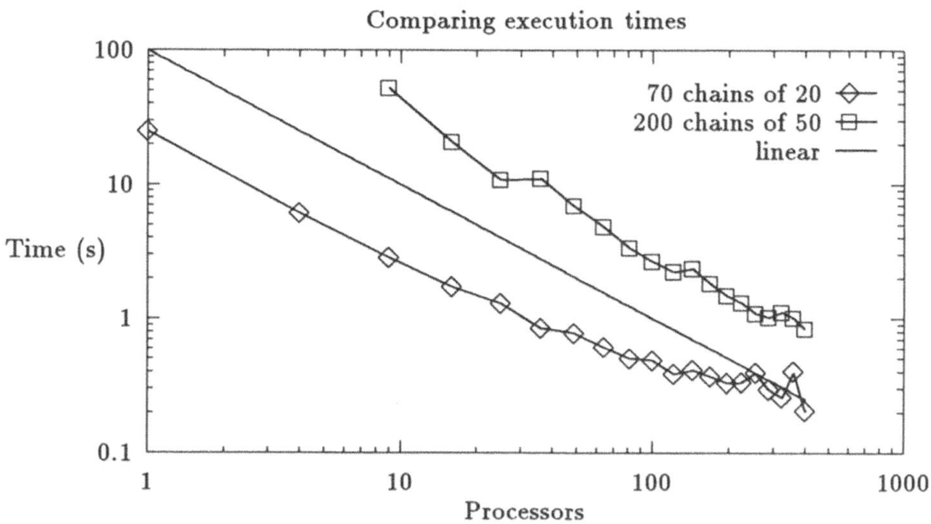

Fig. 4. Execution times

the next section.

5. Nucleation simulations

We have used the parallel implementation described in the previous section to study the nucleation behaviour of linear chains of length 10. These simulations have been performed in collaboration with B.W.H. van Beest, and can be considered as a large-scale follow-up to his simulations of "decane" on sequential machines.

The system consisted of 1000 chains of length 10. The model that we have used to describe the interactions is similar to that of Ref. [34], with the difference that we have chosen a cutoff of $2.5\,\sigma$ in the evaluation of the Lennard-Jones potential. Berendsen's technique [35] has been applied for the loose coupling to an external temperature and pressure bath.

We started by heating the system to a temperature well above its melting point. The result of this heating process is shown in Fig. 5. In this liquid, hardly any ordering can be found.

Then we quenched the liquid to a temperature $T^* = 3.5$. As soon as this temperature was reached, the system started to show crystallization behaviour. The ordering process resulted in a system consisting of a number of large layers wrapped around the periodic boundaries (Fig. 6).

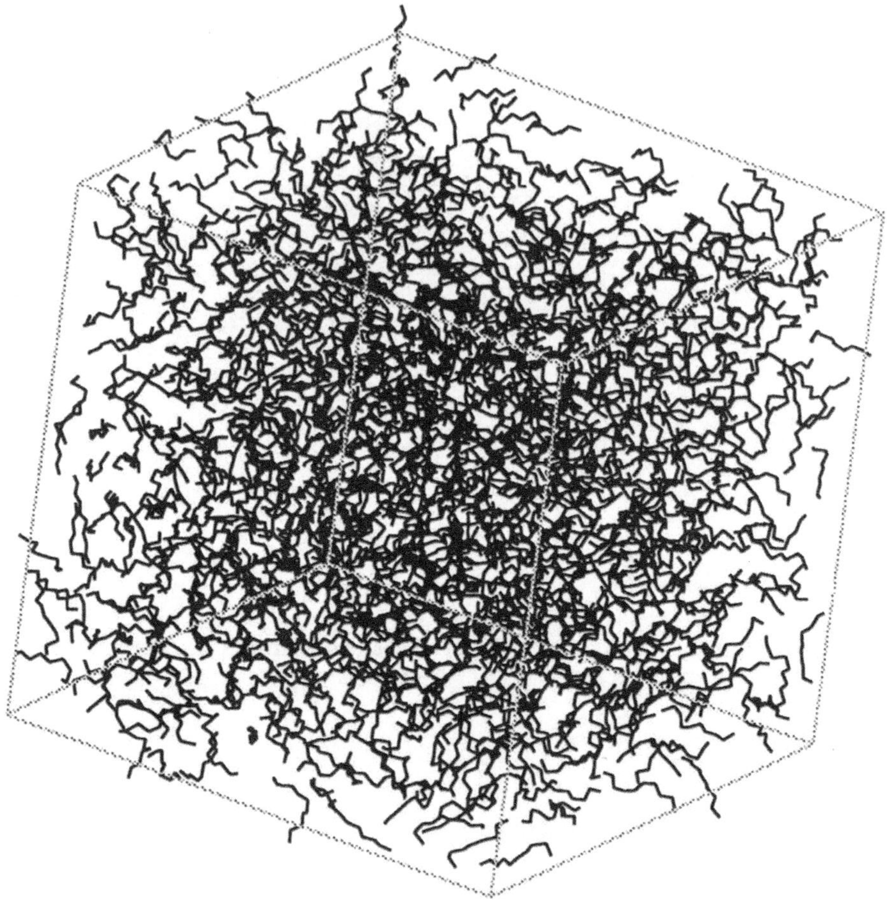

Fig. 5. A configuration at the end of the heating process.

In order to describe these layers, we have introduced the concept of a cluster. If two chains have the same orientation, and the distance between their centres of masses is at most $1.5\,\sigma$, then they belong to the same cluster. A similar algorithm to that used to identify the micelles in a water/oil/surfactant system (see Chapter 12) can now be used to characterize the clusters. For the crystalline configuration, we observed five large clusters and several small ones; in the liquid state, the number of clusters is almost equivalent to the number of chains.

The simulation yielded many other interesting results, which will be presented in a future publication.

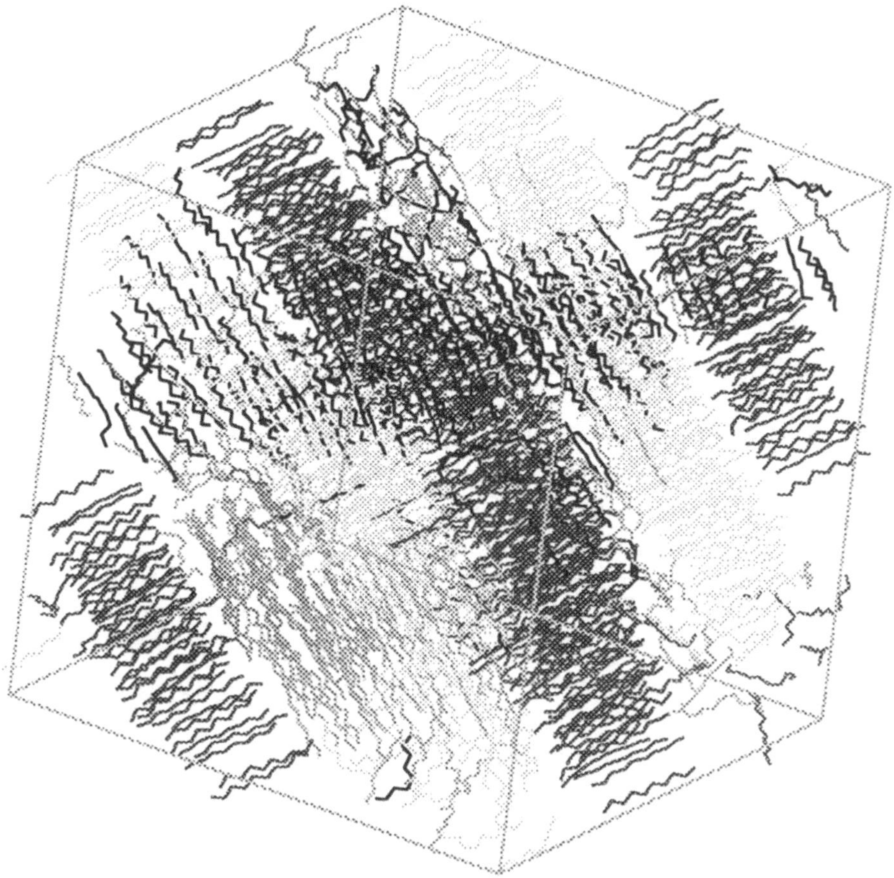

Fig. 6. A configuration just after crystallization.

6. Conclusion

In this Chapter, we have argued that, from the hardware point of view, parallel processor networks are the preferred architectures. We have also emphasized that programs that are to be run on such a network should be scalable.

As an example we have looked at parallel implementations of molecular dynamics, and have shown that for short-range potentials the geometric-parallelism method is the best. Some approaches to the parallel implementation of long-range potentials have been reported in the literature [1, 28], but it is not yet known which method is the most suitable.

Already for small numbers of particles a parallel processor network can compete with conventional supercomputers, but its real power is evident at large system sizes. This feature of parallel processor networks opens new perspectives for the study of large systems, and problems which could not previously be dealt with.

References

[1] H. BEKKER, E. J. DIJKSTRA, and H. J. C. BERENDSEN, Mapping molecular dynamics simulation calculations on a ring architecture, in *Parallel Computing: From Theory to Sound Practice*, edited by W. JOOSEN and E. MILGROM, pp. 268–279, IOS Press, Amsterdam, 1992.

[2] S. CHYNOWETH, U. C. KLOMP, and L. SCALES, *Comput. Phys. Commun.* **62**, 297 (1991).

[3] K. ESSELINK, B. SMIT, and P. A. J. HILBERS, *J. Comput. Phys.* **105** (1993), to appear.

[4] D. FINCHAM, *Molec. Simul.* **1**, 1 (1987).

[5] H. G. PETERSEN and J. W. PERRAM, *Molec. Phys.* **67**, 849 (1989).

[6] M. R. S. PINCHES, D. J. TILDESLEY, and W. SMITH, *Molec. Simul.* **6**, 51 (1991).

[7] S. PLIMPTON and G. HEFFELFINGER, Scalable parallel molecular dynamics on MIMD supercomputers, in *Scalable High Performance Computing Conference SHPCC '92*, pp. 246–251, IEEE Computer Society, IEEE Computer Society Press, 1992.

[8] D. C. RAPAPORT, *Comput. Phys. Commun.* **62**, 217 (1991).

[9] M. P. ALLEN and D. J. TILDESLEY, *Computer Simulation of Liquids*, Oxford Science Publications, 1987.

[10] W. J. DALLY, *A VLSI architecture for concurrent data structures*, Kluwer, Hingham, Massachusetts, 1987.

[11] P. A. J. HILBERS, *Processor Networks and Aspects of the Mapping Problem*, Cambridge University Press, Cambridge, 1991.

[12] C. L. SEITZ, *IEEE Trans. Comput.* **C-33**, 1247 (1984).

[13] A. TREW and G. WILSON, *Past, Present, Parallel: A Survey of Available Parallel Computing Systems*, Springer-Verlag, 1991.

[14] L. D. WITTIE, *IEEE Trans. Comput.* **C–30**, 264 (1981).

[15] S. H. BOKARI, *IEEE Trans. Comput.* **C–30** (1981).

[16] K. M. CHANDY and S. TAYLOR, *An Introduction to Parallel Programming*, Jones and Barlett, 1991.

[17] R. S. COK, *Parallel Programs for the Transputer*, Prentice-Hall, Inc, Englewood Cliffs, New Jersey, 1991.

[18] G. FOX, M. JOHNSON, G. LYZENGA, S. OTTO, J. SALMON, and D. WALKER, *Solving Problems on Concurrent Processors, Volume I*, Prentice-Hall, Inc, Englewood Cliffs, New Jersey, 1988.

[19] D. W. HEERMANN and A. N. BURKITT, *Parallel Algorithms in Computational Science*, Springer-Verlag, Berlin, 1991.

[20] L. D. J. C. LOYENS, *A Design Method for Parallel Programs*, PhD thesis, Eindhoven University of Technology, 1992.

[21] J. J. LUKKIEN, *Parallel Program Design and Generalized Weakest Preconditions*, PhD thesis, University of Groningen, 1991.

[22] J. J. LUKKIEN, Transputer Pascal, a user manual, CS Note 8912, University of Groningen, Dept. Computing Science, P. O. Box 800, 9700 AV Groningen, The Netherlands, 1989.

[23] V. N. RAO and V. KUMAR, Analysis of scalability of parallel algorithms, CS Note, University of Texas at Austin, Dept. of Computer Science, Austin, Texas 78712, 1989.

[24] J. WOO and S. SAHNI, *Journal of Supercomputing* **3**, 209 (1989).

[25] J. L. A. VAN DE SNEPSCHEUT, *What computing is all about*, Springer-Verlag, 1993.

[26] M. OKUDA, D. BROWN, J. H. R. CLARKE, and T. YAMAZAKI, A domain decomposition parallelization strategy for molecular dynamics simulations on distributed memory machines, submitted for publication, 1992.

[27] A. R. C. RAINE, *Molec. Simul.* **7**, 59 (1991).

[28] W. SMITH, *Comput. Phys. Commun.* **67**, 392 (1992).

[29] H. SATO, Y. TANAKA, H. IWAMA, S. KAWAKIKA, M. SAITO, K. MORIKAMI, T. YAO, and S. TSUTSUMI, Parallelization of AMBER molecular dynamics program for the AP1000 highly parallel computer, in *Scalable High Performance Computing Conference SHPCC '92*, pp. 113–120, IEEE Computer Society, IEEE Computer Society Press, 1992.

[30] S. Y. LIEM, D. BROWN, and J. H. R. CLARKE, *Comput. Phys. Commun.* **67**, 261 (1991).

[31] K. ESSELINK and P. A. J. HILBERS, *J. Comput. Phys.* **105** (1993), to appear.

[32] K. ESSELINK and P. A. J. HILBERS, Parallel molecular dynamics on a torus network, in *Scalable High Performance Computing Conference SHPCC '92*, pp. 106–112, IEEE Computer Society, IEEE Computer Society Press, 1992.

[33] P. A. J. HILBERS and K. ESSELINK, Parallel molecular dynamics, in *Parallel Computing: From Theory to Sound Practice*, edited by W. JOOSEN and E. MILGROM, pp. 288–299, IOS Press, Amsterdam, 1992.

[34] D. RIGBY and R. J. ROE, *J. Chem. Phys.* **87**, 7285 (1987).

[35] H. J. C. BERENDSEN, J. P. M. POSTMA, W. F. VAN GUNSTEREN, A. DINOLA, and J. R. HAAK, *J. Chem. Phys.* **81**, 3684 (1984).

Scientific Visualization, a User View

M. Ferrario
Dipartimento di Fisica
Università di Messina
Sezione di Fisica Teorica
C.P. 50
98166 S. Agata di Messina
Italy

ABSTRACT.
Virtually all comprehension in science, technology, and even art, calls on our ability to visualize. The ability to visualize is almost synonymous with understanding. Modern science goes beyond isolated theories, and demands computer simulation to understand real-world situations. Scientific Visualization - a set of software tools coupled with a powerful 3D graphical computer environment - is the eyes through which simulations are viewed. From today's post-processing of data to future tracking and steering of simulations, Scientific Visualization has become an essential part of the research process for the computational scientist.

1. Introduction

In the last decade, as a direct consequence of the supercomputer revolution, we have witnessed the explosion of Computational Science in all scientific fields of research. *"Using the computer as a tool, mathematics as logic, and algorithms as a common language, computational scientists have developed numerical methodology with which to study physical phenomena"* [1]. Given that the primary goal of Computational Science is to simulate models that precisely and completely describe real physical systems (and possibly understand them), with the increase in computational speed it is now possible to run simulations that approach closely the complexity of reality, and therefore it is highly desirable to give a detailed and realistic visual representation of such a *virtual* reality.

However, to restrict Scientific Visualization to this would be unfair. The major problem the computational scientist faces, since the escalation of high-performance supercomputing, is to find a solution to the problem of numerical overload. Supercomputer simulations can now easily produce a billion

M.P. Allen and D.J. Tildesley (eds.), Computer Simulation in Chemical Physics, 497–503.

numbers. *"What knowledge does one absorb or discover from scanning a billion numbers that are produced by a supercomputer? How much data in this form can the researcher comprehend at once? How readily can he or she recognize unanticipated relationships exposed in a complex simulation that is expressed only numerically?"* [1]. Visual exploration of the data helps with the *understanding* of the model. Anyone's visual perception is far more acute than the ability to assimilate the same information expressed only as numerical data. Scientific Visualization becomes a necessary process for the researcher to comprehend the simulation data most completely. *"The process of visualizing acts as a telescope into the simulation"* [1].

The most difficult aspect of Scientific Visualization is the process of inventing the mapping and writing the algorithms that transform abstract data into n-dimensional geometric representations. The most important decisions to be made relate to what visual, geometric, form the data will take. Recent technological advances make it possible to have, available on the scientist's desktop, powerful graphics workstations that provide interactive capabilities, full colour and 3-dimensional rendering, and animation. Similar advances in Computer Graphics mean that techniques are also available, either as developers' tools or fully-functional black box applications, to render and interactively manipulate any sort of 3-dimensional geometrical data.

The job of converting the abstract numerical data into visual, *meaningful*, pictures is left to the computational scientist, who can most benefit from interdisciplinary collaboration with visual artists and computer graphics specialists, interacting *"synergistically to expand the range of options available in the quest for solutions to specific visual communication problems"* [1].

2. Discussion

Historically the term Scientific Visualization was first used in the 1987 ACM SIGGRAPH panel report on "Visualization in Scientific Computing" by McCormick, De Fanti and Brown [2,3]. A recent book [4] provides an excellent up-to-date introductory guide to the field for the computational scientist. Also the journal *Computers in Physics* hosts regularly a Scientific Visualization section, which is edited by R. S. Wolff [5], who, in one of his introductory articles, has rephrased the ancient chinese proverb *"A picture is worth ten thousand words"*, familiar to computer graphics specialists [6], in an interesting and more contemporary way: *"a good visualization is worth several thousands ... "* [7], making explicit reference to the pressing problem nowadays of raising funds for research. However, as the same author points out, there is a great danger related to this aspect: the efforts required to produce good-looking images may distract the scientists from the task of producing correct, numerically-accurate (rather than good-looking) images. It should also be stressed that such tasks should not really be regarded as Scientific

Visualization, but rather as Presentation Graphics. Scientific Visualization, or visual data analysis, is primarily concerned with exploring data and information graphically, as a means of gaining quick understanding and insight into the data. Presentation Graphics is concerned with the communication of information and results that are already understood, while in the Scientific Visualization process we are still seeking to *understand* the data.

2.1. HARDWARE AND SOFTWARE REQUIREMENTS

The principal requirement of the computational scientist for doing visualization is the availability of a desktop/deskside graphics workstation. Depending on the kind of visualization needed, personal computers, such as Macintoshes or even MS-DOS PC's beefed up with enough memory and suitably-large, 1Mpixel, screens, with 8, or possibly 24, color planes, would be enough to render 2-dimensional and 3-dimensional (wireframe) sets of data with sufficient speed and interactive manipulation capabilities. The public domain Scientific Visualization Suite for the Macintosh from NCSA [8] is a good example of what it is possible to do on a machine of this level.

Three-dimensional data sets require dedicated hardware, typically a powerful Unix workstation with a specialized graphics subsystem. Of course it is still possible to do solid modelling on lower-class workstations or high-end PC's. Certainly it would be impossible to have the necessary speed for interactive manipulation without a dedicated graphics processor with 16- or 24-bit Z-buffer to treat in hardware the removal of hidden lines and hidden surfaces; to do proper lighting without a true-colour 24-bit frame buffer; and to render animation without a double-frame video buffer. If a skilled computer graphics specialist can render any kind of scene on a pixel programming base, a computational scientist certainly does not want to get involved with anything more complicated than a high-level graphics development library, such as GL (Silicon Graphics' graphic library) or PHIGS (Programmer Hierarchical Interactive Graphics System, an ANSI and ISO standard). Rather, scientists are more and more interested in full-featured visualization tools such as AVS from Advanced Visual Systems, Iris Explorer from Silicon Graphics, or the Advanced Data Visualizer from Wavefront Technologies. These tools, and more from other vendors, provide full rendering and interactive manipulation capabilities for almost all possible data sets that a computational scientist can think of. They also provide several different rendering techniques, each of which can be used to extract different aspects from the data.

Nevertheless, there is still something left over to the scientist in the Scientific Visualization field. All the above-mentioned tools provide full capabilities for post-processing of data; that is they allow, with some time delay, visualization of the data after all of it is calculated. Even if the visualization may be under interactive control, there is no route back to the source of data. A different approach is growing currently, thanks to the latest advances in high-speed graphics workstations connected with high-speed network links to supercomputer centres. With such set-ups available, simulations can be run in a *tracking*, and possibly, *steering* graphics environment [9, 10]. With *tracking*, the data is visualized while it is being calculated, with little or no time delay. There is an interactive control and a connection with the source of data, with the possibility of stopping the simulation process. With *steering* both visualization and calculation are under interactive control and the calculation can be stopped and restarted after some parameters are modified. In these cases, usually, the graphics and computational processes are distributed over different cpus, a workstation and a supercomputer respectively, and data travels along high-speed networks. A group of researchers active in the field of computer simulation of liquids and solids has started recently a development activity at CECAM, the European Centre for Atomic and Molecular Physics, with the aim of producing a public-domain set of basic tools specifically for visual analysis of simulation. Work is in progress in both fields of post-processing and tracking-steering [11].

Finally, particularly for presentations, good color hardcopy devices are needed. Curiously, this aspect is very often not given enough care, and simple low-quality off-the-screen photographs are the final outcome of very laborious visualization processes. Producing videos is also becoming fashionable, especially for users of supercomputing centres having centralized, dedicated, video recording equipment, while high-quality prints are still rare and expensive. Hopefully, with the rapid growth of the multimedia market, the scientist will soon benefit from reasonable high-quality hardcopy devices.

2.2. DATA REPRESENTATION

Geometrically speaking, computer displays and hardcopy devices are inherently 2-dimensional devices. 3-dimensional images are obtained by projections, hence the need for a dedicated graphics engine to take care of geometrical transformations and hidden-line/hidden-surface removal. Mapping techniques can add dimensionality to the display. This is done mostly using colour coding, but also time and, on the newest workstations, sound can be used. For example, color coding has been used in simple molecular graphics images to give a visual representation of disorder in liquid crystal systems, by rendering each molecule with a color representative of the value of the relevant order parameter [12, 13]. Ordered systems would produce uniformly coloured images, while disordered systems would show up as multicolored *harlequin* pictures [13].

Data can be either scalar, vector or tensor values defined in a multidimensional variable space, usually up to three spatial coordinates plus one time coordinate. While scalar functions of a single variable are routinely visualized using 1-dimensional lines in a 2D chart, there are several different choices for scalar functions of two variables. One can use contour maps (again using 1-dimensional lines in a 2D chart), or colour maps (that is, 2-dimensional polygons in a 2D chart), or height maps (again 2-dimensional polygons but in a 3D chart). Sometimes a degree of redundancy is useful in data representation, for clarity and visual reinforcement. Thus, a color-coded height map, or color-coded contour lines, would enhance the representation of the function. Otherwise, combinations can be used to represent more functions on the same graph, but care should be taken not to overload the image with too much information at one time, which might be confusing.

For functions of three variables, there are basically two different ways of representing scalar data: using isosurfaces, and volume rendering. In the first case, one can use 0-dimensional dot isosurfaces, a 1-dimensional wireframe representation, or 2-dimensional polygon meshes, all in a 3D graph, while in the second case, one has a 3-dimensional voxel (volume element) representation of the data [14]. Volumetric representation is an active field of research in computer graphics and requires massive resources. Since fast algorithms for isosurface calculation and rendering are available, and provided a few isosurfaces are enough to give a clear picture of the data, a proper use of color coding plus transparency may give excellent results [15]. Transparency is a very useful feature, which unfortunately is not always fully supported by graphics hardware.

Vector and tensor data sets require particular techniques for rendering. For vector fields, a useful representation can be made in terms of imaginary particle traces, either with a 0-dimensional dot or a 1-dimensional line representation. Other possible representations use arrows on a 3D grid, with intensity either mapped to the arrow length, or color coded, or both. "Glyphs", special symbols named after hieroglyphs, are also used, in particular for tensor data. The advantage of glyph techniques is that they allow simultaneous rendering of several variables at each grid point. Active research is going on in this field as the design of glyphs, that can successfully convey lots of information in a clear way, is not at all a well-defined matter at present [1].

A few words should be devoted to color coding in general. A computer pseudocolor map is simply a sequential scale of colors that are mapped to a corresponding sequence of numbers, usually referred to as a *color palette*. Historically, color-coded maps have been used in geographical and meteorological maps to visualize quantities like the altitude of mountains, the depth of water, temperature, and other kinds of geographically-distributed properties. Such maps are always accompanied by a legend, a color map that provide a key to a quantitative scale that relates the color to the numerical values of the mapped property. A sequential color scale should always be included on the computer-generated image as a point of reference if the representation is to be a quantitative one, even when *standard* color coding is being used. By standard color coding we refer to either a rainbow color scale or to uniform color scales based on a color intensity mapping. An important point to make is that color coding is most useful when it can be manipulated interactively. The reason for this is that human perception and/or hardware characteristics limit the number of useful different color gradations. User data values can span different orders of magnitude, and, in principle, one does not know which range is the relevant one for the particular phenomenon under study. Assume that one is using an 8-bit color display, which limits the number of possible color gradations to 256. An injudicious choice of the color mapping may result in very flat images with no exposure of the relevant behaviour of the system. Such a problem is overcome if one has the possibility of exploring interactively the wide range of data by changing the mapping that relates colors to numbers. A useful feature is also the possibility of changing the mapping law, for example using a logarithmic rather than a linear mapping. The NCSA Scientific Visualization Suite has embedded, in all applications, interactive manipulation of color palettes and even a tool for building personalized color palettes. A more detailed discussion on usage of color palettes can be found in Ref. [7].

Acknowledgements

We thank Advanced Visual Systems, Hewlett-Packard, IBM and Silicon Graphics for the loan of demonstration material for the duration of the School; John Shelley for demonstrating his own molecular graphics visualization package; and Michel Teschner (Silicon Graphics) for his presentation of molecular and surface rendering tools. The CECAM activity was partially supported by the Italian Consiglio Nazionale delle Ricerche under the P.F. "Sistemi Informatici e Calcolo Parallelo" u.o. "Calcolo Vettoriale e Parallelo in Fisica Statistica". Finally, thanks to Antonio Messina, Marco Ronchetti, John Shelley and Michael Teschner for all the lengthy and interesting discussions on the subject.

References

[1] D. Cox, Scientific Visualization: Supercomputing and Renaissance Teams, talk given at IBM Europe Institute, Computer Graphics, Data Visualization and Animation, 1991.

[2] B. H. McCormick, T. A. DeFanti, and M. D. Brown, *Computer Graphics* **21(6)** (1987), Special Issue.

[3] A. Wolfe, M. Brown, D. Greenberg, M. Keeler, A. R. Smith, and L. Yaeger, *Computers in Physics* **2**, 16 (1988).

[4] R. A. Earnshaw and N. Wiseman, *Introductory Guide to Scientific Visualization*, Springer-Verlag, 1992.

[5] R. S. WOLFF, *Computers in Physics* **4**, 260 (1990).

[6] J. FOLEY and A. VAN DAM, *Fundamentals of Interactive Computer Graphics*, Addison-Wesley, Reading Massachusetts, 1980.

[7] R. S. WOLFF, *Computers in Physics* **5**, 339 (1991).

[8] B. BACON, *Computers in Physics* **3**, 11 (1989).

[9] M. BERNASCHI, M. BRISCOLINI, G. RICHELLI, and F. VALENTINI, 4D-SDV An Advanced Visualization Environment for Three-Dimensional Simulations using Cooperative Processing for Animation, talk given at IBM Europe Institute, Computer Graphics, Data Visualization and Animation, 1991.

[10] L. CALORI, G. FABIANI, and A. MESSINA, *IEEE Computer Graphics and Applications* (1991), in press.

[11] M. FERRARIO, H. JONSSON, and M. RONCHETTI, Scientific Visualization, in *Report of CECAM Activity*, 1992, Besides the three organizers, the researchers involved are: V. Cavecchia, F. DiTolla, D. Faken, A. Messina and J. Shelley.

[12] M. P. ALLEN, Molecular dynamics and ordering near the isotropic-nematic transition, video.

[13] P. PASINI and C. ZANNONI, *International Journal of Modern Physics* (1992), in press. Video shown at the Workshop 'Computer Simulation of Liquid Crystals', Il Ciocco 1991.

[14] R. S. WOLFF, *Computers in Physics* **6**, 421 (1992).

[15] J. C. SHELLEY, M. SPRIK, and M. L. KLEIN, *Langmuir* (1993), in press.

List of Participants

Directors

Prof. M. L. Klein
Univ. of Philadelphia
klein@sg2.chem.upenn.edu

Dr. P. A. Madden
Oxford Univ.
madden@vax.ox.ac.uk

Local Organisers

Prof. G. B. Suffritti
Univ. di Sassari
suffritti@mvchss.cineca.it

Prof. P. Demontis
Univ. di Sassari
demontis@mvchss.cineca.it

Lecturers

Dr. M. P. Allen
Bristol Univ.
m.p.allen@uk.ac.bristol

Prof. D. Coker
Boston Univ.
coker@chem.bu.edu

Prof. K. A. Dawson
Univ. Coll. Dublin
dawson@erin.ucd.ie

Prof. M. Ferrario
Univ. di Messina
ferrario@imeuniv

Prof. D. Frenkel
FOM Inst., Amsterdam
frenkel@amolf.nl

Prof. G. Galli
IRRMA, Lausanne
galli@eldpa.epfl.ch

Dr. P. Hilbers
Shell Research, Amsterdam
hilbers1@ksla.nl

Dr. K. Kremer
IFF, Jülich
iff076@djukfa11

Dr. B. Smit
Shell Research, Amsterdam
smit1@ksla.nl

Prof. M. Sprik
IBM, Zurich
spr@zurlvm1

Prof. D. Tildesley
Univ. of Southampton
chi002@ibm.soton.ac.uk

Prof. F. van Swol
Univ. of Illinois, Urbana-Champaign
vanswol@uiucscs

Participants

Dr. D. Adolf
Univ. of Southampton
chr084@ibm.soton.ac.uk

Prof. J. Alejandre
Univ. Autonoma Metrop., Iztapalapa

Dr. N. G. Almarza
Univ. Complutense de Madrid
almarza@quim.ucm.es

Mr. H. Azzouz
Univ. Pierre et Marie Curie, Paris
azzouz@rox.lptl.jussieu.fr

Mr. A. Bakker
Univ. of Uppsala
albert@kemist.uu.se

Dr. I. Bako
Central Res. Inst. Hungarian Acad. Sci. Budapest
h1203pal@ella.hu

Mr. S. Bernard
Comm. a l'Energie Atomique, CELV, Villeneuve St. Georges

Mr. T. Beutler
ETH, Zurich
tcb@igc.ethz.ch

Mr. T. Biben
Lab. de Physique, l'ENS, Lyon
biben@frensl61.bitnet

Dr. M. Bionducci
Univ. di Cagliari
bionducci@vaxca1.infn.it

Mr. L. Bocquet
Lab. de Physique, l'ENS, Lyon
lbocquet@frensl61.bitnet

Dr. E. Boek
Univ. of Twente, Enschede
boek@utwente.nl

Mr. I. Borzsak
Eotvos Univ. Budapest
ethe102@hueco.uni-wien.ac.at

Prof. C. Caccamo
Univ. di Messina
caccamo@ime.univ

Dr. G. Cardini
Univ. di Firenze
cardini@ifichim.bitnet

Ms. M. do Carmo Lança
New Univ. de Lisboa
mchl@fct.unl.pt

Dr. M. Civi
Ankara Univ.

Mr. M. M. Clark
Oklahoma State Univ. Stillwater
u0580aa@unx.ucc.okstate.edu

Dr. D. Cleaver
Univ. of Southampton
chr085@uk.ac.soton.ibm

Dr. E. de Miguel
Univ. de Sevilla
demiguel@obelix.cica.es

Prof. R. G. Della Valle
Univ. di Venezia
valle@ifichim

Ms. L. Douglas
Univ. of Illinois, Urbana-Champaign
douglas@uiucscs.bitnet

Mr. M. Eldridge
Oxford Univ.
madden@uk.ac.ox.vax

Mr. J. W. Essex
Univ. of Oxford
grb@uk.ac.ox.vax

Mr. J. M. Esteves
Centro de Fisica da Mat. Condens., Lisboa
festeves@ptearn

Dr. I. Etxebarria
Univ. del Pais Vasco, Bilbao
wmbetali@lg.ehu.es

Mr. R. Everaers
IFF KFA, Jülich
ever@iff042.iff.kfa-juelich.de

Mr. M. Fearn
Univ. East Anglia, Norwich
hp003@uk.ac.east-anglia.cpc865

Prof. F. S. Fernandes
Univ. de Lisboa
qccfmssf@ptearn

Mr. S. Fetsko
Univ. of Virginia, Charlottesville
swf2u@virginia.edu

Dr. E. Fois
Univ. di Milano
raimilan@imisiam.mi.cnr.it

Mr. M. Foley
Oxford Univ.
madden@vax.ox.ac.uk

Dr. F. F. M. Freitas
Univ. de Lisboa
qffmf@ptearn

Dr. S. Fritzsche
Leipzig
hlz021@djukfa11

Mr. G. Galassi
Univ. of Southampton
chr059@ibm.soton.ac.uk

Mr. J. C. Gil Montoro
Univ. Complutense de Madrid
juancar@quim.ucm.es

Mr. E. Hernandez
Royal Institution, London
eddy@uk.ac.ri.ricx

Dr. A. Jamshed
Kings College, London
udkj060@elm.cc.kcl.ac.uk

Mr. H. Karaasian
Middle East Tech. Univ. Ankara
a09039@trmetu

Dr. M. Kiselev
Max Planck Inst. für Chemie, Mainz
mpc5@
ibma.ipp-garching.mpg.de

Mr. J. Komorowski
Tech. Univ., Berlin
komo0630@
w418zrz.chem.tu-berlin.de

Mr. G. Kramposthuber
Ludwig Maximilian Univ. Munchen
kramposthuber@
stat.physik.uni-muenchen.dbp.de

Dr. D. Laria
CECAM, Paris
laria@cecam.fr

Dr. A. Luzar
Univ. of California, Berkeley
alenka@gold.cchem.berkeley.edu

Mr. A. Mackie
Cornell Univ., Ithaca
mackie@
nemesis.cheme.cornell.edu

Dr. L. Mariani
Univ. di Perugia

Ms. E. Martin del Rio
Univ. de Sevilla
elvira@obelix.cica.es

Mr. C. Mason
Univ. of Bristol
carl.mason@uk.ac.bristol

Mr. M. McCallum
Univ. of California, Berkeley
mccallum@
lithium.cchem.berkeley.edu

Ms. D. Mei
Boston Univ.
mei@chem.bu.edu

Mr. D. Miguel del Rio
Univ. Complutense de Madrid
Miguel@quim.ucm.es

Ms. G. Mooij
FOM Inst. for Atom. and Mol. Phys., Amsterdam
mooij@amolf.amolf.nl

Mr. P. B. Moore
Boston Univ.
moore@chem.bu.edu

Mr. T. Mounir
Univ. Paris VI

Mr. A. Müller
TH Merseburg
muller@chemie.th-merseburg.de

Mr. C. Mundy
Univ. of California, Berkeley
dawson@ucbcmsa.bitnet

Dr. P. Mura
Univ. di Sassari
mura@mvchss.cineca.it

Prof. G. Orlandi
Univ. di Bologna
mc8a@icineca

Mr. E. O'Toole
Cornell Univ., Ithaca
otoole@setanta.cheme.cornell.edu

Prof. N. Papanicolaou
Univ. of Ioannina
e050b03@grioanun

Ms. C. Pegas
Univ. Pierre et Marie Curie, Paris
gc@frunip62.bitnet

Dr. L. A. T. Pereira Neves
Univ. de Lisboa
qcomputl@ptearn

Dr. S. Prestipino
Univ. di Messina
prestip@imeuniv.bitnet

Dr. S. Quartieri
Univ. di Modena
37984::quartieri

Mr. J. Recht
Cornell Univ., Ithaca
recht@pyladis.cheme.cornell.edu

Mr. D. Rector
Univ. of Illinois, Urbana-Champaign
rector@scsc.scs.uiuc.edu

Dr. P. A. Rejto
Univ. of California, Berkeley
rejto@gold.cchem.berkeley.edu

Dr. A. M. Rigby
Shell Research, Amsterdam
rigby2@ksla.nl

Dr. B. Rosi-Schwartz
Univ. of Reading
spsrosi@cms.am.rdg.ac.uk

Dr. B. Rousseau
CNRS, Univ. Paris Sud

Dr. D. Scharf
Univ. of Philadelphia
scharf@sun0.chem.upenn.edu

Dr. C. Seidel
Max Planck Inst. für Kolloidforschung, Seehof

Dr. A. Sergi
Univ. di Messina
39.857::sergi

Dr. J. Shelley
Univ. of British Columbia
Shelley@theory.chem.ubc.ca

Dr. B. Space
Princeton Univ.
bspace@chemvax.princeton.edu

Mr. M. Spanu
Univ. di Sassari
spanu@mvchss.cineca.it

Dr. M. Stevens
IFF, Jülich
stevens@
iff042.iff.kfa-juelich.de

Prof. A. Tani
Univ. di Pisa
tani@ipisns.ib

Dr. M. Teschner
Silicon Graphics, Basel
micha@basel.sgi.com

Dr. D. Tobias
Univ. of Philadelphia
tobias@sun0.chem.upenn.edu

Dr. A. Torcini
Univ. di Firenze
simulieq@ifiidg.fi.cnr.it

Mr. K. P. Travis
Univ. of Manchester
mcdapkt@uk.ac.mcc.cms

Dr. A. van Buuren
Univ. of Groningen
buuren@rugrcx.rug.nl

Mr. J. van Duijneveldt
Univ. of Utrecht
vduijneveldt@amolf.amolf.nl

Ms. C. van Heusden
Univ. of Cambridge
cmh20@phx.cam.ac.uk

Ms. M. van Leeuwen
Shell Research, Amsterdam
leeuwen4@ksla.nl

Dr. M. L. F. Vega
Univ. de Sevilla
lourdes@sevaxu.cica.es

Mr. H. Voigt
Tech. Univ., Berlin
vogt0433@
w421zrz.physik.tu-berlin.de

Mr. J. Vrabec
Ruhr-Univ., Bochum
johann.fischer@
ruba.rz.ruhr-uni-bochum.dbp.de

Mr. T. Weider
Tech. Univ., Berlin
twei0433@
w421zrz.physik.tu-berlin.de

Mr. M. Wen
Univ. of Philadelphia
mar@sun0.chem.upenn.edu

Mr. M. Wilson
Oxford Univ.
madden@uk.ac.ox.vax

Mr. G. A. Worth
Univ. of Oxford
grc@uk.ac.ox.vax

Dr. M. Yurtsever
Middle East Tech. Univ. Ankara
a07122@trmetu

List of Posters

The posters are listed in the order that they appeared at the meeting.

C. Pegas	Study of molten salt mixtures: MD simulation and thermodynamics.
S. Prestipino	Statistical thermodynamics of four particles on a sphere.
N. Papanicolaou	A molecular dynamics simulation study of the phonon density of states on the Cu(100) surface.
E. Hernandez	Computational studies of hydrocarbon diffusion in zeolites.
I. Borzsak	On the convergence of the Green's entropy expansion.
K. Travis	Frozen distribution sampling and the coupling of torsional motions to transport properties.
A. Torcini	The puzzle of fast sound in liquid water.
D. Cleaver	A practical comparison of histogram and cumulant expansion methods.
C. Mason	MC of hard ellipsoids: cumulant expansions and free energy barriers.
I. Bako	MD and X-ray studies of formic acid.
H. Karaasian	MC simulations of liquids and cluster binary liquid mixtures.
J. C. Gil Montoro	The Voronoi polyhedra as tools for structural analysis.
S. Fritzsche	Some new results on diffusion through small pores: CH_4 in a zeolite.
R. G. Della Valle	MD simulation of glassy silica (SiO_2).
F. F. M. Freitas & F. S. Fernandes	Computer simulation of liquid CH_3Cl.
M. Wen	MD of a self-assembled monolayer.
P. B. Moore & D. Mei	Computer simulation of CS_2: light scattering.
E. de Miguel	Simulation of Gay-Berne and Lennard-Jones mixtures.

H. Voigt	MD simulation of structural relaxation in simple fluids.
C. van Heusden	Classical trajectory calculations with accurate electrostatic potentials.
J. S. van Duijneveldt	Computer simulation of free energy barriers in crystal nucleation.
N. G. Almarza	MC simulation of chemical equilibria.
M. D. Eldridge	Computer simulation of a binary colloidal hard-sphere system.
G. Kramposthuber	Ab initio simulation of charged colloids.
F. Fernandes & L. A. T. Pereira Neves	Phase transitions in ionic clusters a study by molecular dynamics.
P. A. Rejto	A polarizable, dissociable model for H_2O.
E. Martin del Rio & E. de Miguel	Simulation of the liquid-vapour interface in the Gay-Berne model.
G. Cardini	MD simulation of plastic phase I of cyclopentane.
J. Recht	MC simulation of square-well mixtures in the Gibbs ensemble.
E. Boek	MD simulation at interfaces between crystal urea and water.
A. Mackie	Lattice systems and the Gibbs ensemble.
A. Jamshed	MD of polymorphic phase transitions in plastic crystals.
S. Alessandro	Order-disorder transitions in plastic crystals.
A. Bakker	An MD study of aqueous aluminium chloride solution with three-body interactions.
M. Wilson	MD simulations of dense ionic systems including polarization effects.
I. Etxebarria	MD simulation of K_2SeO_4 (phase transitions).
A. Müller	Mixtures of non-polar and dipolar linear molecules.
T. Mounir	Experimental investigations of sorption of C_3H_8 on x-zeolite.

T. Weider	MD simulation of a model ferrofluid.
M. Kiselev	Statistical geometry method for characterising structure fluctuation in solution.
L. Douglas	The role of solvation forces in phase transitions of colloidal systems.
J. Shelley	MD simulation of aqueous ionic micelles.
M. Stevens	MD simulation of polyelectrolytes.
J. Alejandre	Inhomogeneous colloidal suspensions.
G. Galassi	MC studies of linear polymer chains between walls.
A. Torcini	A scaling behaviour for liquid alkali metals.
S. Vrabec	Vapor liquid equilibrium from the NPT test-particle method.
B. Rousseau	A study of homogeneous nucleation in SF_6 clusters.
H. Azzouz	A quantum MD study of proton transfer reactions in asymmetrical H-bonds in solution.
A. Luzar	Structure and dynamics of H-bond networks in DMSO-H_2O mixtures.
D. Adolf	Local polymer motion by Brownian dynamics.
M. Yurtsever	MD simulation of pattern formation in 2D model lipid mixtures.
R. Everaers & K. Kremer	Stretching of a model network: MD used to study polymers with diamond lattice topology.
J. Komorowski	Bond-angle distribution of short 2D chain molecules.
G. Mooij	Direct simulation of phase equilibria in chain molecules.
D. Scharf	Spectra of Li impurity in molecular hydrogen.
M. van Leeuwen	Phase equilibria of polar systems: Stockmayer dipolar soft spheres and dipolar hard spheres.
A. van Buuren	MD simulations of carbohydrate based surfactants at an interface.
C. Seidel	Conformation of a charged chain.
E. O'Toole	Thermodynamic and structural properties of a simple lattice protein model.

B. Rosi-Schwartz	Validation of polymer models with diffraction data.
M. L. F. Vega	A new algorithm for MD in GC ensemble.
E. de Miguel	Nematic, smectic A and the smectic A-B transitions in the Gay-Berne model.
D. Rector	Micelle formation at the fluid-wall interface.
L. Bocquet	Correlation function approach to the dynamics of confined films.
M. Fearn	MD of silicon using bond-order potentials.
B. Space	Excess electronic eigenstates of polarizable noble gases.

Index

GPSR Compliance

The European Union's (EU) General Product Safety Regulation (GPSR) is a set of rules that requires consumer products to be safe and our obligations to ensure this.

If you have any concerns about our products, you can contact us on ProductSafety@springernature.com

In case Publisher is established outside the EU, the EU authorized representative is:

Springer Nature Customer Service Center GmbH
Europaplatz 3
69115 Heidelberg, Germany

Batch number: 10607320

Printed by Printforce, the Netherlands